GPCR MOLECULAR PHARMACOLOGY AND DRUG TARGETING

GPCR MOLECULAR PHARMACOLOGY AND DRUG TARGETING

SHIFTING PARADIGMS AND NEW DIRECTIONS

Edited by

Annette Gilchrist
Department of Pharmaceutical Sciences
Midwestern University, Downers Grove, Illinois

WILEY

A JOHN WILEY & SONS, INC., PUBLICATION

Published by John Wiley & Sons, Inc., Hoboken, New Jersey.
Published simultaneously in Canada.

Library of Congress Cataloging-in-Publication Data:

GPCR molecular pharmacology and drug targeting : shifting paradigms and new directions / edited by Annette Gilchrist.

p. ; cm.

Includes index.

ISBN 978-0-470-30778-6 (cloth)

1. G proteins. 2. Drug targeting. I. Gilchrist, Annette.

[DNLM: 1. Receptors, G-Protein-Coupled–physiology. 2. Drug Delivery Systems. 3. Molecular Structure. QV 38 G725 2010]

QP552.G16G627 2010

612′.015756–dc22

2009052133

10 9 8 7 6 5 4 3 2 1

To Josen and Finn, who arrived somewhere in the middle of all of this ...

CONTENTS

PREFACE

Although G protein-coupled receptors (GPCRs) have been the subject of study since the early days of pharmacology, our understanding of this large family of receptors continues to evolve. As a result, texts that discuss these receptors must constantly be rewritten and revised as emerging concepts are brought forth. Although many texts focus on GPCRs, none provide an overall look at the molecular pharmacology of this important target class. This book presents an up-to-date review on how scientists are thinking about GPCRs.

Early work includes the lock and key model of Fisher, and "receptive substances" to explain the biological actions of exogenous chemicals on cell "receptors," and Clark's occupancy theory that introduced the idea that the effect of an agonist is proportional to the number of occupied receptors. In the 1960s, the concept of intrinsic activity was introduced to explain the observation that not every agonist of a given receptor induced the same maximum effect. Compounds reaching the maximum were referred to as full agonist and other agonists were named partial agonist. This model was later extended to include drug efficacy and the system-independent concept of intrinsic efficacy. The mathematics applied in the models were relatively simple and allowed calculations to be made on affinity and activity. Looking back, it seems remarkable that most of these concepts were developed when little information on the biochemical nature of receptors or the underlying molecular mechanisms involved in signal transduction were available.

I have been working in the field of GPCRs for about 15 years, a time dwarfed by many of the contributing authors, who are well recognized as being experts in the field. In this time period, many of the basic pharmacology concepts for GPCRs have undergone radical changes. Yet, if one looks back, it seems that years before their general acceptance, there was research to support the paradigm shifts. For example, in 1989 Costa and Herz described antagonists with negative intrinsic activity at wild-type δ-opioid receptors (*PNAS 1989 86: 7321-7325*). Many subsequent studies have confirmed that GPCR proteins can signal in an agonist-independent, constitutive manner, and the concept of inverse agonism is now accepted and included in pharmacology textbooks. But in its early days, the concept of inverse agonism was hotly debated. Thus, the way we think about GPCRs, their basal activity, how they get activated, how they communicate the signal across the cell membrane, how they couple to each other as well as to other receptors and other membrane proteins, how their mutations lead to disease, and the many ways in which they can be screened for compounds that modulate them are constantly changing.

This textbook will serve as a resource for any scientists (e.g., pharmacologists, chemists, biologists) investigating GPCRs in both academia and industry (pharmaceutical/biotechnology). The book provides a general overview of GPCRs in terms of their biology and pharmacology, as well as recent developments including structure, deorphanization, dimerization, functional selectivity, and accessory proteins. In addition, it presents detailed methods on how to express, manipulate, and measure GPCRs. It is important to recognize that emerging concepts have shaped not only our understanding of GPCR pharmacology but also the drug discovery process itself. Recognition that most of the compounds initially considered to be competitive antagonists were inverse agonists, and studies indicating that the therapeutic outcome of inverse agonists and neutral antagonists can be different, led many companies to initiate drug discovery efforts specifically aimed at identifying inverse agonists. As a result, it is envisaged that this book on the shifting paradigms to GPCR pharmacology will be an essential resource for scientists working to identify compounds that selectively target members of this diverse family.

The book is organized into 17 chapters; the first seven chapters deal with the evolving pharmacology that surrounds this family, including chapters discussing allosteric modulators, receptor dimerization, GPCR deorphanization, G protein activation, and the Frizzled family. The next six chapters deal with methods to screen GPCRs, and include chapters on cell phenotype, "traditional" biochemical techniques, resonance energy transfer, label-free detection, and approaches for ultra-high-throughput screening. The next three chapters present GPCR structures, new ways to express and crystallize the receptors, as well as computational studies such as modeling and rational drug design. The final chapter describes the use of pharmacological chaperones for diseases caused by GPCR mutations.

Chicago College of Pharmacy ANNETTE GILCHRIST
Midwestern University

CONTRIBUTORS

Ruben Abagyan, Bioinformatics and Systems Biology Graduate Program, University of California, La Jolla, CA

Elfrida R. Benjamin, Amicus Therapeutics, Cranbury, NJ

Magdalena Birker-Robaczewska, Actelion Pharmaceuticals Ltd., Allschwil, Switzerland

Michel Bouvier, Département de Biochimie, Université de Montréal, Montréal, Quebec, Canada

Moritz Bünemann, Department of Pharmacology and Toxicology, Philipps-Universitaet Marburg, Marburg, Germany

Nathalie Clement, Institut Cochin, Université Paris Descartes, Department of Cell Biology, Paris, France

Arthur Christopoulos, Monash Institute of Pharmaceutical Sciences and Department of Pharmacology, Monash University, Victoria, Australia

Maria Teresa Dell'Anno, Department of Psychiatry, Neurobiology, Pharmacology and Biotechnologies, University of Pisa, Pisa, Italy

Maoqing Dong, Department of Molecular Pharmacology and Experimental Therapeutics, Mayo Clinic, Scottsdale, AZ

Richard M. Eglen, PerkinElmer Life and Analytical Sciences, Waltham, MA

James C. Errey, Heptares Therapeutics Limited, BioPark, Welwyn Garden City, Hertfordshire, UK

Michael Freissmuth, Institute of Pharmacology, Center of Biomolecular Medicine and Pharmacology, Medical University of Vienna, Vienna, Austria

John Gatfield, Actelion Pharmaceuticals Ltd., Allschwil, Switzerland

Karen J. Gregory, Monash Institute of Pharmaceutical Sciences and Department of Pharmacology, Monash University, Victoria, Australia

Christian W. Gruber, Institute of Pharmacology, Center of Biomolecular Medicine and Pharmacology, Medical University of Vienna, Vienna, Austria

Ingrid Gsandtner, Institute of Pharmacology, Center of Biomolecular Medicine and Pharmacology, Medical University of Vienna, Vienna, Austria

Kaleeckal G. Harikumar, Department of Molecular Pharmacology and Experimental Therapeutics, Mayo Clinic, Scottsdale, AZ

Stephen J. Hill, The University of Nottingham, Institute of Cell Signalling, School of Biomedical Sciences, Queen's Medical Centre, Nottingham, UK

Carsten Hoffmann, Department of Pharmacology and Toxicology, University of Würzburg, Würzburg, Germany

Nicholas D. Holliday, The University of Nottingham, Institute of Cell Signalling, School of Biomedical Sciences, Queen's Medical Centre, Nottingham, UK

Ralf Jockers, Institut Cochin, Université Paris Descartes, Department of Cell Biology, Paris, France

Vsevolod Katritch, Skaggs School of Pharmacy and Pharmaceutical Sciences, La Jolla, CA

Terry Kenakin, Biochemical Reagents and Assay Development, GlaxoSmithKline Research and Development, Research Triangle Park, NC

Priya Kunapuli, In Vitro Sciences, External Discovery and Preclinical Sciences, Merck & Co., West Point, PA

Christopher Langmead, Heptares Therapeutics Limited, BioPark, Welwyn Garden City, UK

Angélique Levoye, Unité de Pathogénie Virale, Department of Virology, Institut Pasteur, Paris, France

Fiona H. Marshall, Heptares Therapeutics Limited, BioPark, Broadwater Road, Welwyn Garden City, Hertfordshire, UK

Lauren T. May, The University of Nottingham, Institute of Cell Signalling, School of Biomedical Sciences, Queen's Medical Centre, Nottingham, UK

Maria Rosa Mazzoni, Department of Psychiatry, Neurobiology, Pharmacology and Biotechnologies, University of Pisa, Pisa, Italy

Laurence J. Miller, Department of Molecular Pharmacology and Experimental Therapeutics Mayo Clinic, Scottsdale, AZ

Oliver Nayler, Actelion Pharmaceuticals Ltd., Allschwil, Switzerland

Terry Reisine, Independent Consultant, Los Angeles, CA

Patricia René, Département de Biochimie, Université de Montréal, Montréal, Quebec, Canada

Kimberly A. Reynolds, Department of Pharmacology, University of Texas Southwestern Medical Center, Dallas, TX

Michael Sabio, Lundbeck Research USA, Paramus, NJ

Gunnar Schulte, Section of Receptor Biology and Signaling, Department of Physiology and Pharmacology, Karolinska Institutet, Stockholm, Sweden

Patrick M. Sexton, Monash Institute of Pharmaceutical Sciences and Department of Pharmacology, Monash University, Victoria, Australia

John Simms, Monash Institute of Pharmaceutical Sciences and Department of Pharmacology, Monash University, Victoria, Australia

Elodie Tenconi, Institut Cochin, Université Paris Descartes, Department of Cell Biology, Paris, France

Sidney W. Topiol, Lundbeck Research USA, Paramus, NJ

Celine Valant, Monash Institute of Pharmaceutical Sciences and Department of Pharmacology, Monash University, Victoria, Australia

Kenneth J. Valenzano, Amicus Therapeutics, Cranbury, NJ

CHAPTER 1

The Evolution of Receptors: From On–Off Switches to Microprocessors

TERRY KENAKIN

Biochemical Reagents and Assay Development, GlaxoSmithKline Research and Development, Research Triangle Park, NC

... Try to imagine a pharmacology course that made no mention of receptors ...'
—Humphrey Rang, 2006

1.1. INTRODUCTION

The evolution of the receptor concept is traced from the turn of the century where the receptor was an organizational concept to the present day where seven transmembrane (7TM) receptors are considered to be complex control units for cellular function. Drug receptor effect can be quantified without knowledge of molecular mechanism through the operational model of drug action. Alternatively, biochemical knowledge of the receptor as an independent protein unit recognizing both external chemicals and cytosolic protein independently can be used to understand ligand-specific receptor effects and how these might be exploited therapeutically.

1.2. THE RECEPTOR AS AN ON–OFF SWITCH

Historically, the concept of drug receptors began as an abstract idea and continued throughout pharmacology for another 70–80 years as such. Models considering receptors essentially had drugs activate (as a switch) a black box to produce tissue response. It is useful first to consider receptors as pharmacological switches since the theories used to describe drug action in

GPCR Molecular Pharmacology and Drug Targeting: Shifting Paradigms and New Directions, Edited by Annette Gilchrist

these systems lay the foundation for the more mechanistically driven models of receptors in use today. These ideas have formed the basis for quantitative receptor theory on which new information about the biochemistry and structure of receptors has been added. These will then be discussed in terms of specific mechanistic models of 7TM receptor function and how these assist in the quantification of drug effect.

1.3. HISTORICAL BACKGROUND AND CLASSICAL RECEPTOR THEORY

What is considered a "given" today, namely that drugs interact with specific binding sites on the cell membrane called receptors, is an extremely important tenet of pharmacology. As stated by Rang [1], it indeed was "Pharmacology's big idea." The main reason it is such an important cornerstone of pharmacology is that it introduces order into the apparent chaos of physiology. For example, a simple molecule such as epinephrine mediates a myriad of physiological processes. A large subset of these, namely cardiac chronotropy, inotropy, lusitropy, vascular relaxation, lacrimal, pancreatic, and salivary gland secretion, bronchiole, uterine and urinary bladder muscle relaxation, decreased stomach motility, skeletal muscle tremor, and melatonin synthesis are mediated by a small subset of membrane-bound proteins, specifically β_1- and β_2-adrenoceptors. This immediately puts order in the collection of physiological processes in that it gives them a common place to start, namely the interaction of epinephrine with the receptor. This order fits in well with the discipline of medicinal chemistry in that chemists have access to the processes for potential control.

At the turn of the twentieth century, different groups carried out research that caused them to postulate the existence of control points on cells that responded to chemicals, that is, receptors. For example, Paul Ehrlich (1854–1915) carried out studies on agent "606" (salvarsan) for syphilis. His work with dyes and bacteria led him to propose that there are "chemoreceptors" (actually a collection of "amboreceptors," "triceptors," and "polyceptors") on parasites, cancer cells, and microorganisms that could be exploited therapeutically [2]. In Cambridge, John Newport Langley (1852–1926) studied the drug jaborandi (contains the alkaloid pilocarpine) and atropine and concluded that receptors were "switches" that received and generated signals and that these switches could be activated or blocked by specific molecules [2]. However, it is A.J. Clark (1885–1941) who is considered the father of modern receptor pharmacology. Clark was the first to suggest, from studies of acetylcholine and atropine, that a unimolecular interaction occurs between a drug and a "substance on the cell." As he put it [3] ...

> ... it is impossible to explain the remarkable effects observed except by assuming that drugs unite with receptors of a highly specific pattern....

In fact, it was Clark who described pharmacological phenomena in chemical terms [3, 4], a concept readily accepted today, but quite heretical in Clark's time. The prevailing concepts guiding physiology at the turn of the century were rooted in homeopathic theories (i.e., a fundamental theory centered on the surface tension of the cell membrane) like the Arndt–Schulz Law and Weber–Fechner Law [2]. A generally accepted statement to describe physiologic phenomena was simply that "certain phenomena occur frequently."

One of Clark's most valuable contributions to pharmacology was the application of mathematical rules to the behavior of biological systems. Thus, the dose–response curve became the common currency of pharmacology, and its judicious use in the work of Clark and others built the framework for what had become known as "receptor theory," namely the application of simple thermodynamic rules to pharmacological systems.

An early example of mathematics applied to the study of receptors was provided by A.V. Hill, a student of Langley. He expressed the time course of contraction of frog rectus abdominus to the agonist nicotine (N) through an equilibrium concentration–response curve of the form:

$$Y = \frac{N}{k' + kN} - M \tag{1.1}$$

where Y is contraction height, M is threshold, and k′, k are constants. While this work predated the routine use of binding isotherms to receptor work considerably, Hill lost interest in the approach, and it was left to Irving Langmuir, a chemist at General Electric Company in the United States, to devise an equation for the quantification of molecules binding to a surface, in particular, chemicals to metal filaments for light bulbs. Thus, the Langmuir adsorption isotherm quantifies the fraction of the substance bound to a surface (the pharmacological counterpart being receptor, denoted ρ_A) by a molecule [A] as:

$$\rho_A = \frac{[A]}{[A] + K_A} \tag{1.2}$$

where K_A is the ratio of what Langmuir referred to as the "rate of evaporation" of the substance away from the surface (pharmacologically, the rate of offset of the molecule from the receptor) divided by the "rate of condensation" of the molecule toward the surface (pharmacologically, the rate of onset toward the receptor). In pharmacological terms, the specific terminology for molecules that produced such activation is "agonist." While mechanistically, this equation is based on thermodynamic principles governing agonist binding to receptors, operationally, it also defines the universally observed relationship between agonists and the pharmacological responses they induce to tissues and cells. Thus, any observed receptor-mediated response in any tissue can be summarized by a form of the Langmuir isotherm where the fractional maximum given by Equation 1.2 is multiplied by the maximal response observed from the preparation ($Response_A = \rho_A \cdot E_{max}$):

$$Response_A = \frac{[A] \cdot E_{max}}{[A] + EC_{50}} \tag{1.3}$$

where EC_{50} refers to the concentration of agonist A that produces half the maximal response to drug A. It should be noted that Equation 1.3 is written for a system demonstrating a Hill coefficient (in honor of A.V. Hill) of unity. If there is cooperativity in the system (either in the binding of the drug to the receptor [*vide infra*] or in the cellular processes that translate drug binding into cellular response), then Equation 1.3 becomes:

$$Response_A = \frac{[A]^n \cdot E_{max}}{[A]^n + EC_{50}^n} \tag{1.4}$$

where n is the Hill coefficient for the dose–response curve. Figure 1.1 shows data from Clark (effect of acetylcholine on frog heart chronotropy) fit to the Langmuir adsorption isotherm (Eq. 1.3). Irrespective of mechanism, it can be seen that the curve shown in Fig. 1.1 concisely summarizes the data. Thus, the

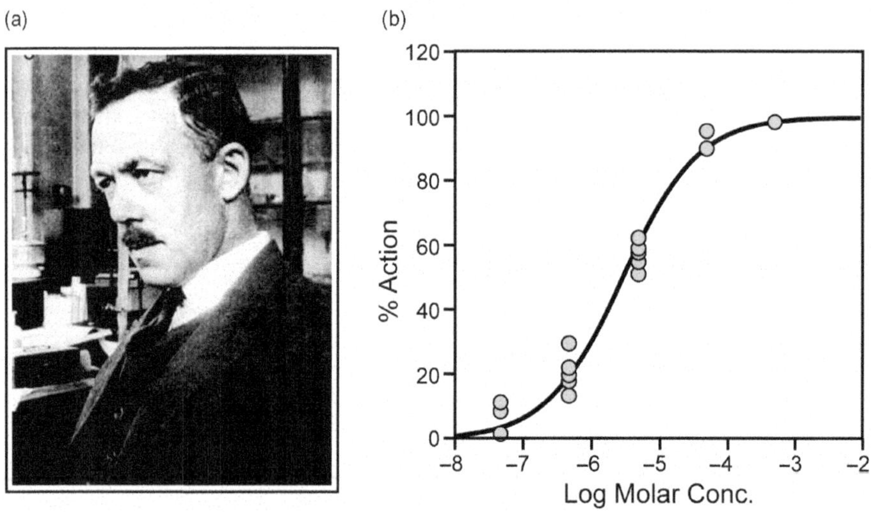

Figure 1.1 (a) Alfred. J. Clark (1885–1941), Professor of Pharmacology at University College London and later Chairman of Pharmacology in Edinburgh. Clark applied chemical laws to biological phenomena and is regarded as the father of receptor pharmacology. (b) Clark's data showing responses of frog heart to acetylcholine. Data points are fit to the Langmuir adsorption isotherm (Eq. 1.3).

16 data points are summarized by a shorthand that states acetylcholine produces a chronotropic effect in the frog heart that begins its effect at 50 nM, is half maximal at 2.8 μM, and is approximately maximum at 500 μM.

The adsorption isotherm furnished an extremely useful method of summarizing and handling dose–response data, but what was still required is a way to relate the observed data to the molecular mechanism of the drugs producing the effect. With no independent estimate of the affinity of the agonist he was using, Clark had to assume a one-to-one correspondence between the molecules of agonist he added to his preparation and the quanta of excitation those molecules gave to the tissue; that is, there was no provision for variance in the "power" of the molecules to induce tissue response. Ariens and Van Rossum, leading a germinal receptor group in Nijmegen, began the process of relating the molecular mechanism of drugs to the observed effects of drugs [5–7]. Thus was introduced the concept of "intrinsic activity," denoted α, as a scaling factor to accommodate the observation that not all agonists produce the maximal response of the preparation. Under these circumstances, Equation 1.4 becomes:

$$\text{Response}_A = \frac{[A]^n \cdot \alpha \cdot E_{max}}{[A]^n + EC_{50}^n} \tag{1.5}$$

where, for $\alpha = 0.5$, this would depict a drug that produced 50% of the tissue maximal response. Intrinsic activity became the first parameter designed to scale observed drug effect with the molecular "power" of an agonist to induce response. While this improved the correspondence between some observed effects of agonists, it was left to R.P. Stephenson, a pharmacologist working in Edinburgh, to extend this process to another level. Stephenson postulated that there was no reason to assume that tissue response was linked to agonist concentration in a linear manner (as was the requirement of the Clark and Ariens treatments). Instead, he postulated the existence of a theoretical parameter he called "stimulus," which is the result of the immediate interaction of the drug with the receptor [8]. This stimulus is imparted to the cell which then processes it in various ways, according to its needs, to yield tissue response. This loosely defined a function (referred to as the stimulus–response function) relating tissue excitation and response. Thus, tissue response was given as:

$$\text{Response}_A = f\left[\frac{[A] \cdot e \cdot E_{max}}{[A] + K_A}\right] \tag{1.6}$$

where e is a term efficacy (used to depict the power of the drug to produce response) and f is the stimulus–response mechanism. The important aspect of Equation 1.6 is that it allows the tissue response to be dissociated from receptor occupancy; experimental data would soon show the importance of that feature of the model.

Stephenson's concept of efficacy was required because of his observation that a series of related alkyltrimethylammonium compounds produced different maximal levels of guinea pig ileal contractions within a similar concentration range [8]. Since the agonist potencies indicated that the compounds had similar affinities for the receptor, Stephenson reasoned that another property of these molecules had to be operative to make them dissimilar in terms of producing muscle contraction; that was the property he termed "efficacy." This concept opened up a completely new way to look at tissue activation through receptors. Specifically, there were no constraints regarding the power of molecules to produce pharmacological response. Technically, powerful agonists could produce maximal tissue response by activating only a portion of the available receptors; the remaining portion would thus be described as being "spare" or not required for the production of maximal response. This offered maximal control to tissue systems since the cellular receptor density (i.e., varying proportions of spare receptors)

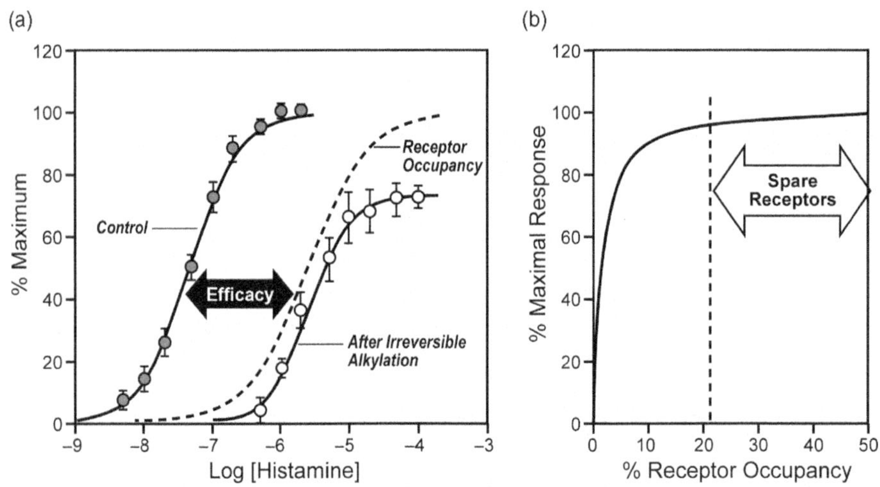

Figure 1.2 Response of guinea pig ileum to contraction by histamine and the relationship between histamine receptor occupancy and tissue response. (a) Contractile responses of guinea pig ileum to histamine. Ordinates: percent maximal contraction to histamine (solid lines) or calculated receptor occupancy from Langmuir adsorption isotherm (Eq. 1.2) for dotted line. Abscissae: logarithm of molar concentrations of histamine. Data shown for control (n = 12) and after alkylation of a portion of the population of histamine receptors with SY-28 (*N*-ethyl-*N*-β-bromoethyl)-1′-naphthylmethylamine) (200 nM exposure for 3 min followed by 3 h wash; n = 8). Data from Reference 11. (b) Calculated relationship between histamine receptor occupancy (dotted line in panel A) and control histamine response. This is the experimentally derived stimulus–response relationship between histamine and guinea pig ileum. Note the abscissal axis for panel B does not extend to complete receptor occupancy and that essentially 100% tissue response is obtained with an 18% histamine receptor occupancy.

could be used by the tissue to control sensitivity to the hormone or neurotransmitter. Experimental evidence to support the existence of spare receptors came with the use of receptor alkylating agents (specifically, β-haloalkylamines) which could irreversibly block portions of the receptor population in any given tissue [9, 10]. It was observed that irreversible removal of large portions of receptor populations rendered tissues less sensitive to agonists but that these agonists still were capable of producing the maximal tissue response. This experimentally defined the shape of stimulus–response relationships as postulated by Stephenson; an example of the process used to do so is shown in Fig. 1.2. These data formed the concepts leading to the present model universally used to depict agonist response in tissues, namely the operational model.

1.4. THE OPERATIONAL MODEL OF DRUG ACTION

While Stephenson's treatment of drug receptor-mediated response can be used to fit data to molecular models, it still utilized efficacy essentially as a fitting parameter tailored to make experimental data fit the model. This arbitrary nature of efficacy led Black and Leff to postulate a new model of drug action based on observed effects of drugs in tissues; they called this approach the operational model of drug action [12]. This model is based on the premise that the efficacy term emerges from an experimentally observed behavior of pharmacological systems, specifically the saturable relationship between receptor stimulation and observed response. An example of the shape of such a relationship is shown in Fig. 1.2b. The hyperbolic shape of this relationship forms the basic premise of this model; the ligand occupied receptor [AR] activates the cellular stimulus–response cascade with a general equilibrium dissociation constant denoted K_E (this is the concentration of [AR] complex producing 50% maximal response):

$$\frac{\text{Response}}{E_{max}} = \frac{[AR]}{[AR] + K_E} \tag{1.7}$$

The more efficient is the process from production of [AR] to response, the smaller is K_E. Substituting mass action for the production of [AR] yields the equation for the operational model:

$$\text{Response} = \frac{[A] \cdot [R_t] \cdot E_{max}}{[A]([R_t] + K_E) + K_A \cdot K_E} \tag{1.8}$$

The constant used to characterize the propensity of a given system and a given agonist to yield response is the ratio $[R_t]/K_E$; this is denoted τ. Substituting for τ yields the working equation for the operational model:

$$\text{Response} = \frac{[A] \cdot \tau \cdot E_{max}}{[A](\tau + 1) + K_A} \tag{1.9}$$

It can be seen that tissue response is now a function of biologically related quantities, namely the receptor density $[R_t]$ and K_E, concentration of occupied (activated) receptor available for interaction with the cellular machinery mediating tissue response.

At this point, it should be pointed out that the unknown nature of the biochemical reactions linking receptor occupancy and tissue response is not an impediment to the system-independent measure of drug activity. This is because of the null method. Thus, when comparisons of agonists are made in the same tissue at equal levels of response, then the impact of the biochemical cascade translating receptor occupancy and tissue response is removed since their effects are the same for both agonists. Under these circumstances, ratios of receptor affinities (i.e., K_A) and/or ratios of efficacies (ratio of τ values) become unique identifiers of the particular agonist–receptor pairs. Up to this point, only agonism has been discussed, but receptor theory also has provided a number of models to describe the antagonism of agonist response. It is worth considering these before discussion of 7TM receptor mechanisms.

1.5. RECEPTOR ANTAGONISM

There are two major mechanisms of receptor antagonism: orthosteric, whereby the antagonist and agonist compete for the same binding site on the receptor, and allosteric, whereby each has their own binding site on the receptor and the interaction between them takes place through a conformational change in the receptor protein. It is important to differentiate these since these respective antagonist types have different behaviors in pharmacological and physiological systems.

All equations for orthosteric molecular interaction can be derived from the integral of the differential equation describing the receptor occupancy by an antagonist with time ($\partial\rho_B/dt$) as a function of time and the competition between agonist [A] and antagonist [B] for receptors [13]:

$$\partial\rho_B/dt = k_2 \frac{[B]}{K_B}\left((1-\rho_B) - \frac{(1-\rho_B)[A]/K_A}{[A]/K_A + 1}\right) - \rho_B \tag{1.10}$$

where ρ_b is the equilibrium receptor occupancy by antagonist and K_A and K_B are the equilibrium dissociation constants of the agonist and antagonist receptor complexes, respectively. Upon integration this yields:

$$\rho_{AB} = ([A]/K_A/([A]/K_A+1))\cdot(1-(\vartheta(1-e^{-k}2^{\Phi t})+\rho_B e^{-k}2^{\Phi t})) \tag{1.11}$$

where:

$$\vartheta = [B]/K_B/([B]/K_B+[A]/K_A+1) \tag{1.12}$$

$$\rho_B = [B]/K_B/([B]/K_B+1) \tag{1.13}$$

$$\Phi = ([B]/K_B+[A]/K_A+1)/([A]/K_A+1) \tag{1.14}$$

For accurate estimation of K_B (1/affinity of the antagonist), there must be enough time elapsed in the experiment for the agonist to re-equilibrate with the antagonist-bound receptors. If there is sufficient time for this to occur (the dissociation rate of the antagonist is rapid such that the agonist can attain correct receptor occupancy) and time/$k_2 > 10$, then Equation 1.11 for receptor occupancy by agonist (ρ_A) reduces to the familiar, and much simpler, equation for simple competitive antagonism presented by Gaddum [14]:

$$\rho_A = \frac{[A]/K_A}{[A]/K_A+[B]/K_B+1} \tag{1.15}$$

This equation predicts the well-known shift to the right of agonist dose–response curves with no diminution of maxima produced by competitive antagonists (note that agonist (ρ_A) will be complete ($\rho_A \to 1$) when $[A] \gg [B]$). On the other hand, if the antagonist has a slow offset, there may not be sufficient time for re-equilibration during the experiments and noncompetitive antagonism may result. Under these circumstances, time/$k_2 < 0.01$ and Equation 1.11 reduces to the Gaddum equation for noncompetitive antagonism:

$$\rho_A = \frac{[A]/K_A}{[A]/K_A(1+[B]/K_B)+[B]/K_B+1} \tag{1.16}$$

Under these conditions, Equation 1.16 predicts that the presence of the antagonist ($[B] \neq 0$) essentially precludes complete receptor occupancy by the agonist (ρ_A always <1 with nonzero values of [B]). This can produce dose–response curves with depressed maxima. It can be seen that competitive (surmountable) and noncompetitive (insurmountable) are only kinetic extremes of the same mechanism of drug action (orthosteric binding of antagonist to the agonist binding site).

The other major mechanism for drug-induced receptor blockade is through allosteric interaction whereby the agonist and antagonist bind to their own sites on the receptor, and the interaction between them occurs through a conformational change in the receptor:

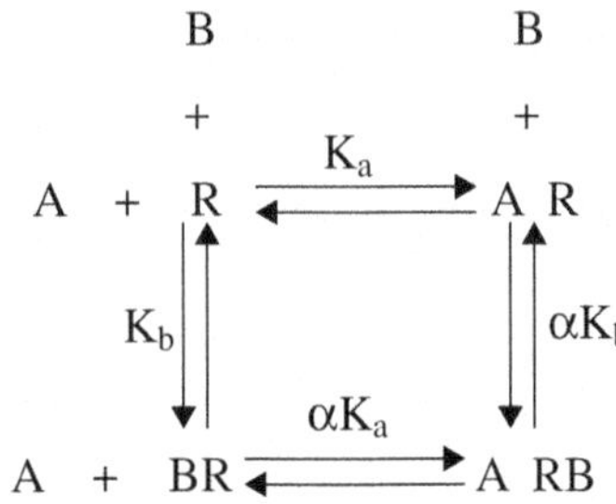

(1.17)

The effect of the modulator on the affinity of the receptor for ligand A is quantified by a factor α (designated the cooperativity constant—the use of this symbol should be differentiated from its use denoting intrinsic activity by Ariens); the affinities of ligands A and B for the receptor are K_a and K_b, respectively. Under these circumstances, the effect of an allosteric modulator on the binding of ligand A is given by [15]:

$$\rho_A = \frac{[A]/K_A(1+\alpha[B]/K_B)}{[A]/K_A(1+\alpha[B]/K_B)+[B]/K_B+1} \qquad (1.18)$$

Equation 1.18 defines the changes in affinity of the receptor for A when the modulator B is bound; these can be positive (i.e., increase affinity when $\alpha > 1$) to yield potentiation of binding or negative (decrease affinity when $\alpha < 1$) to yield antagonism. Unlike orthosteric antagonism, binding is not precluded by a negative allosteric modulator but rather, the affinity of the receptor is reset to a different level. Also, since allosteric effect is mediated by binding of the modulator at a separate site, it is saturable; that is, when the allosteric sites are completely bound by modulator, the effect reaches a maximal limit.

The other major delineation between orthosteric and allosteric effect is that allosteric effects can modulate agonist affinity and efficacy separately. This is because modulators essentially stabilize a new conformation of the receptor. To describe the effect on agonist efficacy, an extended model of allosteric modulation of receptors is required. Thus, the Ehlert allosteric model [15] (scheme 18) is linked to the Black and Leff operational model [12] to yield the following [16–18]:

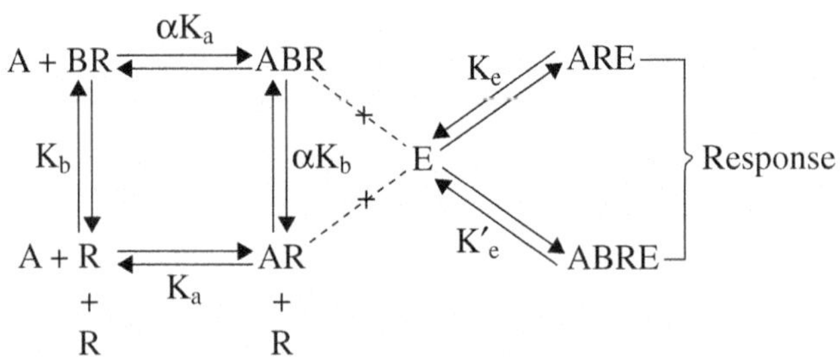

(1.19)

The model for this is a melding of Equations 1.9 and 1.18 to yield [16–18]:

$$\text{Response} = \frac{[A]/K_A\,\tau(1+\alpha\beta[B]/K_B)E_{max}}{[A]/K_A(1+\alpha[B]/K_B+\tau(1+\alpha\beta[B]/K_B))+[B]/K_B+1} \tag{1.20}$$

In this model, the agonist and modulator-bound receptor complex has the potential ability to signal; therefore, β refers to the change in the efficacy of the agonist when the receptor is bound by modulator ($\beta = \tau'/\tau$ where τ' is the efficacy of the agonist when the modulator is bound to the receptor). This permits the model to predict a range of separate effects on affinity and efficacy of the agonist.

While orthosteric effects are preemptive (once the antagonist binds, no effect can be produced by the agonist) whereby there is never a species of protein with both agonist and antagonist bound, allosteric effects are permissive. This latter property means that there are protein species with agonist and antagonist co-binding simultaneously. Under these circumstances, the antagonist (actually more specifically, allosteric modulator) can change the receptor reactivity toward the agonist in a number of ways, that is, ranging from increased to decreased affinity, increased to decreased efficacy. Moreover, these effects can be probe dependent, that is, be different for different agonists [19]. This can lead to interesting effects such as that seen with the N-methyl-D-aspartate (NMDA) receptor antagonist ifenprodil [20]. This drug reduces the efficacy but increases the affinity of the receptor for NMDA; under these circumstances, ifenprodil potency, as an antagonist, increases with increasing concentrations of NMDA; that is, the antagonism increases as the system is more highly driven.

These models of receptor function can be used to characterize agonist and antagonist action using null methods with no knowledge of receptor structure or biochemistry required; that is, the receptor can be viewed as an operational on–off switch. While this was imperative in the early years of receptor theory where no mechanism-based knowledge concerning receptor function was available, the last 20 years of technological advancement has furnished a wealth of information about the structure and function of receptors. This has added greatly to the understanding of drug and receptor function. It is worth considering how knowledge of receptor structure and biochemical function has added to the models used to describe and predict drug action.

1.6. SPECIFIC MODELS OF GPCRs (7TM RECEPTORS)

Improvement in biochemical and structural techniques has provided the capability to characterize receptors and, more importantly, the other cellular bodies with whom they interact. There are two fundamental discoveries that led to our current model of receptor function, namely receptor allosterism and membrane protein translocation (the "mobile receptor hypothesis" [21]). Allosterism was first described for ion channels [22, 23] and enzymes [24] and was subsequently applied to receptors [25–27]. The application of the idea that ligands can specifically bias protein conformation through selective binding (conformational selection [28]) gives a molecular basis for the concept of agonist efficacy (and actually, all allosteric modulation of receptors).

In this scheme, drugs bind to a limited number of preexisting receptor conformations and stabilize those through binding (at the expense of other conformations according to Le Chatelier's principle of an equilibrium responding to perturbation). Selective affinity of the ligand for specific receptor active states creates a bias in the collection of receptor conformations, that is, a ligand with higher affinity for one of the conformations will enrich it through selective binding. Scheme 21, shown below, shows how selective affinity can produce receptor activation: shown are two receptor conformations R and R* controlled by an equilibrium dissociation constant L where L = [R*]/[R]:

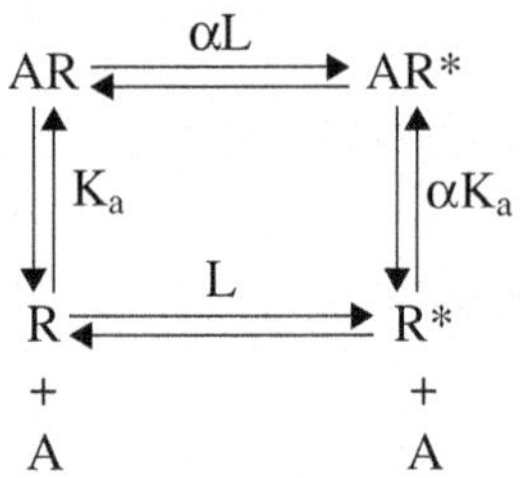

(1.21)

The scheme above shows a ligand A with an affinity (defined as the equilibrium association constant $K_a = k_1/k_2$) of K_a for receptor state R and αK_a for receptor state R*. The factor α denotes the differential affinity of the agonist for R*; that is, α = 10 denotes a 10-fold greater affinity of the ligand for the R* state. Therefore, α (selective affinity) confers the ability of the ligand to alter the equilibrium between R and R*. This can be seen by calculating the amount of R* (both as R* and AR*) present in the system in the absence of A and in the presence of A. The equilibrium expression for ([R*] + [AR*])/ $[R_{tot}]$ where $[R_{tot}]$ is the total receptor concentration given by the conservation equation $[R_{tot}]$ = [R] + [AR] + [R*] + [AR*] is:

$$\rho = \frac{L(1+\alpha[A]/K_A)}{[A]/K_A(1+\alpha L)+1+L} \tag{1.22}$$

In the absence of agonist ([A] = 0), $\rho_0 = L/(1 + L)$ while in the presence of a maximal concentration of ligand (saturating the receptors; $[A] \rightarrow \infty$) $\rho_\infty = (\alpha(1 + L))/(1 + \alpha L)$. Therefore, the effect of the ligand on the proportion of the R* state is given by the ratio ρ_∞/ρ_0. This ratio is given by:

$$\frac{\rho_\infty}{\rho_0} = \frac{\alpha(1+L)}{(1+\alpha L)} \tag{1.23}$$

Equation 23 predicts that if the ligand has an equal affinity for both the R and R* states ($\alpha = 1$) then ρ_∞/ρ_0 will equal unity, and no change in the proportion of R* will result from maximal ligand binding. However, if $\alpha > 1$, then the presence of the conformationally selective ligand will cause the ratio ρ_∞/ρ_0 to be >1, and the R* state will be enriched by the presence of the ligand. R* is a completely new conformation of the receptor which could be more or less sensitive to the endogenous agonist or could signal in its own right (be an agonist). The fact that 7TM receptors are designed to be allosteric in nature (bind small molecule hormones or neurotransmitters in one region of the receptor to cause a change in shape to affect a protein–protein interaction in another part of the receptor) may make 7TM receptors especially prone to allosteric effects by other small molecules.

The other major idea that changed the way in which we view 7TM receptors is the discovery that they float in the lipid membrane and can associate with other membrane-bound proteins to become different species [21]. This confers two special properties on receptors that provide for maximal signal control. The first is that the relative stoichiometries of receptors and interactants can be used by the cell for fine-tuning of signal magnitude and cell sensitivity. The second is that it allows the receptor to function in a much more complicated mode than a simple on–off switch. In a hard-wired mode whereby the activated receptor is mandatorily linked to a single response element, receptor activation is binary in that excitation either is or is not imparted to the cell. In a floating disconnected mode, the receptor can link to a range of couplers giving it the capability to discern which response element it activates in response to which initial stimulus it receives. In short, it becomes a microprocessor.

1.7. THE RECEPTOR AS MICROPROCESSOR: TERNARY COMPLEX MODELS

The floating receptor hypothesis resulted in the prototype model of 7TM receptor action, namely the ternary complex model (first published by DeLean and colleagues [29]). This describes a receptor that, when activated by an agonist, moves laterally in the cell membrane to physically couple to a trimeric subunit referred to as a G protein. Supporting this model were data showing that physical complexes between receptors and G proteins could be isolated after addition of agonist to receptor systems (i.e., References 30, 31). The prevalence of this mechanism (receptor protein that recognize external ligands and transmit signals to cellular GTPase heterotrimers called G proteins to elicit response) led to the pervasive name for 7TM receptors from that period as G protein-coupled receptors (GPCRs).

Receptor behavior that was inconsistent with the ternary complex model was observed nine years after its description; this was the impetus for the publication of the extended ternary complex model [32]. Specifically, Costa and Herz [33] noted that a peptide opioid receptor antagonist selectively reduced the basal level of NG108 cells containing μ opioid receptors and that this behavior was due to the reversal of elevated basal tissue response due to receptors that spontaneously formed an active state. There was no provision for spontaneous non-ligand-dependent activation of receptors in the original ternary complex model; thus, a new model had to be devised: this was the extended ternary complex model.

The extended ternary complex model describes a receptor that can exist in two states, active ($[R_a]$) and inactive ($[R_i]$), named for their ability to activate G proteins [G]) that coexist according to an allosteric constant unique for the receptor type (denoted $L = [R_a]/[R_i]$)—see Fig. 1.3a). The affinity of the ligand for $[R_i]$ is denoted K_a (equilibrium association constant); the ligand has a differential affinity for $[R_a]$ of αK_a. The unbound receptor has an affinity for G protein of K_g; ligands can confer a different affinity of the receptor for G protein denoted γK_g. This model describes response production (elevated concentrations of $[R_a]$ and $[AR_a]$) as a fraction of total receptor species (denoted ρ) as:

$$\rho = \frac{L[G]/K_G(1+\alpha\gamma[A]/K_A)}{[A]/K_A(1+\alpha L(1+\gamma[G]/K_G))+L(1+[G]/K_G)+1} \tag{1.24}$$

where K_A and K_G are equilibrium dissociation constants (reciprocals of association constants). Figure 1.3b shows the effects of changing α on dose–response curves of a system with existing constitutive activity (shown as an elevated basal response which can be reduced by ligands with $\alpha < 1$). Such ligands are referred to as inverse agonists. Formally identical effects are observed with changes in γ values.

The extended ternary complex model gives a vectorial quality to efficacy. As discussed above, efficacy can be described as a selective affinity of the ligand for various receptor states. Thus, $\alpha > 1$ leads to positive agonism while $\alpha < 1$ results in inverse agonism. This model has been referred to as a "two-state" model, probably because of the two unliganded species $[R_i]$ and $[R_a]$. However, this is a misnomer since the model actually describes infinite receptor states when the receptor is ligand bound; that is, the magnitude of γ confers a unique affinity of the receptor for G proteins when the receptor is ligand bound. Under these circumstances, every value of γ defines a new ligand-bound receptor state.

A theoretical shortcoming of the extended ternary complex model is the fact that it allows only the activated receptor to form complexes with G protein. Thermodynamically, there is no reason a priori that all receptor species (active and inactive) cannot bind to G proteins; when this is added to the extended ternary complex model, a cubic model results where the receptors form the species [ARiG] as well as [ARaG]. The resulting model is known as the cubic ternary complex model [34–36]; this is a more rigorous and thermodynamically correct model, but it is more difficult to use since there are a greater number of parameters that cannot be independently estimated.

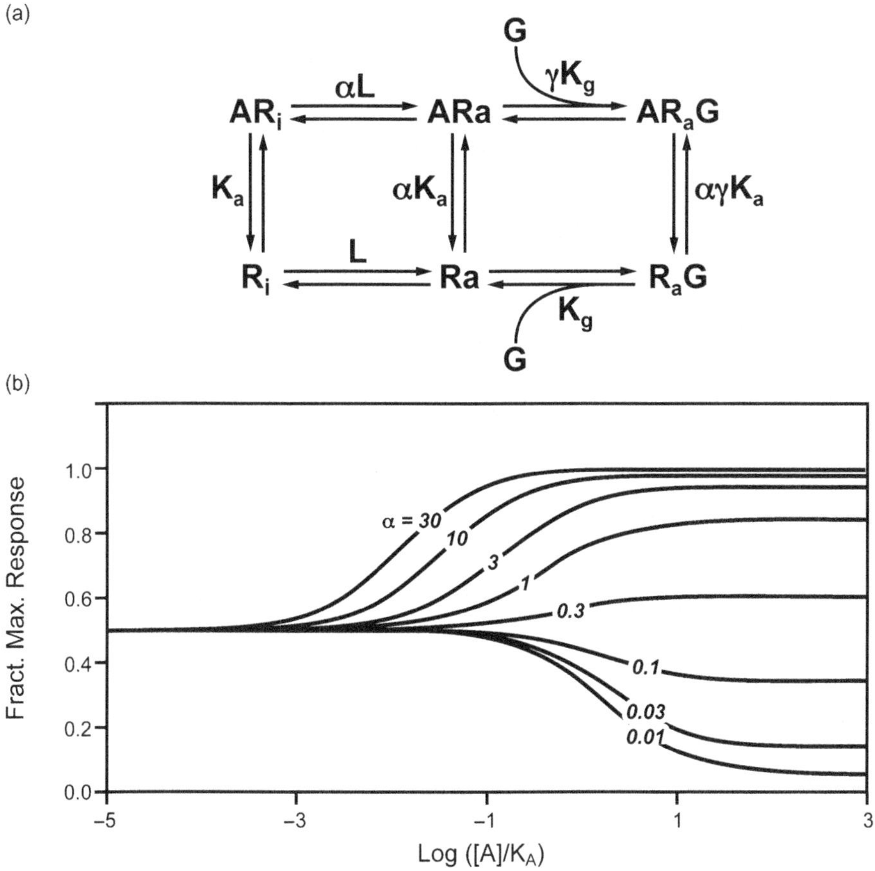

Figure 1.3 The extended ternary complex model for 7TM receptor function [32]. (a) Schematic diagram showing a receptor that can exist in an inactive [Ri] and active [Ra] state; both states can interact with a ligand A and the active state interacts with G protein [G]. A variant is the cubic ternary complex model where the inactive receptor can also interact with G protein [34–36]. (b) Effects of ligands with varying efficacy (α values) producing the response-yielding species [ARaG] according to Equation 1.24. This simulation shows a system with constitutive activity (basal effect = 0.5) and the effect of positive ($\alpha > 1$) and inverse ($\alpha < 1$) agonists.

At this point, the existing models, while being able to accommodate complex receptor function (vide infra), still basically considered receptors as on–off switches. However, the floating nature of the receptor and the ability to form complexes with multiple membrane-bound proteins as well as biochemical studies of receptors produced evidence that multiple interaction of receptors with cytosolic proteins in the cell membrane exist (i.e., see Fig. 1.4). For example, the thyrotropin receptor has the capability of coupling to all four

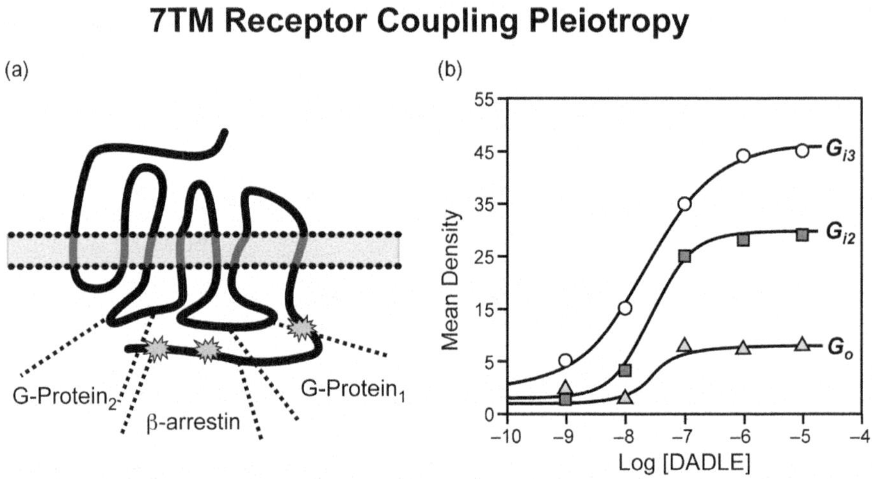

Figure 1.4 Pleitotropic coupling for 7TM receptors. (a) Various regions of the cytosolic loops of receptors can bind to various cellular coupling proteins (i.e., G proteins, β-arrestin). Gray areas represent sites for receptor phosphorylation. (b) Coupling of three G proteins to opioid receptors; ordinates are density measurements reflecting amount of ternary complex (receptor, G protein, and opioid agonist D-Ala^2,D-Leu^5 enkephalin [DADLE]). Data redrawn from Reference 37.

major G protein families (a total of 10 G protein subunits) [38]. In general, the use of recombinant receptor systems made clear that most receptors could couple to multiple G proteins [39–44]; it is not entirely clear to what extent these multiple couplings are artifacts of recombinant overexpression or physiological relevant fine-tuning of receptor signaling.

The preceding discussion has focused on receptor coupling that results in activation of G proteins. Historically, the canonical view of GPCR signaling cascades places receptor activation in a queue preceding receptor phosphorylation, binding of phosphorylated receptor to arrestin-family adapters, uncoupling from G proteins [45], and internalization of those complexes to the endocytotic compartment [46]. The binding to β-arrestin was thought primarily to serve the function of turning off the G protein signal. A major recent development in receptor pharmacology is the discovery of the ability of GPCRs to signal directly through the β-arrestin pathway to activate extracellular signal-regulated kinases (ERKs) through the formation of cytosolic "receptosomes" [47, 48]. Under these circumstances, β-arrestins can be considered to act as multifunctional adapters and scaffolds enabling the recruitment of signaling molecules (i.e., ERK) and other assemblies in an activation-dependent manner [49, 50]. Thus, receptors are now known to produce activation of ERK1/2 pathways via G protein-dependent or G protein-independent pathways [51–53]. G protein versus receptosome signaling differs both from its cellular origins and temporal relationship to ligand stimulation. For example, parathyroid hormone (PTH) produces G protein-dependent and G protein-independent stimulation of ERK1/2. While the G protein component is transient, a temporally distinct long-lasting stimulation is produced by direct stimulation of β-arrestin by the activated receptor. Interestingly, different chemical analogues of PTH have been shown to selectively activate these separate signaling pathways. Thus, while [Trp^{1}]PTHrp-(1–36) selectively produces G protein-mediated ERK1/2 stimulation, [D-Trp^{12},Tyr^{34}]PTH-(7–34) selectively produces β-arrestin-dependent, and G protein-independent, stimulation of ERK1/2 [54].

1.8. RECEPTORS AS BASIC DRUG RECOGNITION UNITS

In lieu of direct biochemical characterization, historically, receptor classification was based on the observed system-independent measurement of agonist potency ratios (PRs) and antagonist equilibrium dissociation constants. This approach depends on the concept that the receptor is the single discerning unit for agonism. Under these circumstances, the affinity and efficacy terms in Equation 1.9 (namely K_A and τ respectively) refer to the specific interaction of a given agonist for a given receptor (irrespective of which cell type response is mediated). Under these circumstances, the magnitude of the relative ratio of potency (PR) for two full agonists (A and B) is a unique identifier of the agonists and receptor type since it is independent of all tissue-based response elements:

$$PR_{AB} = \frac{K_{AA} \cdot (\tau_B + 1)}{K_{AB} \cdot (\tau_A + 1)} \tag{1.25}$$

Deviation of such PR estimates were considered to be presumptive evidence of differences in the receptor (as the minimal recognition unit). The emphasis of classical pharmacology was recognition of chemicals since the response systems usually came as an intact unit (i.e., isolated tissues). However, the independent nature of the receptor and response elements (a "floating" receptor interacting with different free G proteins and other cytosolic proteins) adds another element of recognition, namely the recognition of the response element after agonist binding. The impact of this factor became clear with the use of recombinant receptor systems where the relative stoichiometry of these elements could be varied. In addition, technological advances furnished the means to selectively observe the individual components of cellular response, thereby allowing the measurement of changes in distinct receptor-coupled pathways to agonist activation. These types of systems furnished experimental data that was totally inconsistent with the previously described assumptions concerning PRs of agonists. Specifically, it was observed that the potency of agonists for a single receptor differed when different response cascades, mediated by that same receptor, were measured. For example, for pituitary adenylate cyclase-activating polypeptide (PACAP) receptors, which pleiotropically mediate changes in cyclic adenosine monophosphate (AMP) and inositol phosphate 3 (IP_3), two PACAP peptide fragments ($PACAP_{1-27}$ and $PACAP_{1-38}$) produced elevated cyclic AMP and IP_3 in cells. However, the relative potency of these two agonists for these pathways was reversed [55]. Thus, the relative efficacy of $PACAP_{1-27}$ for cyclic AMP elevation is higher than that for $PACAP_{1-38}$ but lower for elevation of IP_3. In historical terms, this would have implied that the responses to the two agonists were mediated by different receptors. Since only one receptor type was transfected into the cell, this was not an option requiring consideration of alternative ideas. It should be noted that relative ratios of potency need not be reversed to denote functional selectivity since the actual numerical value of the PR accurately depends on relative affinity and efficacy of the agonist for the receptor recognition unit.

A tacit assumption in the historical view of agonist PRs is the idea that all agonists produce a uniform receptor-active state, that is, they all flip the receptor switch in an identical manner. However, there is an abundance of evidence to show that proteins spontaneously produce a myriad of conformations in response to thermal energy [56–59]. In accordance to the concept of conformational selection [28], ligands interact with these collections of conformations (termed "ensembles") and stabilize subsets of them through selectively high affinity [60, 61]. If there were a collection of conformations, then some may have greater affinity for some response elements than others. Under these circumstances, the minimal recognition unit, with respect to the response elements of the cell, would not be the receptor per se but rather the receptor-active state (i.e., ligand-stabilized subset of receptor conformations made after ligand binding). If the active state were the minimal recognition unit, then the experimental results indicating differing PRs for different response pathways can be accommodated. Such differences require that different ligands stabilize different ensembles of receptors to produce different macro-active states interacting with the cell.

The varying PRs observed for PACAP analogs led to a modified model of receptor stimulus whereby stimulus could "traffic" to different portions of a cellular stimulus–response cascade [62]. There has been a large body of evidence since that time to verify this mechanism with many ligands (see References 63–67 for reviews of specific papers), suggesting that many ligands produce different receptor-active states. This is, in fact, consistent with experimental observations. There is another large body of data from a number of experimental approaches that confirm that ligands can stabilize different receptor conformations [68–72]. This idea completes the general notion of 7TM receptors as microprocessors. Thus, a range of transient receptor conformations can be stabilized by different ligands to form varying predominant conformations that then interact with a range of independent cytosolic reactants to produce cellular response.

1.9. RECEPTOR STRUCTURE

In parallel experiments conducted over the past decades, studies on 7TM receptor structure have contributed to the models of function. In particular, point mutation studies have elucidated the various separate regions that interact with different ligands and different response coupling elements. The notion of receptor allostery has attained prominence now that functional, as opposed to binding, high-throughput screens are increasingly being used for new drug discovery. It can be argued that the early use of radioligand binding assays in screening has biased systems to the detection of orthosteric ligands, leaving an impression that allosteric molecules for 7TM receptor are comparably rare [73]. However, 7TM receptors are nature's prototype allosteric protein binding small hormones and neurotransmitters in one region of the protein and changing shape to produce a change in a protein–protein interaction in another part of the protein. Theoretically, it would be predicted that many small molecules would function as drugs in an allosteric manner. As functional high-throughput screens are implemented, a corresponding increase in the number of allosteric modulators are being discovered.

Allosteric molecules can produce immensely powerful receptor-mediated effects. For example, biochemical [74, 75] and structural [76] studies of the chemokine (C-C) motif receptor 5 (CCR5) receptor show a number of small molecule allosteric modulators that can block the interaction of enormous proteins (CCR5 and gp120, the HIV viral coat protein both over 120 Da). These ideas, taken together with the notion that ligands stabilize preferred conformations of receptors to affect response, lead to the notion that the complete surface of the receptor may be considered a potential drug binding active site. With this in mind, structural definition of binding pockets may not be as relevant to 7TM receptors as it is for enzymes.

1.10. FUTURE CONSIDERATIONS

Present concepts of 7TM receptor function suggest a broad range of chemical interventions for possible therapeutic utility. Receptors are now known to have an intrinsic activity (spontaneous formation of active states) that probably is part of the normal sampling of tertiary structures in conformational space. This makes them active control points for cells to limit and increase external signals and internal needs. A useful way to interrogate possible chemical probing of such a system is through the Receptor Probability model described by Onaran and Costa [60, 61]. This model makes no assumptions as to the conformation of the receptor, nor its pharmacological function, but rather describes the probability that a given ligand will produce a conformational bias in the ensemble that may or may not subsequently have biological effect. This opens the concept of efficacy away from a linear idea whereby ligands must produce a set number of effects (i.e., activation, desensitization,

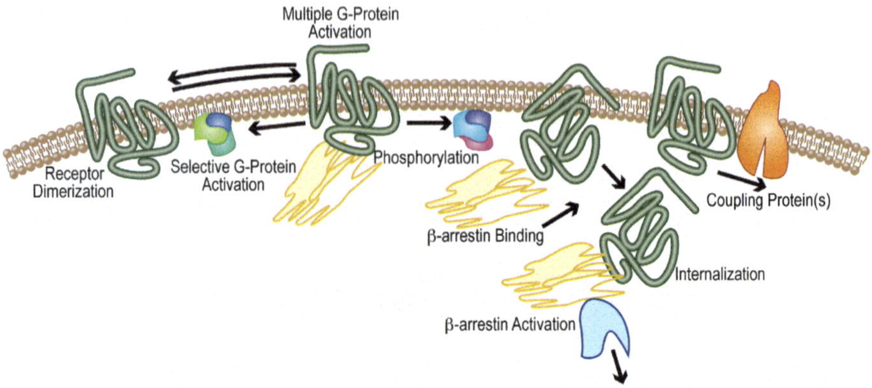

Figure 1.5 Some interactions of 7 transmembrane receptors with cellular components to generate phenotypes of efficacy. Cellular response does not automatically indicate direct interactions of the ligand-bound receptor with some of these processes (i.e., not a universal indicator of molecular efficacy) and ligands can have mixed efficacies, that is, be an inverse agonist at one G protein subtype and a positive agonist at another. This scheme also highlights the potential for cellular signaling through non-G protein pathways (i.e., β-arrestin).

internalization) in order. With the probability model, efficacy can be "collateral" [77] in that ligands can produce subsets of activities. For example, there are a number of antagonists for many receptor types that actively produce receptor internalization without producing receptor activation [78; see also Reference 65].

In general, a rich array of behaviors can be ascribed to 7TM receptors (Fig. 1.5). For example, fine-tuning of external control may be achieved by receptors in the formation of hetero- or homodimers (or higher-order oligomers) or association with other membrane proteins such as receptor activity-modifying proteins (RAMPs). In fact, it is an unfortunate misnomer to refer to 7TM receptors as GPCRs as they are known to couple to a vast other array of signaling proteins in the cell [79–83]. Under these circumstances, 7TM receptors can be considered to be viable active moieties in cells, the activity of which can be altered by agonists, inverse agonists, antagonists, and agents that otherwise alter their disposition (i.e., actively internalize receptors).

Another general idea to emerge is the labile nature of receptors in response to binding. In light of thermodynamic concepts relating to spontaneous formation of receptor conformations, binding should be considered an active, rather than a passive, phenomenon; that is, binding of any ligand to the receptor will change the overall makeup of the conformational ensemble (i.e., according to Eq. 1.23). Simulations of the effects of binding to a random collection of protein conformations show a correlation between affinity and "efficacy," the latter being defined as the ability to change the conformation of the receptor ensemble [84]. This can be made more evident by examination of equations describing collections of protein conformations. Thus, consider a system containing a root inactive receptor conformation R and multiple other conformations labeled R_i where i = 1 to n all controlled by various allosteric constants (denoted L_i). Under these circumstances, the fraction of receptors not in the R state under basal conditions is given by:

$$\rho_{nonR} = \frac{\sum_{i=1}^{n} L_i}{\left(1 + \sum_{i=1}^{n} L_i\right)} \tag{1.26}$$

With affinity of A for R_i as K (equilibrium dissociation constant of ligand–receptor complex is K^{-1}) and affinity of A for each state K_i as $\Psi_i^{-1}K$, the fraction of receptors not in the R state in the presence of a saturating concentration of ligand is given by [66]:

$$\rho_{nonR} = \frac{\sum_{i=1}^{n} L_i + [A]\Big/K \sum_{i=1}^{n} \Psi_i L_i}{[A]/K\left(1 + \sum_{i=1}^{n} \Psi_i L_i\right) + \left(1 + \sum_{i=1}^{n} L_i\right)} \tag{1.27}$$

It can be seen that no change in the fraction of receptors different from the R state will occur in the presence of the ligand only if the affinities of the ligand for every state is the same (equal to K, i.e., $\Psi_{i\ to\ n} = 1$). If the ligand has differential affinity for any of the states R_i, then it will alter the fractional makeup of the ensemble and produce different relative quantities of the specific conformations. Thus, a receptor system, at any given instant, provides a ligand with a choice of conformations to which it can bind; in effect, the ligand enters a "conformational cafeteria." The ligand, in turn, will bind to proportions of states commensurate with the respective affinities it has for each state. However, as these conformations are interconvertible, the proportions of the states will change according to the selective affinities they have for the ligand; that is, the ligand becomes a stabilizing influence toward certain preferred states for which it has the highest affinity. This is an extension of the effects described for two states but highlights the probability of changes in ensembles with increasing conformational states; that is, the more states there are, the higher the probability that a given ligand with macroaffinity for the receptor will alter the relative quantities of the various conformations upon binding.

The rich range of behaviors observed for 7TM receptors, intrinsic or in response to chemicals or cellular reactants, makes these proteins versatile and complex control points for physiological function. This, in turn, gives pharmacologists and medicinal chemists vast opportunity to exploit these complex control units through selective binding to specific tertiary conformations. The complexity of this overall system is both advantageous (high potential) and disadvantageous (difficulty in assessing value). It may be that introduction of new selective ligands into the most complex system of all (patients in the clinic) will be the step that discerns value from extraneous complexity.

REFERENCES

1. Rang, H.P. (2006) The receptor concept: Pharmacology's big idea. *Br J Pharmacol.* 147, S9–S16.
2. Holmstedt, B. Liljestrand, G. (1981) *Readings in Pharmacology*. New York: Raven Press.
3. Clark, A.J. (1937) General pharmacology. In *Heffter's Handbuch Der Experimentellen Pharmakologie*, Ergband 4. Berlin: Springer.
4. Clark, A.J. (1933) *The Mode of Action of Drugs on Cells*. London: Edward Arnold.
5. Ariens, E.J. (1954) Affinity and intrinsic activity in the theory of competitive inhibition. *Arch Int Pharmacodyn Ther.* 99, 32–49.
6. Ariens, E.J. (1964) *Molecular Pharmacology*, Vol. 1. New York: Academic Press.
7. Van Rossum, J.M. (1966) Limitations of molecular pharmacology. Some implications of the basic assumptions underlying calculations on drug-receptor interactions and the significance of biological drug parameters. *Adv Drug Res.* 3, 189–223.
8. Stephenson, R.P. (1956) A modification of receptor theory. *Br J Pharmacol.* 11, 379–393.
9. Furchgott, R.F. (1966) The use of β-haloalkylamines in the differentiation of receptors and in the determination of dissociation constants of receptor-agonist complexes. *Adv Drug Res.* 3, 21–55.
10. Furchgott, R.F. (1972) The classification of adrenoreceptors (adrenergic receptors). An evaluation from the standpoint of receptor theory. In *Handbook of Experimental Pharmacology, Catecholamines*, Vol. 33, ed. H. Blaschko, E. Muscholl, pp. 283–335. Berlin: Springer-Verlag.
11. Kenakin, T.P. (1975) Blockade of histamine-induced contractions of intestinal smooth muscle by β-haloalkylamines. PhD thesis, University of Alberta, Edmonton, Alberta.
12. Black, J.W., Leff, P. (1983) Operational models of pharmacological agonist. *Proc R Soc Lond* [Biol]. 220, 141–162.
13. Paton, W.D.M., Rang, H.P. (1965) The uptake of atropine and related drugs by intestinal smooth muscle of the guinea pig in relation to acetylcholine receptors. *Proc R Soc Lond* [Biol]. 163, 1–44.
14. Gaddum, J.H. (1937) The quantitative effects of antagonistic drugs. *J Physiol Lond.* 89, 7P–9P.
15. Ehlert, F.J. (1988) Estimation of the affinities of allosteric ligands using radioligand binding and pharmacological null methods. *Mol Pharmacol.* 33, 187–194.
16. Kenakin, T.P. (2005) New concepts in drug discovery: Collateral efficacy and permissive antagonism. *Nat Rev Drug Discov.* 4, 919–927.
17. Kenakin, T.P., Jenkinson, S., Watson, C. (2006) Determining the potency and molecular mechanism of action of insurmountable antagonists. *J Pharmacol Exp Ther.* 319, 710–723.

18. Price, M.R., Baillie, G.L., Thomas, A., Stevenson, L.A., Easson, M., Goodwin, R., McLean, A., McIntosh, L., Goodwin, G., Walker, G., Westwood, P., Marrs, J., Thomson, F., Cowley, P., Christopoulos, A., Pertwee, R.G., Ross, R.A. (2005) Allosteric modulation of the cannabinoid CB1 receptor. *Mol Pharmacol.* 68, 1484–1495.
19. Christopolous, A., Kenakin, T.P. (2002) G-protein coupled receptor allosterism and complexing. *Pharmacol Rev.* 54, 323–374.
20. Kew, J.N.C., Trube, G., Kemp, J.A. (1996) A novel mechanism of activity-dependent NMDA receptor antagonism describes the effects of ifenprodil in rat cultured cortical neurons. *J Physiol.* 497, 761–772.
21. Cuatrecasas, P. (1974) Membrane receptors. *Annu Rev Biochem.* 43, 169–214.
22. Katz, B., Thesleff, S. (1957) A study of the "desensitization" produced by acetylcholine at the motor end-plate. *J Physiol.* 138, 63–80.
23. Del Castillo J., Katz B. (1957) Interaction at end-plate receptors between different choline derivatives. *Proc R Soc London* [Biol]. 146, 369–381.
24. Monod, J., Wyman, J., Changeux, J.P. (1965) On the nature of allosteric transitions. *J Biol Chem.* 12, 88–118.
25. Colquhoun, D. (1973) The relationship between classical and cooperative models for drug action. In *Drug Receptors*, ed. H.P. Rang, pp. 149–82. Baltimore, MD: University Park Press.
26. Karlin, A. (1967) On the application of a "plausible model" of allosteric proteins to the receptor for acetylcholine. *J Theor Biol.* 16, 306–320.
27. Thron, C.D. (1973) On the analysis of pharmacological experiments in terms of an allosteric receptor model. *Mol Pharmacol.* 9, 1–9.
28. Burgen, A.S.V. (1966) Conformational changes and drug action. *Fed Proc.* 40, 2723–2728.
29. DeLean, A., Stadel, J.M., Lefkowitz, R.J. (1980) A ternary complex model explains the agonist-specific binding properties of adenylate cyclase coupled β-adrenergic receptor. *J Biol Chem.* 255, 7108–7117.
30. Limbird, L.E., Lefkowitz, R.J. (1978) Agonist induced increase in apparent β-adrenergic receptor size. *Proc Natl Acad Sci U S A.* 75, 228–232.
31. Ross, E.M., Maguire, M.E., Sturgill, T.W., Biltonen, R.L., Gilman, A.G. (1977) Relationship between the β-adrenergic receptor and adenylate cyclase. *J Biol Chem.* 252, 5761–5775.
32. Samama, P., Cotecchia, S., Costa, T., Lefkowitz, R.J. (1993) A mutation-induced activated state of the β_2-adrenergic receptor: Extending the ternary complex model. *J Biol Chem.* 268, 4625–4636.
33. Costa, T., Herz, A. (1989) Antagonists with negative intrinsic activity at δ-opioid receptors coupled to GTP-binding proteins. *Proc Natl Acad Sci U S A.* 86, 7321–7325.
34. Weiss, J.M., Morgan, P.H., Lutz, M.W., Kenakin, T.P. (1996) The cubic ternary complex receptor-occupancy model. I. Model description. *J Theor Biol.* 178, 151–167.
35. Weiss, J.M., Morgan, P.H., Lutz, M.W., Kenakin, T.P. (1996) The cubic ternary complex receptor-occupancy model. II. Understanding apparent affinity. *J Theor Biol.* 178, 169–182.
36. Weiss, J.M., Morgan, P.H., Lutz, M.W., Kenakin, T.P. (1996) The cubic ternary complex receptor-occupancy model. III. Resurrecting efficacy. *J Theor Biol.* 181, 381–397.

37. Roerig, S.C., Loh, H.H., Law, P.Y. (1992) Identification of three separate guanine nucleotide-biding proteins that interact with the δ-opioid receptor in NG108-15 X glioma hybrid cells. *Mol Pharmacol.* 41, 822–831.
38. Laugwitz, K.L., Allgeier, A., Offermanns, S., Spicher, K., Van Sande, J., Dumont, J.E., Schultz, G. (1996) The human thyrotropin receptor: A heptahelical receptor capable of stimulating members of all four G protein families. *Proc Natl Acad Sci U S A.* 93, 116–120.
39. Gudermann, T., Kalbrenner, F., Schultz, G. (1996) Diversity and selectivity of receptor-G protein interaction. *Annu Rev Pharmacol Toxicol.* 36, 429–459.
40. Hermans, E. (2003) Biochemical and pharmacological control of the multiplicity of coupling at G-protein receptors. *Pharmacol Ther.* 99, 25–44.
41. Albert, P.R., Robillard, L. (2002) G-protein specificity: Traffic direction required. *Cell Signal.* 14, 407–418.
42. Burford, N.T., Wang, D.X., Sadee, W. (2000) G-protein coupling of μ-opioid receptors (OP3) elevated basal signaling activity. *Biochem J.* 348, 531–537.
43. Offermanns, S., Wieland, T., Homann, D., Sandmann, J., Bombien, D., Spicher, K., Schultz, G., Jakobs, K.H. (1994) Transfected muscarinic acetylcholine receptors selectively couple to Gi-type proteins and Gq/11. *Mol Pharmacol.* 45, 890–898.
44. Wise, A., Sheehan, M., Rees, S., Lee, M., Milligan, G. (1999) Comparative analysis of the efficacy of A_1 adenosine receptor activation of $G_{i/o}\alpha$ G proteins following coexpression of receptor and G protein and expression of A_1 adenosine receptor $G_{i/o}\alpha$ fusion proteins. *Biochem.* 38, 2272–2278.
45. Kohout, T.A., Lefkowitz, R.J. (2003) Regulation of G protein-coupled receptor kinases and arrestins during receptor desensitization. *Mol Pharmacol.* 63, 9–18.
46. Scott, M.G.H., Benmerah, A., Muntaner, O., Manullo, S. (2002) Recruitment of activated G protein-coupled receptors to pre-existing clathrin-coated pits in living cells. *J Biol Chem.* 277, 3552–3559.
47. DeWire, S.M., Ahn, S., Lefkowitz, R.J., Shenoy, S.K. (2007) β-arrestins and cell signaling. *Annu Rev Physiol.* 69, 483–510.
48. Lefkowitz, R.J., Shenoy, S.K. (2005) Transduction of receptor signals by beta-arrestins. *Science.* 308, 212–217.
49. Terrillon, S., Bouvier, M. (2004) Receptor activity-independent recruitment of β-arrestin reveals specific signaling modes. *EMBO J.* 23, 3950–3961.
50. Luttrell, L.M. (2005) Composition and function of G protein-coupled receptor signalsomes controlling mitogen-activated protein kinase activity. *J Mol Neurosci.* 26, 253–263.
51. Azzi, M., Charest, P.G., Angers, S., Rousseau, G., Kohout, T., Bouvier, M. (2003) β-arrestin-mediated activation of MAPK by inverse agonists reveals distinct active conformations for G-protein-coupled receptors. *Proc Natl Acad Sci U S A.* 100, 11406–11411.
52. Seta, K., Nanmori, M., Modrall, J.G., Neubig, R.R., Sadoshima, J. (2002) AT1 receptor mutant lacking heterotrimeric G-protein coupling activates the Src-Ras-ERK pathway. *J Biol Chem.* 277, 9268–9277.
53. Wei, H., Ahn, S., Shenoy, S.K., Karnick, S.S., Hunyady, L., Luttrell, L.M., Lefkowitz, R.J. (2003) Independent β-arrestin 2 and G-protein-mediated pathways for angiotensin II activation of extracellular signal-regulated kinases 1 and 2. *Proc Natl Acad Sci U S A.* 100, 10782–10787.

54. Gesty-Palmer, D., Chen, M., Reiter, E., Ahn, S., Nelson, C.D., Wang S. (2006) Distinct β-arrestin- and G protein-dependent pathways for parathyroid hormone receptor-stimulated ERK1/2 activation. *J Biol Chem.* 281, 10856–10864.

55. Spengler, D., Waeber, C., Pantaloni, C., Holsboer, F., Bockaert, J., Seeburg, P.H., Journot, L. (1993) Differential signal transduction by five splice variants of the PACAP receptor. *Nature.* 365, 170–175.

56. Fraunfelder, H., Parak, F., Young, R.D. (1988) Conformational substrates in proteins. *Annu Rev Biophys Biophys Chem.* 17, 451–479.

57. Fraunfelder, H., Sligar, S.G., Wolynes, P.G. (1991) The energy landscapes and motions of proteins. *Science.* 254, 1598–1603.

58. Hilser, J., Freire, E. (1997) Predicting the equilibrium protein folding pathway: Structure-based analysis of staphylococcal nuclease. *Proteins.* 27, 171–183.

59. Hilser, J., Dowdy, D., Oas, T.G., Freire, E. (1998) The structural distribution of cooperative interactions in proteins: Analysis of the native state ensemble. *Proc Natl Acad Sci U S A.* 95, 9903–9908.

60. Onaran, H.O., Costa, T. (1997) Agonist efficacy and allosteric models of receptor action. *Ann N Y Acad Sci.* 812, 98–115.

61. Onaran, H.O., Scheer, A., Cotecchia, S., Costa, T.A. (2000) A look at receptor efficacy. From the signaling network of the cell to the intramolecular motion of the receptor. In *The Pharmacology of Functional, Biochemical, and Recombinant Systems Handbook of Experimental Pharmacology*, Vol. 148, ed. T.P. Kenakin, J.A. Angus, pp. 217–280. Heidelberg: Springer.

62. Kenakin, T.P. (1995) Agonist-receptor efficacy II: Agonist-trafficking of receptor signals. *Trends Pharmacol Sci.* 16, 232–238.

63. Kenakin, T.P. (1998) Agonist selective receptor states: The new level of selectivity? *Pharm News.* 5, 20–25.

64. Kenakin, T.P. (2002) Efficacy at G protein coupled receptors. *Annual Rev Pharmacol Toxicol.* 42, 349–379.

65. Kenakin, T.P. (2003) Ligand-selective receptor conformations revisited: The promise and the problem. *Trends Pharmacol Sci.* 24, 346–354.

66. Kenakin, T.P. (2006) Collateral efficacy as pharmacological problem applied to new drug discovery. *Expert Opin Drug Disc.* 1, 635–652.

67. Urban, J.D., Clarke, W.P., von Zastrow, M., Nichols, D.E., Kobilka, B., Weinstein, H., Javitch, J.A., Roth, B.L., Christopoulos, A., Sexton, P.M., Miller, K.J., Spedding, M., Mailman, R.B. (2007) Functional selectivity and classical concepts of quantitative pharmacology. *J Pharmacol Exp Ther.* 320, 1–13.

68. Gether, U., Lin, S., Kobilka, B.K. (1995) Fluorescent labeling of purified β_2-adrenergic receptor: Evidence for ligand specific conformational changes. *J Biol Chem.* 270, 28268–28275.

69. Ghanouni, P., Gryczynski, Z., Steenhuis, J.J., Lee, T.W., Farrens, D. L. Lakowicz, J.R., Kobilka, B.K. (2001) Functionally different agonists produce distinct conformations in G-protein coupling domains of the β2-adrenergic receptor. *J Biol Chem.* 276, 24433–24436.

70. Palanche, T., Ilien, B., Zoffmann, S., Reck, M.P., Bucher, B., Edelstein, S.J., Galzi, J.L. (2001) The neurokinin A receptor activates calcium and cAMP responses through distinct conformational states. *J Biol Chem.* 276, 34853–34861.

71. Swaminath, G., Xiang, Y., Lee, T.W., Steehuis, J., Parnot, C., Kobilka, B.K. (2004) Sequential binding of agonists to the β_2 adrenoceptor: Kinetic evidence for intermediate conformational states. *J Biol Chem.* 279, 686–691.

72. Hruby, V.J., Tollin, G. (2007) Plasmon-waveguide resonance (PWR) spectroscopy for directly viewing rates of GPCR/G-protein interactions and quantifying affinities. *Curr Opin Pharmacol.* 7, 507–514.

73. Rees, S., Morrow, D., Kenakin, T. (2002) PCR drug discovery through the exploitation of allosteric drug binding sites. *Receptors Channels.* 8, 261–268.

74. Watson, C., Jenkinson, S., Kazmierski, W., Kenakin, T.P. (2005) The CCR5 receptor-based mechanism of action of 873140, a potent allosteric non-competitive HIV entry-inhibitor. *Mol Pharmacol.* 67, 1268–1282.

75. Maeda, K., Nakata, H., Koh, Y., Miyakawa, T., Ogata, H., Takaoka, Y., Sbibayama, S., Sagawa, K., Fukushima, D., Moravek, J., Koyanagi, Y., Mitsuya, H. (2004) Spirodiketopiperazine-based CCR5 inhibitor which preserves CC-chemokine/CCR5 interactions and exerts potent activity against R5 human immunodeficiency virus type 1 *in vitro*. *J Virol.* 78, 8654–8662.

76. Kondru, R., Zhang, J., Ji, C., Mirzadegan, T., Rotstein, D., Sankuratri, S., Dioszegi, M. (2008) Molecular interactions of CCR5 with major classes of small-molecule anti-HIV CCR5 antagonists. *Mol Pharmacol.* 73, 789–800.

77. Kenakin, T. (2006) Collateral efficacy as pharmacological problem applied to new drug discovery. *Expert Opin Drug Discov.* 1, 635–652.

78. Gray, J.A., Roth, B.L. (2001) Paradoxical trafficking and regulation of 5-HT_{2A} receptors by agonists and antagonists. *Brain Res Bull.* 56, 441–451.

79. Hall, R.A., Premont, R.T., Leflowitz, R.J. (1999) Heptahelical receptor signaling: Beyond the G protein paradigm. *J Cell Biol.* 145, 927–932.

80. Heuss, C., Gerber, U. (2000) G-protein-independent signaling by G-protein-coupled receptors. *Trends Neurosci.* 23, 469–475.

81. Brady, A.E., Limbird, L.E. (2002) G protein-coupled receptor interacting proteins: Emerging roles in localization and signal transduction. *Cell Signal.* 14, 297–309.

82. Lanier, S.M. (2004) AGS proteins, GPR motifs and the signals processed by heterotrimeric G proteins. *Biol Cell.* 96, 369–372.

83. Hill, S.J. (2006) G-protein-coupled receptors: Past, present and future. *Br J Pharmacol.* 147, S27–S37.

84. Kenakin, T.P., Onaran, O. (2002) The ligand paradox between affinity and efficacy: Can you be there and not make a difference? *Trends Pharmacol Sci.* 23, 275–280.

CHAPTER 2

The Evolving Pharmacology of GPCRs

LAUREN T. MAY, NICHOLAS D. HOLLIDAY, and STEPHEN J. HILL

Institute of Cell Signalling, School of Biomedical Sciences, Queen's Medical Centre, Nottingham, UK

2.1. AGONISTS, NEUTRAL ANTAGONISTS, AND INVERSE AGONISTS

G protein-coupled receptors (GPCRs) represent the largest superfamily of cell surface receptors, with approximately 800 functioning receptor proteins encoded within the human genome. Conserved structural GPCR motifs include seven α-helical transmembrane (TM) domains of approximately 25–35 amino acids, an extracellular N-terminal domain, and an intracellular C-terminal domain. GPCR TM domains are relatively conserved and form the basis of the GPCR classification system of GPCRs into six families [1]. Although some structural motifs are highly conserved, sufficient sequence variability exists within the GPCR superfamily to allow selective accommodation of a vast array of endogenous stimuli, including ions, organic odorants, amines, peptides, proteins, lipids, nucleotides, and photons [1]. In addition, the identification of natural and synthetic compounds found to either mimic or block the actions of endogenous agonists continually increases the diversity of GPCR ligands in the chemical space. The type and strength of interaction a compound has with its receptor is driven both by the chemical structure of the ligand and the nature of the receptor epitopes available within the binding pocket. Classical receptor–ligand interactions define two important system-independent properties of a compound: affinity and intrinsic efficacy.

2.1.1. Affinity and Efficacy

Affinity describes the capacity of the ligand to bind to a receptor and is driven by electrostatic forces, hydrogen bonding, and van der Waals forces.

GPCR Molecular Pharmacology and Drug Targeting: Shifting Paradigms and New Directions, Edited by Annette Gilchrist

Mathematically, the affinity of a ligand is defined as the rate of association divided by the rate of dissociation (k_{on}/k_{off} in Fig. 2.1a). Receptor occupancy is determined both by ligand affinity and concentration, described by the Langmuir adsorption isotherm (rewritten using pharmacological nomenclature; reviewed in Reference 2) as:

$$[AR]/[R_T] = [A]/([A] + K_A)$$

$[R_T]$ represents the density of a given receptor population and [AR] denotes the concentration of agonist (A) receptor complexes. K_A is the equilibrium dissociation constant or alternatively, the reciprocal of the affinity constant. From a plot of fractional occupancy against ligand concentration, the number of receptive binding sites in the preparation is defined by the maximal asymptote and the equilibrium dissociation constant equal to the concentration of ligand required to occupy half of the total receptor population.

Figure 2.1 Pharmacological models of ligand binding. (a) The affinity of a ligand (A) for a receptor (R) is determined both by its rate of association (k_{on}) and dissociation (k_{off}). (b) Left. The two-state model of receptor activation. Middle. The ternary complex model. Right. The allosteric ternary complex model. Within each of these models, the equilibrium dissociation constant of the ligand (A) for the inactive receptor (R) is represented by K_A. Receptors within the two-state model of receptor activation (*left*), can transition between an inactive (R) and active state (R*), defined by the isomerization constant (L); ligand preference for the active receptor is governed by α. The equilibrium dissociation constant of the G protein (G) within the ternary complex model (*middle*) is represented by K_G. The cooperativity between the binding sites when the ternary complex is formed is described by α. The allosteric ternary complex model (*right*) describes orthosteric (A) and allosteric (B) ligand binding, the equilibrium dissociation constants represented as K_A and K_B, respectively. The cooperativity between the two binding sites when the receptor is simultaneously bound by both ligands is described by α. (c) Left. The cubic ternary complex model. Right. The allosteric two-state model. Within these models, the receptor can isomerize between an inactive (R) and an active (R*) state, defined by the isomerization constant (L). Within the cubic ternary complex (*left*), the equilibrium dissociation constant for the ligand (A) and G protein (G) at the inactive receptor (R) are represented by K_A and K_G, respectively. The thermodynamic coupling factor α describes the effect of ligand binding on receptor isomerization, β describes the effect of receptor activation on G protein coupling, γ describes the effect of G protein-coupling on ligand binding, and δ describes the effect of concomitant ligand and G protein-coupling on receptor isomerization. Within the allosteric two-state model (*right*), the equilibrium dissociation constant for the orthosteric ligand (A) and the allosteric ligand (B) at the inactive receptor (R) are represented by K_A and K_B, respectively. The thermodynamic coupling factor γ describes the effect of orthosteric ligand binding on receptor isomerization, β describes the effect of receptor activation on allosteric ligand binding, α describes the effect of allosteric ligand binding on orthosteric ligand binding, and δ describes the effect of concomitant orthosteric and allosteric ligand binding on receptor isomerization.

Intrinsic efficacy (ε), originally described by Furchgott (1966), describes the stimulus (S) exerted by a ligand per receptor molecule, according to the following equation [3]:

$$S = \varepsilon[R_T] \cdot \frac{[A]}{K_A + [A]}$$

Assumptions associated with this model include uniformity in receptor type, the attainment of system equilibrium, and the lack of ligand depletion or receptor desensitization. The ensuing cellular response (E) is generally assumed to be a hyperbolic function (f) of the initial stimulus exerted upon the receptor population. Therefore, E = stimulus/(stimulus + β) where β represents the coupling efficiency of the stimulus to the cellular response. Under these conditions, the cellular response can be expressed as:

$$E = f[([A][R_T]\varepsilon/\beta)/([A](([R_T]\varepsilon/\beta)+1)+K_A)]$$

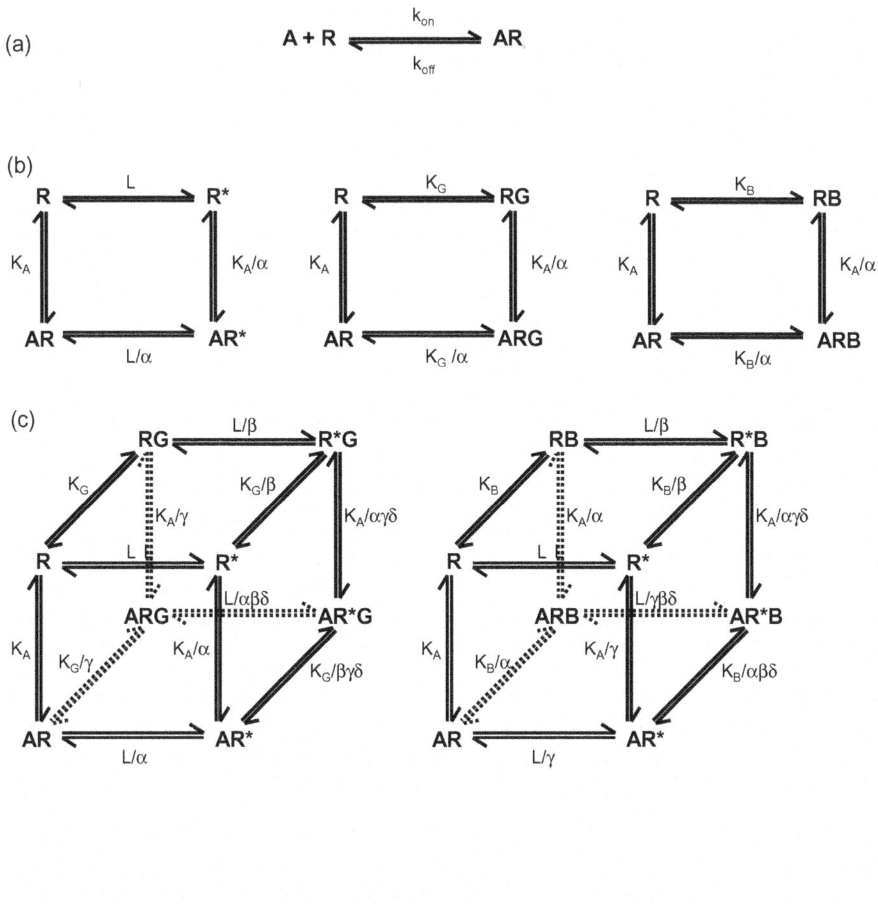

Relative efficacy ($\varepsilon[R_T]/\beta$) of a ligand, together with its affinity K_A, is defined by the application of this equation to experimental data. The stimulus–response coupling component (β) and receptor concentration ($[R_T]$) within the relative efficacy term mean that the relative efficacy observed for a particular ligand is influenced by the coupling efficiency of the effector pathway of interest and the receptor concentration (Fig. 2.2a). A ligand with low positive efficacy may behave as an agonist in a well-coupled system and as an antagonist in a poorly coupled system. A similar relationship is also true in terms of the impact receptor concentration has on the relative efficacy of a ligand. Therefore, in contrast to intrinsic efficacy, which is a system independent property of the ligand, relative efficacy also depends on receptor density and the coupling efficiency of the response which is measured.

An inherent property of the stimulus–response relationship described by Furchgott [3] is the ability of an agonist to impart stimulus to the receptor. Within this context, ligands with positive intrinsic efficacy, agonists, induce a conformational change within the receptor that promotes coupling to intracellular effectors. Partial agonism results from the induction of a receptor conformation that is less effective at coupling to intracellular effectors. Antagonists possess no efficacy, acting only to inhibit agonist-mediated receptor activation. Recent studies using biophysical techniques such as fluorescent spectroscopy and fluorescence resonance energy transfer (FRET) support the suggestion that the conformational changes observed within a GPCR upon ligand binding correlate with the efficacy of the ligand (reviewed in References 4, 5). Fluorescent spectroscopy studies using the human β_2-adrenergic receptor have suggested that agonist binding may behave according to a sequential

Figure 2.2 Ligand directed GPCR signaling. Four examples are illustrated, based on a hypothetical G_S-coupled GPCR with two cellular responses, cAMP accumulation and ERK activation. Ligand 1 reflects the behavior of a classical agonist in each case. Responses in (a) are sufficiently explained by the lower efficacy of ligand 2, which still elicits the same active receptor state as ligand 1. Ligand 2 is a weak partial agonist/neutral antagonist in the cAMP assay, but more efficacious in the ERK assay, because of differences in stimulus coupling efficiency between the two end points (Section 2.1). However, in a well-characterized system with a single receptor population, reversal of the rank orders of potency of ligands 1 and 2 in the two assays (b) requires ligand selection of different active receptor conformations, which directed signaling to either the cAMP or ERK pathway. Equally, the opposing efficacies of ligand 2 in (c) suggest an adoption of a distinct receptor state which couples selectively to the ERK pathway. If the two end points display differing levels of constitutive activity, these responses can also be accounted for by protean agonism (d), in which the receptor conformation stabilized by ligand 2 has lower coupling efficacy than the spontaneously active state. In this instance, ligand 2 will display inverse agonism in a constitutively active system, but agonism in a silent system, even when the same response is measured in both cases.

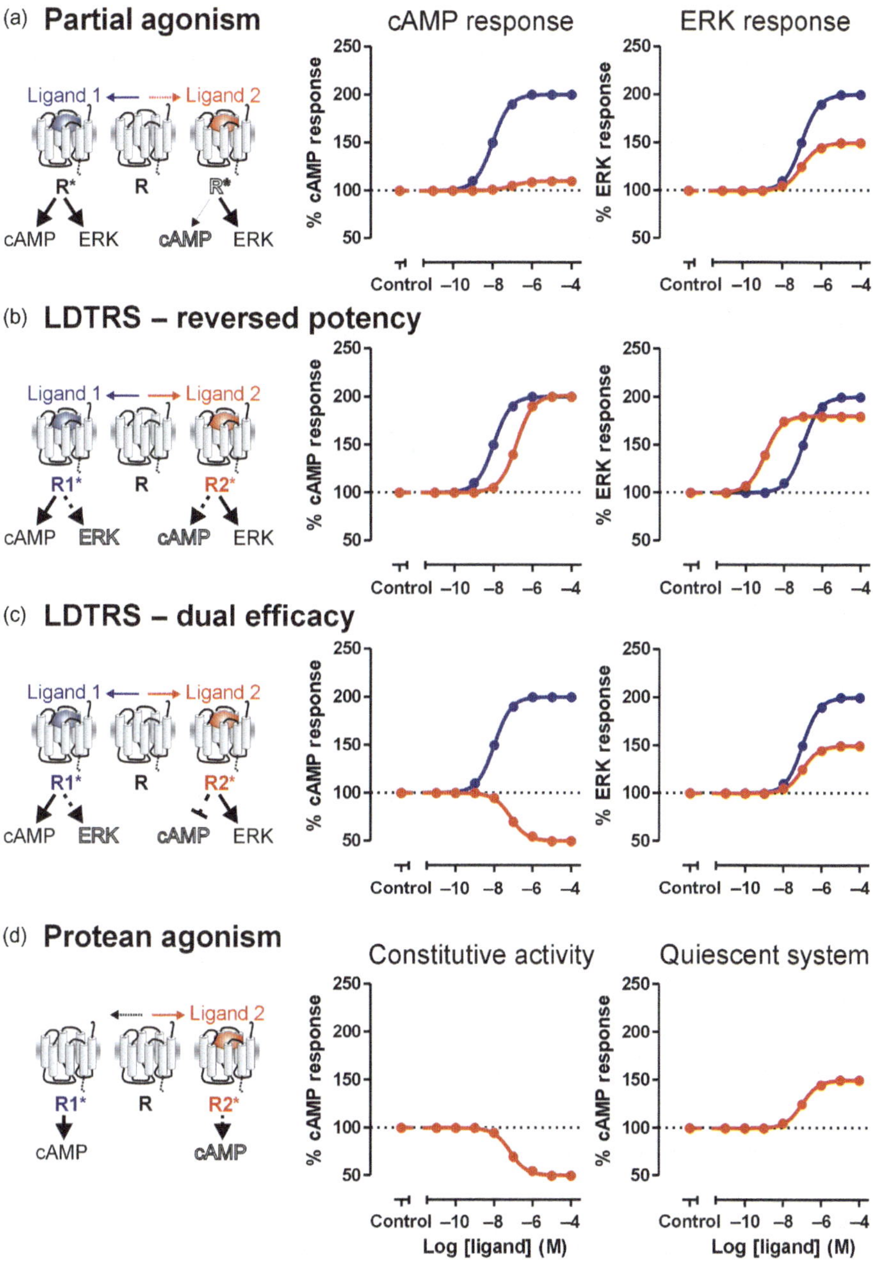
(a) Partial agonism
cAMP response
ERK response
Ligand 1
Ligand 2
R*
R
cAMP
ERK
cAMP
ERK
% cAMP response
% ERK response
250
200
150
100
50
Control –10 –8 –6 –4
(b) LDTRS – reversed potency
Ligand 1
Ligand 2
R1*
R
R2*
cAMP
ERK
cAMP
ERK
(c) LDTRS – dual efficacy
Ligand 1
Ligand 2
R1*
R
R2*
cAMP
ERK
cAMP
ERK
(d) Protean agonism
Constitutive activity
Quiescent system
Ligand 2
R1*
R
R2*
cAMP
cAMP
Log [ligand] (M)
Log [ligand] (M)

model. Within this model, an agonist has the potential to stabilize a number of intermediate receptor conformations, each of which may be linked to a particular intracellular signaling pathway [6]. Kinetic FRET studies using the α_2-adrenoceptor have also suggested that rate and direction of the conformational changes seen within a GPCR upon ligand binding can be used as an index of ligand efficacy. The rate of conformational change was found to be rapid for a full agonist, intermediate for a partial agonist, and slow and in the opposite direction for ligands with negative efficacy, known as inverse agonists [5, 7, 8].

An inverse agonist is defined as a ligand that can decrease the constitutive activity of a receptor, that is, the spontaneous activity of receptor in the absence of an agonist. In most endogenous systems, the level of constitutively active receptors is not sufficient to distinguish an inverse agonist from a neutral antagonist. As a consequence, the phenomenon of constitutive activity evaded detection experimentally prior to the advent of molecular biology and the development of recombinant cell lines. These techniques, in combination with the increased sensitivity of current functional assays, has meant that many ligands originally classified as antagonists have subsequently been shown to decrease constitutive activity and as such, are now referred to as inverse agonists (reviewed in Reference 9). Constitutively active GPCRs have been implicated in important aspects of human physiology and pathophysiology. For example, constitutively active melanocortin MC4 receptors and ghrelin receptors play key roles in determining the set point of hypothalamic neuronal pathways regulating body weight [10, 11]. Indeed, the activity of MC4 receptors is not only influenced by an endogenous neuropeptide agonist (α-melanocyte-stimulating hormone) but also by an endogenous neuropeptide inverse agonist (Agouti-related protein). The pathologies of retinitis pigmentosa and hyperthyroidism are also thought to be a consequence of naturally occurring mutations that cause increased constitutive activity of various GPCRs [9]. In addition, Karposi's sarcoma-associated herpes virus (KSHV) utilizes a constitutively active viral version of the human CXC chemokine receptor subtype 2 to promote cell proliferation and continued viral replication [12, 13].

2.1.2. Pharmacological Models of Agonism, Antagonism, and Inverse Agonism

The simplest models that describe the actions of agonists, antagonists, and inverse agonists are the two state model of receptor activation and the ternary complex model (TCM) [14–16]. Within the two state model of receptor activation (Fig. 2.1b, left), receptors can transition between inactive (R) or constitutively active (R*), an equilibrium which is defined by the isomerization constant (L). The efficacy of a ligand is determined by its preference for binding the active over the inactive form of the receptor (α). The TCM (Fig. 2.1b, middle) describes the actions of a ligand (A) in terms of its affinity for the free receptor (K_A) and the binding cooperativity between the ternary complex (α), defined as the GPCR bound by both ligand and G protein.

Models combining the two state model of agonist action and the TCM have also been developed to describe the effect of a ligand and G protein on a receptor that isomerizes between an inactive and an active state. Two such models are the extended TCM and the more thermodynamically complete version, the cubic TCM (Fig. 2.1c, left). Parameters within the cubic TCM include the affinity, K_A and K_G, of the ligand (A) and G protein (G) for the inactive receptor, respectively, the isomerization constant (L), and the thermodynamic coupling factors α, β, γ, and δ. The factors α and β describe the preference of the ligand and G protein, respectively, for the active over the inactive form of the receptor. The interaction between ligand binding and G protein coupling is described by γ, whereas δ describes the ability of the ternary complex to drive receptor isomerization. Within the cubic TCM, GPCR signaling is thought to emanate from the G protein-bound active form of the receptor. As α describes the preference of a ligand for the active over the inactive form of the receptor, this factor represents the intrinsic efficacy of a ligand. Values of α greater than one will drive the equilibrium toward the active form of the receptor, mediating a subsequent increase in receptor stimulus. In terms of intrinsic efficacy, the value of α for an agonist would be greater than one, for a neutral antagonist equal to one, and for an inverse agonist between zero and one. The intrinsic efficacy of a partial agonist will always be lower than that of a full agonist, reflecting less of a preference for the active over the inactive form of the receptor. In contrast to the TCM, the cubic TCM does not require an agonist to possess positive binding cooperativity with the G protein to mediate receptor signaling. A ligand with neutral G protein binding cooperativity but positive efficacy will still drive an increase in the concentration of active receptor–G protein complexes.

Linkage models such as the extended TCM and the cubic TCM are useful in that they can incorporate the constitutive activity of GPCRs; however, an inherent limitation of this type of model is the requirement to predefine the different types of interacting receptor species. The probabilistic model of GPCR behavior represents an alternative and much more versatile approach that allows unbound GPCRs to fluctuate between a spectrum of receptor conformations, the distribution of which is altered upon ligand binding and/or interactions with other proteins. This model can accommodate a range of GPCR behaviors, and as such differs from the classical idea of GPCR activation resulting in a single, linear signaling pathway [14]. The probabilistic model is also useful for understanding how different agonists can selectively enrich distinct subsets of active receptor conformations, a phenomenon that can lead to ligand-directed trafficking of receptor stimulus (LDTRS) and protean agonism. However, although the probabilistic model can allow qualitative interpretation of the more complex and divergent behaviors of GPCRs, linkage models currently remain fundamental for the quantification of ligand parameters such as affinity and efficacy.

2.2. LDTRS/PROTEAN AGONISM

LDTRS is the concept that different ligands can direct intracellular signaling to particular pathways. It was first introduced by Berg et al. (1998) who showed that the 5-HT_{2C} receptor can couple to two different pathways in Chinese hamster ovary cells (inositol phosphate accumulation and arachidonic acid release) and exhibit a markedly different rank order of efficacies for each response (Fig. 2.2b) [17]. It has been proposed that LDTRS may result from the ability of receptors to spontaneously adopt a variety of different conformations and that a particular ligand can stabilize a certain conformational species to produce a particular biological response [18]. Some insight into how the pharmacology of a particular ligand–receptor interaction can be altered by the nature of the transducing protein (e.g., heterotrimeric G protein alpha subunit) with which the ligand–receptor complex interacts can, however, also be gained from a consideration of allosteric interactions (see also Section 2.4 below). For all intents and purposes, the binding site on the receptor at which the receptor–G protein interaction occurs can be considered to be an allosteric binding site, with the G protein alpha subunit acting as the allosteric regulator via this site (Fig. 2.1c). As a consequence, the binding of the receptor to a particular G protein alpha subunit can change the way in which a particular exogenous ligand can interact with the receptor's orthosteric binding site (at which the endogenous agonist normally interacts).

The potential for ligand-directed trafficking is further complicated by the increasing evidence that GPCRs can elicit downstream signaling responses that are mediated independently of heterotrimeric G proteins [19–21]. For example, it is becoming clear that β-arrestin may have an important role in orchestrating both the location of receptors within particular signaling domains of a cell and their ability to trigger a range of different responses, including the activation of the p42/44 MAP kinase pathway [22–25]. These G protein-independent mechanisms are not limited to those involving β-arrestin, and evidence has accumulated for a role for other signaling and scaffolding proteins, including caveolin-1, A-kinase anchoring proteins, and the Na/K exchange protein [26–29].

Receptor-mediated events within the cell are therefore complex, and there is the possibility that a given receptor can evoke multiple effects via a combination of G protein and non-G protein-mediated events within the same cell. This complexity is nicely illustrated by the human β_2-adrenoceptor, which can signal through both cyclic adenosine monophosphate (AMP) accumulation and extracellular signal-regulated kinases 1 and 2 (ERK1/2) phosphorylation [22, 23, 30]. At this receptor, the prototypical β-blocker propranolol can act as an inverse agonist on Gs-mediated cyclic AMP accumulation, but as an agonist of β-arrestin-mediated ERK1/2 phosphorylation (Fig. 2.2c,d) [22, 23]. Furthermore, evidence is now accumulating to suggest that antagonist affinities may also vary depending upon the agonist and signaling response used to screen for them [30].

The concept that certain agonists can reverse their effects under different physiological situations was suggested by Kenakin (2001) in his discussion of protean agonism [31]. His hypothesis was that some agonists (protean agonists) may produce an active receptor conformation that has a lower efficacy than the spontaneously active conformation (in the absence of agonist) that is responsible for constitutive receptor activity. As a consequence, in a constitutively active system where there is a significant proportion of spontaneously active receptors, activation by a protean agonist would lead to a ligand–receptor conformation that has lower efficacy than the constitutively active (agonist-free) receptor conformation. In this situation, the protean agonist would manifest itself as inverse agonist. In a quiescent system (i.e., no constitutive activity), the same ligand would produce an agonist effect by virtue of changing the predominant resting state of the receptor to the more active protean conformation. Thus, in a quiescent system, the protean ligand acts as an agonist and, in a constitutively active system, the ligand is an inverse agonist. Protean agonism has been suggested to explain the *in vitro* and *in vivo* actions of proxyfan at the histamine H_3-receptor (which appears to be constitutively active under physiological conditions; Fig. 2.2d) [32].

2.3. MOLECULAR MECHANISMS OF GPCR LIGAND BINDING

Of the six GPCR structural classes, three families (A–C) are responsible for recognition of endogenous neurotransmitters and hormones [1]. Rhodopsin-like receptors, also called family A or family 1 (Fig. 2.3), form the overwhelming majority of ~670 members, including ~390 odorant receptors. In this class, 1–3TM amino acid residues per helix are highly conserved, motifs which include the DRY motif at the base of TM3, the NPXXY motif in TM7, and single proline residues which generate kinks in TM domains 5 and 6 (Fig. 2.3a; [1]). This conservation is central to the integrity of the GPCR helix bundle, and for the transition between inactive and active conformations. There is substantial diversity in the remaining amino acid sequences of individual family A receptors, particularly in the extracellular loop (ECL) and TM domains. This should not be surprising, because the ligand binding sites in rhodopsin-like receptors must generate specificity for the greatest chemical array of ligands of all the GPCR classes. These not only include many agonists that are small molecules, such as amines, nucleotides, or lipids, but also peptides and large glycoproteins.

2.3.1. Rhodopsin-Like Receptor Binding Sites

The classical model of ligand binding involves a pocket formed within the receptor helix bundle, now illuminated by the crystal stuctures of inverse agonist-bound rhodopsin (covalently attached 11-*cis*-retinal; [33]), β-adrenoceptors (bound to cyanopindolol or carazolol; [34, 35]), and the A_{2A}

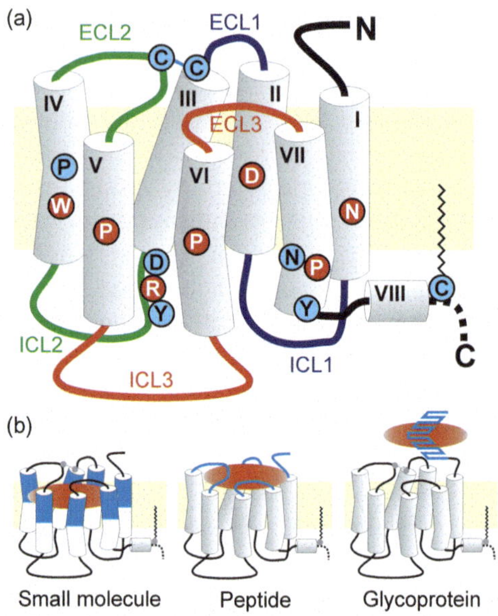

Figure 2.3 (a) A schematic representation of a typical rhodopsin-like GPCR. The transmembrane helices (I–VII) are arranged counterclockwise (viewed from the extracellular side), beginning with the extracellular N-terminus and including three extracellular and intracellular loops (ECL, ICL). There are invariant amino acids in each transmembrane domain (white on red) that provide a reference point for all rhodopsin-like receptors in the Ballesteros–Weinstein numbering system (X.50). Other highly conserved motifs mentioned in the text are also indicated (black on blue). (b) Different modes of binding for rhodopsin-like GPCRs are summarized by the binding "pockets" occupied by different sizes of agonist (red) and the broad regions of receptor involved in ligand interaction (blue).

adenosine receptor [36]. For example, Fig. 2.4 highlights important residues involved in adrenaline binding to the β_2-adrenoceptor, localized to the upper half of different TM domains. Asp113 (3.32) forms an ionic bond with the adrenaline secondary amine, while the ligand phenol hydroxyl groups are recognized by TM5 Ser204 (5.43) and Ser207 (5.46) [37, 38]. Additional van der Waal's contacts are made between the aromatic ring itself and Phe290 (6.52) [39], while the adrenaline β-hydroxyl is likely to form a hydrogen bond with Asn312 (7.39) [34]. All β-ligands contain the same aliphatic backbone with amine and β-hydroxyl groups as adrenaline, and the corresponding points of interaction with the β_2-adrenoceptor, such as Asp113, remain important. In contrast, the accommodation of the aromatic groups in the binding pocket is different, leading to agonist or inverse agonist compounds of varying efficacy [40]. Residues lining the amine GPCR subtype binding pockets are highly conserved (e.g., between β_1 vs. β_2-adrenoceptors; [35, 41]), driven by the chemical similarity of the endogenous ligands—for example, an equivalent TM3

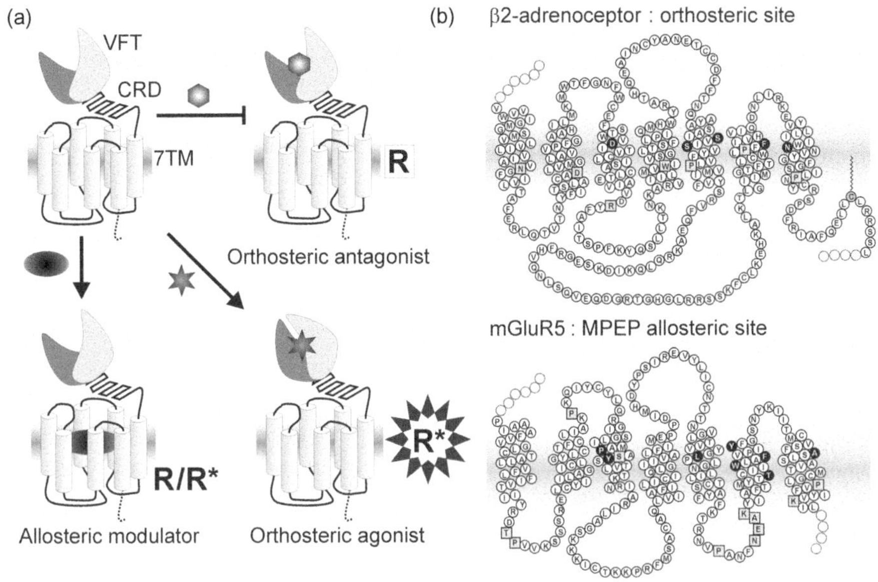

Figure 2.4 (a) Othosteric and allosteric ligands at class C GPCRs. Orthosteric agonists bind and close the N-terminal Venus Fly trap (VFT) domain, connected to the hepta-helical bundle (7TM) via the cysteine-rich domain (CRD). Orthosteric antagonists interact with the same VFT site but prevent agonist occupancy and VFT closure, preventing receptor activation. In contrast, allosteric class C GPCR ligands bind to a separate intramembrane pocket within the 7TM bundle, and may influence receptor function in conjunction with orthosteric ligands and/or independently. (b) A comparison of the key transmembrane residues (white on black) forming the orthosteric site of the β2-adrenoceptor (for adrenaline; [34]) and the allosteric site for the modulator MPEP at the mGluR5 receptor (rat sequence; [62]). Conserved TM residues (squares) for family A receptors (see Fig. 2.3a) and family C GPCRs are also indicated.

aspartate residue recognizes the agonist amine in all acetylcholine, adrenaline, dopamine, histamine, and 5-HT receptors. Thus, while the compact binding space of these receptors has historically proved most tractable for drug discovery, there are limitations to the selectivity that can be achieved when targeting only the orthosteric binding site.

An intramembrane binding pocket also forms the predominant recognition domain for a few peptide GPCRs, particularly for smaller ligands such as Angiotensin II at the AT1 receptor [42]. Invariably, larger peptides require binding epitopes on the extracellular surface of the receptor (Fig. 2.3b). For example, neuropeptide Y binding to the Y1 receptor represents a halfway house, in which the C-terminal end of the peptide agonist inserts superficially into the helix bundle pocket while the majority of the agonist contacts ECL residues [43]. It is also clear that peptides can act as agonists even if they *only* bind the extracellular domains. The relatively small 10 amino acid peptide substance P is recognized exclusively by motifs in the N-terminus and ECLs of the NK1 receptor [44–46]. Indeed, intramembrane residues first described as direct participants in substance P binding now appear to prevent general receptor conformational changes associated with agonism, at a site distal to the binding domain [47]. For the protease-activated receptors, a tethered peptide agonist (6 amino acids) is revealed by enzymatic cleavage of the receptor N-terminus and binds only ECL 2 [48]. The largest glycoprotein hormones (e.g., follicle-stimulating hormone [FSH]) use a large N-terminal domain in their receptors for high-affinity interaction (Fig. 2.3b) [1]. The FSH receptor binding surface is formed by a concave leucine rich repeat, which has been studied by crystallography of the N-terminal ectodomain in complex with the hormone [49]. The mechanisms by which glycoprotein binding elicits receptor activation are still open to debate. One model suggests that once associated with the N-terminus, the hormone is presented to other recognition epitopes in the receptor extracellular or TM domains. This model shares features with the proposed mode of binding for class B GPCRs, a select group of peptide receptors which are also distinguished by a large structured N-terminal binding domain [50]. It is also possible that the N-terminal binding site is the only direct interaction between the protein hormone and the receptor. In this case, the resultant changes in this ectodomain (particularly in the hinge region at the junction with TM1) may either directly lead to receptor activation, or remove the inhibitory constraints suppressing constitutive activity [49].

2.3.2. Ligand Recognition in Class C Receptors

The class C receptor family includes neuronal metabotropic receptors for amino acid transmitters (glutamate, gamma-aminobutyric acid (GABA)), as well as members which can sense small molecules such as calcium ions or sugars. Despite their compact ligands, these receptors are distinguished by a very large N-terminus that forms the exclusive orthosteric binding domain (Fig. 2.4a) [51]. Part of the N-terminus forms a two-lobed structure related in evolution to bacterial periplasmic binding proteins, involved in the transport of similar small molecules. Ligands such as glutamate are recognized by hydrophilic interaction with amino acid residues in both lobes. Their binding leads to the closure of the lobes around the agonist molecule at the interface, commonly referred to as the Venus flytrap (VFT) mechanism. This process, and the converse prevention of VFT domain closure by orthosteric antagonists, has been extensively studied by crystallography of the isolated N-terminal ectodomain (e.g., mGluR1; [52, 53]). With the exception of $GABA_B$ receptors, the VFT domain is followed by a second structured N-terminal region, called the cysteine-rich domain. This participates in the transmission of the agonist-induced conformational changes in the N-terminus (VFT closure) to the 7TM domains responsible for G protein coupling [54]. The ability of class C VFT domains to form constitutive dimers is also an essential structural element in the activation process, as discussed below (2.5.1).

2.3.3. Molecular Mechanisms of Rhodopsin-Like Receptor Activation

With notable exceptions (such as those outlined in Section 2.2 above), GPCRs translate the binding of their endogenous agonists into high-fidelity intracellular signals. This implies a consistent molecular interface between an active receptor and its preferred downstream partner, in the presence of many competing effectors. The requirements for activating a particular class of heterotrimeric G proteins, for example Gs, must therefore be similar, but Gs-coupled rhodopsin-like receptors include a full range of binding modes (Fig. 2.3b), from β-adrenoceptors for adrenaline to the FSH glycoprotein receptors. How can very different structural ligand binding sites be reconciled with the need to produce reproducible active receptor conformations, recognized by the same G protein?

Based on mutagenesis and other biophysical data, the cytoplasmic ends of the helix bundle in rhodopsin-like receptors are held close together in the inactive conformation. The widely conserved helix residues participate in this interaction, such as the NPXXY motif at the bottom of TM7 forming hydrophobic interactions with TM1 and TM2 residues. The DRY motif at the cytoplasmic end of TM3 was also originally suggested to form an ionic lock with a TM6 aspartate residue. Receptor activation involves the breakage of these interhelix links, resulting in the outward movement of the inner ends of the TM6 and TM7 from the rest of the bundle, and the exposure of a cytoplasmic cleft which is recognized by the G protein [45]. A recent model (the "toggle-switch") suggests that because the individual TM domains are relatively rigid, the outward helical movement on the cytoplasmic face of the receptor is driven by movement of the helices toward each other on the extracellular side after agonist activation [55]. In constitutively active GPCRs, this switch movement can occur spontaneously, for example, following mutation of the DRY inactivating ionic lock [56], through the absence of constraining influences conferred by ECLs (melanocortin receptors; [57]) or the presence of a self-contained intramolecular interaction that brings the outer ends of the TM helices into close proximity (the ghrelin receptor family; [58]). The toggle-switch is also an attractive mechanism conceptually, because it is then possible for agonists to stabilize the same see-saw action in independent ways for different receptors—through an upper membrane binding pocket above the fulcrum for amine receptors, or by restraining the extracellular ends of the helices for peptide receptors. It also allows prediction and subsequent engineering of new agonist binding sites, for example, for metal ions in the upper TM regions of the β_2-adrenoceptor and the NK1 receptor [59, 60]. More recently, a toggle-switch mechanism in the constitutively active ghrelin receptor was supported by studying the efficacy of a core peptide ligand, which could be reversed by receptor mutations which altered its vertical position in the binding pocket. As predicted, more superficial binding resulted in agonism, while a deeper location generated inverse agonist properties [61].

Despite the ability of the toggle-switch mechanism to predict much experimental data, it remains a model based on indirect evidence of the changes that occur when 7TM receptors are activated. The increasing availability of 7TM receptor crystal structures may now challenge some long-held assumptions. The most surprising finding to date is that with the exception of the original rhodopsin structure [33], the TM3 DRY motif is too far away from TM6 in the β-adrenoceptor or A_{2A} adenosine receptor structure to form the "ionic lock" proposed as a central tenet of the inactive receptor state [34–36]. Although these receptors were crystallized in an "inactive" antagonist-bound conformation, in each case, the aspartate of the DRY motif interacts with closer residues in intracellular loop 2. Refinement in our understanding of 7TM receptor activation will require resolution of these key differences between high-resolution structural data in artificial crystals and indirect approaches in a more physiological environment.

2.4. TWO GPCR LIGANDS BINDING AT ONCE—CONCEPT OF ALLOSTERISM

Despite sharing a common seven transmembrane (7TM)-spanning structure, GPCRs are malleable proteins that, as described in Section 2.3, have varying modes of endogenous ligand binding. In some cases, it appears that the endogenous, or orthosteric, binding pocket on one GPCR may represent a second, alternative, binding site on a different GPCR. This phenomenon has been observed for family C GPCRs where, in contrast to that for endogenous agonists, the binding pocket for a number of potent and selective synthetic ligands appears to reside exclusively within the heptahelical TM domains. Moreover, receptor mutagenesis studies have revealed that the location of this binding pocket correlates well with that for agonists and antagonists of rhodopsin-like GPCRs (Fig. 2.4) [45, 51, 62]. Any additional receptor binding sites that are topographically distinct from that of the orthosteric binding site are referred to as allosteric binding sites and the ligands that recognize them, allosteric modulators. GPCRs can be said to be natural allosteric proteins in that they require allosteric interactions to transmit stimulus from an extracellular agonist to an intracellular effector. However, in addition to intracellular proteins such as G proteins and β-arrestin, GPCRs can also be allosterically regulated by extracellular ions such as Zn^{2+}, Na^{2+}, and Ca^{2+}, endogenous peptides, endogenous lipids, autoantibodies, and synthetic ligands (reviewed in Reference 63).

2.4.1. Classes of Allosteric Modulators

Similar to that of agonists and inverse agonists, allosteric modulators stabilize a select subset of receptor conformations. Allosteric modulators can alter the binding and/or functional properties of orthosteric ligands and, in some cases, when an active receptor conformation is stabilized, can have efficacy in their own right. As such, in addition to the various classes of orthosteric ligands, there are a number of different categories of allosteric ligands [64–66]. At equilibrium, positive and negative allosteric modulators enhance or inhibit orthosteric ligand binding and/or function, respectively, neutral allosteric modulators have no effect, and allosteric agonists mediate receptor activation in the absence of an orthosteric ligand. However, the properties of an allosteric ligand can rarely be confined to a single category. For example, at the M_4 muscarinic acetylcholine receptor (mAChR), the allosteric ligand LY2033298 can increase basal receptor activity, enhance agonist binding, and inhibit antagonist binding. As such, LY2033298 behaves as an allosteric agonist, as a positive allosteric modulator, and as a negative allosteric modulator [67]. The contrasting properties of LY2033298 highlight the probe dependence of allosteric modulators, that is, the ability of the orthosteric ligand used in the experiment to define the direction and degree of modulation mediated by an allosteric ligand.

2.4.2. Pharmacological Models of Allosteric Interactions

The classical form of the TCM suggests that the probe dependence of the G protein defines the ability of the orthosteric ligand to promote the formation of the ternary complex and as such, the efficacy of an orthosteric ligand (Fig. 2.1b, middle). Within this model, the G protein has positive cooperativity with an orthosteric agonist, negative cooperativity with an orthosteric inverse agonist, and neutral cooperativity with an orthosteric antagonist. The TCM can also be used as a more general model to describe any allosteric interaction between two binding sites on a receptor. The allosteric ternary complex model (ATCM) is the simplest mechanistic framework that can quantify the binding properties of an allosteric modulator (Fig. 2.1c, right). The difference underlying the two models is that in contrast to the classical TCM, the ATCM makes no assumptions with regards to the efficacy of either ligand, purely modeling ligand binding. At the level of binding, allosteric modulators are described by two parameters, affinity and cooperativity. In contrast to affinity, which is a system-independent property of an allosteric modulator, cooperativity is dependent on the orthosteric ligand present, and is described as the ratio of the affinity of the free receptor to that of the occupied receptor. The cooperativity between an orthosteric and allosteric ligand is always reciprocal, and therefore, a single cooperativity factor can be used to describe the effect of an orthosteric ligand on allosteric binding and vice versa [68–71]. According to the ATCM, the fractional occupancy of the orthosteric ligand is governed by the following equation:

$$\rho_A = \frac{\frac{[A]}{K_A}}{\frac{[A]}{K_A} + \frac{\left(1+{[B]}/{K_B}\right)}{\left(1+{\alpha[B]}/{K_B}\right)}}$$

Within this model, K_A and K_B denote the equilibrium dissociation constants of the orthosteric ligand (A) and the allosteric ligand (B), respectively, and α denotes the binding cooperativity. A value of α greater than one describes positive cooperativity, less than one describes negative cooperativity, and equal to one describes neutral cooperativity. Highly negative cooperative interactions occur when α approaches zero. In these instances, the allosteric interaction between the allosteric modulator and orthosteric ligand can become indistinguishable from those of a competitive nature [72].

The ATCM is useful when describing allosteric interactions at the binding level; however, when assessing receptor function, a number of allosteric modulators have been identified whose properties do not follow the predictions of the ATCM. For example, at the human cannabinoid CB_1 receptors, Org27569 displays positive cooperativity in terms of binding of the orthosteric agonist CP55940, but negative cooperativity in terms of efficacy [73]. Under such circumstances, a more complex model is required. The allosteric two-state model (ATSM) is an extension of the ATCM that allows for allosteric effects on orthosteric ligand efficacy and allosteric agonism (Fig. 2.1c, right) [74]. The ATSM describes orthosteric (A) and allosteric (B) ligand binding in terms of their affinity for the free and bound active (R*) and inactive form (R) of the receptor. Within this model, α is analogous to that in the ATCM; β and γ describe the preference of A and B, respectively, for R* over R, and as such, their intrinsic efficacy and δ describes the activation cooperativity between the two ligands when they are simultaneously bound, forming the ternary complex. The ATSM separates the modulation of an allosteric ligand into independent effects on orthosteric ligand binding (α) and efficacy (δ). Another important property of this model is that it allows allosteric ligands to possess efficacy in their own right. Allosteric ligands have been identified that can modulate receptor activity in the absence of orthosteric ligand. For example, the well-characterized allosteric modulator, alcuronium, has been shown to decrease the constitutive activity of the M_2 mAChR [75]. In contrast, the highly selective 2-amino-3-benzoylthiophene compounds that act as allosteric enhancers of agonist binding to the adenosine A_1 receptor have been shown to enhance constitutive receptor activity from an allosteric binding site [76]. Similar to the effects that increasing the constitutive activity has on agonist affinity (reviewed in Reference 16), the efficacy of an allosteric modulator will alter the equilibrium between the different receptor states and, as a result, alter the affinity of the orthosteric agonists and inverse agonists. Allosteric agonists will enrich the active form of the receptor, enhancing the affinity of orthosteric agonists but decreasing the affinity of orthosteric inverse agonists. Allosteric inverse agonists will drive the equilibrium toward the inactive receptor and mediate the reciprocal effect to that of an allosteric agonist. This modulation of orthosteric affinity is driven by γ and therefore is independent to that defined by the cooperativity factors α and δ. The overall modulation mediated by the allosteric ligand will be a combination of all three cooperativity factors. While remaining a useful tool to explain qualitative effects of an allosteric modulator on receptor and/or orthosteric ligand efficacy, the complex nature of the ATSM has meant that it remains elusive in terms of obtaining parameter estimates from experimental data.

2.4.3. Advantages of Allosteric Ligands

To date, GPCR pharmacology has primarily been focused on the discovery and characterization of ligands that target the endogenous agonist binding site. This approach to drug discovery has yielded many highly efficacious and therapeutically useful compounds. However, for some GPCRs, the identification of compounds with the desired pharmacological properties remains elusive. A common obstacle is the development of a ligand with sufficient selectivity between subtypes of the same receptor family. The evolutionary pressure to accommodate a common endogenous ligand has meant that the orthosteric binding site is often highly conserved between receptor subtypes. In contrast, allosteric binding sites do not accommodate a common endogenous ligand and as a consequence, generally display much greater sequence divergence between receptor subtypes [77]. Therefore, targeting the allosteric binding site may represent a more successful strategy to obtain highly subtype selective ligands. The family of mAChRs (M_1–M_5) has high sequence homology within the orthosteric binding site [78]. However, mAChRs can also be allosterically regulated by a wide range of structurally diverse ligands through multiple allosteric binding sites [79–82]. A number of muscarinic ligands that demonstrate subtype selectivity either at the level of function or binding have been shown to have an allosteric mode of action. Allosteric modulators such as gallamine, alcuronium, and C_7/3-phth have considerably higher affinity for the M_2 mAChR as compared with the M_5 mAChR. This binding selectivity is driven by residues that are not conserved between the receptor subtypes such as the acidic ^{172}EDGE^{175}, which is specific to the second ECL of the M_2 mAChR [83–85]. An alternative approach for obtaining subtype selectivity is to use a bivalent ligand, the allosteric portion of which confers selectivity, whereas the orthosteric portion mediates the desired pharmacology. At the M_2 mAChR, bivalent ligands have been developed using a hybrid of the potent nonselective orthosteric agonist, iperoxo, and fragments of bis(ammonio) alkane-type allosteric modulators. These compounds were full agonists at the M_2 mAChR and displayed a higher affinity for the M_2 mAChR when compared against the M_1 and M_3 mAChR [86].

Another advantage of allosteric modulators is that they have the potential to conserve the spatial and temporal signaling profile of the endogenous agonist [87]. Under normal physiological conditions, the action of an endogenous agonist is generally intermittent and is highly regulated. In contrast, exogenous agonists mediate continual receptor activation where and when they are present. Allosteric modulators that lack intrinsic efficacy have no effect on receptor function until the endogenous agonist is present. The magnitude and direction of the modulation imparted by the allosteric ligand will represent the combination of the cooperativity between ligand binding, efficacy, and activation. This aspect of allosteric modulators is particularly important when the therapeutic objective is a chronic enhancement of the action of an endogenous agonist.

2.5. GPCR DIMERIZATION

Current understanding of ligand–GPCR interactions was built upon a signaling unit consisting of one receptor protein and one effector, such as the heterotrimeric G protein. However, from the birth of GPCRs as a receptor family, there were hints that these receptors might exist as multimers, of which the simplest case is a dimer [88]. Three advances in the past decade accelerated wider interest in GPCRs as dimeric proteins. First, the puzzle of inactive cloned $GABA_B$ receptors was solved by recognition that both $GABA_{B1}$ and $GABA_{B2}$ subunits were required for a functioning heterodimer [89, 90]. Later, atomic force microscopy revealed an elegant array of paired rhodopsin receptors in native rod outer segments [91], albeit in membranes that contain the highest physiological concentration of any known GPCR. Finally, bioluminescence resonance energy transfer (BRET) and fluorescence resonance energy transfer (FRET) techniques have been extensively applied to investigate the distance between individual GPCR monomers. Close apposition (<10 nm) between bioluminescent/fluorescent donor and acceptor-labeled proteins allows the donor energy to be transferred to the acceptor, resulting in increased longer wavelength emission. With some exceptions [92, 93], significant BRET/FRET signals between a wide array of GPCR monomers support constitutive receptor homo- and heterodimerization in transfected cells [88, 94–96]. Such techniques support organized multimeric GPCR complexes but do not resolve whether dimerization is an essential prerequisite for ligand binding and signaling. Here, we consider the extent to which the protomers in a dimer must interact to activate downstream effectors, and whether communication between homo- and heterodimer ligand binding sites has the potential to generate novel pharmacology.

2.5.1. Dimerization Is Essential for Class C Receptor Function

The modular ligand binding VFT domains in class C receptors exhibit a rigid dimerization interface between the first lobes, consolidated in mGluR and Ca-sensing receptors by a cys–cys disulfide bond [53, 97] (for a review, see Reference 51). Constitutive N-terminal dimerization can be supported by other contacts, for example, within the C terminus. The coiled–coil interaction between the C-terminal tails of the $GABA_B$ subunits masks the endoplasmic reticulum retention signal of $GABA_{B1}$ and allows surface delivery of the heterodimer [89, 90], while C-terminal domains of glutamate receptors can prevent inappropriate heterodimerization [98].

The VFT dimer interface provides a rotational anchor, in which domain closure after agonist binding shortens the distance between the C termini of each module, bringing the attached respective 7TM domains closer together [99]. FRET studies support such a rearrangement after activation of mGluR1 [100]. Intersubunit communication is critical for $GABA_B$ receptor activation since $GABA_{B1}$ subunits do not function alone even when the C tail retention sequence is mutated to allow plasma membrane expression [89], while the $GABA_{B2}$ subunit lacks the conserved VFT domain GABA binding site [101]. Moreover, receptor mutations that prevent $GABA_{B1/B2}$ G protein signaling are effective only when applied to the $GABA_{B2}$ and not the $GABA_{B1}$ protomer [102, 103]. $GABA_{B2}$ also contains the binding site for the allosteric modulator CGP7630 [104]. Such evidence led to the proposal that binding of GABA to a single site in $GABA_{B1}$ *trans*-activates its $GABA_{B2}$ partner, responsible for G protein signaling. However, the stark division between "agonist binding" and "coupling" $GABA_B$ subunits is a simplification. The presence of the nonbinding $GABA_{B2}$ VFT domain still facilities closure of the $GABA_{B1}$ VFT and thereby enhances its affinity for GABA [105–107]. Equally, the presence of the $GABA_{B1}$ 7TM domain in the $GABA_{B1/B2}$ heterodimer improves G protein coupling efficiency, and *cis*-activation of chimeric GABA receptor consisting of efficient binding ($GABA_{B1}$ VFT) and activation ($GABA_{B2}$ 7TM) domains can only occur on coexpression with a $GABA_{B2}$ VFT/$GABA_{B1}$ 7TM chimera. Thus, the critical element in $GABA_B$ receptor activation is not specific complementation between the "binding" and "activation" monomers. Rather, it is the formation of the $GABA_B$ dimer itself, which allows global rearrangement of the 7TM domains to be driven by closure of the linked VFT modules.

This activation theme runs through other class C receptors, including taste receptor heterodimers [108] and even mGluR homodimers, in which the orthosteric VFT binding sites are identical. Recently, an elegant investigation into the mechanisms of mGluR5 activation made use of the $GABA_B$ C tail quality control system, to express defined mGluR5 dimers at the cell surface with complementary mutations in one or other of the protomers [54]. These prevented glutamate binding in the VFT, G protein activation, or dissociated the coupling between the VFT and the 7TM domain in the same subunit; in addition, the binding site for the mGluR1 allosteric modulator Ro-01-6128 was introduced into one mGluR5 promoter, so that it could be locked in an active conformation. This study indicated first that the binding of glutamate to one VFT domain was sufficient for a response. Second, following glutamate binding, only one 7TM domain in the dimer reached an active conformation at a given time, but this activated protomer could either be in the opposing one (*trans*) from the ligand-bound VFT or in the same subunit (*cis*). In either case, the activation mechanism required communication between receptor protomers. As for $GABA_B$ receptors, the formation of mGluR5 dimers is therefore integral to their function.

2.5.2. Is Dimerization Required for Class A GPCR Activation?

While the 7TM domains of class A and class C receptors have little overall sequence homology, they exhibit similar topology and share motifs important in GPCR activation [109], and class C receptors expressed without VFT domains behave like class A GPCRs [110]. Moreover, both classes of GPCR must activate heterotrimeric G proteins, and the larger cytoplasmic footprint of a receptor dimer is proposed to accommodate both $G\alpha$ and $G\beta\gamma$ subunits more easily [111, 112], though this model has been questioned [113]. With BRET/FRET experiments indicating that class A GPCRs can form multimeric complexes, it is easy to assume that dimerization must be a fundamental theme in the signaling of all GPCRs. Yet the critical N-terminal VFT domains are unique to class C GPCRs. Is the dimer-based activation mechanism similarly specific, or do the VFT domains tether individual protomers to enhance the formation of a universal GPCR "active dimer"?

Experimental proof of class A dimers through mutagenesis is challenging because the monomers interact within highly conserved hydrophobic TM domains, even for the glycoprotein GPCRs with large N-termini [114]. Cross-linking experiments indicate an interface between helices IV and V of opposing monomers in opsin, 5HT, and dopamine GPCRs [115–117], with additional formation of higher-order oligomers (e.g., through helix I; [115, 117]). Targeting the interface by mutagenesis, or isolated expression of the relevant isolated helices, inhibits dimer formation and associated GPCR maturation and signaling [96, 111, 118–121], but side effects on appropriate folding of the individual proteins cannot be excluded. Complementation of coexpressed nonbinding and noncoupling GPCRs, akin to the $GABA_B$ receptor, can restore function through dimerization [114, 122, 123], but swapping individual domains in these receptors might also achieve a signaling monomeric unit. Importantly, these approaches (in common with most BRET/FRET experiments; [95]) do not measure proportions of monomeric versus multimeric receptors. The most compelling case for functional relevance of class A GPCR dimers derives from recombinant leukotriene BLT_1 receptors in an artificially reconstituted system. Cross-linking and other approaches indicate that LTB_4 agonist increases the proportion of BLT_1 dimers and that these complex with a single heterotrimeric G protein [111]. In addition, heterodimeric BLT_1 receptors, in which one subunit is mutated to reduce LTB_4 affinity or eliminate its binding altogether [124, 125], indicate that a single bound agonist promotes G protein activation by the dimer. Switching on the first protomer excludes activation of the second, reflecting the proposed mechanism for class C receptors [54], but only *cis* activation of the same BLT_1 monomer binding agonist occurs [124, 125]. Even in these experiments, a role of BLT_1 monomers in living cells is not excluded, because the monomer/dimer ratio of the artificially purified receptors is significantly influenced by solubilization with different detergents [125].

If GPCR dimers are obligatory, then monomeric receptor preparations should be nonfunctional. In fact, reconstitution of β_2-adrenoceptors in lipid particles, as verified monomers, preserves both G protein coupling and classical ligand pharmacology [126]. Rhodopsin and neurotensin NTS1 receptors also signal as individual proteins, using a variety of methods to obtain receptor micelles [127–129]. When stoichiometry is altered to obtain dimeric receptor preparations, there is a reduction in G protein coupling efficiency [128, 129], supporting exclusive activation of only one receptor protomer per dimer. Clearly, reconstitution studies can demonstrate that class A GPCR monomers are signaling competent, and indicate that dimers may play a modulatory rather than a central role.

2.5.3. Influence of Receptor Dimers on Binding Studies

A model in which receptor dimers activate one G protein predicts that the two orthosteric binding sites will not behave identically. In this arrangement, binding of the G protein heterotrimer is asymmetric (Gα contacts one monomer, and the Gβγ subunit the second) and thus, its allosteric effects on ligand interaction differ between the two monomers. Agonist binding to the first orthosteric site will stabilize G protein association with the dimer in a defined orientation and thereby influence ligand affinity at the second site. If, as expected, allosteric stabilization of an agonist–receptor complex occurs through Gα, this explains activation of only one receptor protomer per dimer and predicts negative cooperative binding of orthosteric ligands, leading also to a single bound ligand per dimer [54, 111, 124, 130]. Indeed, ligand binding assays can suggest this cooperativity, for example, through differences in receptor number measured by saturation binding of two ligands [131] and alterations in the shape of saturation and competition isotherms [114, 132]. For a limited number of GPCRs, the best evidence for negative cooperativity is the enhancement of radioligand off rates by excess unlabeled ligand [114, 130, 132–134]. Effects on dissociation kinetics cannot be accounted for by heterogeneous receptor populations, and a significant proportion of receptor dimers must be present in the membrane for their observation. Even here it is necessary to be cautious, particularly as most of the current studies use peptide or protein ligands where binding may be a multistep process—similar results could be obtained if ligands compete for independent parts of a single large orthosteric "pocket." Additional complications include the fact that some GPCR ligands in fact display positive cooperativity [129, 132, 135], and both types of cooperativity may be transient and dependent on the effector [136]. For example, the presence of G protein eliminates the positive cooperativity between neurotensin binding sites in the NTS1 receptor dimer [129]. However, wider application of kinetic experiments to a spectrum of GPCR ligands would help define the extent of cooperative interactions between orthosteric sites and, by implication, the prevalence of GPCR dimers [136].

Eventually, a conceptual framework involving GPCR dimers may also provide new insight into the pharmacology of both allosterism and dual efficacy. For example, the mapped binding sites on the class A ghrelin receptor for orthosteric (ghrelin) and allosteric ligands (L-692,429) overlap significantly, yet they must bind simultaneously to exert their actions [10, 137]. An attractive explanation, among others, is for each ligand to bind separate protomers in a ghrelin receptor dimer, analogous to class C allosteric interactions [54, 138]. This is particularly the case if negative cooperativity ensures the second protomer site does not bind ghrelin [137]. We also know little about the mode of binding of other effectors to a potential receptor dimer, such as the arrestins. A popular model, based on inactive (and therefore noninteracting) structures of rhodopsin and arrestin, suggests that the two arrestin binding domains associate with different GPCR monomers in 2:1 stoichiometry [112]. However, the most detailed investigations on the binding of radiolabeled arrestins to different receptors [139] or the interactions between rhodopsin and arrestin in rods [140] provide convincing data on the association of one GPCR with one arrestin molecule. Here, the allosteric influence of arrestin on a GPCR dimer will be symmetric, and in marked contrast to the effects of G proteins, agonist binding to *both* ligand binding sites should be stabilized. Immediately, it becomes apparent that ligand binding to GPCR dimer stabilized by a G protein (with negative cooperativity between sites) is qualitatively different from a symmetric ternary complex containing arrestin, providing another mechanism for ligand-directed signaling to associated pathways. Resolution of such issues requires more work on the stoichiometry of GPCR interaction with different effectors, including a vital role for classical pharmacology techniques.

2.5.4. GPCR Heterodimerization

For certain class C GPCRs, such as the $GABA_B$ receptor and taste receptors, the evidence for heterodimerization is overwhelming. However, there is no reported heterodimerization among different metabotropic glutamate receptors, despite several closely related receptor subtypes, coexpressed in the same neurons [51, 98]. Ability to form class C dimers is clearly a tightly regulated phenomenon. What is then to be made of FRET studies which indicate numerous interacting combinations of different class A GPCRs [141] supported by more recent techniques detecting heterodimers using bimolecular fluorescence complementation [142, 143]? Often, these investigations suggest that closely related GPCRs, such as the β_1/β_2-adrenoceptors and CCR2/CCR5 chemokine receptors, form homo- and heterodimers with equal efficiency [95, 144]. If the same range of heterodimers existed in physiology, would the original (and successful) definitions of GPCR subtypes using *in vitro* pharmacology have been possible?

To stringently test the relevance of GPCR heterodimers, strict criteria should accompany their acceptance as a new pharmacological entity. In the first instance, identification needs to be supported in native cells, in addition to heterologous systems. For example, co-immunoprecipitation and cooperativity in ligand binding and function have indicated μ/δ or κ/δ opioid, D_1/D_2 dopamine, and A_1/A_{2A} adenosine receptor heterodimers *in vivo* [145–148]. Second, many heterodimers are indicated by novel patterns of signaling or regulation when the different GPCRs are expressed together. Thus, D_1/D_2 receptor coexpression results in Gq/11 activation, compared to Gs or Gi coupling for the individual dopamine receptors [147]. In addition, coexpression of μ/δ opioid receptors produces a unique pattern of arrestin-mediated ERK signaling [149]. Moreover, trafficking of orexin-1 receptors is altered by coexpression of the CB1 cannabinoid receptor, and can then be modulated by cannabinoid ligands [150]. These experiments demonstrate close interaction between GPCR subtypes, with physiological relevance and the potential for novel regulation by drugs. Yet, the studies are limited in that they do not distinguish between a new pharmacologically distinct heterodimer and cross-talk between closely complexed receptors in the same microdomains.

A cornerstone for the proof of heterodimerization should therefore be binding experiments, to show that communication exists between different orthosteric binding sites. An important thermodynamic consideration is that any effects should be reciprocal. If the binding of a selective ligand to one protomer affects selective ligand binding at the second, then the converse should also be the case. Surprisingly, this is manifestly true for some heterodimer combinations (e.g., CCR5/CCR2; [130]) but not for others. For example, an entirely new pharmacology is generated for kappa and delta opioid receptor heterodimers, with eliminated affinity for the selective agonists at either parent receptor [151]. In such cases, the integrity of the individual protomer orthosteric sites cannot be maintained, and domain swapping between monomers might generate pharmacologically distinct binding sites in the heterodimer. It is then imperative to show that this does not result from GPCR overexpression, but is physiologically relevant, as appears to be case for native κ/δ opioid heterodimers [148]. For heterodimers formed by two protomers which retain their individual characteristics, the acceleration of dissociation kinetics provides the strongest validation of communicating binding sites [114, 130]. For CCR5/CCR2 receptor heterodimers, as for the respective homodimers, there is strong negative cooperativity, which suggests only one ligand occupies the heterodimer [130]. Here, cross-talk between the binding of different ligands can certainly generate a new structure activity relationship, and the presentation of a different dimer cytoplasmic interface has potential (though unproven) to interact uniquely with different effectors. If such careful definitions of GPCR heterodimers are repeated more widely, there is certainly scope for new avenues in drug discovery in the future.

2.6. FUTURE PERSPECTIVES

The complexity of GPCR signaling and the potential for them to form homo- and heterodimeric structures that interact with heterotrimeric G proteins and other signaling proteins in different stoichiometric complexes within the same cells means that we need to give greater consideration to the temporal and spatial aspects ligand–receptor interactions. The recent development of fluorescent agonists and antagonists alongside high-resolution confocal imaging techniques and biophysical approaches (e.g., [152–154]) may provide the stimulus for significant progress in this area.

ACKNOWLEDGMENTS

L.T.M. is a National Health and Medical Research Council (NHMRC) of Australia Training (Postdoctoral) Fellow, N.D.H. received funding support from the Medical Research Council (MRC) and the Royal Society, and S.J.H. received funding support from the Biotechnology and Biological Sciences Research Council (BBSRC) and the MRC.

REFERENCES

1. Lagerstrom, M.C., Schioth, H.B. (2008) Structural diversity of G protein-coupled receptors and significance for drug discovery. *Nat Rev Drug Discov.* 7, 339–357.
2. Kenakin, T. (2004) *A Pharmacology Primer: Theory, Application, and Methods*. San Diego, CA: Elsevier Academic Press.
3. Furchgott, R.F. (1966) The use of β-haloalkylamines in the differentiation of receptors and in the determination of dissociation constants of receptor-agonist complexes. In *Advances in Drug Research*, ed. H. Blaschko, E. Muscholl, pp. 21–55. London and New York: Academic Press.
4. Kobilka, B.K., Deupi, X. (2007) Conformational complexity of G-protein-coupled receptors. *Trends Pharmacol Sci.* 28, 397–406.
5. Lohse, M.J., Nikolaev, V.O., Hein, P., Hoffmann C., Vilardaga, J.P., Bunemann, M. (2008) Optical techniques to analyze real-time activation and signaling of G-protein-coupled receptors. *Trends Pharmacol Sci.* 29, 159–165.
6. Swaminath, G., Xiang, Y., Lee, T.W., Steenhuis, J., Parnot, C., Kobilka, B.K. (2004) Sequential binding of agonists to the β2 adrenoceptor: Kinetic evidence for intermediate conformational states. *J Biol Chem.* 279, 686–691.
7. Nikolaev, V.O., Hoffmann, C., Bunemann, M., Lohse, M.J., Vilardaga, J.P. (2006) Molecular basis of partial agonism at the neurotransmitter α2A-adrenergic receptor and Gi-protein heterotrimer. *J Biol Chem.* 281, 24506–24511.
8. Vilardaga, J.P., Steinmeyer, R., Harms, G.S., Lohse, M.J. (2005) Molecular basis of inverse agonism in a G protein-coupled receptor. *Nat Chem Biol.* 1, 25–28.
9. Seifert, R. Wenzel-Seifert, K. .(2002) Constitutive activity of G-protein-coupled receptors: Cause of disease and common property of wild-type receptors. *Naunyn Schmiedebergs Arch Pharmacol.* 366, 381–416.
10. Holst, B., Brandt, E., Bach, A., Heding, A., Schwartz, T.W. (2005) Nonpeptide and peptide growth hormone secretagogues act both as ghrelin receptor agonist and as positive or negative allosteric modulators of ghrelin signaling. *Mol Endocrinol.* 19, 2400–2411.

11. Nijenhuis, W.A., Oosterom, J., Adan, R.A. (2001) AgRP(83-132) acts as an inverse agonist on the human-melanocortin-4 receptor. *Mol Endocrinol.* 15, 164–171.
12. Arvanitakis, L., Geras-Raaka, E., Varma, A., Gershengorn, M.C., Cesarman, E. (1997) Human herpesvirus KSHV encodes a constitutively active G-protein-coupled receptor linked to cell proliferation. *Nature.* 385, 347–350.
13. Smit, M.J., Vink, C., Verzijl, D., Casarosa, P,. Bruggeman, C.A., Leurs, R. (2003) Virally encoded G protein-coupled receptors: Targets for potentially innovative anti-viral drug development. *Curr Drug Targets.* 4, 431–441.
14. Kenakin, T. (1996) The classification of seven transmembrane receptors in recombinant expression systems. *Pharmacol Rev.* 48, 413–463.
15. Leff, P. (1995) The two-state model of receptor activation. *Trends Pharmacol Sci.* 16, 89–97.
16. Lefkowitz, R.J., Cotecchia, S., Samama, P., Costa, T. (1993) Constitutive activity of receptors coupled to guanine nucleotide regulatory proteins. *Trends Pharmacol Sci.* 14, 303–307.
17. Berg, K.A., Maayani, S., Goldfarb, J., Scaramellini, C., Leff, P., Clarke, W.P. (1998) Effector pathway-dependent relative efficacy at serotonin type 2A and 2C receptors: Evidence for agonist-directed trafficking of receptor stimulus. *Mol Pharmacol.* 54, 94–104.
18. Kenakin, T. (2002) Drug efficacy at G protein-coupled receptors. *Annu Rev Pharmacol Toxicol.* 42, 349–379.
19. Luttrell, L.M., Roudabush, F.L., Choy, E.W., Miller, W.E., Field, M.E., Pierce, K.L., Lefkowitz, R.J. (2001) Activation and targeting of extracellular signal-regulated kinases by beta-arrestin scaffolds. *Proc Natl Acad Sci U S A.* 98, 2449–2454.
20. Seta, K., Nanamori, M., Modrall, J.G. , Neubig, R.R., Sadoshima, J. (2002) AT1 receptor mutant lacking heterotrimeric G protein coupling activates the Src-Ras-ERK pathway without nuclear translocation of ERKs. *J Biol Chem.* 277, 9268–9277.
21. Wei, H., Ahn, S., Shenoy, S.K., Karnik, S.S., Hunyady, L., Luttrell, L.M., Lefkowitz, R.J. (2003) Independent beta-arrestin 2 and G protein-mediated pathways for angiotensin II activation of extracellular signal-regulated kinases 1 and 2. *Proc Natl Acad Sci U S A.* 100, 10782–10787.
22. Azzi, M., Charest, P.G., Angers, S., Rousseau, G., Kohout, T., Bouvier, M., Pineyro, G. (2003) Beta-arrestin-mediated activation of MAPK by inverse agonists reveals distinct active conformations for G protein-coupled receptors. *Proc Natl Acad Sci U S A.* 100, 11406–11411.
23. Baker, J.G., Hall, I.P., Hill, S.J. (2003) Agonist and inverse agonist actions of β-blockers at the human β2-adrenoceptor provide evidence for agonist-directed signaling. *Mol Pharmacol.* 64, 1357–1369.
24. Gesty-Palmer, D., Chen, M., Reiter, E., Ahn, S., Nelson, C.D., Wang, S., Eckhardt, A.E., Cowan, C.L., Spurney, R.F., Luttrell, L.M., Lefkowitz, R.J. (2006) Distinct β-arrestin- and G protein-dependent pathways for parathyroid hormone receptor-stimulated ERK1/2 activation. *J Biol Chem.* 281, 10856–10864.
25. Wisler, J.W., DeWire, S.M., Whalen, E.J., Violin, J.D., Drake, M.T., Ahn, S., Shenoy, S.K., Lefkowitz, R.J. (2007) A unique mechanism of β-blocker action: Carvedilol stimulates β-arrestin signaling. *Proc Natl Acad Sci U S A.* 104, 16657–16662.

26. Fraser, I.D., Cong, M., Kim, J., Rollins, E.N., Daaka, Y., Lefkowitz, R.J., Scott, J.D. (2000) Assembly of an A kinase-anchoring protein-beta(2)-adrenergic receptor complex facilitates receptor phosphorylation and signaling. *Curr Biol.* 10, 409–412.

27. Gines, S., Ciruela, F., Burgueno, J., Casado, V., Canela, E.I., Mallol, J., Lluis, C., Franco, R. (2001) Involvement of caveolin in ligand-induced recruitment and internalization of A(1) adenosine receptor and adenosine deaminase in an epithelial cell line. *Mol Pharmacol.* 59, 1314–1323.

28. Hall, R.A., Premont, R.T., Chow, C.W., Blitzer, J.T., Pitcher, J.A., Claing, A., Stoffel, R.H., Barak, L.S., Shenolikar, S., Weinman, E.J., Grinstein, S., Lefkowitz, R.J. (1998) The β2-adrenergic receptor interacts with the Na^+/H^+-exchanger regulatory factor to control Na^+/H^+ exchange. *Nature.* 392, 626–630.

29. Ostrom, R.S., Insel, P.A. (2004) The evolving role of lipid rafts and caveolae in G protein-coupled receptor signaling: Implications for molecular pharmacology. *Br J Pharmacol.* 143, 235–245.

30. Baker, J.G., Hill, S.J. (2007) Multiple GPCR conformations and signalling pathways: Implications for antagonist affinity estimates. *Trends Pharmacol Sci.* 28, 374–381.

31. Kenakin, T. (2001) Inverse, protean, and ligand-selective agonism: Matters of receptor conformation. *FASEB J.* 15, 598–611.

32. Gbahou, F., Rouleau, A., Morisset, S., Parmentier, R., Crochet, S., Lin, J.S., Ligneau, X., Tardivel-Lacombe, J., Stark, H., Schunack, W., Ganellin, C.R., Schwartz, J.C., Arrang, J.M. (2003) Protean agonism at histamine H3 receptors in vitro and in vivo. *Proc Natl Acad Sci U S A.* 100, 11086–11091.

33. Palczewski, K., Kumasaka, T., Hori, T., Behnke, C.A., Motoshima, H., Fox, B.A., Le Trong, I., Teller, D.C., Okada, T., Stenkamp, R.E., Yamamoto, M., Miyano, M. (2000). Crystal structure of rhodopsin: A G protein-coupled receptor. *Science.* 289, 739–745.

34. Rosenbaum, D.M., Cherezov, V., Hanson, M.A., Rasmussen, S.G., Thian, F.S., Kobilka, T.S., Choi, H.J., Yao, X.J., Weis, W.I., Stevens, R.C., Kobilka, B.K. (2007) GPCR engineering yields high-resolution structural insights into β2-adrenergic receptor function. *Science.* 318, 1266–1273.

35. Warne, T., Serrano-Vega, M.J., Baker, J.G., Moukhametzianov, R., Edwards, P.C., Henderson, R., Leslie, A.G., Tate, C.G., Schertler, G.F. (2008) Structure of a β1-adrenergic G-protein-coupled receptor. *Nature.* 454, 486–491.

36. Jaakola, V.P., Griffith, M.T., Hanson, M.A., Cherezov, V., Chien, E.Y., Lane, J.R., Ijzerman, A.P., Stevens, R.C. (2008) The 2.6 angstrom crystal structure of a human A2A adenosine receptor bound to an antagonist. *Science.* 322, 1211–1217.

37. Strader, C.D., Candelore, M.R., Hill, W.S., Sigal, I.S., Dixon, R.A. (1989) Identification of two serine residues involved in agonist activation of the β-adrenergic receptor. *J Biol Chem.* 264, 13572–13578.

38. Strader, C.D., Sigal, I.S., Candelore, M.R., Rands, E., Hill, W.S., Dixon, R.A. (1988). Conserved aspartic acid residues 79 and 113 of the β-adrenergic receptor have different roles in receptor function. *J Biol Chem.* 263, 10267–10271.

39. Shi, L., Liapakis, G., Xu, R., Guarnieri, F., Ballesteros, J.A., Javitch, J.A. (2002) Beta2 adrenergic receptor activation. Modulation of the proline kink in transmembrane 6 by a rotamer toggle switch. *J Biol Chem.* 277, 40989–40996.

40. Swaminath, G., Deupi, X., Lee, T.W., Zhu, W., Thian, F.S., Kobilka, T.S., Kobilka, B. (2005) Probing the β2 adrenoceptor binding site with catechol reveals differences in binding and activation by agonists and partial agonists. *J Biol Chem.* 280, 22165–22171.
41. Baker, J.G., Proudman, R.G., Hawley, N.C., Fischer, P.M., Hill, S.J. (2008) Role of key transmembrane residues in agonist and antagonist actions at the two conformations of the human β1-adrenoceptor. *Mol Pharmacol.* 74, 1246–1260.
42. Clement, M., Martin, S.S., Beaulieu, M.E., Chamberland, C., Lavigne, P., Leduc, R., Guillemette, G., Escher, E. (2005) Determining the environment of the ligand binding pocket of the human angiotensin II type I (hAT1) receptor using the methionine proximity assay. *J Biol Chem.* 280, 27121–27129.
43. Sjodin, P., Holmberg, S.K., Akerberg, H., Berglund, M.M., Mohell, N., Larhammar, D. (2006) Re-evaluation of receptor-ligand interactions of the human neuropeptide Y receptor Y1: A site-directed mutagenesis study. *Biochem J.* 393, 161–169.
44. Fong, T.M., Huang, R.R., Strader, C.D. (1992) Localization of agonist and antagonist binding domains of the human neurokinin-1 receptor. *J Biol Chem.* 267, 25664–25667.
45. Gether, U. (2000) Uncovering molecular mechanisms involved in activation of G protein-coupled receptors. *Endocr Rev.* 21, 90–113.
46. Gether, U., Johansen, T.E., Snider, R.M., Lowe III, J.A., Nakanishi, S., Schwartz, T.W. (1993) Different binding epitopes on the NK1 receptor for substance P and non-peptide antagonist. *Nature.* 362, 345–348.
47. Rosenkilde, M.M., Cahir, M., Gether, U., Hjorth, S.A., Schwartz, T.W. (1994) Mutations along transmembrane segment II of the NK-1 receptor affect substance P competition with non-peptide antagonists but not substance P binding. *J Biol Chem.* 269, 28160–28164.
48. Lerner, D.J., Chen, M., Tram, T., Coughlin, S.R. (1996) Agonist recognition by proteinase-activated receptor 2 and thrombin receptor: Importance of extracellular loop interactions for receptor function. *J Biol Chem.* 271, 13943–13947.
49. Fan, Q.R., Hendrickson, W.A. (2005) Structure of human follicle-stimulating hormone in complex with its receptor. *Nature.* 433, 269–277.
50. Neumann, J.M., Couvineau, A., Murail, S., Lacapere, J.J., Jamin, N., Laburthe, M. (2008) Class-B GPCR activation: Is ligand helix-capping the key? *Trends Biochem Sci.* 33, 314–319.
51. Pin, J.P., Kniazeff, J., Liu, J., Binet, V., Goudet, C., Rondard, P., Prezeau, L. (2005) Allosteric functioning of dimeric class C G-protein-coupled receptors. *FEBS J.* 272, 2947–2955.
52. Kunishima, N., Shimada, Y., Tsuji, Y., Sato, T., Yamamoto, M., Kumasaka, T., Nakanishi, S., Jingami, H., Morikawa, K. (2000) Structural basis of glutamate recognition by a dimeric metabotropic glutamate receptor. *Nature.* 407, 971–977.
53. Tsuchiya, D., Kunishima, N., Kamiya, N., Jingami, H., Morikawa, K. (2002) Structural views of the ligand-binding cores of a metabotropic glutamate receptor complexed with an antagonist and both glutamate and Gd3+. *Proc Natl Acad Sci U S A.* 99, 2660–2665.
54. Brock, C., Oueslati, N., Soler, S., Boudier, L., Rondard, P., Pin, J.P. (2007) Activation of a dimeric metabotropic glutamate receptor by intersubunit rearrangement. *J Biol Chem.* 282, 33000–33008.

55. Schwartz, T.W., Frimurer, T.M., Holst, B., Rosenkilde, M.M., Elling, C.E. (2006) Molecular mechanism of 7TM receptor activation—A global toggle switch model. *Annu Rev Pharmacol Toxicol.* 46, 481–519.
56. Scheer, A., Fanelli, F., Costa, T., De Benedetti, P.G., Cotecchia, S. (1996) Constitutively active mutants of the alpha 1B-adrenergic receptor: Role of highly conserved polar amino acids in receptor activation. *EMBO J.* 15, 3566–3578.
57. Holst, B., Schwartz, T.W. (2003) Molecular mechanism of agonism and inverse agonism in the melanocortin receptors: Zn(2+) as a structural and functional probe. *Ann N Y Acad Sci.* 994, 1–11.
58. Holst, B., Holliday, N.D., Bach, A., Elling, C.E., Cox, H.M., Schwartz, T.W. (2004) Common structural basis for constitutive activity of the ghrelin receptor family. *J Biol Chem.* 279, 53806–53817.
59. Elling, C.E., Frimurer, T.M., Gerlach, L.O., Jorgensen, R., Holst, B., Schwartz, T.W. (2006) Metal ion site engineering indicates a global toggle switch model for seven-transmembrane receptor activation. *J Biol Chem.* 281, 17337–17346.
60. Holst, B., Elling, C.E., Schwartz, T.W. (2000) Partial agonism through a zinc-ion switch constructed between transmembrane domains III and VII in the tachykinin NK(1) receptor. *Mol Pharmacol.* 58, 263–270.
61. Holst, B., Mokrosinski, J., Lang, M., Brandt, E., Nygaard, R., Frimurer, T.M., Beck-Sickinger, A.G., Schwartz, T.W. (2007) Identification of an efficacy switch region in the ghrelin receptor responsible for interchange between agonism and inverse agonism. *J Biol Chem.* 282, 15799–15811.
62. Malherbe, P., Kratochwil, N., Zenner, M.T., Piussi, J., Diener, C., Kratzeisen, C., Fischer, C., Porter, R.H. (2003) Mutational analysis and molecular modeling of the binding pocket of the metabotropic glutamate 5 receptor negative modulator 2-methyl-6-(phenylethynyl)-pyridine. *Mol Pharmacol.* 64, 823–832.
63. May, L.T., Leach, K., Sexton, P.M., Christopoulos, A. 2007. Allosteric modulation of G protein-coupled receptors. *Annu Rev Pharmacol Toxicol.* 47, 1–51.
64. Langmead, C.J., Christopoulos, A. (2006) Allosteric agonists of 7TM receptors: Expanding the pharmacological toolbox. *Trends Pharmacol Sci.* 27, 475–481.
65. May, L.T., Christopoulos, A. (2003) Allosteric modulators of G-protein-coupled receptors. *Curr Opin Pharmacol.* 3, 551–556.
66. Neubig, R.R., Spedding, M., Kenakin, T., Christopoulos, A. (2003) International Union of Pharmacology Committee on Receptor Nomenclature and Drug Classification. XXXVIII. Update on terms and symbols in quantitative pharmacology. *Pharmacol Rev.* 55, 597–606.
67. Chan, W.Y., McKinzie, D.L., Bose, S., Mitchell, S.N., Witkin, J.M., Thompson, R.C., Christopoulos, A., Lazareno, S., Birdsall, N.J., Bymaster, F.P., Felder, C.C. (2008) Allosteric modulation of the muscarinic M4 receptor as an approach to treating schizophrenia. *Proc Natl Acad Sci U S A.* 105, 10978–10983.
68. Christopoulos, A. (2000) Quantification of allosteric interactions at G protein-coupled receptors using radioligand binding assays. In *Current Protocols in Pharmacology*, pp. 1.22.21–1.22.40. New York: Wiley and Sons.
69. Ehlert, F.J. (1988) Estimation of the affinities of allosteric ligands using radioligand binding and pharmacological null methods. *Mol Pharmacol.* 33, 187–194.
70. Lazareno, S., Birdsall, N.J. (1995) Detection, quantitation, and verification of allosteric interactions of agents with labeled and unlabeled ligands at G protein-

coupled receptors: Interactions of strychnine and acetylcholine at muscarinic receptors. *Mol Pharmacol.* 48, 362–378.

71. Stockton, J.M., Birdsall, N.J., Burgen, A.S., Hulme, E.C. (1983) Modification of the binding properties of muscarinic receptors by gallamine. *Mol Pharmacol.* 23, 551–557.
72. Kenakin, T. (1997) *Pharmacologic Analysis of Drug-Receptor Interaction.* Philadelphia: Lippincott-Raven.
73. Price, M.R., Baillie, G.L., Thomas, A., Stevenson, L.A., Easson, M., Goodwin, R., McLean, A., McIntosh, L., Goodwin, G., Walker, G., Westwood P., Marrs, J., Thomson, F., Cowley, P., Christopoulos, A., Pertwee, R.G., Ross R.A. (2005) Allosteric modulation of the cannabinoid CB1 receptor. *Mol Pharmacol.* 68, 1484–1495.
74. Hall, D.A. (2000) Modeling the functional effects of allosteric modulators at pharmacological receptors: An extension of the two-state model of receptor activation. *Mol Pharmacol.* 58, 1412–1423.
75. Zahn, K., Eckstein, N., Trankle, C., Sadee, W., Mohr, K. (2002) Allosteric modulation of muscarinic receptor signaling: Alcuronium-induced conversion of pilocarpine from an agonist into an antagonist. *J Pharmacol Exp Ther.* 301, 720–728.
76. Bruns, R.F., Fergus, J.H. (1990) Allosteric enhancement of adenosine A1 receptor binding and function by 2-amino-3-benzoylthiophenes. *Mol Pharmacol.* 38, 939–949.
77. Christopoulos, A., Kenakin, T. (2002) G protein-coupled receptor allosterism and complexing. *Pharmacol Rev.* 54, 323–374.
78. Wess, J. (1993) Mutational analysis of muscarinic acetylcholine receptors: Structural basis of ligand/receptor/G protein interactions. *Life Sci.* 53, 1447–1463.
79. Birdsall, N.J., Lazareno, S. (2005) Allosterism at muscarinic receptors: Ligands and mechanisms. *Mini Rev Med Chem.* 5, 523–543.
80. Christopoulos, A., Lanzafame, A., Mitchelson, F. (1998) Allosteric interactions at muscarinic cholinoceptors. *Clin Exp Pharmacol Physiol.* 25, 185–194.
81. Ellis, J., Huyler, J., Brann, M.R. (1991) Allosteric regulation of cloned m1-m5 muscarinic receptor subtypes. *Biochem Pharmacol.* 42, 1927–1932.
82. Tucek, S., Proska, J. (1995) Allosteric modulation of muscarinic acetylcholine receptors. *Trends Pharmacol Sci.* 16, 205–212.
83. Voigtlander, U., Johren, K., Mohr, M., Raasch, A., Trankle, C., Buller, S., Ellis, J., Holtje, H.D., Mohr, K. (2003) Allosteric site on muscarinic acetylcholine receptors: Identification of two amino acids in the muscarinic M2 receptor that account entirely for the M2/M5 subtype selectivities of some structurally diverse allosteric ligands in N-methylscopolamine-occupied receptors. *Mol Pharmacol.* 64, 21–31.
84. Leppik, R.A., Miller, R.C., Eck, M., Paquet, J.L. (1994) Role of acidic amino acids in the allosteric modulation by gallamine of antagonist binding at the m2 muscarinic acetylcholine receptor. *Mol Pharmacol.* 45, 983–990.
85. Gnagey, A.L., Seidenberg, M., Ellis, J. (1999) Site-directed mutagenesis reveals two epitopes involved in the subtype selectivity of the allosteric interactions of gallamine at muscarinic acetylcholine receptors. *Mol Pharmacol.* 56, 1245–1253.
86. Antony, J., Kellershohn, K., Mohr-Andra, M., Kebig, A., Prilla, S., Muth, M., Heller, E., Disingrini, T., Dallanoce, C., Bertoni, S., Schrobang, J., Trankle, C., Kostenis, E., Christopoulos, A., Holtje, H.D., Barocelli, E., De Amici, M.,

Holzgrabe, U., Mohr, K. (2009) Dualsteric GPCR targeting: A novel route to binding and signaling pathway selectivity. *FASEB J.* 23, 442–450.

87. Birdsall, N.J., Lazareno, S., Matsui, H. (1996) Allosteric regulation of muscarinic receptors. *Prog Brain Res.* 109, 147–151.
88. Milligan, G., Bouvier, M. (2005) Methods to monitor the quaternary structure of G protein-coupled receptors. *FEBS J.* 272, 2914–2925.
89. Margeta-Mitrovic, M., Jan, Y.N., and Jan, L.Y. (2000) A trafficking checkpoint controls GABA(B) receptor heterodimerization. *Neuron.* 27, 97–106.
90. White, J.H., Wise, A., Main, M.J., Green, A., Fraser, N.J., Disney, G.H., Barnes, A.A., Emson, P., Foord, S.M., Marshall, F.H. (1998) Heterodimerization is required for the formation of a functional GABA(B) receptor. *Nature.* 396, 679–682.
91. Fotiadis, D., Liang, Y., Filipek, S., Saperstein, D.A., Engel, A., Palczewski, K. (2003) Atomic-force microscopy: Rhodopsin dimers in native disc membranes. *Nature.* 421, 127–128.
92. James, J.R., Oliveira, M.I., Carmo, A.M., Iaboni, A., Davis, S.J. (2006) A rigorous experimental framework for detecting protein oligomerization using bioluminescence resonance energy transfer. *Nat Methods.* 3, 1001–1006.
93. Meyer, B.H., Segura, J.M., Martinez, K.L., Hovius, R., George, N., Johnsson, K., Vogel, H. (2006) FRET imaging reveals that functional neurokinin-1 receptors are monomeric and reside in membrane microdomains of live cells. *Proc Natl Acad Sci U S A.* 103, 2138–2143.
94. Maurel, D., Comps-Agrar, L., Brock, C., Rives, M.L., Bourrier, E., Ayoub, M.A., Bazin, H., Tinel, N., Durroux, T., Prezeau, L., Trinquet, E., Pin, J.P. (2008) Cell-surface protein-protein interaction analysis with time-resolved FRET and snap-tag technologies: Application to GPCR oligomerization. *Nat Methods.* 5, 561–567.
95. Mercier, J.F., Salahpour, A., Angers, S., Breit, A., Bouvier, M. (2002) Quantitative assessment of β1- and β2-adrenergic receptor homo- and heterodimerization by bioluminescence resonance energy transfer. *J Biol Chem.* 277, 44925–44931.
96. Salahpour, A., Angers, S., Mercier, J.F., Lagace, M., Marullo, S., Bouvier, M. (2004) Homodimerization of the beta2-adrenergic receptor as a prerequisite for cell surface targeting. *J Biol Chem.* 279, 33390–33397.
97. Jiang, Y., Minet, E., Zhang, Z., Silver, P.A., Bai, M. (2004) Modulation of interprotomer relationships is important for activation of dimeric calcium-sensing receptor. *J Biol Chem.* 279, 14147–14156.
98. Remelli, R., Robbins, M.J., McIlhinney, R.A. (2008) The C-terminus of the metabotropic glutamate receptor 1b regulates dimerization of the receptor. *J Neurochem.* 104, 1020–1031.
99. Muto, T., Tsuchiya, D., Morikawa, K., Jingami, H. (2007) Structures of the extracellular regions of the group II/III metabotropic glutamate receptors. *Proc Natl Acad Sci U S A.* 104, 3759–3764.
100. Tateyama, M., Abe, H., Nakata, H., Saito, O., Kubo, Y. (2004) Ligand-induced rearrangement of the dimeric metabotropic glutamate receptor 1alpha. *Nat Struct Mol Biol.* 11, 637–642.
101. Kniazeff, J., Galvez, T., Labesse, G., Pin, J.P. (2002) No ligand binding in the GB2 subunit of the GABA(B) receptor is required for activation and allosteric interaction between the subunits. *J Neurosci.* 22, 7352–7361.

102. Havlickova, M., Prezeau, L., Duthey, B., Bettler, B., Pin, J.P., Blahos, J. (2002) The intracellular loops of the GB2 subunit are crucial for G-protein coupling of the heteromeric gamma-aminobutyrate B receptor. *Mol Pharmacol.* 62, 343–350.

103. Robbins, M.J., Calver, A.R., Filippov, A.K., Hirst, W.D., Russell, R.B., Wood, M.D., Nasir, S., Couve, A., Brown, D.A., Moss, S.J., Pangalos, M.N. (2001) GABA(B2) is essential for G-protein coupling of the GABA(B) receptor heterodimer. *J Neurosci.* 21, 8043–8052.

104. Binet, V., Goudet, C., Brajon, C., Le Corre, L., Acher, F., Pin, J.P., Prezeau, L. (2004) Molecular mechanisms of GABA(B) receptor activation: New insights from the mechanism of action of CGP7930, a positive allosteric modulator. *Biochem Soc Trans.* 32, 871–872.

105. Galvez, T., Duthey, B., Kniazeff, J., Blahos, J., Rovelli, G., Bettler, B., Prezeau, L., Pin, J.P. (2001) Allosteric interactions between GB1 and GB2 subunits are required for optimal GABA(B) receptor function. *EMBO J.* 20, 2152–2159.

106. Liu, J., Maurel, D., Etzol, S., Brabet, I., Ansanay, H., Pin, J.P., Rondard, P. (2004) Molecular determinants involved in the allosteric control of agonist affinity in the GABAB receptor by the GABAB2 subunit. *J Biol Chem.* 279, 15824–15830.

107. Rondard, P., Huang, S., Monnier, C., Tu, H., Blanchard, B., Oueslati, N., Malhaire, F., Li, Y., Trinquet, E., Labesse, G., Pin, J.P., Liu, J. (2008) Functioning of the dimeric GABA(B) receptor extracellular domain revealed by glycan wedge scanning. *EMBO J.* 27, 1321–1332.

108. Xu, H., Staszewski, L., Tang, H., Adler, E., Zoller, M., Li, X. (2004) Different functional roles of T1R subunits in the heteromeric taste receptors. *Proc Natl Acad Sci U S A.* 101, 14258–14263.

109. Binet, V., Duthey, B., Lecaillon, J., Vol, C., Quoyer, J., Labesse, G., Pin, J.P., Prezeau, L. (2007) Common structural requirements for heptahelical domain function in class A and class C G protein-coupled receptors. *J Biol Chem.* 282, 12154–12163.

110. Goudet, C., Gaven, F., Kniazeff, J., Vol, C., Liu, J., Cohen-Gonsaud, M., Acher, F., Prezeau, L., Pin, J.P. (2004) Heptahelical domain of metabotropic glutamate receptor 5 behaves like rhodopsin-like receptors. *Proc Natl Acad Sci U S A.* 101, 378–383.

111. Baneres, J.L., Parello, J. (2003) Structure-based analysis of GPCR function: Evidence for a novel pentameric assembly between the dimeric leukotriene B4 receptor BLT1 and the G-protein. *J Mol Biol.* 329, 815–829.

112. Fotiadis, D., Jastrzebska, B., Philippsen, A., Muller, D.J., Palczewski, K., Engel, A. (2006) Structure of the rhodopsin dimer: A working model for G-protein-coupled receptors. *Curr Opin Struct Biol.* 16, 252–259.

113. Chabre, M., le Maire, M. (2005) Monomeric G-protein-coupled receptor as a functional unit. *Biochemistry.* 44, 9395–9403.

114. Urizar, E., Montanelli, L., Loy, T., Bonomi, M., Swillens, S., Gales, C., Bouvier, M., Smits, G., Vassart, G., Costagliola, S. (2005) Glycoprotein hormone receptors: Link between receptor homodimerization and negative cooperativity. *EMBO J.* 24, 1954–1964.

115. Guo, W., Urizar, E., Kralikova, M., Mobarec, J.C., Shi, L., Filizola, M., Javitch, J.A. (2008) Dopamine D2 receptors form higher order oligomers at physiological expression levels. *EMBO J.* 27, 2293–2304.

116. Kota, P., Reeves, P.J., Rajbhandary, U.L., Khorana, H.G. (2006) Opsin is present as dimers in COS1 cells: Identification of amino acids at the dimeric interface. *Proc Natl Acad Sci U S A.* 103, 3054–3059.

117. Mancia, F., Assur, Z., Herman, A.G., Siegel, R., Hendrickson, W.A. (2008) Ligand sensitivity in dimeric associations of the serotonin 5HT2c receptor. *EMBO Rep.* 9, 363–369.

118. Carrillo, J.J., Lopez-Gimenez, J.F., Milligan, G. (2004) Multiple interactions between transmembrane helices generate the oligomeric alpha1b-adrenoceptor. *Mol Pharmacol.* 66, 1123–1137.

119. Harikumar, K.G., Pinon, D.I., Miller, L.J. (2007) Transmembrane segment IV contributes a functionally important interface for oligomerization of the Class II G protein-coupled secretin receptor. *J Biol Chem.* 282, 30363–30372.

120. Hebert, T.E., Moffett, S., Morello, J.P., Loisel, T.P., Bichet, D.G., Barret, C., Bouvier, M. (1996) A peptide derived from a beta2-adrenergic receptor transmembrane domain inhibits both receptor dimerization and activation. *J Biol Chem.* 271, 16384–16392.

121. Lopez-Gimenez, J.F., Canals, M., Pediani, J.D., Milligan, G. (2007) The alpha1b-adrenoceptor exists as a higher-order oligomer: Effective oligomerization is required for receptor maturation, surface delivery, and function. *Mol Pharmacol.* 71, 1015–1029.

122. Bakker, R.A., Dees, G., Carrillo, J.J., Booth, R.G., Lopez-Gimenez, J.F., Milligan, G., Strange, P.G., Leurs, R. (2004) Domain swapping in the human histamine H1 receptor. *J Pharmacol Exp Ther.* 311, 131–138.

123. Maggio, R., Vogel, Z., Wess, J. (1993) Coexpression studies with mutant muscarinic/adrenergic receptors provide evidence for intermolecular "cross-talk" between G-protein-linked receptors. *Proc Natl Acad Sci U S A.* 90, 3103–3107.

124. Damian, M., Martin, A., Mesnier, D., Pin, J.P., Baneres, J.L. (2006) Asymmetric conformational changes in a GPCR dimer controlled by G-proteins. *EMBO J.* 25, 5693–5702.

125. Damian, M., Mary, S., Martin, A., Pin, J.P., Baneres, J.L. (2008) G protein activation by the leukotriene B4 receptor dimer. Evidence for an absence of trans-activation. *J Biol Chem.* 283, 21084–21092.

126. Whorton, M.R., Bokoch, M.P., Rasmussen, S.G., Huang, B., Zare, R.N., Kobilka, B., Sunahara, R.K. (2007) A monomeric G protein-coupled receptor isolated in a high-density lipoprotein particle efficiently activates its G protein. *Proc Natl Acad Sci U S A.* 104, 7682–7687.

127. Whorton, M.R., Jastrzebska, B., Park, P.S., Fotiadis, D., Engel, A., Palczewski, K., Sunahara, R.K. (2008) Efficient coupling of transducin to monomeric rhodopsin in a phospholipid bilayer. *J Biol Chem.* 283, 4387–4394.

128. Ernst, O.P., Gramse, V., Kolbe, M., Hofmann, K.P., Heck, M. (2007) Monomeric G protein-coupled receptor rhodopsin in solution activates its G protein transducin at the diffusion limit. *Proc Natl Acad Sci U S A.* 104, 10859–10864.

129. White, J.F., Grodnitzky, J., Louis, J.M., Trinh, L.B., Shiloach, J., Gutierrez, J., Northup, J.K., Grisshammer, R. (2007) Dimerization of the class A G protein-coupled neurotensin receptor NTS1 alters G protein interaction. *Proc Natl Acad Sci U S A.* 104, 12199–12204.

130. Springael, J.Y., Le Minh, P.N., Urizar, E., Costagliola, S., Vassart, G., Parmentier, M. (2006) Allosteric modulation of binding properties between units of chemokine receptor homo- and hetero-oligomers. *Mol Pharmacol.* 69, 1652–1661.

131. Armstrong, D., Strange, P.G. (2001) Dopamine D2 receptor dimer formation: Evidence from ligand binding. *J Biol Chem.* 276, 22621–22629.

132. Albizu, L., Balestre, M.N., Breton, C., Pin, J.P., Manning, M., Mouillac, B., Barberis, C., Durroux, T. (2006) Probing the existence of G protein-coupled receptor dimers by positive and negative ligand-dependent cooperative binding. *Mol Pharmacol.* 70, 1783–1791.

133. Limbird, L.E., Lefkowitz, R.J. (1976) Negative cooperativity among beta-adrenergic receptors in frog erythrocyte membranes. *J Biol Chem.* 251, 5007–5014.

134. Svendsen, A.M., Vrecl, M., Ellis, T.M., Heding, A., Kristensen, J.B., Wade, J.D., Bathgate, R.A., De Meyts, P., Nohr, J. (2008) Cooperative binding of insulin-like peptide 3 to a dimeric relaxin family peptide receptor 2. *Endocrinology.* 149, 1113–1120.

135. Wreggett, K.A., Wells, J.W. (1995) Cooperativity manifest in the binding properties of purified cardiac muscarinic receptors. *J Biol Chem.* 270, 22488–22499.

136. Durroux, T. (2005) Principles: A model for the allosteric interactions between ligand binding sites within a dimeric GPCR. *Trends Pharmacol Sci.* 26, 376–384.

137. Schwartz, T.W., Holst, B. (2007) Allosteric enhancers, allosteric agonists and ago-allosteric modulators: Where do they bind and how do they act? *Trends Pharmacol Sci.* 28, 366–373.

138. Binet, V., Brajon, C., Le Corre, L., Acher, F., Pin, J.P., Prezeau, L. (2004) The heptahelical domain of GABA(B2) is activated directly by CGP7930, a positive allosteric modulator of the GABA(B) receptor. *J Biol Chem.* 279, 29085–29091.

139. Gurevich, V.V., Dion, S.B., Onorato, J.J., Ptasienski, J., Kim, C.M., Sterne-Marr, R., Hosey, M.M., Benovic, J.L. (1995) Arrestin interactions with G protein-coupled receptors. Direct binding studies of wild type and mutant arrestins with rhodopsin, beta 2-adrenergic, and m2 muscarinic cholinergic receptors. *J Biol Chem.* 270, 720–731.

140. Hanson, S.M., Gurevich, E.V., Vishnivetskiy, S.A., Ahmed, M.R., Song, X., Gurevich, V.V. (2007) Each rhodopsin molecule binds its own arrestin. *Proc Natl Acad Sci U S A.* 104, 3125–3128.

141. Milligan, G., Smith, N.J. (2007) Allosteric modulation of heterodimeric G-protein-coupled receptors. *Trends Pharmacol Sci.* 28, 615–620.

142. Briddon, S.J., Gandia, J., Amaral, O.B., Ferre, S., Lluis, C., Franco, R., Hill, S.J., Ciruela, F. (2008) Plasma membrane diffusion of g protein-coupled receptor oligomers. *Biochim Biophys Acta.* 1783, 2262–2268.

143. Vidi, P.A., Chemel, B.R., Hu, C.D., Watts, V.J. (2008) Ligand-dependent oligomerization of dopamine D(2) and adenosine A(2A) receptors in living neuronal cells. *Mol Pharmacol.* 74, 544–551.

144. El-Asmar, L., Springael, J.Y., Ballet, S., Andrieu, E.U., Vassart, G., Parmentier, M. (2005) Evidence for negative binding cooperativity within CCR5-CCR2b heterodimers. *Mol Pharmacol.* 67, 460–469.

145. Ciruela, F., Casado, V., Rodrigues, R.J., Lujan, R., Burgueno, J., Canals, M., Borycz, J., Rebola, N., Goldberg, S.R., Mallol, J., Cortes, A., Canela, E.I.,

Lopez-Gimenez, J.F., Milligan, G., Lluis, C. Cunha, R.A., Ferre, S., Franco, R. (2006) Presynaptic control of striatal glutamatergic neurotransmission by adenosine A1-A2A receptor heteromers. *J Neurosci.* 26, 2080–2087.

146. Gomes, I., Gupta, A., Filipovska, J., Szeto, H.H., Pintar, J.E., Devi, L.A. (2004) A role for heterodimerization of mu and delta opiate receptors in enhancing morphine analgesia. *Proc Natl Acad Sci U S A.* 101, 5135–5139.
147. Rashid, A.J., So, C.H., Kong, M.M., Furtak, T., El-Ghundi, M., Cheng, R., O'Dowd, B.F., George, S.R. (2007) D1-D2 dopamine receptor heterooligomers with unique pharmacology are coupled to rapid activation of Gq/11 in the striatum. *Proc Natl Acad Sci U S A.* 104, 654–659.
148. Waldhoer, M., Fong, J., Jones, R.M., Lunzer, M.M., Sharma, S.K., Kostenis, E., Portoghese, P.S., Whistler, J.L. (2005) A heterodimer-selective agonist shows in vivo relevance of G protein-coupled receptor dimers. *Proc Natl Acad Sci U S A.* 102, 9050–9055.
149. Rozenfeld, R., Devi, L.A. (2007) Receptor heterodimerization leads to a switch in signaling: Beta-arrestin2-mediated ERK activation by mu-delta opioid receptor heterodimers. *FASEB J.* 21, 2455–2465.
150. Ellis, J., Pediani, J.D., Canals, M., Milasta, S., Milligan, G. (2006) Orexin-1 receptor-cannabinoid CB1 receptor heterodimerization results in both ligand-dependent and -independent coordinated alterations of receptor localization and function. *J Biol Chem.* 281, 38812–38824.
151. Jordan, B.A., Devi, L.A. (1999) G-protein-coupled receptor heterodimerization modulates receptor function. *Nature.* 399, 697–700.
152. Briddon, S.J., Hill, S.J. (2007) Pharmacology under the microscope: The use of fluorescence correlation spectroscopy to determine the properties of ligand-receptor complexes. *Trends Pharmacol Sci.* 28, 637–645.
153. Briddon, S.J., Middleton, R.J., Cordeaux, Y., Flavin, F.M., Weinstein, J.A., George, M.W., Kellam, B., Hill, S.J. (2004) Quantitative analysis of the formation and diffusion of A1-adenosine receptor-antagonist complexes in single living cells. *Proc Natl Acad Sci U S A.* 101, 4673–4678.
154. Cordeaux, Y., Briddon, S.J., Alexander, S.P., Kellam, B., Hill, S.J. (2008) Agonist-occupied A3 adenosine receptors exist within heterogeneous complexes in membrane microdomains of individual living cells. *FASEB J.* 22, 850–860.

CHAPTER 3

The Emergence of Allosteric Modulators for G Protein-Coupled Receptors

KAREN J. GREGORY, CELINE VALANT, JOHN SIMMS, PATRICK M. SEXTON, and ARTHUR CHRISTOPOULOS

Monash Institute of Pharmaceutical Sciences and Department of Pharmacology, Monash University, Victoria, Australia

3.1. INTRODUCTION

Guanine nucleotide-binding protein (G protein)coupled receptors (GPCRs) are integral membrane proteins that mediate the actions of a vast array of endogenous stimuli and represent the major targets for approximately 30% of all medicines on the market [1–3]. Due to their ubiquitous cell surface expression, GPCRs are very tractable drug targets. However, GPCR-based drug discovery programs, in common with those of many other targets, suffer from a high attrition rate. This is likely due to two major reasons: (1) an insufficient mechanistic understanding of the relationship between common biological screening assays of GPCR behavior and the actual GPCR therapeutic end point that is being targeted in the drug discovery program; (2) a paucity of highly GPCR subtype-selective ligands. With respect to the former, the growing characterization of the *in vivo* physiological roles of GPCRs via knockout animals in recent years, as well as a realization of the need to compare multiple indices of GPCR function when screening for drug effects, is promising to make some headway in guiding more informed approaches to appropriate drug screening. With respect to the selectivity issue, this is possibly due to the fact that traditional drug discovery has targeted GPCR orthosteric sites, that is, the natural binding sites for the receptors' endogenous ligands, in

GPCR Molecular Pharmacology and Drug Targeting: Shifting Paradigms and New Directions, Edited by Annette Gilchrist

the search for novel molecules that either mimic or block the actions of the endogenous ligands. Because orthosteric sites are generally well conserved across different subtypes of a given receptor, chemotypes that target such binding pockets may be expected to lack selectivity if they primarily utilize attachment points common to each subtype of GPCR. However, it is now recognized that many GPCRs possess topographically distinct, allosteric binding sites, and that ligands that bind to these sites offer tremendous potential for more selective and/or effective therapies than conventional orthosteric ligands.

3.2. FOUNDATIONS OF ALLOSTERIC RECEPTOR THEORY

The development of allosteric theory was basically driven by two interrelated concepts: the first being the discovery of cooperativity in the binding of some ligands to their target proteins, and the second being the realization that proteins naturally explore multiple conformations and that conformational changes can be transmitted across protein structures such that the binding of a ligand could affect the conformation of a distant binding pocket recognized by another ligand. The idea that certain proteins could bind more than one ligand was notably described early last century for the enzyme, hemoglobin, which can simultaneously bind up to four molecules of oxygen [4], and subsequently extended to other enzymes and ligand-gated ion channels [5, 6]. In most known instances of cooperativity, the binding of one ligand to its site has the potential to alter the affinity for subsequent equivalents of the same ligand at the remaining (unoccupied) sites [7]. In the case of hemoglobin, for instance, this corresponds to a 200-fold increase in the affinity of the fourth oxygen molecule bound compared to the first [8]. Around the middle of the last century, important studies were also underway that led to the concept of global protein conformational changes, whereby proteins were postulated to isomerize between multiple conformations that can show different affinities for ligands and/or different functional properties [9, 10]. In such a scheme, the process of ligand binding can be viewed as an event that biases certain protein conformational states toward those that show higher affinity for the ligand, at the expense of other possible protein states.

The term "allosteric" was first coined by Monod and colleagues in studies of enzyme inhibition [11, 12]. They noted that inhibitors that were structurally diverse from the enzyme substrate were likely to act at alternate binding sites that were somehow conformationally linked to that of the substrate binding pocket. The authors defined allosteric proteins on the following criteria: an oligomeric architecture, cooperativity in binding, and the ability of ligand binding to preferentially select for certain protein conformations that exist in equilibrium. This strict definition of allosteric proteins has since been relaxed somewhat, with the term being used to describe multisite interactions on proteins, irrespective of whether or not they are oligomeric, and/or to describe the transition of a protein between different conformational states, regardless of whether or not they possess multiple binding sites.

Although the phenomenon was initially studied on enzyme and ion channel systems, it should be noted that GPCRs are naturally allosteric proteins. Specifically, they are designed to recognize extracellular stimuli and then to transmit them across biological membranes to impart the signal to their cognate G protein(s) and other intracellular effectors. Thus, the endogenous orthosteric ligand binds to one site on the receptor that is usually extracellularly accessible, while the G protein and any other accessory proteins interact with interfaces that are present inside the cell, and therefore topographically distinct from the binding site of the endogenous ligand. Moreover, there is a wealth of evidence that the binding of G proteins, and other types of GPCR–protein interactions, can influence the conformational state of the receptor, as evidenced by altered ligand affinity and/or signaling [13–17]. Given this tremendous conformational plasticity, it is perhaps not surprising that many small molecules are now being discovered with increasing prevalence that can bind to sites other than the orthosteric site on a GPCR to promote conformational changes that have a profound effect on the binding and/or function of orthosteric ligands.

3.3. MODELS FOR UNDERSTANDING THE EFFECTS OF ALLOSTERIC MODULATORS

The recognition that GPCRs can possess binding sites in addition to the orthosteric site has necessitated the development of mathematical models to facilitate descriptions and understanding of allosteric drug–receptor interactions. The binding of an allosteric ligand to its site will change the conformation of the receptor, which means that the "geography" of the orthosteric binding pocket and any other potential receptor–ligand/protein interfaces, may also change. As a consequence, the binding affinity and/or signaling efficacy of the orthosteric ligand can be modulated, either in a positive or negative manner—hence the common use of the term "allosteric modulator" to encompass different types of allosteric ligands. The simplest allosteric GPCR model is one in which the binding of an allosteric ligand modulates only the affinity of the orthosteric ligand; this simple model is referred to herein as the allosteric ternary complex model (ATCM; Fig. 3.1a). Within the framework of the ATCM, the interaction is governed by the concentration of each ligand, the equilibrium dissociation constants of the orthosteric and allosteric ligands (K_A and K_B, respectively) for their binding to their sites on the unoccupied receptor, and the "cooperativity factor", α, which is a measure of the magnitude and direction of the allosteric interaction between the two conformationally linked sites. A value of $\alpha < 1$ (but greater than 0) indicates negative cooperativity, such that the binding of an allosteric ligand inhibits the binding of the orthosteric ligand. Values of $\alpha > 1$ indicate positive cooperativity, such that the

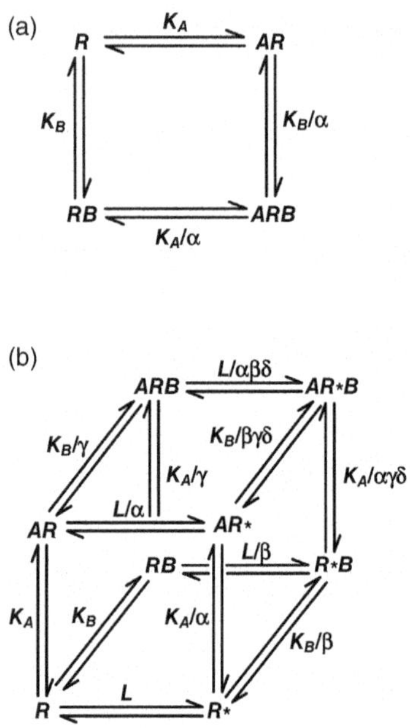

Figure 3.1 Allosteric models of GPCR-ligand interactions. (a) The allosteric ternary complex model (ATCM), which describes the interaction between an orthosteric ligand, A, and allosteric modulator, B, in terms of their respective equilibrium dissociation constants (K_A, K_B) and a cooperativity factor, α, which denotes the magnitude and direction of the allosteric effect on ligand binding affinity. (b) The allosteric two state model (ATSM), which describes allosteric modulator effects on affinity, efficacy, and the distribution of the receptor between active (R*) and inactive (R) states, in terms of distinct conformations selected by ligands according to their cooperativity factors for the different states.

allosteric ligand promotes the binding of orthosteric ligand, whereas values of $\alpha = 1$ indicate neutral cooperativity, that is, no net change in binding affinity at equilibrium. Because the two sites are conformationally linked, the allosteric interaction is reciprocal at equilibrium, in that the orthosteric ligand will modulate the binding of the allosteric ligand in the same manner and to the same extent.

The simple ATCM describes the effect of the modulator only in terms of changes in orthosteric ligand affinity, and vice versa. However, there is no a priori reason why the conformational change engendered by an allosteric modulator in the GPCR does not perturb signaling efficacy in addition to, or independently of, any effects on orthosteric ligand binding affinity. In order to incorporate allosteric modulation of efficacy as well as affinity, the simple ATCM has been extended into an allosteric two-state model (ATSM) [18]. This model (Fig. 3.1b) describes GPCR function in terms of: (a) the ability of the receptor to constitutively isomerize between active (R*) and inactive (R) states, as determined by the isomerization constant, L; (b) the ability of orthosteric and allosteric ligands to modify this transition between states, that is, to act as either agonists or inverse agonists, which is governed by the parameters α and β; (c) the ability of each ligand to allosterically modulate the binding affinity of the other, governed by the "binding cooperativity" parameter, γ; and (d) the ability of either ligand to modulate the transition to an active receptor state when both ligands are bound, governed by the "activation cooperativity" parameter, δ. Although the model is more complex due to the increased number of parameters, it has proven very useful in conceptualizing divergent allosteric modulator effects on ligand affinity versus efficacy, which are being recognized with increasing prevalence in many studies [19].

More recently, it has become accepted that GPCRs can adopt multiple active and inactive conformations beyond the simple "R and R*" paradigm. This realization has a number of significant consequences, including the discovery that different ligands can preferentially stabilize unique GPCR conformations, each associated with its own profile of receptor behaviors, to the relative exclusion of other states. The phenomenon has been variously termed "stimulus-trafficking," "biased agonism," "collateral efficacy," "ligand-directed signaling," and "functional selectivity," among others [20]. Because allosteric ligands can themselves stabilize GPCR conformations that change the reactivity of the receptor toward orthosteric ligands and/or its host cellular environment, there is a significant potential for allosteric modulators to engender stimulus-bias and functional selectivity [21]. This is likely to have a significant impact on drug discovery programs in the coming years.

3.4. TYPES OF ALLOSTERIC MODULATORS AND THEIR PROPERTIES

Allosteric modulators have a vast array of potential effects. For instance, they can enhance orthosteric ligand affinity and/or efficacy, inhibit orthosteric ligand affinity and/or efficacy, or even act as allosteric agonists/inverse agonists in their own right (Fig. 3.2a). Moreover, some of these properties can change dramatically, depending on the nature of the orthosteric ligand with which the modulator is interacting. Thus, it is not surprising that there have been a number of phenomenological descriptors, some more confusing and/or cryptic than others, applied to allosteric modulators. It should also be noted that there is no reason why a modulator could not express more than one of these properties concomitantly, for example, agonism (positive or inverse) together with enhancement or inhibition of orthosteric ligand binding/function [19, 22]. For example, alcuronium, an allosteric modulator of muscarinic acetylcholine receptors (mAChRs), is an inverse (allosteric) agonist, an allosteric enhancer of the affinity of the orthosteric antagonist, [^{3}H]*N*-methylscopolamine, an allosteric inhibitor of the affinity of the orthosteric antagonist, [^{3}H]quinuclidinyl benzilate, and an allosteric inhibitor of the efficacy of the orthosteric

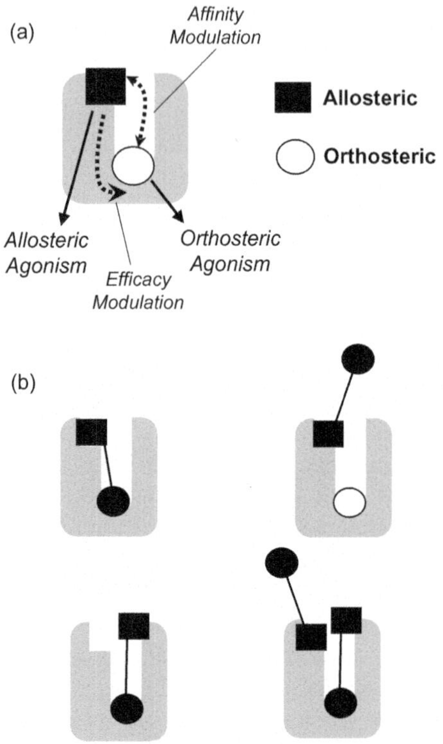

Figure 3.2 Modes of allosteric or bitopic ligand effects. (a) An allosteric modulator, by binding to a topographically distinct site, can modulate the affinity and/or signaling efficacy of an orthosteric ligand. In addition, the modulator may be able to engender receptor agonism/inverse agonism in its own right. It is also possible for a given allosteric modulator to express more than one of these properties. (b) Possible modes of engagement of a receptor's orthosteric and allosteric sites by a bitopic ligand, either in the absence or presence of an orthosteric ligand.

agonist, pilocarpine, at the M_2 mAChR [23, 24]. Different types of cooperativity between an allosteric modulator and orthosteric ligands at a given GPCR is referred to as "probe dependence," highlighting that the type of allosteric modulation can change depending on the nature of the orthosteric ligand that is used as a reference probe of receptor binding or function; alcuronium is an excellent example of the probe-dependent nature of allosteric interactions.

In addition to displaying probe dependence, allosteric interactions can result in strikingly different behaviors for the *same* pair of interacting ligands, depending on the nature of the assay format utilized to detect their activity. For instance, the allosteric modulator, ORG27569, is an enhancer of the binding affinity of the orthosteric agonist, CP55940 at cannabinoid CB_1 receptors, but an inhibitor of CP55940 signaling efficacy, resulting in noncompetitive antagonism of function despite an increase in agonist potency [25]. A similar observation has been made at M_2 muscarinic receptors for the interaction between the allosteric modulator, naphmethonium and the orthosteric agonist, pilocarpine [26]. There are two important considerations that arise from observations such as these. First, this type of noncompetitive antagonism of function is similar to what has previously been referred to as "use-dependent blockade," in that the antagonist actually works better in the presence of increasing amounts of agonist tone; because the modulator increases the affinity of the agonist, the agonist will also increase the affinity of the modulator, which would then bind with this higher affinity to abolish the signaling of the agonist. Use-dependent antagonism of GPCRs is simply not achievable in an orthosteric setting. The second important consideration is that, depending on the assay format (e.g., binding versus function), a different result will be obtained and thus, there is a risk of misinterpreting the pharmacology of the modulator if a range of assays is not utilized. It should be noted that none of the assays are to be considered "incorrect" in this regard; rather, they each provide a window into the conformational states that the allosteric modulator promotes.

Currently, it remains to be determined whether a single ligand phenotype (modulator only) or a combination of both modulator and agonist properties is the optimal approach to treating different diseases with GPCR allosteric ligands. Most likely, different therapies will benefit from one phenotype relative to another. However, an alternative means of exploiting the agonist/modulator paradigm is via the generation of "bitopic ligands," that is, hybrid molecules composed of both an orthosteric moiety and an allosteric moiety (Fig. 3.2b). One such example is the muscarinic receptor agonist, McN-A-343, which displays high efficacy at the M_1 and M_4 subtypes of muscarinic receptor, but low efficacy and functional selectivity at the M_2 muscarinic subtype; we have recently discovered that the latter is due to the fact that McN-A-343 is composed of an orthosteric agonist, trimethylammonium, coupled to a negative allosteric modulator of agonist efficacy, 3-chlorophenyl carbamate [27]. Presumably, the higher efficacy of the McN-A-343 at other muscarinic receptor subtypes reflects differences in the allosteric binding pockets of those subtypes. This finding raises the interesting question as to whether other functionally selective GPCR agonists also represent hitherto unappreciated bitopic orthosteric/allosteric ligands. A logical extension of this approach is to actively engineer such bitopic ligands, as has been successfully demonstrated recently for novel hybrid molecules that selectively target the M_2 muscarinic receptor [28]; an advantage of this approach is the ability to take a nonselective orthosteric agonist and engender subtype and functional selectivity in its actions by attaching an appropriately selective allosteric moiety.

Irrespective of phenotype, the most obvious advantage of allosteric ligands is the potential for greater receptor subtype selectivity, as allosteric sites need not have evolved to accommodate the endogenous ligand [29]. An additional advantage of allosteric modulators that have no agonistic activity in the absence of orthosteric ligand is the ability to retain the spatial and temporal aspects of normal (physiological) receptor function; the modulator would only exert an effect where and when the endogenous neurotransmitter or hormone is present. Furthermore, modulators with limited cooperativity will have an in-built "ceiling" level to their effect, suggesting that they may be potentially safer than orthosteric ligands if administered in very large doses [19]. Perhaps the most important challenge in this field, however, currently remains the ability to detect and quantify the myriad of possible allosteric effects that can arise when two ligands simultaneously occupy a GPCR.

3.5. DETECTION AND QUANTIFICATION OF ALLOSTERIC INTERACTIONS

Although cell-based functional assays have largely surpassed radioligand binding assays as primary screens for allosteric GPCR modulators, there are advantages and disadvantages to both types of assays when measuring allosteric modulator effects, and ideally, a combination of such experiments should be used where possible.

3.5.1. Radioligand Binding Assays

Equilibrium Binding Assays Detection of allosteric interactions using an equilibrium binding assay relies on the ability of a putative allosteric ligand to alter the binding affinity of a radiolabeled orthosteric ligand to either increase or decrease the specific binding of the orthosteric probe. The most common assay in this regard is to perform a modulator titration curve against a single fixed concentration of orthosteric radioligand. According to the simple ATCM, the maximum fractional binding (FB) of the orthosteric ligand is a function of the cooperativity of the interaction, and is given by [30]:

$$FB = \frac{\alpha([A] + K_A)}{\alpha[A] + K_A}$$

From this relationship, it can be seen that an allosteric ligand exhibiting very high negative cooperativity ($\alpha \rightarrow 0$) can completely inhibit the specific binding of the radioligand and therefore be misinterpreted as a competitive orthosteric ligand. If the negative cooperativity is not too high, then the interaction will deviate from the expectations of competition due to the fact that appreciable ternary complex (ARB, Fig. 3.1a) will still be detected at high concentrations of modulator, resulting in a nonzero degree of specific radioligand binding (Fig. 3.3). Thus, one means of differentiating allosteric antagonism from competitive antagonism is to utilize large concentrations ($>K_A$) of orthosteric radioligand as a probe.

A second issue that needs to be considered in the conduct and interpretation of equilibrium binding assays is that an allosteric modulator with a weak

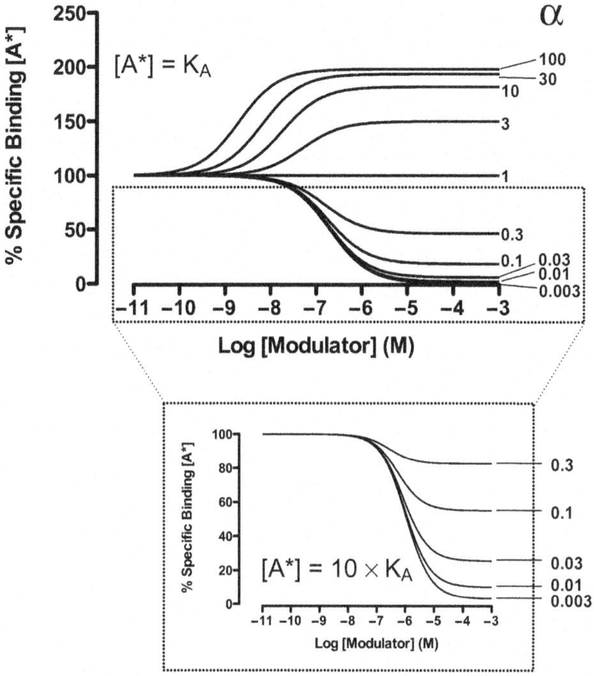

Figure 3.3 Effect of degree of cooperativity on the detection of allosteric modulation in a radioligand binding assay. The simulations illustrate the effects of increasing concentrations of an allosteric modulator ($K_B = 10^{-7}$M) under conditions of different degrees of positive or negative cooperativity (α values indicated in the figure) on the percent specific binding of an orthosteric radioligand, denoted A*. Note that for high degrees of negative cooperativity (boxed region, top panel), the interaction becomes difficult to distinguish from competitive antagonism. However, increasing the concentration of radioligand (bottom panel) can unmask the allosteric effect.

cooperativity factor ($\alpha \approx 1$) may not be detected due to a lack of sensitivity of the method in this instance; this is common for allosteric modulators that exert their effects predominantly on receptor signaling rather than binding. Utilizing another orthosteric ligand belonging to a separate chemical class may overcome this, assuming that the allosteric modulator exhibits a different degree of binding cooperativity with the new probe. In general, however, radioligand binding assays are most useful if the modulator behaves according to the simple ATCM with a reasonable degree of either positive or negative cooperativity. There are a few instances where high-affinity allosteric radioligands have been developed, for example, for M_2 muscarinic receptors [31] or metabotropic glutamate receptors (mGluRs) [32], but by and large, the equilibrium binding assay using a radioactive allosteric ligand to screen for other allosteric ligands that act at the same site is not yet a routine method of detection of allosteric modulators.

Dissociation/Association Kinetic Binding Assays An alternative application of radioligand binding assays to detect and quantify allosteric interactions involves the analysis of allosteric effects on the rates of dissociation or association of the orthosteric radioligand with the GPCR of interest. Kinetic binding assays can be used individually or in tandem with equilibrium binding assays; if an allosteric modulator has the ability to change the affinity of an orthosteric ligand, then this will be manifested either as a change in the association and/or dissociation rates of the orthosteric probe on a modulator-occupied receptor. In general, the measurement of association kinetics is problematic with respect to interpretation because competitive (orthosteric) ligands will also alter the apparent association rate of the radioligand simply by delaying the time taken for the radiolabeled orthosteric probe to reach equilibrium. In contrast, the only way that the dissociation rate of a pre-equilibrated radioligand–receptor complex can be modified is if the interacting ligand binds to another site on this complex to change the receptor conformation prior to the radioligand dissociating. Thus, radioligand dissociation kinetic assays represent a useful means of detecting and validating an allosteric mode of action. Furthermore, under certain conditions, these assays can also be used to quantify the allosteric effect in terms of the ATCM [33, 34]. An additional advantage of kinetic binding assays is that they have the potential to detect allosteric ligands that possess neutral binding cooperativity ($\alpha = 1$) at equilibrium, if the mechanistic basis of the neutral cooperativity is due to the modulator altering orthosteric ligand association and dissociation rates to the same extent [34]. However, there are a number of situations where the utility of dissociation kinetic assays is limited. First, when the allosteric interaction is primarily manifested by an alteration in the association, rather than dissociation, rate of the orthosteric ligand. Second, under conditions of very high negative cooperativity, the affinity of the modulator for the radioligand-bound receptor may be sufficiently low such that, unless impractically high concentrations of modulator are utilized, there is no perturbation of dissociation kinetics. Third, a dissociation kinetic assay will fail to detect allosterism if the effect of the modulator is primarily on orthosteric ligand efficacy rather than affinity.

Another consequence of the effects of some allosteric modulators on orthosteric ligand binding kinetics is that the interaction between the modulator and radiolabeled orthosteric probe may not reach equilibrium over the time course of the assay, leading to complex behaviors that are not accommodated by the simple equilibrium ATCM [33, 35, 36].

3.5.2. Functional Assays

Although radioligand binding assays provide a direct means for detecting and quantifying allosteric interactions at GPCRs, they are limited in that they can only measure allosteric changes in radioligand affinity. In order to fully define the pharmacology of an allosteric ligand, it is necessary to consider its effects in a functional context, which can reveal allosteric effects on signaling efficacy as well as whether or not the modulator possesses agonism (either positive or inverse) in its own right. Furthermore, it is more frequently possible to use the endogenous ligand in a functional assay, relative to a binding assay, to directly probe the influence of the allosteric ligand on the natural agonist response.

As with the radioligand assay, the simplest allosteric effect in the context of a functional assay occurs when only the affinity of an orthosteric ligand for its receptor changes. In this scenario, the orthosteric agonist concentration–response curve is shifted to the left by an allosteric enhancer while an allosteric inhibitor will shift the curve to the right; no effects are expected on either the maximum or basal system responses. According to the ATCM, the translocation of concentration–response curves in this situation will approach a limit as determined by the cooperativity, α, with the orthosteric agonist, beyond which increasing concentrations of modulator will be unable to further shift the agonist curve. In this respect, the ATCM can be readily applied to functional interaction data to determine the affinity of the allosteric modulator and the cooperativity between the interacting ligands. For allosteric inhibitors that have very high negative cooperativity with the orthosteric agonist tested, this limit may not be reached over the concentration range utilized. Such allosteric modulators may be mistaken for competitive antagonists. Indeed, this is predicted in the ATCM ($\alpha \to 0$), as described above.

If the presence of an allosteric modulator results in a change in the maximal agonist effect (E_{max}) or the basal responsiveness of the system, then this is indicative of orthosteric efficacy modulation, in the case of the former, or allosteric agonism, in the case of the latter. The allosteric two-state model (Fig. 3.1b) accommodates efficacy modulation and agonism in addition to affinity modulation. However, the model is not readily amenable to fitting experimental data due to the large number of parameters. An alternative approach has recently been presented, based on combining the operational model of agonism [37] with the ATCM [21, 25] to yield an operational model of allosterism:

$$E = \frac{E_m(\tau_A[A](K_B + \alpha\beta[B]) + \tau_B[B]K_A)^n}{([A]K_B + K_AK_B + K_A[B] + \alpha[A][B])^n + (\tau_A[A](K_B + \alpha\beta[B]) + \tau_B[B]K_A)^n}$$

where E is the effect, and A, B, K_A, and K_B are as defined for the ATCM (Fig. 3.1a) above. As for the simple ATCM, allosteric modulation of binding affinity in the operational model of allosterism is governed by the cooperativity factor α. Allosteric modulation of orthosteric ligand efficacy is incorporated into the model by the introduction of another (empirical) parameter, β, which scales from zero to infinity and denotes the magnitude by which the allosteric modulator modifies the stimulus generated by the orthosteric agonist on the ARB ternary complex. The parameters τ_A and τ_B relate to the ability of the orthosteric and allosteric ligands, respectively, to promote receptor activation. Both τ_A and τ_B incorporate the intrinsic efficacy of each ligand, the total density of receptors, and the efficiency of stimulus–response coupling. The

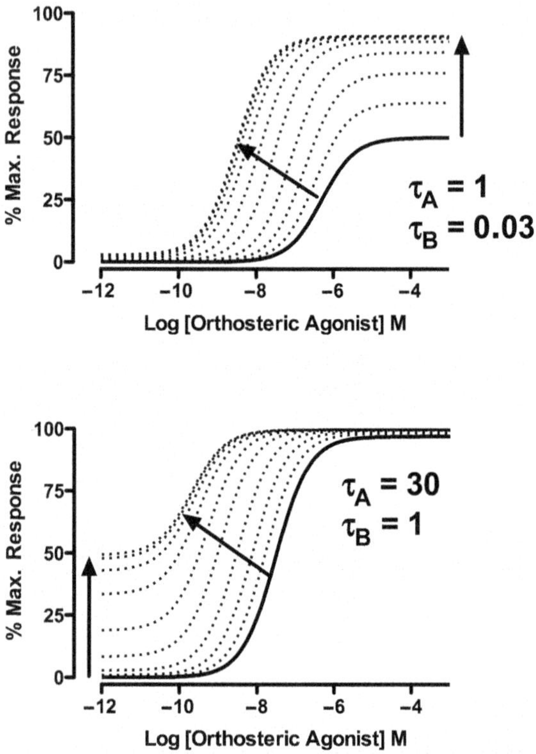

Figure 3.4 Effects of cellular stimulus–response coupling on the manifestation of allosteric modulation and agonism. Simulations based on an operational model of allosterism (see text) for a system with low coupling efficiency or receptor expression (top panel) or high coupling efficiency or receptor expression (low panel). For all simulations, the following parameter values were used: E_m = 100, n = 1, K_A = 10^{-6} M, K_B = 10^{-7} M, α = 30, β = 10. The allosteric modulator, B, was present at concentrations ranging from 0.3 nM to 3 μM. Despite the same pair of ligands being tested at the same receptor, strikingly different apparent behaviors are imposed by the stimulus–response machinery of the host system, which can be modeled by changes in the parameter, τ, of the operational model. Data adapted from Reference 21.

parameters E_m and n denote the maximal possible system response and the slope factor of the transducer function that links occupancy to response, respectively [21].

In addition to the modulator K_B value, the operational model of allosterism describes modulation as being mediated by two parameters, α and β, which can vary for each and every set of interacting ligands at a GPCR. Theoretically, however, these should not change for a given set of ligands and GPCR between different assays of GPCR function. An important caveat to this is the potential for pathway-specific modulation; allosteric modulator-engendered functional selectivity would be indicated by assay-dependent changes in the value of the β parameter.

With respect to the τ (agonism) parameters of the model, because these are determined not only by the intrinsic properties of each ligand but also by the biological system under investigation, they can change between different assay systems (Fig. 3.4). Analyzing an effector pathway that has low stimulus–response coupling efficiency or in a cellular background with very low levels of receptor expression can yield a low τ_B value for an agonist and, as such, its efficacy may not be discernible; if the compound is allosteric, the interaction will manifest primarily as a change in the potency and/or maximal response to orthosteric ligand, with no effect on basal signaling. In the case of receptor overexpression or high stimulus–response coupling efficiency, and subsequently high τ_B values, the allosteric ligand efficacy will substantially increase the basal responsiveness of the assay system, and may also shift the orthosteric agonist potency. However, enhancement of the maximal orthosteric agonist response may not be evident, as a GPCR that has high coupling efficiency may already be approaching the maximal possible response of the entire cellular system (E_m) for any agonist being tested [21]. These are important considerations in terms of designing screening strategies for allosteric ligand-based drug discovery programs, interpreting the pharmacology of putative allosteric ligands and also translating research from recombinant systems to tissues and beyond.

3.6. SOME EXAMPLES OF GPCR ALLOSTERIC MODULATORS

3.6.1. Small Molecule Allosteric Modulators

Clearly, the most obvious potential for exploiting allosteric sites on GPCRs is with respect to the discovery of novel small molecule chemotypes, which traditionally are seen as the most suitable agents amenable for drug development. In recent years, small molecule allosteric modulators have been discovered to act at all three major families of GPCRs (see Table 3.1 for some representative examples). While many of these have remained pharmacological tools, they have highlighted the applicability of allosteric modulation to GPCR-targeted

TABLE 3.1 Representative Allosteric Modulators of GPCRs

Family A	
Adenosine A_1	PD 81,723; LUF 5484
Adenosine A_{2A}	Amilorides
Adenosine A_3	VU5455; VU8504; DU124183
Adrenoceptor α_1	Amilorides; benzodiazepines; conopeptide r-TIA
Adrenoceptor α_{2A}, α_{2B}	Amilorides
Adrenoceptor α_{2D}	Agmatine
Adrenoceptor β_2	Zinc

TABLE 3.1 (*Continued*)

Cannabinoid CB_1	Org 27569; Org 27759; Org 29647; PSNCBAM-1; RTI-371
Chemokine CXCR1	Reparixin; SCH 527123; AZ Cmpds. A & B
Chemokine CXCR2	Reparixin; SCH 527123; SB 656933; AZ Cmpds. A & B
Chemokine CXCR3	IP-10; I-TAC
Chemokine CXCR4	RSVM, ASLW; prichosanthin; plerixafor
Chemokine CCR1	BX-471; CP-481715; UCB35625
Chemokine CCR3	UCB35625; TAK779;
Chemokine CCR5	Trichosanthin; AK602; AK530; TAKK220; TAK779; SCH 351125; ancriviroc; vicriviroc; aplaviroc; maraviroc
Dopamine D_1	Zinc
Dopamine D_2	Amilorides; zinc; L-prolyl-L-leucylglycinamide
Endothelin ET_A	Aspirin, sodium salicylate
FFA2 (GPR43)	Phenylacetamides 1 and 2
Follicle stimulating hormone	BMS compounds 2–7
GnRh receptor	Furan-1
Growth hormone secretagogue	L-692,429; GH-releasing peptide 6
Luteinizing hormone	Org 41841; [^{3}H]Org 43553
Muscarinic M_1–M_5	Gallamine, alcuronium, brucine, W84, C_7/3-phth; WIN 62,577; AC-42; thiochrome; MT7; MT3; staurosporine; tacrine; McN-A-343; LY2033298
Neurokinin NK_1	Heparin
Opioid μ, δ	Cannabidiol
Purine $P2_{Y1}$	2,2′-pyridylsatogen tosylate
Serotonin $5HT_{1B/1D}$	5HT moduline
Serotonin $5HT_{2A}$, $5HT_7$	Oleamide
Serotonin $5HT_{2C}$	Oleamide; PNU-69176E
Family B	
CRF1 receptor	Antalarmin; NBI 35965; DMP696; NBI 27914
CGRP receptor	BIBN4096BS
Glucagon	Bay27-9955; L-168,049
GLP1 receptor	NN compounds 1–6; T-0632
Family C	
Calcium sensing receptor	Fendeline; Cinacalcet; NPS 467; NPS 568; L-amino acids; NPS 2143; Calhex 231
$GABA_B$	CGP7930; CGP13501; GS39783
Glutamate $mGluR_1$	(-)-CPCCOEt; Ro 67-7476, Ro 01-6128; BAY36-7620; [^{3}H]R214127; NPS 2390; EM-TBPC; *cis*-64a; JNJ 16259685
Glutamate $mGluR_2$	LY 487379; Biphenyl-indanone A; LY-181837; Ro 67-6221
Glutamate $mGluR_4$	SIB-1893, MPEP; (-)-PHCCC; VU0155041; VU0080421
Glutamate $mGluR_5$	MPEP; MTEP; DFB, DmeoB, DCB; CPPHA; CDPPB; ADX-47273
Glutamate $mGluR_7$	AMN082; MMPIP

drug discovery. The following sections briefly discuss some exemplar models of modulators from each of the main GPCR families.

Family A GPCRs The mAChR family is arguably one of the most well-characterized Family A GPCRs that possess at least one (and likely two) small molecule allosteric binding site(s). Indeed, the full spectrum of allosteric ligand phenotypes have been described for this family, from prototypical allosteric inhibitors and enhancers, such as gallamine and alcuronium, to allosteric agonists, such as LY2033298 (3-amino-5-chloro-6-methoxy-4-methyl-thieno[2,3-b] pyridine-2-carboxylic acid cyclopropylamide) and bitopic ligands, such as McN-A-343 [23, 27, 38, 39]. Examples of the entire range (and complexities) of allosteric behaviors are also evident for the muscarinic receptor family, including positive and inverse agonism through the allosteric site, probe dependence of allosteric interactions, subtype selectivity driven through cooperativity rather than affinity, and differential allosteric modulation of efficacy and affinity of the same orthosteric ligand [40]. In this regard, the muscarinic receptors have been widely utilized as a model system in the development of assay techniques and application of theoretical models, such as the ATCM and the operational model of allosterism, to experimental data. In addition, the location of the allosteric site of muscarinic receptors utilized by prototypical allosteric modulators has been well characterized via numerous structure–function studies identifying the importance of regions such as the second extracellular loop and the top of transmembrane (TM) domain 7 for allosteric drug interaction.

From a therapeutic viewpoint, the M_1 and M_4 subtypes have more recently been identified as desirable targets for allosteric modulation. Recent drug discovery efforts have uncovered highly functionally selective M_1 mAChR agonists, such as AC-42 (4-n-butyl-1-[4-2(2-methylphenyl)-4-oxo-1-butyl]-piperidine) [41], 77-LH-28-1 (1-[3-(4-butyl-1-piperidinyl) propyl]-3,3-dihydro-2(1H)-quinolinone), AC-260584 (4-[3-(butylpiperidin-1-yl)propyl]-7-fluoro-4H-benzo[1,4]oxasin-3-one) [42, 43], and TBPB ((1-(1′-2-methylbenzyl)-1,4′-bipiperin-4-yl)-1Hbenzo[*d*]imidazole-1(3H)-one) [44], which have been suggested to mediate at least some of their effects via allosteric mechanisms. For example, the mode of action of AC-42 has been characterized using a combination of mutagenesis, signaling, binding, and dissociation kinetic assays, which have suggested that this compound operates at a distinct location to the orthosteric site [41–43, 45]. *N*-desmethlyclozapine is another putative allosteric agonist identified for the M_1 mAChR [46]. A truly M_1 selective allosteric agonist could be used to treat cognitive function and behavioral symptoms of Alzheimer's disease, with an allosteric mode of action overcoming the lack of selectivity and subsequent dose-limiting side effects associated with orthosteric agonists. For the M_4 mAChR subtype, recent studies have reported M_4-selective positive allosteric modulators, LY2033298 and VU10010 (3-amino-*N*-(4-chlorobenzyl)-4,6-dimethylthieno[2,3-b]pyridine-2-carboxamide) [39, 47], with the exciting potential of providing a novel approach to treating schizophrenia. Mechanistic studies of these ligands indicated that they exert their effects at the M_4 mAChR by increasing the affinity of ACh and its coupling to G proteins. Interestingly, LY2033298 has also been shown to restore the function of ACh at a mutant muscarinic receptor with dramatically impaired orthosteric site functionality [48], highlighting the potential to use allosteric ligands to rescue function of receptors with impaired orthosteric pockets and/or signaling.

The adenosine family of GPCRs is another noteworthy example where many studies have focused on the potential for selective allosteric modulators. Indeed, the A_1 adenosine receptor subtype was the first GPCR for which positive allosteric modulators were reported [49]. A_1 allosteric enhancers have been pursued as an avenue for drug development in the treatment of a number of pathologies including neurological, cardiac, sleep, immune, and inflammatory disorders as well as cancer. Initial modulators were generated from an aroylthiophene chemical scaffold, which, although showing promise, exhibited factors that made it unattractive for further development [50]. Better compounds have since been developed from 6-aryl-8H-indeno[1,2-d]thiazol-2-ylamines, including IDTA-3x [51]. In addition, VUF5455 (4-methoxy-*N*-[7-methyl-3-(2-pyridinyl)-1-isoquinolinyl] benzamide) and DU124183 (2-cyclopentyl-4-phenylamino-1H-imidazo[4,5-c] quinoline) have been developed as selective, nanomolar potency, enhancers for the adenosine A_3 receptor [52, 53]. All these compounds, however, appear to exhibit a mixed orthosteric/allosteric binding mode, which may have an impact on their use therapeutically.

With respect to current therapeutic utility of allosteric modulators, the chemokine receptor family is worth highlighting because maraviroc (UK-427,857), an allosteric antagonist of CCR5 receptors, was approved for use by the FDA in 2007 for the treatment of HIV-1 infection [54]. The chemokine receptor family is also of interest in that they possess numerous endogenous ligands and probably multiple allosteric sites, given the recent discovery of at least one class of modulator of the CXCR2 receptor with an intracellular site of action [55]. Because of the potential for probe-dependence of allosteric interactions, best practice would be to assess potential allosteric ligands against the full array of endogenous orthosteric ligands for any given chemokine receptor subtype. Indeed, a recent study has shown that a number of metal ion chelator complexes that bind allosterically at the CC-chemokine receptor 1, can simultaneously enhance the binding of one endogenous ligand (CC-chemokine 3), while acting as an antagonist for another (CC-chemokine 5) at the same receptor [56]. Clearly, the phenomenon of probe dependence is an important consideration with respect to the development of allosteric ligands from a drug discovery perspective.

Family B GPCRs Family B GPCRs include a number of peptide hormone GPCRs such as the corticotropin releasing factor (CRF) and glucagon-like peptide 1 (GLP1) receptors. Furthermore, this family of GPCRs is characterized by a relatively large N-terminal domain, which comprises most of the orthosteric binding site. Traditionally, this class of receptors has been recalcitrant to drug discovery targeting the diffuse pharmacophore associated with the peptide orthosteric site. However, a number of small molecules have recently emerged that act allosterically, an example of which is found with molecules targeting the CRF-1 receptor. Evidence for the allosteric nature of these molecules has been derived from both pharmacological studies that highlighted behavior inconsistent with the predictions of simple orthosteric competition, and from site-directed mutagenesis studies that demonstrated these compounds bound to residues within the TM regions of the CRF1 receptor rather than the N-terminal domain [57]. To date, the majority of CRF-1 receptor allosteric antagonists are used as pharmacological tools to study anxiety, depression, and irritable bowel syndrome. However, recently a series of tetraazaacenaphthylenes have shown promising results in preclinical studies and a Phase II clinical trial of major depressive disorder [58–60].

Novo Nordisk recently reported the first series of allosteric agonist/modulators, exemplified by compound 2, which specifically targets the GLP-1R [61]. In addition to its own agonist activity, compound 2 was able to increase the affinity of the endogenous ligand for the receptor while not affecting the maximal cyclic adenosine monophosphate (cAMP) response in functional studies. The molecular mechanisms for the effects of compound 2 are, to date, unknown. While there have been relatively few studies on other Family B GPCRs, these findings suggest that the allosteric approach represents a viable path forward for discovering small molecules directed against these receptors.

Family C GPCRs Family C GPCRs have traditionally proved most amenable to allosteric potentiation, and represent the first family of GPCR for which an allosteric modulator was approved and marketed as a novel therapeutic. Cinacalcet (NPS-1493) is a positive allosteric modulator of the calcium-sensing receptor (CaR) and represents an oral calcimimetic for the treatment of secondary hyperparathyroidism in patients with chronic kidney disease [62, 63]. Cinacalcet interacts with the TM regions of the CaR to potentiate the actions of calcium, which binds in the large N-terminal, "Venus flytrap" (VFT) orthosteric binding domain [64]. Interestingly, the removal of the VFT from the CaR converts another positive allosteric modulator, calindol, into an allosteric agonist that exhibits synergy with additional calcium binding sites located within the remaining seven transmembrane (7TM) helical bundle [64]. This highlights not only the interaction between the allosteric and orthosteric sites on these receptors, but also the inherent complexity in some of these systems that must be considered in drug discovery efforts.

Allosteric modulators of mGluRs are also the subject of continued research because of their emerging therapeutic potential for a range of psychiatric and neurological disorders such as pain, anxiety, cognition, Parkinson's disease, and schizophrenia [65]. Positive allosteric modulators have recently been described for group I (mGluR1 and mGluR5), group II (mGluR2), and group

III (mGluR4) mGluRs. Relative to classical mGluR agonists, these molecules offer improved selectivity, chemical tractability, and may reduce receptor desensitization [65]. Negative allosteric modulators of mGluRs have also been described [66].

3.6.2. Proteins as Allosteric Modulators

Over the last decade, our understanding of GPCR structure and function has been challenged by the discovery that these receptors are able to form homo- and hetero-oligomers with other proteins, which have the capacity to affect several aspects of their function, including trafficking, cell surface expression, and receptor pharmacology [19, 67]. Each protein interacting with a receptor, which modifies its function, can be considered as an allosteric modulator. As mentioned earlier, the classic example of this phenomenon is the interaction of the G protein itself with a receptor. Importantly, there is an ever-growing list of "accessory proteins" that have been shown to interact with GPCRs and modulate their binding and/or functional properties.

Perhaps the key exemplar model of a GPCR-protein interaction that has a profound influence on receptor phenotype is that of the receptor activity modifying proteins (RAMPs), a family of three, single-pass, TM proteins (RAMP1, RAMP2, and RAMP3) [68]. Hetero-oligomerization of the calcitonin-like receptor (CLR) with RAMPs is required for the cell surface expression of this GPCR and also dictates its pharmacology [68]. The sensitivity of the CLR to the peptide agonists, CGRP, or adrenomedullins varies accordingly with its association with either RAMP1 or RAMPs 2 or 3, respectively. It is likely that the differences in ligand binding affinity at the CLR can, at least partially, be attributed to allosteric regulation of the receptor by RAMPs [69]. Thus, RAMPs may induce conformational changes in the CLR that alter its specificity to form either a CGRP or adrenomedullin receptor. Furthermore, in addition to their association with the CLR, RAMPs can modulate the pharmacology of the closely related calcitonin receptor (CTR), engendering high affinity amylin receptors that each have distinct agonist and antagonist pharmacology depending upon which RAMP is present [70, 71]. RAMPs also interact with, and modulate the function of, other Family B GPCRs, including those receptors for vasoactive intestinal peptide/pituitary adenylate cyclase-activating peptide (VPAC). Family B GPCRs couple predominantly through $G_{\alpha s}$ to induce the downstream formation of cAMP, but can also couple to other G proteins at higher agonist concentrations. Coexpression of RAMP2 with the VPAC1R causes a specific augmentation of IP_3 production, presumably downstream of $G_{\alpha q}$, without affecting cAMP responses [72]. As agonist potency was not altered, it has been speculated that this may be due to compartmentalization of the RAMP2/VPAC1R complex to a G_q-enriched microenvironment. The ability of RAMPs to modulate both ligand selectivity, as in the case of the CLR/CTR, and G protein coupling profiles, suggests these proteins have important allosteric roles in regulating GPCR function. Furthermore, considering RAMPs only exhibit 30% sequence identity, the accessory proteins themselves may provide a basis for selective drugs, especially when the interaction surface between the RAMP and the respective GPCR(s) is also taken into account. Development of screening assays that measure RAMP/GPCR interaction could lead to discovery and development of a novel family of drugs that act as allosteric regulators of GPCRs through RAMPs [73].

There are numerous other examples of accessory and scaffolding proteins that interact with GPCRs to influence expression, membrane localization, or perturb the intracellular signaling events arising from GPCR activation. For example, it is well-known that regulators of G protein signaling (RGS) proteins affect GPCR/G protein coupling [74]. Interactions with accessory proteins that result in GPCR organization into membrane microdomains, such as the caveolins, can direct GPCR/G protein coupling to those G proteins colocalized within the same microdomain. As a result, agonists that have a higher affinity for that particular GPCR/G protein-coupled state may display increased efficacy. Many GPCRs are known to directly interact with proteins, such as arrestins and PDZ domain-containing proteins, which result in the activation of intracellular signaling cascades that are distinct from those arising from G protein coupling. In this respect, the GPCR-protein interaction modulates ligand efficacy, and there are many instances of ligands that have efficacy for these alternate signaling cascades and no efficacy for classical G protein-mediated signaling [75–78].

In addition to interactions with accessory and scaffolding proteins, GPCRs are also known to form homo- and hetero-oligomers with each other, the existence of which are likely to influence the affinity and/or efficacy of both orthosteric and allosteric ligands [79]. However, relatively few studies have investigated experimentally whether cooperative binding/allosteric modulation occurs across such a complex. Allosteric modulation between receptor partners in a complex has been shown spectroscopically using the leukotriene 1B receptor, where the binding of a ligand to one member of the homodimer resulted in the conformational change in the other receptor [80]. With respect to the heterodimeric gamma-amino butyric acid $(GABA)_B$ receptor or some of the taste receptor complexes, orthosteric ligand binding has been shown to occur on one receptor partner, while allosteric modulators bind the other member of the heterodimer [81, 82]. Further studies exploring this phenomenon utilized chimeras between the thyrotropin receptor (TSHR) and the lutropin receptor (LTR) [83]. When expressed separately, the receptors displayed a specific type of pharmacology, that is, the TSH receptor bound TSH but not human chorionic gonadotropin (hCG; a ligand of the lutropin receptor), whereas the LTR bound specifically hCG but not TSH. However, when the LTR and the TSH receptors were coexpressed, hCG was able to compete for radiolabeled TSH. The allosteric nature of the heterologous competition was confirmed with kinetic dissociation experiments that suggested binding of hCG to the LTR induced a conformational change in both units of the heterodimer that resulted in the release of ^{125}I-TSH from the TSH receptor [83].

A rapid dissociation of prebound radiolabeled TSH upon addition of unlabelled TSH was also monitored on cells expressing only the TSHR that suggested negative binding cooperativity applies to receptor homodimers as well. The presence of two binding sites with different affinities was confirmed by saturation binding assays using radiolabeled TSH, which identified a low-affinity binding site in addition to the classical high-affinity site [83]. Experiments on the TSH receptor in a native cellular background also confirmed the presence of two binding sites acting allosterically, in which one site expressed a negative binding cooperativity for the other [83].

Using a similar approach, strong negative binding cooperativity has also been identified in homo- and heterodimers of the chemokine receptors (CCR) CCR2 and CCR5 [84]. CCR5-specific ligands, unable to compete for the binding of CCR2-specific ligands on cells expressing CCR2 alone, efficiently prevented binding when the two receptors were coexpressed. This effect was repeated with CCR2 selective ligands modulating the binding of CCR5 specific ligands, but only when both receptors were expressed. Similar coexpression studies with an array of Family A GPCRs, including the μ and δ opioid receptors or the orexin1 and cannabinoid CB_1 receptors, have shown that coexpression results in altered ligand pharmacology [85, 86].

The molecular mechanisms underlying allosteric modulation of binding sites within receptor dimers are not known, but are likely to involve structural rearrangement of both receptor units. This has been shown with the dopamine D_2 receptor in which the contact interface between dimers alters with the activation state of the receptor [87]. However, full activation of a receptor is not mandatory for allosteric modulation, since negative binding cooperativity is also observed with antagonists [66].

The existence of binding cooperativity between GPCRs in a complex raises the question of the stoichiometry of ligand binding. For receptors displaying a strong negative binding cooperativity, only a single ligand is expected to bind to a dimmer [84]. However, in other receptors such as the mGluR, the complex may bind two ligands simultaneously [88]. More recently, experiments have also indicated that the binding cooperativity in the vasopressin V1a receptor dimer could be either positive or negative, depending on the ligand studied [89]. Although intriguing, the increasing wealth of data describing cooperativity in the binding of ligands to GPCRs highlights the fact that the properties of dimers will differ according to the receptor studied, the ligand investigated, and possibly other factors such as interactions with other proteins in the signaling complex such as RAMPs.

3.7. CONCLUDING REMARKS

Allosteric interactions at GPCRs, whether mediated by small molecules or other proteins, can result in profound perturbations in ligand binding and GPCR function. Targeting small molecule allosteric sites on GPCRs holds much promise in the field of drug discovery. However, a continuing challenge to the field is the ability to detect, validate, and quantify allosteric ligand behaviors. In this respect, appropriately designed screening strategies are required to overcome potential confounding issues such as probe-dependence and differential modulation of affinity and efficacy. Regardless, compared to orthosteric sites, allosteric sites can accommodate a greater diversity of chemotypes and as such offer the potential for the development of therapeutics with greater subtype selectivity. For allosteric modulators, there is also the prospect of maintaining spatial and temporal aspects of GPCR function. Exploiting a bitopic binding mode is an alternative avenue for the development of selective agonists with high affinity, but perhaps also pathway-specific efficacy. Conversely, this approach could be used to develop highly selective antagonists and inverse agonists, thus contributing significantly to the pharmacologist's armamentarium and hopefully, to the clinician's therapeutic options.

REFERENCES

1. Drews, J. (2000) Drug discovery: A historical perspective. *Science*. 287, 1960–1964.
2. Hopkins, A.L., Groom, C.R. (2002) The druggable genome. *Nat Rev Drug Discov*. 1, 727–730.
3. Overington, J.P., Al-Lazikani, B., Hopkins, A.L. (2006) How many drug targets are there? *Nat Rev Drug Discov*. 5, 993–996.
4. Bohr, C., Hasselbalch, K.A., Krogh, A. (1904). Uber einen in biologischen Beziehung wichtigen Einfluss den die Kohlen saurespannung des Blutes auf dessen Sauerstoffbindung ubt. *Skand Arch Physiol*. 16, 401–412.
5. Galzi, J.L., Revah, F., Bouet, F., Menez, A., Goeldner, M., Hirth, C., Changeux, J.P. (1991) Allosteric transitions of the acetylcholine receptor probed at the amino acid level with a photolabile cholinergic ligand. *Proc Natl Acad Sci U S A*. 88, 5051–5055.
6. Sigel, E., Buhr, A. (1997) The benzodiazepine binding site of GABAA receptors. *Trends Pharmacol Sci*. 18, 425–429.
7. Edsall, J.T. (1980) Hemoglobin and the origins of the concept of allosterism. *Fed Proc*. 39, 226–235.
8. Holt, J.M., Ackers, G.K. (1995) The pathway of allosteric control as revealed by hemoglobin intermediate states. *FASEB J*. 9, 210–218.
9. Wyman, J., Allen, D.W. (1951) The problem of heme interactions in hemoglobin and the basis for the Bohr effect. *J Polymer Sci*. 7, 499–518.
10. Weber, G. (1972) Ligand binding and internal equilibria in proteins. *Biochemistry*. 11, 864–878.
11. Monod, J., Changeux, J.P., Jacob, F. (1963) Allosteric proteins and cellular control systems. *J Mol Biol*. 6, 306–329.
12. Monod, J., Wyman, J., Changeux, J.P. (1965) On the nature of allosteric transitions: A plausible model. *J Mol Biol*. 12, 88–118.

13. Lefkowitz, R.J., Mullikin, D., Caron, M.G. (1976) Regulation of β-adrenergic receptors by guanyl-5′-yl imidodiphosphate and other purine nucleotides. *J Biol Chem.* 251, 4686–4692.
14. De Lean, A., Stadel, J.M., Lefkowitz, R.J. (1980) A ternary complex model explains the agonist-specific binding properties of the adenylate cyclase-coupled beta-adrenergic receptor. *J Biol Chem.* 255, 7108–7117.
15. Sibley, D.R., De Lean, A., Creese, I. (1982) Anterior pituitary dopamine receptors: Demonstration of interconvertible high and low affinity states of the D-2 dopamine receptor. *J Biol Chem.* 257, 6351–6361.
16. Sexton, P.M., Morfis, M., Tilakaratne, N., Hay, D.L., Udawela, M., Christopoulos, G., Christopoulos, A. (2006) Complexing receptor pharmacology: Modulation of family B G protein-coupled receptor function by RAMPs. *Ann N Y Acad Sci.* 1070, 90–104.
17. Whorton, M.R., Bokoch, M.P., Rasmussen, S.G., Huang, B., Zare, R.N., Kobilka, B., Sunahara, R.K. (2007) A monomeric G protein-coupled receptor isolated in a high-density lipoprotein particle efficiently activates its G protein. *Proc Natl Acad Sci U S A.* 104, 7682–7687.
18. Hall, D.A. (2000) Modeling the functional effects of allosteric modulators at pharmacological receptors: An extension of the two-state model of receptor activation. *Mol Pharmacol.* 58, 1412–1423.
19. May, L.T., Leach, K., Sexton, P.M., Christopoulos, A. (2007) Allosteric modulation of g protein-coupled receptors. *Annu Rev Pharmacol Toxicol.* 47, 1–51.
20. Urban, J.D., Clarke, W.P., von Zastrow, M., Nichols, D.E., Kobilka, B.K., Weinstein, H., Javitch, J.A., Roth, B.L., Christopoulos, A., Sexton, P., Miller, K., Spedding, M., Mailman, R.B. (2007) Functional selectivity and classical concepts of quantitative pharmacology. *J Pharmacol Exp Ther.* 320, 1–13.
21. Leach, K., Sexton, P.M., Christopoulos, A. (2007) Allosteric GPCR modulators: Taking advantage of permissive receptor pharmacology. *Trends Pharmacol Sci.* 28, 382–389.
22. Schwartz, T.W., Holst, B. (2007) Allosteric enhancers, allosteric agonists and ago-allosteric modulators: Where do they bind and how do they act? *Trends Pharmacol Sci.* 28, 366–373.
23. Jakubik, J., Bacakova, L., el-Fakahany, E.E., Tucek, S. (1995) Subtype selectivity of the positive allosteric action of alcuronium at cloned M1-M5 muscarinic acetylcholine receptors. *J Pharmacol Exp Ther.* 274, 1077–1083.
24. Zahn, K., Eckstein, N., Trankle, C., Sadee, W., Mohr, K. (2002) Allosteric modulation of muscarinic receptor signaling: Alcuronium-induced conversion of pilocarpine from an agonist into an antagonist. *J Pharmacol Exp Ther.* 301, 720–728.
25. Price, M.R., Baillie, G.L., Thomas, A., Stevenson, L.A., Easson, M., Goodwin, R., McLean, A., McIntosh, L., Goodwin, G., Walker, G., Westwood, P., Marrs, J., Thomson, F., Cowley, P., Christopoulos, A., Pertwee, R.G., Ross, R.A. (2005) Allosteric modulation of the cannabinoid CB1 receptor. *Mol Pharmacol.* 68, 1484–1495.
26. Jager, D., Schmalenbach, C., Prilla, S., Schrobang, J., Kebig, A., Sennwitz, M., Heller, E., Tränkle, C., Holzgrabe, U., Höltje, HD, Mohr, K. (2007) Allosteric small molecules unveil a role of an extracellular E2/transmembrane helix 7 junction for G protein-coupled receptor activation. *J Biol Chem.* 282, 34968–34976.

27. Valant, C., Gregory, K.J., Hall, N.E., Scammells, P.J., Lew, M.J., Sexton, P.M., Christopoulos, A. (2008) A novel mechanism of G protein-coupled receptor functional selectivity. Muscarinic partial agonist McN-A-343 as a bitopic orthosteric/allosteric ligand. *J Biol Chem*. 283, 29312–29321.

28. Antony, J., Kellershohn, K., Mohr-Andrä, M., Kebig, A., Prilla, S., Muth, M., Heller, E., Disingrini, T., Dallanoce, C., Bertoni, S., Schrobang, J., Tränkle, C., Kostenis, E., Christopoulos, A., Höltje, H.D., Barocelli, E., De Amici, M., Holzgrabe, U., Mohr, K. (2009) Dualsteric GPCR targeting: A novel route to binding and signaling pathway selectivity. *FASEB J*. 23, 442–450.

29. Christopoulos, A. (2002) Allosteric binding sites on cell-surface receptors: Novel targets for drug discovery. *Nat Rev Drug Discov*. 1, 198–210.

30. Christopoulos, A., Kenakin, T. (2002) G protein-coupled receptor allosterism and complexing. *Pharmacol Rev*. 54, 323–374.

31. Trankle, C., Mies-Klomfass, E., Cid, M.H., Holzgrabe, U., Mohr, K. (1998) Identification of a [3H]ligand for the common allosteric site of muscarinic acetylcholine M2 receptors. *Mol Pharmacol*. 54, 139–145.

32. Lavreysen, H., Janssen, C., Bischoff, F., Langlois, X., Leysen, J.E., Lesage, A.S. (2003) [3H]R214127: A novel high-affinity radioligand for the mGlu1 receptor reveals a common binding site shared by multiple allosteric antagonists. *Mol Pharmacol*. 63, 1082–1093.

33. Lazareno, S., Birdsall, N. (1995) Detection, quantitation, and verification of allosteric interactions of agents with labeled and unlabeled ligands at G protein-coupled receptors: Interactions of strychnine and acetylcholine at muscarinic receptors. *Mol Pharmacol*. 48, 362–378.

34. Kostenis, E., Mohr, K. (1996) Two-point kinetic experiments to quantify allosteric effects on radioligand dissociation. *Trends Pharmacol Sci*. 17, 280–283.

35. Proska, J., Tucek, S. (1994) Mechanisms of steric and cooperative actions of alcuronium on cardiac muscarinic acetylcholine receptors. *Mol Pharmacol*. 45, 709–717.

36. Avlani, V., May, L.T., Sexton, P.M., Christopoulos, A. (2004) Application of a kinetic model to the apparently complex behavior of negative and positive allosteric modulators of muscarinic acetylcholine receptors. *J Pharmacol Exp Ther*. 308, 1062–1072.

37. Black, J.W., Leff, P. (1983) Operational models of pharmacological agonism. *Proc R Soc Lond B Biol Sci*. 220, 141–162.

38. Clark, A.L., Mitchelson, F. (1976) The inhibitory effect of gallamine on muscarinic receptors. *Br J Pharmacol*. 58, 323–331.

39. Chan, W.Y., McKinzie, D.L., Bose, S., Mitchell, S.N., Witkin, J.M., Thompson, R.C. Christopoulos, A., Lazareno, S., Birdsall, N.J.M., Bymaster, F.P., Felder, C.C. (2008) Allosteric modulation of the muscarinic M4 receptor as an approach to treating schizophrenia. *Proc Natl Acad Sci U S A*. 105, 10978–10983.

40. Gregory, K.J., Sexton, P.M., Christopoulos, A. (2007) Allosteric modulation of muscarinic acetycholine receptors. *Curr Neuropharm*. 5, 157–167.

41. Spalding, T.A., Trotter, C., Skjaerbaek, N., Messier, T.L., Currier, E.A., Burstein, E.S., Li, D., Hacksell, U., Brann, M.R. (2002) Discovery of an ectopic activation site on the M(1) muscarinic receptor. *Mol Pharmacol*. 61, 1297–1302.

42. Langmead, C.J., Fry, V.A., Forbes, I.T., Branch, C.L., Christopoulos, A., Wood, M.D., Herdon, H.J. (2006) Probing the molecular mechanism of interaction between 4-n-butyl-1-[4-(2-methylphenyl)-4-oxo-1-butyl]-piperidine (AC-42) and the muscarinic M(1) receptor: Direct pharmacological evidence that AC-42 is an allosteric agonist. *Mol Pharmacol.* 69, 236–246.

43. Spalding, T.A., Ma, J.N., Ott, T.R., Friberg, M., Bajpai, A., Bradley, S.R., Davis, R.E., Brann, M.R., Burstein, E.S. (2006) Structural requirements of transmembrane domain 3 for activation by the M1 muscarinic receptor agonists AC-42, AC-260584, clozapine, and N-desmethylclozapine: Evidence for three distinct modes of receptor activation. *Mol Pharmacol.* 70, 1974–1983.

44. Jones, C.K., Brady, A.E., Davis, A.A., Xiang, Z., Bubser, M., Tantawy, M.N., Kane, A.S., Bridgers, T.M., Kennedy, J.P., Bradley, S.R., Peterson, T.E., Ansari, M.S., Baldwin, R.M., Kessler, R.M., Deutch, A.Y., Lah, J.J., Levey, A.I., Lindsley, C.W., Conn, P.J. (2008) Novel selective allosteric activator of the M1 muscarinic acetylcholine receptor regulates amyloid processing and produces antipsychotic-like activity in rats. *J Neurosci.* 28, 10422–10433.

45. Lebon, G., Langmead, C.J., Tehan, B.G., Hulme, E.C. (2009) Mutagenic mapping suggests a novel binding mode for selective agonists of M1 muscarinic acetylcholine receptors. *Mol Pharmacol.* 75, 331–341.

46. Sur, C., Mallorga, P.J., Wittmann, M., Jacobson, M.A., Pascarella, D., Williams, J.B., Brandish, P.E., Pettibone, D.J., Scolnick, E.M., Conn, P.J. (2003) N-desmethylclozapine, an allosteric agonist at muscarinic 1 receptor, potentiates N-methyl-D-aspartate receptor activity. *Proc Natl Acad Sci U S A.* 100, 13674–13679.

47. Shirey, J.K., Xiang, Z., Orton, D., Brady, A.E., Johnson, K.A., Williams, R., Ayala, J.E., Rodriguez, A.L., Wess, J., Weaver, D., Niswender, C.M., Conn, P.J. (2008) An allosteric potentiator of M4 mAChR modulates hippocampal synaptic transmission. *Nat Chem Biol.* 4, 42–50.

48. Nawaratne, V., Leach, K., Suratman, N., Loiacono, R.E., Felder, C.C., Armbruster, B.N., Roth, B.L., Sexton, P.M., Christopoulos, A. (2008) New insights into the function of M4 muscarinic acetylcholine receptors gained using a novel allosteric modulator and a DREADD (designer receptor exclusively activated by a designer drug). *Mol Pharmacol.* 74, 1119–1131.

49. Bruns, R.F., Fergus, J.H. (1990) Allosteric enhancement of adenosine A1 receptor binding and function by 2-amino-3-benzoylthiophenes. *Mol Pharmacol.* 38, 939–949.

50. Joshi, E., Chordia, M.D., Macdonald, T.L., Linden, J., Olsson, R. (2004) Regioselective oxidation of 2-amino-3-aroyl-4,5-dialkylthiophenes by DMSO. *Bioorg Med Chem Lett.* 14, 929–933.

51. Chordia, M.D., Zigler, M., Murphree, L.J., Figler, H., Macdonald, T.L., Olsson, R.A., Linden, J. (2005) 6-aryl-8H-indeno[1,2-d]thiazol-2-ylamines: A1 adenosine receptor agonist allosteric enhancers having improved potency. *J Med Chem.* 48, 5131–5139.

52. Gao, Z.G., Van Muijlwijk-Koezen, J.E., Chen, A., Muller, C.E., Ijzerman, A.P., Jacobson, K.A. (2001) Allosteric modulation of A(3) adenosine receptors by a series of 3-(2-pyridinyl)isoquinoline derivatives. *Mol Pharmacol.* 60, 1057–1063.

53. Gao, Z.-G., Kim, S.G., Soltysiak, K.A., Melman, N., Ijzerman, A.P., Jacobson, K.A. (2002) Selective allosteric enhancement of agonist binding and function at human A3 adenosine receptors by a series of imidazoquinoline derivatives. *Mol Pharmacol.* 62, 81–89.

54. Dorr, P., Westby, M., Dobbs, S., Griffin, P., Irvine, B., Macartney, M., Mori, J., Rickett, G., Smith-Burchnell, C., Napier, C., Webster, R., Armour, D., Price, D., Stammen, B., Wood, A., Perros, M. (2005) Maraviroc (UK-427,857), a potent, orally bioavailable, and selective small-molecule inhibitor of chemokine receptor CCR5 with broad-spectrum anti-human immunodeficiency virus type 1 activity. *Antimicrob Agents Chemother.* 49, 4721–4732.

55. Andrews, G., Jones, C., Wreggett, K.A. (2008) An intracellular allosteric site for a specific class of antagonists of the CC chemokine G protein-coupled receptors CCR4 and CCR5. *Mol Pharmacol.* 73, 855–867.

56. Jensen, P.C., Thiele, S., Ulven, T., Schwartz, T.W., Rosenkilde, M.M. (2008) Positive versus negative modulation of different endogenous chemokines for CC-chemokine receptor 1 by small molecule agonists through allosteric versus orthosteric binding. *J Biol Chem.* 283, 23121–23128.

57. Hoare, S.R., Brown, B.T., Santos, M.A., Malany, S., Betz, S.F., Grigoriadis, D.E. (2006) Single amino acid residue determinants of non-peptide antagonist binding to the corticotropin-releasing factor1 (CRF1) receptor. *Biochem Pharmacol.* 72, 244–255.

58. Zobel, A.W., Nickel, T., Kunzel, H.E., Ackl, N., Sonntag, A., Ising, M., Holsboer, F. (2000) Effects of the high-affinity corticotropin-releasing hormone receptor 1 antagonist R121919 in major depression: The first 20 patients treated. *J Psychiatr Res.* 34, 171–181.

59. Kehne, J., De Lombaert, S. (2002) Non-peptidic CRF1 receptor antagonists for the treatment of anxiety, depression and stress disorders. *Curr Drug Targets CNS Neurol Disord.* 1, 467–493.

60. Hoare, S.R., Fleck, B.A., Gross, R.S., Crowe, P.D., Williams, J.P., Grigoriadis, D.E. (2008) Allosteric ligands for the corticotropin releasing factor type 1 receptor modulate conformational states involved in receptor activation. *Mol Pharmacol.* 73, 1371–1380.

61. Knudsen, L.B., Kiel, D., Teng, M., Behrens, C., Bhumralkar, D., Kodra, J.T., Holst, J.J., Jeppesen, C.B., Johnson, M.D., de Jong, J.C., Jorgensen, A.S., Kercher, T., Kostrowicki, J., Madsen, P., Olesen, P.H., Petersen, J.S., Poulsen, F., Sidelmann, U.G., Sturis, J., Truesdale, L., May, J., Lau, J. (2007) Small-molecule agonists for the glucagon-like peptide 1 receptor. *Proc Natl Acad Sci U S A.* 104, 937–942.

62. Lindberg, J.S., Culleton, B., Wong, G., Borah, M.F., Clark, R.V., Shapiro, W.B., Roger, S.D., Husserl, F.E., Klassen, P.S., Guo, M.D., Albizem, M.B., Coburn, J.W. (2005) Cinacalcet HCl, an oral calcimimetic agent for the treatment of secondary hyperparathyroidism in hemodialysis and peritoneal dialysis: A randomized, double-blind, multicenter study. *J Am Soc Nephrol.* 16, 800–807.

63. Poon, G. (2005) Cinacalcet hydrochloride (Sensipar). *Proc (Bayl Univ Med Cent).* 18, 181–184.

64. Petrel, C., Kessler, A., Dauban, P., Dodd, R.H., Rognan, D., Ruat, M. (2004) Positive and negative allosteric modulators of the Ca^{2+}-sensing receptor interact within overlapping but not identical binding sites in the transmembrane domain. *J Biol Chem.* 279, 18990–18997.

65. Shipe, W.D., Wolkenberg, S.E., Williams Jr., D.L., Lindsley, C.W. (2005) Recent advances in positive allosteric modulators of metabotropic glutamate receptors. *Curr Opin Drug Discov Devel.* 8, 449–457.

66. Rodriguez, A.L., Nong, Y., Sekaran, N.K., Alagille, D., Tamagnan, G.D., Conn, P.J. (2005) A close structural analog of 2-methyl-6-(phenylethynyl)-pyridine acts as a neutral allosteric site ligand on metabotropic glutamate receptor subtype 5 and blocks the effects of multiple allosteric modulators. *Mol Pharmacol.* 68, 1793–1802.

67. Milligan, G. (2006) G-protein-coupled receptor heterodimers: Pharmacology, function and relevance to drug discovery. *Drug Discov Today.* 11, 541–549.

68. McLatchie, L.M., Fraser, N.J., Main, M.J., Wise, A., Brown, J., Thompson, N., Solari, R., Lee, M.G., Foord, S.M. (1998) RAMPs regulate the transport and ligand specificity of the calcitonin-receptor-like receptor. *Nature.* 393, 333–339.

69. Hilairet, S., Foord, S.M., Marshall, F.H., Bouvier, M. (2001) Protein-protein interaction and not glycosylation determines the binding selectivity of heterodimers between the calcitonin receptor-like receptor and the receptor activity-modifying proteins. *J Biol Chem.* 276, 29575–29581.

70. Christopoulos, G., Perry, K.J., Morfis, M., Tilakaratne, N., Gao, Y., Fraser, N.J., Main, M.J., Foord, S.M., Sexton, P.M. (1999) Multiple amylin receptors arise from receptor activity-modifying protein interaction with the calcitonin receptor gene product. *Mol Pharmacol.* 56, 235–242.

71. Hay, D.L., Christopoulos, G., Christopoulos, A., Poyner, D.R., Sexton, P.M. (2005) Pharmacological discrimination of calcitonin receptor: Receptor activity-modifying protein complexes. *Mol Pharmacol.* 67, 1655–1665.

72. Christopoulos, A., Christopoulos, G., Morfis, M., Udawela, M., Laburthe, M., Couvineau, A., Kuwasako, K., Tilakaratne, N., Sexton, P.M. (2003) Novel receptor partners and function of receptor activity-modifying proteins. *J Biol Chem.* 278, 3293–3297.

73. Sexton, P.M., Poyner, D.R., Simms, J., Christopoulos, A., Hay, D.L. (2009) Modulating receptor function through RAMPs: Can they represent drug targets in themselves? *Drug Discov Today.* 14, 413–419.

74. Abramow-Newerly, M., Ming, H., Chidiac, P. (2006) Modulation of subfamily B/R4 RGS protein function by 14-3-3 proteins. *Cell Signal.* 18, 2209–2222.

75. Bisello, A., Chorev, M., Rosenblatt, M., Monticelli, L., Mierke, D.F., Ferrari, S.L. (2002) Selective ligand-induced stabilization of active and desensitized parathyroid hormone type 1 receptor conformations. *J Biol Chem.* 277, 38524–38530.

76. Blanpain, C., Vanderwinden, J.M., Cihak, J., Wittamer, V., Le Poul, E., Issafras, H., Stangassinger, M., Vassart, G., Marullo, S., Schlndorff, D., Parmentier, M., Mack, M. (2002) Multiple active states and oligomerization of CCR5 revealed by functional properties of monoclonal antibodies. *Mol Biol Cell.* 13, 723–737.

77. Wei, H., Ahn, S., Shenoy, S.K., Karnik, S.S., Hunyady, L., Luttrell, L.M., Lefkowitz, R.J. (2003) Independent beta-arrestin 2 and G protein-mediated pathways for angiotensin II activation of extracellular signal-regulated kinases 1 and 2. *Proc Natl Acad Sci U S A.* 100, 10782–10787.

78. Gesty-Palmer, D., Chen, M., Reiter, E., Ahn, S., Nelson, C.D., Wang, S., Eckhardt, A.E., Cowan, C.L., Spurney, R.F., Luttrell, L.M., Lefkowitz, R.J. (2006) Distinct beta-arrestin- and G protein-dependent pathways for parathyroid hormone receptor-stimulated ERK1/2 activation. *J Biol Chem.* 281, 10856–10864.

79. Milligan, G., Smith, N.J. (2007) Allosteric modulation of heterodimeric G-protein-coupled receptors. *Trends Pharmacol Sci.* 28, 615–620.

80. Mesnier, D., Baneres, J.L. (2004) Cooperative conformational changes in a G-protein-coupled receptor dimer, the leukotriene B(4) receptor BLT1. *J Biol Chem.* 279, 49664–49670.
81. Binet, V., Brajon, C., Le Corre, L., Acher, F., Pin, J.P., Prezeau, L. (2004) The heptahelical domain of GABA(B2) is activated directly by CGP7930, a positive allosteric modulator of the GABA(B) receptor. *J Biol Chem.* 279, 29085–29091.
82. Xu, H., Staszewski, L., Tang, H., Adler, E., Zoller, M., Li, X. (2004) Different functional roles of T1R subunits in the heteromeric taste receptors. *Proc Natl Acad Sci U S A.* 101, 14258–14263.
83. Urizar, E., Montanelli, L., Loy, T., Bonomi, M., Swillens, S., Gales, C., Bouvier, M., Smits, G., Vassart, G., Costagliola, S. (2005) Glycoprotein hormone receptors: Link between receptor homodimerization and negative cooperativity. *EMBO J.* 24, 1954–1964.
84. Springael, J.Y., Le Minh, P.N., Urizar, E., Costagliola, S., Vassart, G., Parmentier, M. (2006) Allosteric modulation of binding properties between units of chemokine receptor homo- and hetero-oligomers. *Mol Pharmacol.* 69, 1652–1661.
85. Gomes, I., Jordan, B.A., Gupta, A., Trapaidze, N., Nagy, V., Devi, L.A. (2000) Heterodimerization of mu and delta opioid receptors: A role in opiate synergy. *J Neurosci.* 20, RC110.
86. Hilairet, S., Bouaboula, M., Carriere, D., Le Fur, G., Casellas, P. (2003) Hypersensitization of the Orexin 1 receptor by the CB1 receptor: Evidence for cross-talk blocked by the specific CB1 antagonist, SR141716. *J Biol Chem.* 278, 23731–23737.
87. Guo, W., Shi, L., Filizola, M., Weinstein, H., Javitch, J.A. (2005) Crosstalk in G protein-coupled receptors: Changes at the transmembrane homodimer interface determine activation. *Proc Natl Acad Sci U S A.* 102, 17495–17500.
88. Kniazeff, J., Bessis, A.S., Maurel, D., Ansanay, H., Prezeau, L., Pin, J.P. (2004) Closed state of both binding domains of homodimeric mGlu receptors is required for full activity. *Nat Struct Mol Biol.* 11, 706–713.
89. Albizu, L., Balestre, M.N., Breton, C., Pin, J.P., Manning, M., Mouillac, B., Barberis, C., Durroux, T. (2006) Probing the existence of G protein-coupled receptor dimers by positive and negative ligand-dependent cooperative binding. *Mol Pharmacol.* 70, 1783–1791.

CHAPTER 4

Receptor-Mediated G Protein Activation: How, How Many, and Where?

INGRID GSANDTNER, CHRISTIAN W. GRUBER, and MICHAEL FREISSMUTH

Institute of Pharmacology, Center of Biomolecular Medicine and Pharmacology, Medical University of Vienna, Vienna, Austria

For the past 20 years, the G protein cycle has been understood in considerable detail: the G protein cycles between an inactive GDP-bound conformation and an active GTP-bound conformation. By virtue of its guanine nucleotide exchange factor (GEF) activity, the active (agonist-liganded) receptor operates as the switch, which turns on the signal transduction process [1, 2]. The intrinsic GTPase of the G protein α subunit functions as the timed turn-off switch, which is—in most instances—assisted by the GTPase activating protein (GAP) activity of a family of proteins known as regulators of G protein signaling (RGS) proteins [3]. Snapshots exist for these reactions, which allow for extracting several mechanistic details of the underlying reactions at atomic resolution [4]. However, one reaction in this cycle has remained elusive, namely how the receptor-mediated GDP release is brought about. There are three layers, at which this problem has been addressed. These can be referred to as (1) the mechanical problem = receptor-induced movements within the G protein that allow for the formation of a GDP exit path, (2) the dimer problem = the nature of the receptor species that contacts the G protein, and (3) the signalosome problem = anisotropic distribution of receptor–G protein complexes within the membrane and the resulting higher order of organization into signalosomes/signalplexes.

GPCR Molecular Pharmacology and Drug Targeting: Shifting Paradigms and New Directions, Edited by Annette Gilchrist

4.1. THE MECHANICAL PROBLEM—THREE DIFFERENT SOLUTIONS

The seven transmembrane helices (TM1–7) represent the conserved hydrophobic core of G protein-coupled receptors (GPCRs). This is a versatile scaffold, which allows for the many modes of ligand-dependent activation. There are many variations on the theme [5], because the ligand may be prebound (as in rhodopsin), may bind primarily to an extracellular domain (e.g., the fly trap domain of metabotropic glutamate receptors), to the extracellular face of the hydrophobic core (as in peptide receptors), or within the hydrophobic core (as in receptors for biogenic amines, e.g., adrenergic receptors). Regardless of the details, the fundamental problem is to relay the agonist-induced conformational change via the helical arrangement to the intracellular side, which—from the viewpoint of the G protein—is the business side of the receptor.

GDP is deeply buried within the G protein α subunit. Heterotrimeric G protein α subunits are larger in size than the small monomeric RAS-like G proteins because they have an "extra" helical domain. In addition, the G protein βγ dimer covers a large surface of the G protein α subunit and physically blocks the exit of GDP. This also explains why βγ dimers increase the affinity for GDP of those G protein α subunits that have measurable spontaneous GPD release rates [6, 7]. Finally, as can be seen from Fig. 4.1a, GPCRs cannot contact the GDP binding pocket directly: the nucleotide binding pocket embedded in the GTPase domain is about 30 Å away from the interface between receptor and G protein. This distance precludes a direct contact site between intracellular loops of the receptor and GDP binding pocket. This is also true for squid rhodopsin [8], which is shown in Fig. 4.1a. This most recent addition to the available structures of GPCRs (mammalian rhodopsin in several conformations and the human β_2-adrenergic receptor) is remarkable for several reasons, not the least of which is the extended protrusion of transmembrane helices 2 and 3 (TM2 and TM3) into the cytoplasm. But it is evident from Fig. 4.1a that—in spite of this long intracellular loop—there is not any conceivable way how to arrange the G protein heterotrimer and the receptor to allow for a direct contact between receptor and GDP binding pocket. Several GPCRs have long C-termini; in an extended conformation, these may readily reach into the vicinity of the GDP binding pocket. However, it can be convincingly argued that the long C-termini are irrelevant to the basic mechanism of G protein activation because GPCRs with short C-termini also efficiently activate G proteins. In addition, in those instances, where this issue has been examined, truncation of a long terminus does not affect G protein activation (see, e.g., Reference 9) provided that it does not affect the structure of helix 8 (the proximal portion of the C-terminus adjacent to the seventh transmembrane helix (TM7) [10].

Therefore, it is generally accepted that receptors act "at a distance": they must somehow relay the conformational signals arising within the hydrophobic core via their G protein interaction surface to the guanine nucleotide binding pocket to trigger GDP release. Three solutions have been proposed:

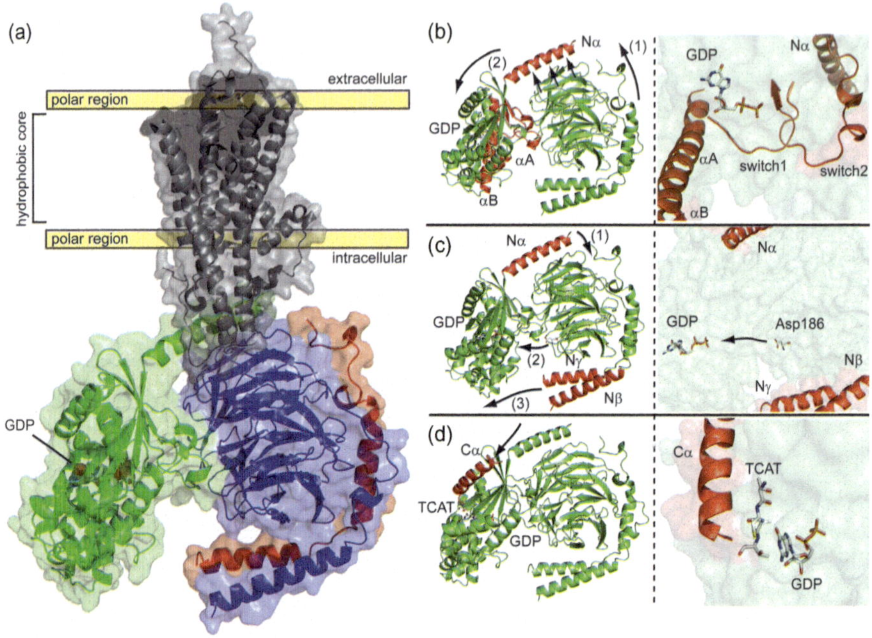

Figure 4.1 Mechanism of G protein activation. (a) Interface of a G protein-coupled receptor, such as squid rhodopsin (shown as grey Ribbon model, Protein Data Bank (PDB) ID *2ZIY*; Reference 8) and a G protein heterotrimer, such as transducin (PDB ID *1GOT*; Reference 13). The transducin heterotrimer is composed of the nucleotide-binding G_t α-subunit (shown as green, transparent space fill/ribbon model) and the G_t β (blue) and γ (red) subunits. Despite the long C-terminal cytoplasmic domain of squid rhodopsin, the nucleotide GDP (shown as sticks representation) is still too far away to be in direct contact with the receptor. This raises the question of how receptors can activate G proteins to release GDP. Three models have been proposed, which provide different solutions to the mechanical problem (b–d): (b) The "lever-arm" model: (1) the receptor uses N-terminal Nα helix of Gα (red) as a lever to pull Gβγ away from Gα (indicated by arrows). (2) This motion invokes movements in parts of the helical region of Gα (αA and αB, red) and the switch regions 1 and 2 (red) that causes the release of GDP. The right panels show the zoomed region near GDP. (c) "Gear-shift" model: (1) the receptor pushes Nα toward the Gβγ subunits. (2) This motions brings Gβ Asp186 (shown as stick representation; motion represented by the arrow in the blowup) in close proximity to GDP, which (3) pushes the coiled-coil helices Nβ and Nγ (red) toward the helical domain of Gα to invoke GDP release. (d) "C-terminal latch" model: the interaction of the receptor with the C-terminus of Gα causes a rigid movement of the adjacent (C-terminal) helix α5 (Cα; red, indicated by an arrow), which is translated to the TCAT motif (stick representation) and triggers the release of GDP.

two of these models (termed "lever-arm" and "gear-shift" models) assume that GDP release is contingent on an active involvement of Gβγ. Both models postulate a conceptual analogy between Gβγ and GEFs of small G proteins and ribosomal elongation effectors (EF-Tu): GEFs engage a large surface of the small G protein (or of EF-Tu) and are in close vicinity to residues stabilizing GDP binding; the same is true for Gβγ. In the third model referred to as "C-terminal latch" model, Gβγ provides a receptor docking site, but is not actively involved in the release process per se.

4.1.1. The Lever-Arm Model

This model was the first to be proposed based on the structures of various G protein heterotrimers as well as insights from (disease-causing) mutations [11]. In the lever-arm model (Fig. 4.1b), the GDP exit pathway is positioned in the area formed by the so-called switch regions 1 and 2 (switch regions are those portions of G_α that differ in conformation between the inactive GDP-liganded and the GTP-liganded active form). The model posits that Gβγ is the go-between that opens the GDP exit pathway in response to receptor activation: Gβγ covers a hydrophobic cavity between switches 1 and 2 and contacts the N-terminus of Gα [12, 13]. The lever in the model proposed is this N-terminal helix (Nα in Fig. 4.1b). This lever is operated by the activated receptor which displaces (and rotates) Gβγ relative to Gα. The resulting traction causes movement in switch 1 and in particular switch 2 and thus allows for GDP exit. The extensive contacts of Gβγ and Gα stabilize the GDP-free state of Gα, which is notorious for its instability. In fact, the ternary complex (HRG) of agonist (H)-liganded receptor (R) and guanine nucleotide-free G protein heterotrimer (G) is amazingly stable. This model predicts that a manipulation that favors tilting of Gα in the heterotrimer ought to facilitate G protein activation. This prediction has been verified by using a mutant of $G\alpha_s$ (the α subunit that stimulates adenylyl cyclase isoforms): if the N-terminal helix of $G\alpha_s$ is shortened by one helical turn (i.e., truncated by four amino acids) on its end adjacent to the RAS-like domain, activation of $G\alpha_s$ by Gβγ is facilitated [14]. However, it still requires an inductive leap to extrapolate this effect of Gβγ to the action of the receptor. In fact, in the presence of high Mg^{2+} concentrations, Gβγ facilitates GDP release from Gα by a mechanism that is poorly understood [6]. It is not clear if this action has any relevance to the GEF action of GPCRs.

4.1.2. The "Gear-Shift" Model

This model [15] also assumes an active role of Gβγ in mediating GDP release, but proposes a different GDP exit pathway. As mentioned above, the helical domain caps the guanine nucleotide binding site. As in the lever-arm model, the gear shift posits that the receptor engages the N-terminal α-helix of Gα (Nα in Fig. 4.2c). However, the receptor presses Gβγ downward; the βγ dimer moves as a rigid body and it is in particular the N-terminal end of Gγ (Nγ in

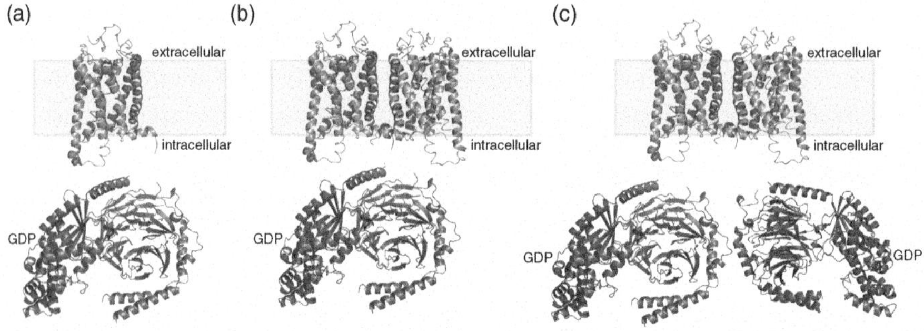

Figure 4.2 Receptor dimerization. (a) One receptor interacts with one G protein resulting in 1:1 complex of GPCR and G protein: a single monomeric G protein-coupled receptor, represented by a GPCR such as the β_2 adrenergic receptor or bovine rhodopsin monomer (shown as grey, membrane spanning Ribbon model, Protein Data Bank [PDB] ID, *2RHI361*; ref. 54), interacts with a single G protein, that is, heterotrimeric transducin (shown as grey Ribbon model, PDB ID *1GOT*; ref. 13), to activate the G protein and hence release GDP. This schematic rendering orients the receptor in a manner consistent with the "lever-arm" and "gear-shift" models of G protein activation. (b) Two receptors dimerize with one G protein: a GPCR dimer, indicated by the rhodopsin dimer, interacts with a single G protein (here, the transducin heterotrimer). The receptor dimer is positioned over the G protein in a manner consistent with the "sequential fit" model of G protein activation, where one receptor engages the receptor contact sites on Gβγ and the second receptor triggers the C-terminal latch (note, though, that in the sequential fit model, one receptor moiety sequentially interacts with the two contact sites). (c) Two receptors and two G proteins: a GPCR dimer, indicated by the rhodopsin dimer, interacts simultaneously with two G proteins (here: heterotrimeric transducin). This model is difficult to conceptualize because the two G proteins are sterically hindered from interacting with the receptor dimer simultaneously.

Fig. 4.2c), which displaces the helical domain. Accordingly, helical and RAS-like domains are pried apart to create an exit pathway for GDP.

Switch 1 is connected to the helical domain via helix F; thus both, the lever-arm and the gear-shift models, require a movement in switch 1. Receptor-induced movements have been observed in switch 1 and switch 2 by electron spin resonance of appropriately labeled $G\alpha_{i-1}$ [16, 17]. The observations are consistent with a change in the mobility (decline in the mobility of switch 1 residues, [17]; increase in the mobility of switch 2 residues). However, the data do not allow for differentiating between these two models. Neither the lever-arm, nor the gear-shift model account for the fact that receptor-mimetic peptides (e.g., D2N, see below) do not require Gβγ or the N-terminus of Gα to activate GDP release. Similarly, it is possible to observe an—albeit inefficient—interaction between Gα and receptor in the absence of Gβγ [18, 19]. Last, but not least, a given receptor has multiple contact sites; on Gβ, Gγ, and the N- and C-termini as well as the on the α4/β6 segment of Gα. A glance at

Fig. 4.1a illustrates that a monomeric receptor has a foot size that is not commensurate with the shoe size of its biochemical footprint [20]. In other words, it appears unlikely that a (monomeric) receptor can bind all contact sites simultaneously.

4.1.3. The "C-Terminal Latch" Model

As mentioned above, the footprint of the receptor is larger than its foot size. The C-terminal latch hypothesis assumes that the receptor screens the contacts sites and when properly positioned activates the latch, which is the C-terminal end of Gα [21]. Thus, this model builds on and extends the "sequential fit" model [22, 23]. The interaction between receptor and cognate G protein(s) can be divided into two steps: (1) a docking process and (2) an activation process. During docking, the receptor encounters Gβγ, and the consecutive interaction with Gα is responsible for GDP release. A crucial finding in support of this model is the fact that binding of Gγ decreases binding of Gα in a dose-dependent manner and vice versa; according to the sequential fit model, Gβγ is crucial for a primarily receptor–G protein interaction but dispensable for GDP release [23]. This model accounts for several observations: (1) the receptor may not only discriminate between Gα subunits in the heterotrimer but may also preferentially interact with specific Gβ and Gγ isoforms. This was originally observed by electrophysiological recordings in cells treated with antisense oligonucleotides to deplete specific subunits [24, 25]. However, analogous findings have also been recapitulated by biochemical approaches [26, 27]. (2) A sequential fit permits the activated receptor to sample the nature of the G protein, which it collides with, and thus allows for kinetic proofreading. In many (but not all, see below) instances, the interactions between receptors and G proteins are consistent with collision coupling; that is, the activated receptor takes a random walk and collides with its cognate G protein, which results in activation. However, in this scenario, the receptor must also encounter noncognate G proteins. The sequential fit scenario allows for a step in which the nature of the G protein can be verified and the irrelevant G proteins rejected by kinetic proofreading, while the cognate G protein is engaged in a productive high-affinity interaction [28]. (3) A sequential fit model accounts for the fact that the peptide D2N efficiently activates $G\alpha_{i-1}$, $G\alpha_o$, and other related subunits in the absence of Gβγ [21]. D2N comprises the N-terminal end of the third intracellular loop of the D_2 dopamine receptor [29] and represents one of the active sites on the D_2 receptor [30]. (4) Last, but not least, the sequential fit model provides a plausible solution to the foot size/shoe size problem mentioned above.

The C-terminal latch model posits that, after the initial Gβγ-dependent docking, the receptor is positioned for the crucial second step, that is, engagement of the C-terminus of Gα (Fig. 4.1d). The fact that the receptor requires the very C-terminus of Gα is beyond doubt. The C-terminus is connected via the α5 helix (marked in red in Fig. 4.1d) to the α5/β6 loop. This loop contacts the guanine base via the TCAT motif (Fig. 4.1d, blowup upon the right hand side). The receptor is thought to trigger a structural rearrangement in the C-terminus, which is propagated via helix α5 to the guanine nucleotide binding pocket (Fig. 4.1d). Several lines of evidence support this model. (1) Mutations, which interfere with hydrophobic packing of the helix, accelerate GDP release [31]. (2) Mutations which disrupt the helix (e.g., introducing proline or glycine residues) preclude efficient receptor-dependent activation of the mutated Gα, while complex formation between receptor and G protein is preserved [32, 33]. In contrast, insertions, which maintain a rigid helix α5, do not affect receptor-dependent activation [33]. (3) Finally, electron paramagnetic resonance measurements of spin-labeled mutants indicate that the mobility of the flexible C-terminal end of Gα is reduced upon binding to the receptor, consistent with this point being the site of engagement. The changes in mobility of spin-labeled residues that lie within the helix α5 are consistent with a rigid body movement of the helix, which is rotated and displaced toward the β6 strand [34]. The resulting movement has been proposed to translate into a rearrangement of the TCAT motif, which contacts the guanine base. It is worth noting that an instructive mutation resides in this region: addition of a hydroxyl group (A351/366S in $G\alpha_s$—numbering dependent on the splice variant; the homologous position is A326 in $G\alpha_{i-1}$) causes a gain of function in testis (resulting in precocious puberty) because the fraction of GTP-bound protein at steady state increases. The latter is defined by the ratio of $[k_{cat,GTP}/(k_{cat,GTP} + k_{off,GDP})]$ and is thus less than 5% in a given G protein heterotrimer but approaches 50% and more, if the off rate of GDP release is close to or exceeds k_{cat} for GTP hydrolysis. In the rest of the body, the mutation is associated with a loss of function because of its enhanced thermal lability (resulting in the phenotype of pseudohypoparathyroidism) [35]. Originally, the extra hydroxyl group arising from the serine substitution was thought to cause a steric clash with the purine ring of GDP. However, experimental results suggest this does not appear to be the case and indicate that subtle changes within the TCAT motif in this region suffice to greatly accelerate GDP release [36].

4.1.4. Are the Three Models Mutually Exclusive?

The previous description focused on the differences between the models; yet, it is obvious that none of the model suffices to explain all experimental observations. Some of the shortcomings of the lever-arm model and the gear-shift model have been pointed out above (see Section 4.1.2.). The explanatory power of the C-terminal latch hypothesis also has its limitations: for instance, it ignores a role for Gβγ in opening the guanine nucleotide exit pathway and therefore does not take into account the evidence suggesting that Gβγ may prime Gα by inducing a preactivated conformation [37]. The wasp venom mastoparan is the prototypical receptomimetic peptide; mastoparan has been reported to affect the circular dichroism spectrum of $G\alpha_{i-1}$ in a manner consistent with a reduced helical content. This observation has been interpreted as evidence for mastoparan-induced melting of helix α5 [38]. However, this interpretation is incompatible with the currently proposed version of the C-terminal latch model, that is, rigid body motion of helix α5. Mastoparan was generally accepted to engage the C-terminus of Gα because its action was blocked by pertussis toxin pretreatment [39]. Pertussis toxin ADP-ribosylates a cysteine residue four amino acids removed from the C-terminus on Gα subunits of the $G_i/G_o/G_t$ family. Pertussis toxin also abrogates the action on Gα of D2N [29] and of receptomimetic lipoamines [40]. Binding of D2N and mastoparan is mutually exclusive [29], and the action of D2N is blocked by an antibody to the Gα C-terminus [21]. This has led to the proposal that cationic receptomimetic compounds, including D2N, bind to and exert their action via the C-terminal latch of Gα [21]. More recently, the position of D2N was examined in a crystal comprising $G\alpha_{i-1}$ and a second activating peptide KB-752 [41]. Like D2N, KB-752 has some GEF activity and, when combined, the action of the two peptides is potentiated. Those residues of D2N, which were traceable in the crystal structure, did not contact the C-terminus but rather were found in an adjacent position, namely in a cleft between the helix α4 and the β6 strand. Similarly, KB-752 was found in the pocket close to the switch 2 region, a position normally occluded by Gβγ [41]. It is not clear, however, why this crystal did not reveal evidence for a GDP-free form of Gα. This shortcoming may be remedied by studying D2N and KB-752 bound to the A326S-mutant of $G\alpha_{i-1}$ [36]. Nevertheless, and at the very least, the complex of $G\alpha_{i-1}$, D2N, and KB-752 highlights that the lever-arm and C-terminal latch models are not mutually exclusive; it is also conceivable that some aspects of the gear-shift model represent a better approximation of reality than the other two models.

Finally, it is evident that the bulk of the observations have been made with the pair $G\alpha_{i-1}$ (or $G\alpha_t$) and rhodopsin, given the lack of a better alternative, and the resulting insights are often extrapolated to all receptors and G proteins. This generalization is likely to be an oversimplification as receptors differ widely in the sequence of their intracellular loops. Moreover, a comparison of the structure of rhodopsin and of the β_2-adrenergic receptor reveals substantial differences. For example, in the inactive rhodopsin, there is an ionic lock between the ERY motif at the (cytoplasmic) bottom of TM3 and an arginine residue at the bottom of TM6 (E247 in rhodopsin) which clamps rhodopsin in the ground state [42]. In contrast, in the β_2-adrenergic receptor, the (homologous DRY motif-based) lock was seen in an open position, although the receptor was also trapped in its inactive conformation by the antagonist carazol (which acts as an inverse agonist like the 11-*cis* retinal chromophore in rhodopsin) [43, 44]. Thus, it is not surprising that the relative importance of contact sites on $G\alpha_{i-1}$ differ for individual receptors [45].

4.2. RECEPTOR MONOMERS–DIMERS–OLIGOMERS: ONE SIZE FITS ALL?

While GPCR monomers were the conceptual norm until the mid 1990s, receptor dimers have become very fashionable [46]. If one ignores the possible existence of higher-order oligomers (see below), three possibilities are conceivable in principle (Fig. 4.2): (1) the classical model: one single receptor protein encounters one G protein (Fig. 4.2a); (2) two receptors form a functional dimer and engage one G protein (Fig. 4.2b)—this is invoked to understand signaling in homodimeric receptors and heterodimeric receptors, which couple to a single cognate G protein; (3) two receptors form a dimer and each of them couples to a distinct G protein (Fig. 4.2c)—this model must be considered in proposed heterodimers of receptors, which couple to distinct sets of G proteins, for example, a heterodimer comprising a G_s- and a G_i-coupled receptor.

4.2.1. Evidence for GPCR Dimers

There is overwhelming evidence to support the existence of GPCR dimers: $GABA_B$ receptors are obligatory dimers, because they are not exported to the cell surface unless both receptors are present [47–49]. Taste perception provides another instructive example [50]: Of the five different known taste perceptions (sweet, bitter, sour, salty, umami—the typical taste of the amino acids monosodium glutamate and aspartate), three are mediated by G protein-coupled receptors, referred to as T_1 receptors for sweet and umami (and further divided into T_1R_1, T_1R_2, and T_1R_3) and T_2 receptors for bitter. The perception of sweet is contingent on the presence of T_1R_2 and T_1R_3. In contrast, mice rendered genetically deficient in either T_1R_1 or T_1R_2 do not respond to the amino acids monosodium glutamate and aspartate. These two examples are formal proof for receptor dimerization, but they do not prove that the receptors work as dimers. It can be argued that oligomer formation is only required for endoplasmic reticulum (ER) export because ER retention motifs are rendered inaccessible in the dimer [51] or because oligomerization favors the assembly of the ER export machinery (as observed with other membrane proteins; see Reference 52). However, using an ingenuous strategy of coexpressing appropriately ligand-binding and G protein-binding deficient receptors, Hlavackova et al. showed that the dimer also signals [53]. In other words, the data support the hypothetical arrangement of a receptor dimer contacting a single G protein heterotrimer depicted in Fig. 4.2b. This type of arrangement is not only aesthetically pleasing, it also provides for a solution to the foot size/shoe size problem alluded to in Sections 4.1.2. and 41.3 above [20] and obviates the need to invoke a two-step sequential fit mechanism; one receptor undertakes the docking part to $G\beta\gamma$ and thus positions $G\alpha$ to allow for triggering the C-terminal latch by the second receptor.

4.2.2. GPCR Dimers Are Not Universally Required as Prerequisites for G Protein Activation

Importantly, though, these three examples are from the G branch of GPCRs (according to the GRAFS nomenclature [5]), which comprises the metabotropic glutamate (mGluR), calcium sensing (CaSR), and $GABA_B$ receptors. The relevance of dimerization for G protein activation is much less clear, if the other GPCR subfamilies are examined. This is, in particular, true for the largest, that is, the R branch of rhodopsin-like receptors. The controversy is most readily evident with the eponymous member of the branch, namely rhodopsin itself. Rhodopsin dimers have been visualized in crystals [54] and observed by atomic force microscopy in native disk membranes [55], but it is not clear if they are meaningful for understanding the mechanism of G protein activation:

1. Under fully dark adapted conditions, a single photon—captured by a single chromophore in a single rhodopsin—is likely to activate a rod to elicit the primary visual response. Given this exquisite sensitivity and the powerful amplification, it is a priori difficult to conceptualize a role of dimeric rhodopsin, unless the second rhodopsin in the dimer need not be active. This model, however, implies precoupling, that is, prebinding of transducin G_t to (inactive dark) adapted rhodopsin. This is inconceivable because rhodopsin is in (up to ~12-fold) excess over transducin; precoupling would preclude the catalytic action of rhodopsin required for signal amplification and make it impossible for rhodopsin to operate under the range of illuminations encountered during a given diurnal cycle [56].
2. Although rhodopsin may assemble to dimers and higher-order arrays in rod outer segments [55], this does not a priori prove that it is the dimeric form that activates the G protein. In fact, the dimer/monomer discussion has not only inspired insightful reviews [57], but has also stimulated ingenuous experiments designed to test the catalytic prowess of monomeric rhodopsin: if solubilized in the detergent dodecyl maltoside, rhodopsin fulfills all criteria of a monomerically dispersed protein. In this monomeric form, rhodopsin activates transducin with catalytic perfection, that is, in a diffusion-limited manner [58]. The rate of activation in solution is lower than that observed in disc membranes but this can be accounted for by the restricted diffusional freedom/directed orientation imposed by the membrane rather than any impairment of monomeric rhodopsin. This interpretation is supported by observations on rhodopsin inserted into nanodiscs of defined size [59] and high-density lipoprotein (HDL) particles of ~10 nm [60]. Under these conditions, the lipid particles are either too small to contain more than one rhodopsin [60] or their size can be defined to allow for insertion of one or two rhodopsin molecules [61]. The results are unequivocal: a single rhodopsin molecule suffices to efficiently activate transducin. In fact, when inserted into the nanodisc, the second rhodopsin does not enhance catalysis: the turnover number is halved, indicating that it does not contribute to any appreciable extent to transducin activation [61]. Similarly, transducin only stabilizes one rhodopsin molecule in the metarhodopsin II (MII) state

(the equivalent of the ternary HRG complex). These data may be interpreted to justify a dimeric arrangement as outlined in Fig. 4.2b), with one active MII state stabilized by G_t and the second rhodopsin as an inactive scaffold. However, the interpretation of the experiment is limited by the fact that the orientation of rhodopsin dimer is not known. If it is random, there are only 50% dimers with parallel orientation and only these dimers can bind transducin. In fact, earlier experiments showed that—at saturating G protein levels—the number of receptors trapped as HRG complexes in reconstituted vesicles approached only 50% of that observed in detergent solutions [62]; in contrast, in fusion proteins, where Gα is directly tethered to the receptor, virtually all receptors are capable of forming high-affinity complexes [28]. This clearly indicates that receptor orientation can be limiting. Taken together, these data show that monomeric rhodopsin is perfectly capable of activating transducin. An analogous conclusion can also be drawn for the β_2-adrenergic receptor, which—when confined to an HDL particle as monomer—activates its cognate G protein G_s very efficiently [61]. This is remarkable because fluorescence resonance energy transfer (FRET) microscopy and bioluminescence resonance energy transfer (BRET) recordings have implied that, in living cells, the β_2-adrenergic receptors are found in various (homo- and hetero-) dimeric arrangements (for review, see Reference 46).

At the very least, one would have to concede that—at the current stage—explanations that invoke a universal role of dimers are not parsimonious. The conclusion is justified that receptor dimers (or higher-order oligomers) appear to be dispensable for G protein activation by those rhodopsin branch members, where it has been subjected to rigorous testing.

3. Rhodopsin and its bleached version opsin differ in their sensitivity to thermal denaturation. As mentioned above, a dimeric model (similar to that illustrated in Fig. 4.2b) is conceivable, but it implies a mixed active/inactive arrangement. Thus, if the state of rhodopsin is sampled by differential scanning, the presence of dimers in rod (outer segment) disk membranes ought to be evident from a change in the thermal denaturation curve. This has been examined recently [63]: the data provided little evidence for the presence of dimeric rhodopsin, leading to the conclusion that the bulk of rhodopsin in the disc membrane is a monomer. It is generally accepted that dimeric arrangements visualized by X-ray crystallography only provide circumstantial evidence for the state of the protein in a living cell. In contrast, atomic force microscopy can, in principle, sample the state(s) of a protein in its native environment. Thus, the observations of rhodopsin dimers/higher-order oligomers by atomic force microscopy [55] cannot be easily dismissed, because it must reflect a biologically relevant property of the protein. It has been pointed out that the mica surface (employed in atomic force microscopy) depletes

the membranes of lipids [57]. The rhodopsin concentration in the disk membrane is very high (>25,000 molecules/μm^2), and the ratio of phospholipid to rhodopsin is about 1:65 [64]; accordingly, the surrounding lipid anulus can only comprise a few layers of phospholipids. Any manipulation, which reduces the solvation of the hydrophobic segment of rhodopsin (including phospholipid depletion and hydrophobic mismatch), is likely to promote oligomeric assemblies [65, 66]. Thus, the monomer/dimer equilibrium may reflect local packing and may be relevant *in vivo*, when GPCRs are present in high concentrations, for example, disc membranes and synapses. But these dimers or higher-order oligomers are not necessary for G protein activation.

4.2.3. Dimers May Allow for Conformational Switches Underlying Receptor Cross-Talk and Other Forms of Allosterism

At first glance, the model depicted in Fig. 4.2c does not seem to have any specific merit in providing a parsimonious explanation. With a little imagination, it is possible to draw ever-more complicated arrangements, in which arrays of receptors are confronted with tubes of G proteins of various stoichiometries. The arrangement in Fig. 4.2c specifically suffers from the drawback that it does not solve the foot size/shoe size problem [20]. But, it does have the advantage that it can account for mutually antagonistic cross-talk in heterodimeric receptors. Such a finding has been observed for the A_{2A} adenosine receptor and the D_2 dopamine receptor, which can assemble into a heterodimer [67, 68]. This dimer formation is thought to be of interest for the treatment of Parkinson's disease, because the mutual inhibition of the two receptors is predicted to have an impact on movement control by the corpus striatum: activation of the A_{2A} receptor impedes D_2 receptor coupling and is thus thought to aggravate the symptoms of Parkinson's disease; conversely, A_{2A} antagonists are likely to facilitate voluntary movements [69]. Mutual inhibition can be readily rationalized upon inspection of Fig. 4.2c: in a dimeric arrangement, activation of one partner and the resulting recruitment of the cognate G protein are likely to sterically preclude recruitment of a different G protein to the other receptor in the dimer. It is worth noting, though, that mutual antagonism is not always observed; D_2 receptors and A_{2A} receptors can synergize under appropriate conditions [70]. Steric hindrance may also underlie the observation that homodimers and heterodimers formed by the G_s-coupled TSH (thyroid stimulating hormone) and LH/hCG (luteinizing hormone/human chorionic gonadotropin) receptors show negative cooperativity of agonist binding [71]. A tug-of-war can be envisaged at high ligand concentration: agonist (H for hormone) binding is stabilized in the ternary HRG complex; at high ligand concentrations, when all binding sites are occupied, only one receptor can be trapped within the HRG complex, while the second receptor must bind the agonist/hormone with lower affinity. In other words, the receptors must compete for the G protein during the reaction trajectory, and this may give rise to negative cooperativity.

Heterodimers have also been shown to change the pharmacological specificity of the complex. The most impressive example is a heterodimer composed of κ- and δ-opioid receptors. This complex can be selectively activated by the agonist 6′-guanidinonaltrindole, which is only weakly active at each individual receptor (or on heterodimers comprising the μ-opiod receptor), and this selective action can also be seen *in vivo*: the compound causes spinal analgesia upon intrathecal injection but not central analgesia upon intracerebroventricular injection consistent with the observations that the two receptors are coexpressed in the spinal cord but not in central nuclei relevant for pain perception and relief such as the periaqueductal gray matter [72]. In the cycle of G protein activation and deactivation, the agonist-liganded receptor confers information to the G protein, but there also must be a signal that is transferred back to the agonist binding site in the receptor. This is conspicuously evident from the stabilization high-affinity agonist binding in the ternary complex (HRG, see above). This mechanism must obviously also operate with the dimer-selective agonist, but it is not clear whether the backward signal is transferred from one G protein to one or two binding sites. In other words, the dimeric receptor may engage a single G protein as depicted in Fig. 4.2b: a sequential fit scenario is conceivable, in which the low intrinsic activity of 6′-guanidinonaltrindole on the δ-opioid receptor may suffice for positioning the G protein to become more fruitfully activated by the agonist liganded κ-opioid receptor. The otherwise low efficacy of 6′-guanidinonaltrindole is then enhanced, making it a full agonist. Alternatively, 6′-guanidinonaltrindole may be bound to both receptors, and the G protein may signal back to the two receptors to trap 6′-guanidinonaltrindole with high affinity in the two binding pockets. This alternative is less likely in view of the observation that dimeric TSH, LH/hCG, and FSH receptors show negative cooperativity [71]. Other scenarios are even less likely, that is, that a complex depicted in Fig. 4.2c (one κ- and one δ-opioid receptor—each liganded to 6′-guanidinonaltrindole) is specifically stabilized by two G protein heterotrimers. Finally, it is worth pointing out that many receptors are endowed with binding sites distinct from that of their canonical ligand (the orthosteric site; [73, 74]). Binding to these other (allosteric) sites is also likely to be affected in dimeric forms of receptors; for it is difficult to envisage why dimerization would only affect the orthosteric ligand (which it does in many ways other than those outlined here; for review, see Reference 75).

4.3. CORRALS, FENCES, RAFTS—ARE THERE PRIVILEGED PLACES FOR GPCR ACTIVATION?

4.3.1. The Actin Cytoskeleton Confines GPCRs by Several Mechanisms

The cycle of receptor-dependent G protein activation and GTPase-induced deactivation allows for amplification: as long as an agonist is present (and as long as desensitization has not been triggered) one receptor molecule can activate a large number of G protein molecules by collision coupling. This, however, requires that the receptor has unlimited access to G proteins; as mentioned above, the situation is epitomized by rod-mediated vision, where a single photon can excite a single rhodopsin, and this suffices to create an optic impression, because rhodopsin activates a large number of transducin molecules. This is, however, not the norm, because in many instances, receptors are subject to restricted collision coupling—a receptor does not have access to all its cognate G proteins: restricted collision coupling may arise from:

1. fencing by the actin cytoskeleton (Fig. 4.3a);
2. picketing by other transmembrane proteins, which are tethered to the cortical actin (Fig. 4.3b); and
3. direct tethering of a receptor to a scaffolding protein (for review, see Reference 76), which is *per se* directly or indirectly linked to the cytoskeleton (Fig. 4.3c).

These mechanisms are readily grasped by intuition and they can actually be shown to operate on GPCRs: single dye tracking reveals that the trajectories of receptors are limited to small areas for prolonged periods of times, suggesting that a given receptor is confined to a small compartment; however, there are instances where the receptor can apparently switch compartment by an event, which is referred to as hop diffusion [77]. It is worth noting that agonist stimulation of many GPCRs is predicted to affect the apparent rate of diffusion—in particular to increase the probability of hop diffusion. Several mechanisms can be envisaged: upon agonist binding, receptors may signal to the β isofoms of phospholipase C either by (1) generating GTP-bound $G\alpha_q$ family members ($G\alpha_{q/11/14/15/16}$) or (2) by liberating enough free Gβγ dimers. Activated phospholipase C will consume PIP_2 (phosphatidylinositol-4,5-bisphosphate) and thus reduce its availability. This phospholipid is essential for recruiting many actin-binding proteins (e.g., ezrin, radixin, moesin—via their FERM domain; N-WASP via a polybasic motif) to the inner leaflet of the plasma membrane. (3) Gβγ dimers may also activate the γ isoform of PI3-kinase, which again consumes PIP_2 (to generate PIP_3, phosphatidylinositol-3,4,5-trisphosphate). Again, this causes redistribution of cytoskeletal proteins and scaffolding proteins (which may interact with PIP_2 and/or PIP_3 via their pleckstrin homology domain). (4) Activation of G_{12}/G_{13}-coupled receptors engages an exchange factor of the p115RhoGEF family, which relay the signal via the small G protein RhoA to the actin cytoskeleton. Last, but not least, the actin cytoskeleton has been proposed to stabilize specialized lipid domains, which arise from the different miscibility of lipids [78, 79] and which are commonly referred to as lipid rafts (see below).

The term "restricted collision coupling" may have a negative connotation because it implies a limited range of G protein activation. It is evident though that fencing (Fig. 4.3a) and picketing (Fig. 4.3b) can—in principle—allow

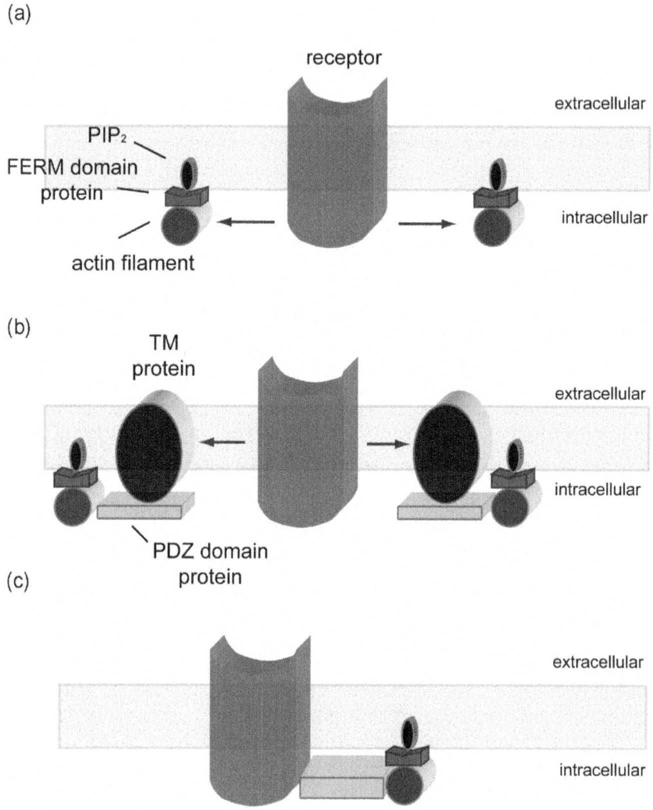

Figure 4.3 Mechanisms for local confinement of GPCRs. (a) "corralling/fencing model": a GPCR (indicated as dark grey, U-shaped rectangle) is restricted in its mobility (indicated with arrows) within the phospholipid bilayer (shown in light grey) by membrane-bound actin filaments (indicated as grey cylinders). Actin is attached to the membrane via FERM domain proteins (F for 4.1 protein, E for ezrin, R for radixin, M for moesin; shown as grey "M"-shaped rectangles). These bind PIP_2 (phosphatidylinositol 4,5 bisphosphate; shown as small dark oval) embedded in the inner leaflet of the plasma membrane. (b) "fencing/picketing model": the GPCR is restricted in its mobility by transmembrane (TM) proteins (indicated as big dark oval), which proteins are attached to the membrane-bound actin complex (PIP_2/FERM protein/actin complexes) via their associated PDZ domain-containing scaffolding proteins. Lateral motion causes the GPCR to collide with the other TM proteins or the lipid shell surrounding these. (c) spatial confinement by direct interaction: a receptor is restricted in its mobility within the membrane by direct binding to a scaffolding (PDZ domain-containing) protein which is tethered to membrane-bound actin filaments.

for a concentration of signaling components and thus speed up signal transmission. The most efficient means of concentrating signaling components, however, is the binding of GPCRs, effectors, and regulators to scaffolding proteins. The paradigmatic example is the InaD mutation which greatly limits signaling by rhodopsin in the drosophila eye (hence the acronym, Inactivation No-After Potential-D) [80]. The affected gene encodes a scaffolding protein with 5 PDZ domains, which assembles rhodopsin, phospholipase C (insect rhodopsin signals via G_q), protein kinase C, the *TRP* channel (a nonselective cation channel; *TRP* = transient receptor potential), and other components into a large complex, termed signalosome or signalplex (for review, see Reference 81).

4.3.2. Cholesterol-Rich Domains and Lipid Rafts

Originally, the cell membrane was conceptualized as an isotropic lipid bilayer, which allowed for free diffusion of all constituents therein. This model was metaphorically referred to as a fluid mosaic of freely and randomly diffusing proteins embedded in phospholipids [82]. Obviously, this model ignores the compartmentalization induced by the cortical actin (Fig. 4.3), which as mentioned above, may also affect the distribution of lipids [78]. In addition, it does not account for the fact that lipids are not equally distributed over the various cell membrane compartments, although these compartments all communicate with each other. Cholesterol, for instance, is enriched in the plasma membrane and in endosomal vesicles, but it is synthesized in the endoplasmic reticulum. It is also clear that the outer and inner leaflet of the membrane do not have the same lipid composition: for example, sphingolipids are enriched in the outer leaflet. These gradients must be actively maintained; this fact is generally best appreciated for phosphatidylserine, which is conspicuously absent from the outer leaflet unless cells undergo apoptosis. Thus, the asymmetric distribution of lipids is actively maintained. In addition, lipids have been proposed to segregate and self-organize based on their different miscibility. Because the saturated lipids of sphingomyelins are more prone to associate with cholesterol, membrane areas may arise that are enriched in these lipids and the platforms—termed lipid rafts—were proposed to attract specific proteins [83]. Cholesterol-rich domains may be further stabilized by the cytoskeleton (see above) and by caveolin [84]. These platforms (rafts and caveolae) have attracted much interest in the field of signal transduction, because many key signaling molecules are modified with lipid moieties (e.g., farnesyl or geranylgeranoyl thioethers on the C-termini of RAS-like small G proteins or myristoyl amide bonds on the N-termini of nonreceptor tyrosine kinases of the SRC family).

Because many GPCRs are palmitoylated (on a cysteine residue at the end of helix 8 in the C-terminus) and several G protein α subunits are myristoylated and palmitoylated, they were proposed to be attracted into lipid rafts. Accordingly, there is a long list of reports, which examined the role of lipid rafts and cholesterol-rich domains in signaling by GPCRs: most reports emphasize a model, where G protein activation preferentially takes place in lipid rafts/cholesterol-rich domains (for review, see References 85, 86). It is not clear, though, why this should be the case; the original reconstitution experiments (of purified receptors and purified G proteins) employed lipid mixtures that were devoid of cholesterol and found robust G protein activation [87] and transfer of the signal to the effector [88]. Similarly, when expressed in *Escherichia coli* and reconstituted with G proteins, GPCRs efficiently interact with their cognate G proteins [89–91], although the inner membrane of gram-negative bacteria is devoid of cholesterol.

Lipid rafts are thought to be too small to directly visualize them; their proposed size varies between 5 and 100 nm and is thus below the limit of optical resolution imposed by Abbé's limit. Similarly, caveolae are not present in all cells; the phenotype of mice genetically deficient in caveolin is very mild (for review, see Reference 84). Thus, the evidence for a role of these putative lipid rafts is only circumstantial. There have been two popular approaches used to examine the role of lipid rafts: (1) the (re)distribution of receptors and G proteins into detergent-resistant membranes and (2) the sensitivity of signaling events to cholesterol extraction. For (1), sucrose density gradient fractionation of cell membranes provide a means of resolving two surface membrane fractions. The lipid enriched light fraction is thought to contain the lipid rafts. Similarly, a proportion of membrane proteins cannot be extracted by Triton X-100, and this detergent-resistant fraction again is thought to represent lipid rafts. With (2), cells are pretreated with the polyene antibiotic filipin 3 and the caging compound methyl-β-cyclodextrin, which sequesters and depletes cholesterol, respectively, and thus disrupt lipid rafts/cholesterol-rich domains. The rationale underlying these approaches have been questioned [92]: typically cells are homogenized in ice-cold buffers and density gradient centrifugation is done at low temperature, which per se has a strong effect on the miscibility of lipids and thus on the diffusion of receptors [77]. Similarly, addition of Triton X-100 may per se affect the miscibility of lipids and thus induce segregation of lipids, which was initially not present [93]. Proteins may also be resistant to detergent extraction, because they are tightly bound to the cytoskeleton.

Extraction of cholesterol may have effects other than disrupting lipid rafts. In fact, like all other membrane proteins, GPCRs affect the shape of the membrane, because for any individual membrane, protein solvation of the hydrophobic transmembrane segment may be more readily accomplished in one type of lipid. This will result in selective attraction of lipids to the anular protein lipid boundary [94, 95]. Cholesterol enhances the stability of rhodopsin but reduces the efficacy of G protein activation [96, 97]. Cholesterol may interfere with rhodopsin activation largely by steric hindrance: the protein expansion that is caused by transition of metarhodopsin I (MI) into the active form MII is impeded by dense packing of cholesterol, which renders the phospholipid bilayer more rigid. Rod outer segments are heterogeneous with respect to their cholesterol content. Disc membranes in the rod outer segments are continuously replenished from evaginations of the plasma membrane at the base of the outer segment. Within approximately 10 days, the disc membranes reach the top of the outer segment where they are engulfed by phagocytosis by the surrounding pigment epithelium. Whereas the ratio of protein to phospholipd remains constant during migration of the disc to the apical tip, cholesterol content falls. In newly formed disc membranes at the base of the outer segment, cholesterol accounts for 30% of the lipid mass; mature discs residing on the tip of the rod outer segment are depleted in cholesterol (with a ratio of phospholipid to cholesterol = 20:1; [64]). From a teleological perspective, this change in cholesterol is useful; a high level of cholesterol may not only stabilize rhodopsin, but it may also prevent its premature activation: photons may also activate those rhodopsin molecules that are still en route to the rod outer segment [98]. These effects of cholesterol do neither require the formation of specific cholesterol-enriched domains nor the presence of a cholesterol binding site on rhodopsin. However, an electron density compatible with cholesterol has been found to be trapped in both crystals of rhodopsin [99] and of the β_2-adrenergic receptor [100], raising the possibility that cholesterol may also regulate receptor function by directly binding to the hydrophobic core.

If GPCRs other than rhodopsin are examined, depletion of cholesterol does not uniformly affect signaling; in some instances, G protein activation is enhanced, while in others, it is abrogated (reviewed in Reference 85). In addition, in at least one receptor, the presence or absence of cholesterol, may affect signaling in a more subtle way than all-or-none: cholesterol depletion abrogates the ability of the A_{2A}-adenosine receptor to activate G_s (the cognate G protein) and hence cAMP accumulation, but it does not interfere with G protein-independent recruitment of ARNO (the exchange factor for the small G protein ARF6) and the resulting stimulation of mitogen activated protein (MAP) kinase [68]. Because many GPCRs recruit more than one signaling cascade, it is likely that analogous effects will also be observed with other receptors. In the lipid raft model, one would be inclined to ascribe the differential effects of cholesterol extraction on the A_{2A} adenosine receptor to the segregation of signaling components, with G protein-dependent signaling contingent on cholesterol-rich domains and ARNO recruitment being independent of membrane compartmentalization. However, at the current stage, it is not possible to determine whether cholesterol extraction alters receptor-dependent G protein activation through disruption of lipid rafts or other cholesterol-rich domains by changing the ability of the lipid bilayer to accommodate conformational changes in an activated GPCR, or by removing cholesterol that is tightly bound to the GPCR.

This section has focused on factors extrinsic to the G protein cycle that may allow for organizing signaling components. However, it should be kept in mind that the reaction kinetics of the G protein cycle allow for self-organization to emerge without any need for additional factors: if a G_q-coupled receptor is allowed to interact with a G protein that is subject to rapid deactivation by the GAP activity residing in an RGS protein and in the phospholipase Cβ (the G_q-regulated) effector, a stable tightly associated complex emerges, which allows for very rapid cycling [101].

ACKNOWLEDGMENTS

Work from the authors' laboratory was supported by the Austrian Science Fund (FWF). I.G. and C.W.G. are recipients of a DOC-fFORTE fellowship of the Austrian Academy of Sciences and a Lise-Meitner stipend (M-1086) of the FWF, respectively; C.W.G. is also supported by a grant from the European Union (FP7: IRG 230970).

REFERENCES

1. Oldham, W.M., Hamm, H.E. (2008) Heterotrimeric G protein activation by G-protein-coupled receptors. *Nat Rev Mol Cell Biol.* 9, 60–71.
2. Johnston, C.A., Siderovski, D.P. (2007) Receptor-mediated activation of heterotrimeric G-proteins: Current structural insights. *Mol Pharmacol.* 72, 219–230.
3. Hollinger, S., Hepler, J.R. (2002) Cellular regulation of RGS proteins: Modulators and integrators of G protein signaling. *Pharmacol Rev.* 54, 527–559.
4. Sprang, S.R. (1997) G protein mechanisms: Insights from structural analysis. *Annu Rev Biochem.* 66, 639–678.
5. Fredriksson, R., Lagerström, M.C., Lundin, L.G., Schiöth, H.B. (2003) The G-protein-coupled receptors in the human genome form five main families. Phylogenetic analysis, paralogon groups, and fingerprints. *Mol Pharmacol.* 63, 1256–1272.
6. Higashijima, T., Ferguson, K.M., Sternweis, P.C., Smigel, M.D., Gilman, A.G. (1986) Effects of Mg^{2+} and the βγ-subunit complex on the interactions of guanine nucleotides with G proteins. *J Biol Chem.* 262, 762–766.
7. Graziano, M.P., Freissmuth, M., Gilman, A.G. (1989) Expression of $G_{s\alpha}$ in *Escherichia coli*: Purification and characterization of two forms of the protein. *J Biol Chem.* 264, 409–418.
8. Murakami, M., Kouyama, T. (2008) Crystal structure of squid rhodopsin. *Nature.* 453, 363–367.
9. Klinger, M., Kuhn, M., Just, H., Stefan, E., Palmer, T., Freissmuth, M., Nanoff, C. (2002) Removal of the carboxy terminus of the A_{2A}-adenosine receptor blunts constitutive activity: Differential effect on cAMP accumulation and MAP kinase stimulation. *Naunyn Schmiedeberg's Arch Pharmacol.* 366, 287–298.
10. Pankevych, H., Korkhov, V., Freissmuth, M., Nanoff, C. (2003) Truncation of the A_1-adenosine receptor reveals distinct roles of the membrane-proximal carboxy terminus in receptor folding and G protein coupling. *J Biol Chem.* 278, 30283–30293.
11. Iiri, T., Farfel, Z., Bourne, H.R. (1998) G-protein diseases furnish a model for the turn-on switch. *Nature.* 394, 35–38.

12. Wall, M.A., Coleman, D.E., Lee, E., Iñiguez-Lluhi, J.A., Posner, B.A., Gilman, A.G., Sprang, S.R. (1995) The structure of the G protein heterotrimer Giα-1β1γ2. *Cell.* 83, 1047–1058.

13. Lambright, D.G., Sondek, J., Bohm, A., Skiba, N.P., Hamm, H.E., Sigler, P.B. (1996) The 2.0 Å crystal structure of a heterotrimeric G protein. *Nature.* 379, 311–319.

14. Rondard, P., Iiri, T., Srinivasan, S., Meng, E., Fujita, T., Bourne, H.R. (2001) Mutant G protein α subunit activated by Gβγ: A model for receptor activation? *Proc Natl Acad Sci U S A.* 98, 6150–6155.

15. Cherfils, J., Chabre, M. (2003) Activation of G-protein Gα subunits by receptors through Gα-Gβ and Gα-Gγ interactions. *Trends Biochem Sci.* 28, 13–17.

16. Van Eps, N., Oldham, W.M., Hamm, H.E., Hubbell, W.L. (2006) Structural and dynamical changes in an α-subunit of a heterotrimeric G protein along the activation pathway. *Proc Natl Acad Sci U S A.* 103, 16194–16199.

17. Oldham, W.M., Van Eps, N., Preininger, A.M., Hubbell, W.L., Hamm, H.E. (2007) Mapping allosteric connections from the receptor to the nucleotide-binding pocket of heterotrimeric G proteins. *Proc Natl Acad Sci U S A.* 104, 7927–7932.

18. Freissmuth, M., Schütz, W., Linder, M.E. (1991) Interactions of the bovine brain A_1-adenosine receptor with recombinant G protein α-subunits: Selectivity for $rG_{i\alpha\text{-}3}$. *J Biol Chem.* 266, 17778–17783.

19. Herrmann, R., Heck, M., Henklein, P., Hofmann, K.P., Ernst, O.P. (2006) Signal transfer from GPCRs to G proteins: Role of the G alpha N-terminal region in rhodopsin-transducin coupling. *J Biol Chem.* 281, 30234–30241.

20. Hamm, H.E. (2001) How activated receptors couple to G proteins. *Proc Natl Acad Sci U S A.* 98, 4819–4821.

21. Nanoff, C., Koppensteiner, R., Yang, Q., Fuerst, E., Ahorn, H., Freissmuth, M. (2006) The carboxyl terminus of the Gα-subunit is the latch for triggered activation of heterotrimeric G proteins. *Mol Pharmacol.* 69, 397–405.

22. Kisselev, O.G., Meyer, C.K., Heck, M., Ernst, O.P., Hofmann, K.P. (1999) Signal transfer from rhodopsin to the G-protein: Evidence for a two-site sequential fit mechanism. *Proc Natl Acad Sci U S A.* 96, 4898–4903.

23. Herrmann, R., Heck, M., Henklein, P., Henklein, P., Kleuss, C., Hofmann, K.P., Ernst, O.P. (2004) Sequence of interactions in receptor-G protein coupling. *J Biol Chem.* 279, 24283–24290.

24. Kleuss, C., Scherübl, H., Hescheler, J., Schultz, G., Wittig, B. (1992) Different β-subunits determine G-protein interaction with transmembrane receptors. *Nature.* 358, 424–426.

25. Kleuss, C., Scherübl, H., Hescheler, J., Schultz, G., Wittig, B. (1993) Selectivity in signal transduction determined by γ-subunits of heterotrimeric G proteins. *Science.* 259, 832–834.

26. McIntire, W.E., MacCleery, G., Garrison, J.C. (2001) The G protein β-subunit is a determinant in the coupling of Gs to the β_1-adrenergic and A_{2a}-adenosine receptors. *J Biol Chem.* 276, 15801–15809.

27. Kühn, B., Christel, C., Wieland, T., Schultz, G., Gudermann, T. (2002) G-protein βγ-subunits contribute to the coupling specificity of the β2-adrenergic receptor to G_s. *Naunyn Schmiedeberg's Arch Pharmacol.* 365, 231–241.

28. Waldhoer, M., Wise, A., Milligan, G., Freissmuth, M., Nanoff, C. (1999) Kinetics of ternary complex formation with fusion proteins composed of the A_1-adenosine receptor and G protein α-subunits. *J Biol Chem.* 274, 30571–30579.
29. Voss, T., Wallner, E., Csernilofsky, A.P., Freissmuth, M. (1993) An amphipathic α-helical structure does not predict the ability of receptor-derived peptides to interact with G proteins. *J Biol Chem.* 268, 4637–4642.
30. Bofill-Cardona, E., Kudlacek, O., Yang, Q., Ahorn, H., Freissmuth, M., Nanoff, C. (2000) Binding of calmodulin to the D_2-dopamine receptor reduces receptor signaling by arresting the G protein activation switch. *J Biol Chem.* 275, 32672–32680.
31. Marin, E.P., Krishna, A.G., Sakmar, T.P. (2001) Rapid activation of transducin by mutations distant from the nucleotide-binding site: Evidence for a mechanistic model of receptor-catalyzed nucleotide exchange by G proteins. *J Biol Chem.* 276, 27400–27405.
32. Marin, E.P., Krishna, A.G., Sakmar, T.P. (2002) Disruption of the α5 helix of transducin impairs rhodopsin-catalyzed nucleotide exchange. *Biochemistry.* 41, 6988–6994.
33. Natochin, M., Moussaif, M., Artemyev, N.O. (2001) Probing the mechanism of rhodopsin-catalyzed transducin activation. *J Neurochem.* 77, 202–210.
34. Oldham, W.M., Van Eps, N., Preininger, A.M., Hubbell, W.L., Hamm, H.E. (2006) Mechanism of the receptor-catalyzed activation of heterotrimeric G proteins. *Nat Struct Mol Biol.* 13, 772–777.
35. Iiri, T., Herzmark, P., Nakamoto, J.M., van Dop, C., Bourne, H.R. (1994) Rapid GDP release from Gsα in patients with gain and loss of endocrine function. *Nature.* 371, 164–168.
36. Posner, B.A., Mixon, M.B., Wall, M.A., Sprang, S.R., Gilman, A.G. (1998) The A326S mutant of Giα-1 as an approximation of the receptor-bound state. *J Biol Chem.* 273, 21752–21758.
37. Abdulaev, N.G., Ngo, T., Zhang, C., Dinh, A., Brabazon, D.M., Ridge, K.D., Marino, J.P. (2005) Heterotrimeric G-protein α-subunit adopts a "preactivated" conformation when associated with βγ-subunits. *J Biol Chem.* 280, 38071–38080.
38. Tanaka, T., Kohno, T., Kinoshita, S., Mukai, H., Itoh, H., Ohya, M., Miyazawa, T., Higashijima, T., Wakamatsu, K. (1998) Alpha helix content of G protein α-subunit is decreased upon activation by receptor mimetics. *J Biol Chem.* 273, 3247–3252.
39. Higashijima, T., Uzu, S., Nakajima, T., Ross, E.M. (1988) Mastoparan, a peptide toxin from wasp venom, mimics receptors by activating GTP-binding regulatory proteins (G proteins). *J Biol Chem.* 263, 6491–6494.
40. Breitweg-Lehmann, E., Czupalla, C., Storm, R., Kudlacek, O., Schunack, W., Freissmuth, M., Nürnberg, B. (2002) Activation and inhibition of G proteins by lipoamines. *Mol Pharmacol.* 61, 628–636.
41. Johnston, C.A., Siderovski, D.P. (2007) Structural basis for nucleotide exchange on Gα-i subunits and receptor coupling specificity. *Proc Natl Acad Sci U S A.* 104, 2001–2006.
42. Palczewski, K., Kumasaka, T., Hori, T., Behnke, C.A., Motoshima, H., Fox, B.A., Le Trong, I., Teller, D.C., Okada, T., Stenkamp, R.E., Yamamoto, M., Miyano, M. (2000) Crystal structure of rhodopsin: A G protein-coupled receptor. *Science.* 289, 739–745.

43. Cherezov, V., Rosenbaum, D.M., Hanson, M.A., Rasmussen, S.G., Thian, F.S., Kobilka, T.S., Choi, H.J., Kuhn, P., Weis, W.I., Kobilka, B.K., Stevens, R.C. (2007) High-resolution crystal structure of an engineered human β2-adrenergic G protein-coupled receptor. *Science*. 318, 1258–1265.

44. Rasmussen, S.G., Choi, H.J., Rosenbaum, D.M., Kobilka, T.S., Thian, F.S., Edwards, P.C., Burghammer, M., Ratnala, V.R., Sanishvili, R., Fischetti, R.F., Schertler, G.F., Weis, W.I., Kobilka, B.K. (2007) Crystal structure of the human beta2 adrenergic G-protein-coupled receptor. *Nature*. 450, 383–387.

45. Slessareva, J.E., Ma, H., Depree, K.M., Flood, L.A., Bae, H., Cabrera-Vera, T.M., Hamm, H.E., Graber, S.G. (2003) Closely related G-protein-coupled receptors use multiple and distinct domains on G-protein α-subunits for selective coupling. *J Biol Chem*. 278, 50530–50536.

46. Pin, J.P., Neubig, R., Bouvier, M., Devi, L., Filizola, M., Javitch, J.A., Lohse, M.J., Milligan, G., Palczewski, K., Parmentier, M., Spedding, M. (2007) International Union of Basic and Clinical Pharmacology. LXVII. Recommendations for the recognition and nomenclature of G protein-coupled receptor heteromultimers. *Pharmacol Rev*. 59, 5–13.

47. Kaupmann, K., Malitschek, B., Schuler, V., Heid, J., Froestl, W., Beck, P., Mosbacher, J., Bischoff, S., Kulik, A., Shigemoto, R., Karschin, A., Bettler, B. (1998) $GABA_B$-receptor subtypes assemble into functional heteromeric complexes. *Nature*. 396, 683–687.

48. Jones, K.A., Borowsky, B., Tamm, J.A., Craig, D.A., Durkin, M.M., Dai, M., Yao, W.J., Johnson, M., Gunwaldsen, C., Huang, L.Y., Tang, C., Shen, Q., Salon, J.A., Morse, K., Laz, T., Smith, K.E., Nagarathnam, D., Noble, S.A., Branchek, T.A., Gerald, C. (1998) $GABA_B$ receptors function as a heteromeric assembly of the subunits $GABA_BR1$ and $GABA_BR2$. *Nature*. 396, 674–679.

49. White, J.H., Wise, A., Main, M.J., Green, A., Fraser, N.J., Disney, G.H., Barnes, A.A., Emson, P., Foord, S.M., Marshall, F.H. (1998) Heterodimerization is required for the formation of a functional $GABA_B$ receptor. *Nature*. 396, 679–682.

50. Chandrashekar, J., Hoon, M.A., Ryba, N.J., Zuker, C.S. (2006) The receptors and cells for mammalian taste. *Nature*. 444, 288–294.

51. Margeta-Mitrovic, M., Jan, Y.N., Jan, L.Y. (2000) A trafficking checkpoint controls $GABA_B$ receptor heterodimerization. *Neuron*. 27, 97–106.

52. Farhan, H., Freissmuth, M., Sitte, H.H. (2006) Oligomerization of neurotransmitter transporters: A ticket from the endoplasmic reticulum to the plasma membrane. *Handb Exp Pharmacol*. 175, 233–249.

53. Hlavackova, V., Goudet, C., Kniazeff, J., Zikova, A., Maurel, D., Vol, C., Trojanova, J., Prézeau, L., Pin, J.P., Blahos, J. (2005) Evidence for a single heptahelical domain being turned on upon activation of a dimeric GPCR. *EMBO J*. 24, 499–509.

54. Salom, D., Lodowski, D.T., Stenkamp, R.E., Le Trong, I., Golczak, M., Jastrzebska, B., Harris, T., Ballesteros, J.A., Palczewski, K. (2006) Crystal structure of a photo-activated deprotonated intermediate of rhodopsin. *Proc Natl Acad Sci U S A*. 103, 16123–16128.

55. Liang, Y., Fotiadis, D., Filipek, S., Saperstein, D.A., Palczewski, K., Engel, A. (2003) Organization of the G protein-coupled receptors rhodopsin and opsin in native membranes. *J Biol Chem*. 278, 21655–21662.

56. Sokolov, M., Lyubarsky, A.L., Strissel, K.J., Savchenko, A.B., Govardovskii, V.I., Pugh Jr., E.N., Arshavsky, V.Y. (2002) Massive light-driven translocation of transducin between the two major compartments of rod cells: A novel mechanism of light adaptation. *Neuron.* 34, 95–106.

57. Chabre, M., le Maire, M. (2005) Monomeric G-protein-coupled receptor as a functional unit. *Biochemistry.* 44, 9395–9403.

58. Ernst, O.P., Gramse, V., Kolbe, M., Hofmann, K.P., Heck, M. (2007) Monomeric G protein-coupled receptor rhodopsin in solution activates its G protein transducin at the diffusion limit. *Proc Natl Acad Sci U S A.* 104, 10859–10864.

59. Bayburt, T.H., Leitz, A.J., Xie, G., Oprian, D.D., Sligar, S.G. (2007) Transducin activation by nanoscale lipid bilayers containing one and two rhodopsins. *J Biol Chem.* 282, 14875–14881.

60. Whorton, M.R., Jastrzebska, B., Park, P.S., Fotiadis, D., Engel, A., Palczewski, K., Sunahara, R.K. (2008) Efficient coupling of transducin to monomeric rhodopsin in a phospholipid bilayer. *J Biol Chem.* 283, 4387–4394.

61. Whorton, M.R., Bokoch, M.P., Rasmussen, S.G., Huang, B., Zare, R.N., Kobilka, B., Sunahara, R.K. (2007) A monomeric G protein-coupled receptor isolated in a high-density lipoprotein particle efficiently activates its G protein. *Proc Natl Acad Sci U S A.* 104, 7682–7687.

62. Freissmuth, M., Selzer, E., Schütz, W. (1991) Interactions of the purified bovine brain A_1-adenosine receptor with G-proteins: Reciprocal modulation of agonist and antagonist binding. *Biochem J.* 276, 651–656.

63. Edrington, T.C., Bennett, M.P., Albert, A.D. (2008) Calorimetric studies of bovine rod outer segment disk membranes support a monomeric unit for both rhodopsin and opsin. *Biophys J.* 95, 2859–2866.

64. Boesze-Battaglia, K., Hennessey, T., Albert, A.D. (1989) Cholesterol heterogeneity in bovine rod outer segment disk membranes. *J Biol Chem.* 264, 8151–8155.

65. Periole, X., Huber, T., Marrink, S.J., Sakmar, T.P. (2007) G protein-coupled receptors self-assemble in dynamics simulations of model bilayers. *J Am Chem Soc.* 129, 10126–10132.

66. Botelho, A.V., Huber, T., Sakmar, T.P., Brown, M.F. (2006) Curvature and hydrophobic forces drive oligomerization and modulate activity of rhodopsin in membranes. *Biophys J.* 91, 4464–4477.

67. Canals, M., Marcellino, D., Fanelli, F., Ciruela, F., de Benedetti, P., Goldberg, S.R., Neve, K., Fuxe, K., Agnati, L.F., Woods, A.S., Ferré, S., Lluis, C., Bouvier, M., Franco, R. (2003) Adenosine A_{2A} dopamine D_2 receptor-receptor heteromerization: Qualitative and quantitative assessment by fluorescence and bioluminescence energy transfer. *J Biol Chem.* 278, 46741–46749.

68. Charalambous, C., Gsandtner, I., Keuerleber, S., Milan-Lobo, L., Kudlacek, O., Freissmuth, M., Zezula, J. (2008) Restricted collision coupling of the A_{2A} receptor revisited: Evidence for physical separation of two signaling cascades. *J Biol Chem.* 283, 9276–9288.

69. Fuxe, K., Ferré, S., Canals, M., Torvinen, M., Terasmaa, A., Marcellino, D., Goldberg, S.R., Staines, W., Jacobsen, K.X., Lluis, C., Woods, A.S., Agnati, L.F., Franco, R. (2005) Adenosine A_{2A} and dopamine D_2 heteromeric receptor complexes and their function. *J Mol Neurosci.* 26, 209–220.

70. Kudlacek, O., Just, H., Korkhov, V.M., Vartian, N., Klinger, M., Pankevych, H., Yang, Q., Nanoff, C., Freissmuth, M., Boehm, S. (2003) The human D_2 dopamine receptor synergizes with the A2A adenosine receptor to stimulate adenylyl cyclase in PC12 cells. *Neuropsychopharmacology*. 28, 1317–1327.
71. Urizar, E., Montanelli, L., Loy, T., Bonomi, M., Swillens, S., Gales, C., Bouvier, M., Smits, G., Vassart, G., Costagliola, S. (2005) Glycoprotein hormone receptors: Link between receptor homodimerization and negative cooperativity. *EMBO J*. 24, 1954–1964.
72. Waldhoer, M., Fong, J., Jones, R.M., Lunzer, M.M., Sharma, S.K., Kostenis, E., Portoghese, P.S., Whistler, J.L. (2005) A heterodimer-selective agonist shows in vivo relevance of G protein-coupled receptor dimers. *Proc Natl Acad Sci U S A*. 102, 9050–9055.
73. May, L.T., Leach, K., Sexton, P.M., Christopoulos, A. (2007) Allosteric modulation of G protein-coupled receptors. *Annu Rev Pharmacol Toxicol*. 47, 1–51.
74. Leach, K., Sexton, P.M., Christopoulos, A. (2007) Allosteric GPCR modulators: Taking advantage of permissive receptor pharmacology. *Trends Pharmacol Sci*. 28, 382–389.
75. Springael, J.Y., Urizar, E., Costagliola, S., Vassart, G., Parmentier, M. (2007) Allosteric properties of G protein-coupled receptor oligomers. *Pharmacol Ther*. 115, 410–418.
76. Bockaert, J., Fagni, L., Dumuis, A., Marin, P. (2004) GPCR interacting proteins (GIP). *Pharmacol Ther*. 103, 203–221.
77. Suzuki, K., Ritchie, K., Kajikawa, E., Fujiwara, T., Kusumi, A. (2005) Rapid hop diffusion of a G-protein-coupled receptor in the plasma membrane as revealed by single-molecule techniques. *Biophys J*. 88, 3659–3680.
78. Liu, A.P., Fletcher, D.A. (2006) Actin polymerization serves as a membrane domain switch in model lipid bilayers. *Biophys J*. 91, 4064–4070.
79. Chichili, G.R., Rodgers, W. (2007) Clustering of membrane raft proteins by the actin cytoskeleton. *J Biol Chem*. 282, 36682–36691.
80. Tsunoda, S., Sierralta, J., Sun, Y., Bodner, R., Suzuki, E., Becker, A., Socolich, M., Zuker, C.S. (1997) A multivalent PDZ-domain protein assembles signalling complexes in a G-protein-coupled cascade. *Nature*. 388, 243–249.
81. Wang, T., Montell, C. (2007) Phototransduction and retinal degeneration in *Drosophila*. *Pflugers Arch*. 454, 821–847.
82. Singer, S.J., Nicolson, G.L. (1972) The fluid mosaic model of the structure of cell membranes. *Science*. 175, 720–731.
83. Simons, K., Ikonen, E. (1997) Functional rafts in cell membranes. *Nature*. 387, 569–572.
84. Parton, R.G., Richards, A.A. (2003) Lipid rafts and caveolae as portals for endocytosis: New insights and common mechanisms. *Traffic*. 4, 724–738.
85. Chini, B., Parenti, M. (2004) G-protein coupled receptors in lipid rafts and caveolae: How, when and why do they go there? *J Mol Endocrinol*. 32, 325–338.
86. Insel, P.A., Head, B.P., Patel, H.H., Roth, D.M., Bundey, R.A., Swaney, J.S. (2005) Compartmentation of G-protein-coupled receptors and their signalling components in lipid rafts and caveolae. *Biochem Soc Trans*. 33, 1131–1134.
87. Cerione, R.A., Codina, J., Benovic, J.L., Lefkowitz, R.J., Birnbaumer, L., Caron, M.G. (1984) The mammalian β_2-adrenergic receptor: Reconstitution of functional

interactions between pure receptor and pure stimulatory nucleotide binding protein of the adenylate cyclase system. *Biochemistry*. 23, 4519–4525.

88. May, D.C., Ross, E.M., Gilman, A.G., Smigel, M.D. (1985) Reconstitution of catecholamine-stimulated adenylate cyclase activity using three purified proteins. *J Biol Chem*. 260, 15829–15833.
89. Freissmuth, M., Selzer, E., Marullo, S., Schütz, W., Strosberg, A.D. (1991) Expression of two human β-adrenergic receptors in *Escherichia coli*: Functional interaction with two forms of the stimulatory G protein. *Proc Natl Acad Sci U S A*. 88, 8548–8552.
90. Bertin, B., Freissmuth, M., Breyer, R., Schütz, W., Strosberg, A.D., Marullo, S. (1992) Functional expression of the human 5-HT1a receptor in *Escherichia coli*: Ligand binding properties and interaction with recombinant G protein α-subunits. *J Biol Chem*. 267, 8200–8206.
91. Jockers, R., Linder, M., Hohenegger, M., Nanoff, C., Bertin, B., Strosberg, A.D., Marullo, S., Freissmuth, M. (1994) Species differences in the G protein selectivity of the human and bovine A_1-adenosine receptor. *J Biol Chem*. 269, 32077–32084.
92. Munro, S. (2003) Lipid rafts: Elusive or illusive? *Cell*. 115, 377–388.
93. Heerklotz, H. (2002) Triton promotes domain formation in lipid raft mixtures. *Biophys J*. 83, 2693–2701.
94. Pitman, M.C., Grossfield, A., Suits, F., Feller, S.E. (2005) Role of cholesterol and polyunsaturated chains in lipid-protein interactions: Molecular dynamics simulation of rhodopsin in a realistic membrane environment. *J Am Chem Soc*. 127, 4576–4577.
95. Grossfield, A., Feller, S.E., Pitman, M.C. (2006) A role for direct interactions in the modulation of rhodopsin by omega-3 polyunsaturated lipids. *Proc Natl Acad Sci U S A*. 103, 4888–4893.
96. Niu, S.L., Mitchell, D.C., Litman, B.J. (2002) Manipulation of cholesterol levels in rod disk membranes by methyl-β-cyclodextrin: Effects on receptor activation. *J Biol Chem*. 277, 20139–20145.
97. Niu, S.L., Mitchell, D.C., Litman, B.J. (2001) Optimization of receptor-G protein coupling by bilayer lipid composition II: Formation of metarhodopsin II-transducin complex. *J Biol Chem*. 276, 42807–42811.
98. Albert, A.D., Boesze-Battaglia, K. (2005) The role of cholesterol in rod outer segment membranes. *Prog Lipid Res*. 44, 99–124.
99. Ruprecht, J.J., Mielke, T., Vogel, R., Villa, C., Schertler, G.F. (2004) Electron crystallography reveals the structure of metarhodopsin I. *EMBO J*. 23, 3609–3620.
100. Hanson, M.A., Cherezov, V., Griffith, M.T., Roth, C.B., Jaakola, V.P., Chien, E.Y., Velasquez, J., Kuhn, P., Stevens, R.C. (2008) A specific cholesterol binding site is established by the 2.8 Å structure of the human β_2-adrenergic receptor. *Structure*. 16, 897–905.
101. Turcotte, M., Tang, W., Ross, E.M. (2008) Coordinate regulation of G protein signaling via dynamic interactions of receptor and GAP. *PLoS Comput Biol*. 4, e1000148.

CHAPTER 5

Molecular Pharmacology of Frizzleds—with Implications for Possible Therapy

GUNNAR SCHULTE

Section of Receptor Biology and Signaling, Department of Physiology and Pharmacology, Karolinska Institutet, Stockholm, Sweden

5.1. INTRODUCTION

WNTs, a family of secreted lipoglycoproteins, originally identified as mammary oncogenes, bind and activate their receptors, the Frizzleds. This review will present WNT–Frizzled interaction, Frizzled pharmacology, and signaling pathways—both basic paradigms and novel developments. The aim is to summarize the current status of our knowledge and point at apparent gaps. The involvement of Frizzleds in embryogenesis, adult physiology, and some diseases, renders these receptors a suitable target for drug development and therapy. The increasing understanding of molecular aspects of the WNT/Frizzled signaling system will support the discovery of drug targets as well as the advancement of therapeutic applications targeting FZD signal transduction. Currently, promising lines of therapy are evolving, for example, cancer therapy, treatment of cardiovascular and inflammatory diseases, and stem cell-based replacement therapy for neurodegenerative conditions such as Parkinson's disease.

5.2. FRIZZLEDS AS WNT RECEPTORS

5.2.1. Frizzleds—The Discovery

About 20 years ago, the product of *int-1*, a viral mammary oncogene, was observed to resemble that of the *Drosophila melanogaster* segment polarity

GPCR Molecular Pharmacology and Drug Targeting: Shifting Paradigms and New Directions, Edited by Annette Gilchrist

gene *wingless* [1]. This led to the discovery of the WNT (acronym of wingless and int-1) family of lipoglycoproteins [2, 3]. In parallel, the gene product of the *frizzled* locus in *Drosophila melanogaster* was determined to code for a protein with seven potential transmembrane domains and a rather large N-terminus, possibly a G protein-coupled receptor (GPCR) [4]. A mutation in this gene leads to disturbed cuticular patterns of bristles that lend the fly a somewhat "frizzled" appearance, suggesting that *frizzled* and *wingless* play related roles in regulation of cell polarity (apparent effects on cuticular cell polarity have also been reported in a FZD_6 knockout mouse [5]). It was first surmised [6] and then shown that *Drosophila* Frizzled2 functions as Wingless receptor, which clearly implied that the Frizzled family of seven transmembrane spanning proteins are WNT receptors [7, 8].

5.2.2. The Frizzled Family

The Frizzled family of receptors [for recent reviews, see References 9, 10] includes 10 mammalian isoforms. Along with the related receptor Smoothened (SMO) [11], Frizzleds were recently classified by the International Union of Basic and Clinical Pharmacology (IUPHAR) as a novel family of GPCRs [12] and taken up into the IUPHAR's database of GPCRs and list of 7TM receptors (see also www.iuphar.org). The recommended nomenclature for the mammalian Frizzleds 1–10 is FZD_{1-10}.

Based on homology studies, FZDs were originally described as secretin receptor-like proteins [13] but more recently have been grouped together with bitter taste2 receptors [14]. The latter analysis indicated that human FZDs share from 20% to 40% identity, with closer identity in four different clusters, notably $FZD_{1,2,7}$ (75% identity), $FZD_{5,8}$ (70% identity), $FZD_{4,9,10}$, and $FZD_{3,6}$ (50% identity). One of the most conserved regions between all the FZDs is the N-terminal cysteine-rich domain (CRD) linked to the receptor core by a highly divergent linker ranging from about 40–100 aa. Further, the receptors contain 7 hydrophobic stretches of 20–25 aa resembling the transmembrane helices I–VII as well as a short C-terminus in the range of 25–200 aa [8]. The C-terminus contains two domains important for binding of the PDZ (*P*SD-95/*d*isc large/*Z*O-1 homologous) ligand (see Fig. 5.1): (1) a highly conserved KTxxxW sequence, which is necessary for FZD signal transduction, serves a docking site for the PDZ domain of the central phosphoprotein Dishevelled (DVL) and (2) a rather well-conserved terminal PDZ ligand domain that can interact with many intracellular proteins, for example, postsynaptic density proteins [9].

Other structural features of classical GPCRs, which are implicated in receptor function and G protein coupling are the presence of charged amino acids, such as arginine residues both on the C- and N-terminal side of the third intracellular loop are present in FZDs. Further, several FZDs ($FZD_{3,4,5,6,8,9,10}$) contain cysteine residues, putative targets for palmytoylation, in the C-terminus, which are required for membrane anchorage and formation

human FZD_1:
Intracellular loop 1
FZD_1 – 344-DMRRFSYPERP-354
Intracellular loop 2
FZD_1 – 424-SLTWFLAAGMKWGHEAIEANSQ-445
Intracellular loop 3
FZD_1 – 511-VSLFRIRTIMKHDGTKTEKLEKLMVR-536
C-terminus
FZD_1 – 623-SG**KTLNSW**RKFYTRLTNSKQGETTV-647

human FZD_2:
Intracellular loop 1
FZD_2 – 269-DMQRFRYPERP-279
Intracellular loop 2
FZD_2 – 349-SLTWFLAAGMKWGHEAIEANSQ-370
Intracellular loop 3
FZD_2 – 436-VSLFRIRTIMKHDGTKTEKLERLMVR-461
C-terminus
FZD_2-541-SG**KTLHSW**RKFYTRLTNSRHGETTV -565

human FZD_3:
Intracellular loop 1
FZD_3 – 227-DVTRFRYPERP-237
Intracellular loop 2
FZD_3 – 310-TWFLAAVPKWGSEAIEKKA-328
Intracellular loop 3
FZD_3 – 396-SLNRVRIEIPLEKENQDKLVKFMIR-420
C-terminus
FZD_3-499-GSK**KTCFEW**ASFFHGRRKKEIVNESRQVL
QEPDFAQSLLRDPNTPIIRKSRGTSTQGTSTHASSTQ
LAMVDDQRSKAGSIHSKVSSYHGSLHRSRDGRYTPC
SYRGMEERLPHGSMSRLTDHSRHSSSHRLNEQSR
HSSIRDLSNNPMTHITHGTSMNRVIEEDGTSA-666

human FZD_4:
Intracellular loop 1
FZD_4 – 244-DSSRFSYPERP-254
Intracellular loop 2
FZD_4 – 324-TLTWFLAAGLKWGHEAIEMHS-344
Intracellular loop 3
FZD_4 – 411-VALFKIRSNLQKDGTKTDKLERLMVK-436
C-terminus
FZD_4-499-**KTLHTW**QKCSNRLVNSGKVKREKRGNGW
VKPGKGSETVV -537

human FZD_5:
Intracellular loop 1
FZD_5 – 260-DMERFRYPERP-270
Intracellular loop 2
FZD_5 – 337-SLTWFLAAGMKWGNEAIAGYAQ-358
Intracellular loop 3
FZD_5 – 424-VSLFRIRSVIKQGGTKTDKLEKLMIR-449
C-terminus
FZD_5-522-WSG**KTVESW**RRFTSRCCCRPRRGHKSG
GAMAAGDYPEASAALTGRTGPPGPAATYHKQVS
LSHV-585

human FZD_6:
Intracellular loop 1
FZD_6 – 223-DVRRFRYPERP-233
Intracellular loop 2
FZD_6 – 306-TWFLAAGRKWSCEAIEQKA-324
Intracellular loop 3
FZD_6 – 392-SLNHVRQVIQHDGRNQEKLKKFMIR-416
C-terminus
FZD_6-465-GSK**KTCTEW**AGFFKRNRKRDPISESRRVLQ
ESCEFFLKHNSKVKHKKKHYKPSSHKLKVISKSMGTST
GATANHGTSAVAITSHDYLGQETLTEIQTSPETSMREV
KADGASTPRLREQDCGEPASPAASISRLSGEQVDGKG
QAGSVSESARSEGRISPKSDITDTGLAQSNNLQVPSSS
EPSSLKGSTSLLVHPVSGVRKEQGGGCHSDT-706

human FZD_7:
Intracellular loop 1
FZD_7 – 278-DMRRFSYPERP-288
Intracellular loop 2
FZD_7 – 358-SLTWFLAAGMKWGHEAIEANSQ-379
Intracellular loop 3
FZD_7 – 445-VSLFRIRTIMKHDGTKTEKLEKLMVR-470
C-terminus
FZD_7-550-SG**KTLQSW**RRFYHRLSHSSKGETAV-574

human FZD_8:
Intracellular loop 1
FZD_8 – 297-STFLIDMERFKYPERP-312
Intracellular loop 2
FZD_8 – 418-SLTWFLAAGMKWGNEAIAGYSQY-439
Intracellular loop 3
FZD_8 – 505-VSLFRIRSVIKQQDGPTKTHKLEKLMIR-532
C-terminus
FZD_8-606-SG**KTLESW**RSLCTRCCWASKGAAVGGGAG
ATAAGGGGGPGGGGGGGPGGGGGPGGGGGSLYSDV
STGLTWRSGTASSVSYPKQMPLSQV-694

human FZD_9:
Intracellular loop 1
FZD_9 – 251-LTFLLEPHRFQYPERP-366
Intracellular loop 2
FZD_9 – 337-TWFLAAGKKWGHEAIEAHG-355
Intracellular loop 3
FZD_9 –422-VALFHIRKIMKTGGTNTEKLEKLMVK-447
C-terminus
FZD_9-530-SS**KTFQTW**QSLCYRKIAAGRARAKACRAPG
SYGRGTHCHYKAPTVVLHMTKTDPSLENPTHL -591

human FZD_{10}:
Intracellular loop 1
FZD_{10} – 247-LTFLIDPARFRYPERP-262
Intracellular loop 2
FZD_{10}– 333-TWFLAAGKKWGHEAIEANS-351
Intracellular loop 3
FZD_{10} – 415-SGFVALFHIRRVMKTGGENTDKLEKLMVR-443
C-terminus
FZD_{10}-524-TS**KTLQSW**QQVCSRRLKKKSRRKPASVITSGG
IYKKAQHPQKTHHGKYEIPAQSPTCV-581

Figure 5.1 Amino acid sequence of the intracellular domains of human FZD_{1-10}. SwissProt accession numbers: FZD_1: Q9UP38; FZD_2: Q14332; FZD_3; Q9NPG1; FZD_4: Q9ULV1; FZD_5: Q13467; FZD_6: O60353; FZD_7: O75084; FZD_8: Q9H461; FZD_9: O00144; FZD_{10}: Q9ULW2. Putative phosphorylation sites are marked as underlined letters in the amino acid sequences (determined with the MiniMotif Miner software). The internal KTxxxW-DVL binding sequence is highlighted in bold, and the terminal PDZ ligand domains are marked in light gray.

of a putative helix 8 between the helix VII and the cysteine [15]. Palmitoylation of GPCR at their C terminus was shown to be important for G protein coupling, receptor phosphorylation, and agonist-induced desensitization and downregulation [16].

In summary, the general architecture of the proteins indeed suggests that the FZDs belong to the superfamily of GPCRs [9, 17] even though other well-conserved GPCR features are lacking or are less well conserved such as the DRY motif at the C-terminus of the third intracellular loop.

5.2.3. Frizzled Ligands

As mentioned above, WNTs were well-known before FZDs were identified as their receptors. With the establishment of the WNT receptor concept, the development of tools such as recombinant FZD-CRDs, and the discovery of the endogenous soluble FZD-like proteins (sFRPs), led to identification of a series of FZD-interacting proteins that were capable of regulating FZD activity and WNT-mediated effects [9, 18]. Most strikingly, R-spondin [19] and Norrin [20] were identified as selective FZD ligands interacting with the CRD and mediating intracellular signaling.

5.2.4. WNT-Frizzled Interactions

In mammals, 10 FZDs and 19 WNTs have been identified. This allows a large number of putative ligand–receptor pairs and raises the question of what structural determinants govern WNT/FZD selectivity [21]. In *Drosophila melanogaster*, 4 FZDs (Fz, DFz2, 3, 4) and 5 WNTs (wingless, Wg; DWNT2, 3, 4, 8) were described, and their interaction profiles have been determined [22] and suggest a rather specific pattern of FZD–WNT interactions in higher organisms. So far, very few WNTs have been purified and extensively tested in their active form [23, 24]. Because of this, hardly any quantitative pharmacological information on affinity constants, binding kinetics or dose–effect relationships is available [22, 25, 26]. In the cited studies, affinities of WNTs to FZD-CRDs were found to be in the lower nanomolar range both in cell-free and cell-based assays. For example, the interaction between *Xenopus* WNT-8 and mouse FZD_8-CRD showed a dissociation constant $K_d = 8\,nM$ [25]. When discussing the role of the CRD for FZD binding to WNTs, it is important to mention results indicating that the CRD might actually not be required for WNT signal transmission [27]. Thus, CRD binding of WNTs could serve merely as a means of recruiting WNTs to the receptor, whereas WNT interaction with parts of the FZD extracellular loops and core region is necessary for signal transduction [27, 28] and establishment of a high-affinity ligand–receptor complex. Another level of complexity was added by a study reporting the dimerization of FZD even in the absence of the CRD [29].

The availability of purified and active WNTs galvanized the research field: these tools allowed more controlled stimulation of WNT pathways as well as a more quantitative analysis of dose–effect relationships [30, 31]. This is important because the broad expression of FZDs in various cell types renders a pharmacological and cell-based analysis of single WNT–FZD pairs difficult. In fact, we have not been able to identify a suitable mammalian cellular system without endogenous FZD expression. On the contrary, many cell types express multiple isoforms as well as co-receptors (unpublished observations).

5.2.5. Intracellular Posttranslational Modifications

The intracellular domains, that is, the intracellular loop $i_{1\text{-}3}$ as well as the C-terminus of FZDs, are putative sites for both protein–protein interaction [9] and posttranslational modifications (see Table 5.1; Figs. 5.1 and 5.2). Even though all FZDs contain plenty of putative phosphorylation sites—mainly for serine/threonine kinases but also some tyrosine kinases (as determined by the MiniMotif Miner software [32])—only one study actually reports DVL-dependent phosphorylation of $XFZD_3$, at serine 576 [33]. Phosphorylation at this site and other unidentified sites in the $XFZD_3$ C-terminus play a role in downregulation of $XFZD_3$ signaling. Analysis of the C-terminal sequence of $XFZD_3$ with the MiniMotif Miner suggests a casein kinase 1 site at serine 576 that requires a phosphorylated serine in position 573. Ser573 resembles a 90 kDa ribosomal S6 kinase (RSK)-phosphorylation site motif.

As shown in Table 5.1, all human FZDs are predicted to have many putative phosphorylation sites, which suggests a high degree of modulation through possible WNT-induced feedback, for example, through CK1/2 (casein kinase 1 and 2), GSK3 (glycogen synthase kinase 3), CamKII (Ca^{2+}- and calmodulin-dependent kinase II), and PKC (Ca^{2+}-dependent protein kinase). In addition, there may also be cross-talk with other signaling pathways, such as various kinases involved in cell cycle regulation, for example, polo-like kinase (PLK), p70S6 kinase, p90 S6 ribosomal kinase (RSK), ataxia telangiectasia mutated (ATM) kinase, cyclic AMP-dependent protein kinase (PKA), phosphorylase kinase, EGF (epidermal growth factor) receptor, and Janus kinase (JAK) signaling. Prediction of the putative phosphorylation sites is based on published consensus sites for kinases. The list remains to be confirmed, but it is far from complete and leaves out G protein-coupled receptor kinases (GRKs), which are known to regulate GPCR phosphorylation, thus altering their dynamics and activity [34].

Of special interest are the putative phosphorylation sites that appear frequently not only in the DVL-binding sequence KTxxxW, but also those in the terminal PDZ ligand domain. Phosphorylation in either of these regions might have an effect on protein–protein interaction between FZD and DVL, or between FZD and other proteins. Thus, it will be exciting to watch future developments as we learn more about the roles of various kinases, FZD phosphorylation sites, and possible cross-talk for WNT-induced receptor signaling, dynamics, and desensitization [35].

TABLE 5.1 Summary of Kinases (according to MiniMotif Miner) with Putative Phosphorylation Sites in the Intracellular Domains of FZD_{1-10}

Human FZDs	PKA	PKC	PKG	CK1	CK2	CamKII	RSK	p70 S6 kinase	ERK 1/2	ATM kinase
FZD_1_i1	√		√		√	√	√			
FZD_1_i2										√
FZD_1_i3		√			√	√	√ (×2)			
FZD_1_c-term		√ (×2)								
FZD_2_i1										
FZD_2_i2										√
FZD_2_i3		√			√	√	√			
FZD_2_c-term	√	√ (×2)		√			√			
FZD_3_i1										
FZD_3_i2							√			
FZD_3_i3										
FZD_3_c-term	√ (×6)	√ (×4)		√ (×7)	√ (×2)	√ (×5)	√ (×6)	√	√	√ (×2)
FZD_4_i1	√	√		√	√					
FZD_4_i2										
FZD_4_i3		√			√		√ (×2)			
FZD_4_c-term		√ (×2)								
FZD_5_i1										
FZD_5_i2										
FZD_5_i3		√			√	√	√			
FZD_5_c-term	√ (×2)	√ (×4)	√	√		√ (×2)	√ (×3)			
FZD_6_i1										
FZD_6_i2	√		√			√	√			
FZD_6_i3										
FZD_6_c-term	√ (×2)	√ (×8)		√ (×8)	√ (×7)	√	√ (×5)		√	
FZD_7_i1	√ (×2)		√		√	√	√			
FZD_7_i2										√
FZD_7_i3		√			√	√	√ (×2)			
FZD_7_c-term	√	√ (×3)		√ (×2)	√		√			
FZD_8_i1										
FZD_8_i2										√
FZD_8_i3		√				√	√ (×2)			
FZD_8_c-term		√ (×3)		√ (×6)		√	√			√
FZD_9_i1										
FZD_9_i2										
FZD_9_i3		√			√					
FZD_9_c-term	√	√		√	√		√			

JAK2	GSK-3alpha	GSK-3beta	PLK	CDK	Phosphorylase kinase	ABL	EGFR	TPK-IIB/p38Syk	CSK
√			√						
√			√						
	√ (×11)	√ (×3)		√	√ (×2)				
	√								
			√		√				
	√ (×7)		√				√		
√									
√	√				√				
√			√						

TABLE 5.1 (*Continued*)

Human FZDs	PKA	PKC	PKG	CK1	CK2	CamKII	RSK	p70 S6 kinase	ERK 1/2	ATM kinase
FZD_{10}_i1										
FZD_{10}_i2										
FZD_{10}_i3		√			√					
FZD_{10}_ c-term	√	√ (×3)	√	√			√ (×2)			

(i1: intracellular loop 1; i2: intracellular loop 2; i3: intracellular loop 3; c-term: C-terminus). √ indicates a single, whereas numbers in parentheses indicate multiple phosphorylation sites. Kinases: PKA, cyclic AMP-dependent protein kinase; PKC, Ca^{2+}-dependent protein kinase; PKG, cyclic GMP-dependent protein kinase; CK1, casein kinase 1; CK2, casein kinase 2; CamKII, Ca^{2+}- and calmodulin-dependent protein kinase; RSK, 90 kDa ribosomal S6 kinases; ERK1/2, extracellular signal-regulated protein kinase1/2; ATM kinase, Ataxia telangiectasia-mutated kinase; JAK2, Janus kinase 2 (autophosphorylation site); GSK3α/β, glycogen synthase kinase 3α/β; CDK, cyclin-dependent kinases; ABL, product of the c-abl oncogene; EGFR, epidermal growth factor receptor; tyrosine protein kinase TPK-IIB/p38Syk; CSK, carboxy-terminal Src kinase.

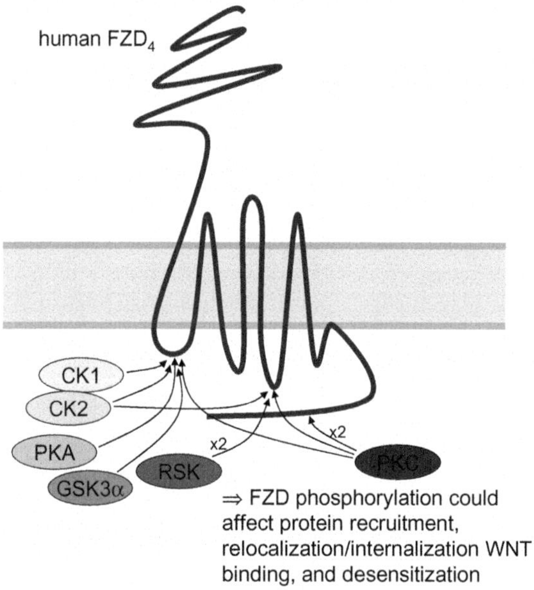

Figure 5.2 Schematic view of human FZD_4 containing putative phosphorylation sites on intracellular domains. For abbreviations, see Table 5.1.

5.3. FRIZZLED SIGNALING

Signaling through FZDs is mainly subdivided into β-catenin-dependent [36] and β-catenin-independent pathways [37]. These branches are also known as canonical and noncanonical WNT pathways, respectively. However, the terminology refers to the historical chronology of the discovery of the pathways, rather than indicating that β-catenin is the WNT-default pathway. Indeed, β-catenin signaling can be activated by other GPCRs unrelated to FZD (e.g.,

JAK2	GSK-3alpha	GSK-3beta	PLK	CDK	Phosphorylase kinase	ABL	EGFR	TPK-IIB/ p38Syk	CSK
	√				√	√		√	√

TABLE 5.2 Summary of β-Catenin-Independent FZD Signaling

Signaling Path	Cellular System	Activating WNT	Reference
WNT/cGMP/Ca^{2+}	Mouse F9 teratocarcinoma	WNT-5A	[75]
WNT/IPs	Mouse F9 teratocarcinoma	WNT-3A (IP_5)	[116]
	Osteoblasts (murine ST2)	WNT-3A	[117]
WNT/RAC-1	Xenopus HEK293	WNT-1	[118]
WNT/RHO	Xenopus HEK293	WNT-1	[118]
WNT/ROR2/CDC42	Xenopus	WNT-5A	[41]
WNT/RAP1	Xenopus HEK293	WNT-8	[119]
WNT/PKA/CREB	mouse presomitic mesoderm cells	WNT-1; WNT-3A; WNT-7A	[120]
WNT/mTOR	Various mammalian cells	WNT-; WNT-3; WNT-10B	[121]
WNT/RYK	Drosophila 293T cells	WNT-1; WNT-3A	[122, 123]

Reference 38). Therefore, I will use the nomenclature β-catenin-dependent and β-catenin-independent pathways. The β-catenin-independent pathways are highly diverse [37] and are named according to the specific molecules involved in signaling (see Table 5.2). This chapter will not go into detail with planar cell polarity (PCP) signaling, referring to WNT signaling responsible for the uniform polarization and orientation of epithelial cells along the apical-basal axis as typically investigated in *Drosophila* [39]. In vertebrates, the lengthening and narrowing of a tissue is referred to as convergent extension

regulating movement of polarized cells during embryonic development. PCP signaling and regulation of convergent extension movements are governed by PCP core proteins and multiple signaling pathways including classical players such as DVL, as well as RHO-like GTPases and JNK [39–41].

5.3.1. β-Catenin-Dependent Signaling

WNTs were originally subdivided into WNT-1-like and WNT-5A-like groups on the basis of differential transforming potential in a mammary tumor cell line. WNT-1-like WNTs, such as WNT-3A, mediate β-catenin-dependent signaling. WNT-3A, one of the purified and commercially available WNTs, generally activates β-catenin signaling in various mammalian cell types. In *Xenopus* models, β-catenin signaling can be studied by the anterior-posterior axis duplication assay [42]. On the molecular level, WNTs interact with both FZDs and their co-receptors low-density lipoprotein receptor-related protein (LRP5/6) to promote signaling that results in inhibition of a constitutively active destruction complex, which by GSK3-dependent phosphorylation, primes β-catenin to degradation in the proteasome [36]. The composition of the destruction complex is not yet completely understood, but the most important players are GSK3, axin, and APC (adenomatous polyposis coli). Upstream of this inhibition of the degradation complex, WNT-stimulation activates the central phosphoprotein DVL, which is dependent on casein kinases [31, 43] and the scaffold protein β-arrestin [44]. DVL activity is regulated by phosphorylation, subcellular localization, and degradation, and DVL is a central relay station of many—if not all—WNT/FZD signaling pathways [45]. DVL activation status can be monitored by an electrophoretic mobility shift, which is dependent on DVL phosphorylation. The shifted band on an immunoblot is therefore often referred to as phosphorylated and shifted DVL (PS-DVL; [31, 46]). The inhibition of the degradation complex is apparently accomplished through a collaborative communication through WNT-bound FZDs and LRPs [47]. Interestingly, there is evidence that WNT-3A-induced β-catenin signaling is composed of a rapid, low-dose and a slower, high-dose pathway, mediated through cooperation between LRPs and FZDs [31] where the rapid, LRP-dependent pathway is independent of PS-DVL formation.

Even though this WNT signaling route is being characterized in ever greater detail, important gaps need to be filled. For example, it is still unclear how the WNT-induced FZD/LRP complex communicates, or when it is required; which LRP kinases other than GSK3 are involved and how they are recruited [47, 48]. Moreover, the regulatory role of DVL for LRP phosphorylation and the inhibition of GSK3 still remain obscure.

5.3.2. β-Catenin-Independent Signaling

Historically, any signaling pathway activated by WNTs that did not lead to the stabilization of β-catenin was coined "noncanonical". Initially, that would only have included what is known in *Drosophila* as the PCP pathway and WNT-induced Ca^{2+} signaling. Recent development in the field of WNT signaling shows that β-catenin-independent signaling actually comprises a vast array of signaling pathways, indicating that the term "noncanonical" is a gross oversimplification [37]. Thus, it is more informative to distinguish the individual signaling branches by referring to the signaling components involved as identified by the original studies (see Table 5.2). In general β-catenin independent signaling regulates major reorganizations during embryonic development, such as tissue polarization and convergent extension movements [40]. Again, the *Xenopus laevis* embryo is a suitable model organism for the study of β-catenin-independent WNT signaling using, for example, Keller explants to investigate tissue elongation and constriction as a measure of convergent extension. In addition, polarized tissues exist not only in the fly but obviously also in vertebrate organisms, such as frog, fish, or mammals [37, 39]. The molecular signaling routes as summarized in Table 5.2 and in Semenov et al. [37] regulate crucial cellular processes, such as gene transcription, proliferation, differentiation, and cytoskeletal reorganization that rely on cell movements and polarization, to name but a few. Surprisingly, even though many 7 transmembrane spanning receptors have been shown to activate mitogen-activated protein kinases (MAPK) through diverse G protein-dependent and -independent mechanisms [49, 50], only very few reports on WNT-induced regulation of extracellular signal-regulated kinases, a major proliferative pathway, have appeared [51].

β-catenin-independent signaling also appears as a regulator of β-catenin-dependent events, even though the precise mechanisms of this cross-talk are not clear yet [30, 52]. In the original report, WNT-5A was shown to decrease WNT-3A signaling to β-catenin by an increase in the GSK3-dependent β-catenin degradation [52]. Recent evidence implicates a crucial role of the atypical receptor tyrosine kinase ROR2 for the WNT-5A-mediated inhibition of β-catenin signaling, which is of special relevance in cells that do not express ROR2 but respond to both WNT-3A and WNT-5A.

5.3.3. Intracellular Scaffolds (DVL and β-arrestin)

DVL has been mentioned previously several times as a central signaling relay for both β-catenin-dependent and -independent signaling. In addition to the physical interaction with FZDs through the highly conserved internal PDZ ligand domain KTxxxW [53], DVL can interact with many proteins, such as axin, kinases, GTPases, guanine nucleotide exchange factors, and structural proteins [45, 54, 55], justifying the term signaling scaffold. DVL is composed of a DEP, a PDZ and a DIX domain: this molecular structure underlines DVL's versatility as a binding partner for many structurally unrelated proteins [45]. Regarding DVL localization in the cells, it has become clear that several distribution patterns are possible and that they are dynamically interchangeable depending on the degree of DVL expression, FZD expression, DVL kinase activity, and WNT stimulation (Fig. 5.3).

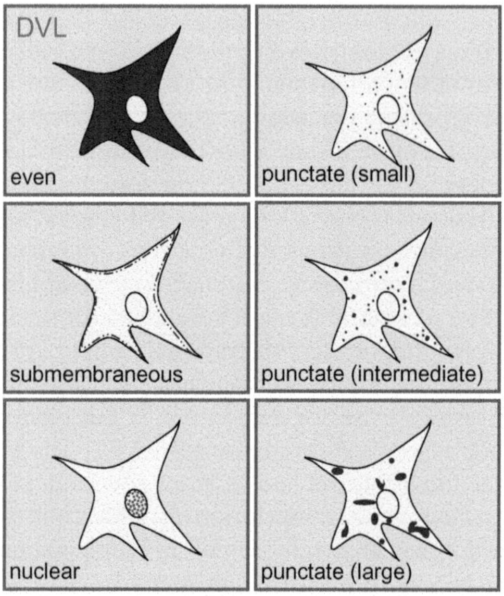

Figure 5.3 Schematic view of the subcellular distribution of dishevelled (DVL). Exogenously expressed DVL in mammalian cells can be observed either evenly distributed throughout the cytosol or in DVL aggregates of various sizes. When coexpressed with FZDs, DVL is recruited to the plasma membrane. In addition, nuclear DVL was reported to mediate WNT signaling. For details, see text.

In an excess of FZD, DVL is recruited to the receptor and shows a submembraneous distribution [56–58]. This pattern can also be achieved by WNT-overexpression in *Drosophila* or *Xenopus* [59, 60]. Exogenously overexpressed DVL commonly leads to a punctate pattern in the cytosol representing dynamic DVL–DVL aggregates of various sizes (e.g., References 30, 61, 62). However, this pattern has not clearly been established yet for endogenously expressed DVL proteins. DVL has also been observed as an evenly distributed cytosolic protein and a recent study [30] relating PS-DVL formation to DVL subcellular distribution showed that under conditions promoting PS-DVL (CK1ε overexpression, WNT stimulation) DVL is predominantly evenly distributed, whereas with conditions promoting the unshifted/unphosphorylated form of DVL, the punctate distribution is preferred. The results suggest that DVL located in punctate represents the inactive form, while the evenly distributed DVL represent the active phosphorylated pool. Additionally, nuclear DVL regulating β-catenin-dependent responses, such as T cell factor (TCF)/lymphoid enhancer factor (LEF)-directed transcription and axis duplication in *Xenopus* embryos was observed under conditions where nuclear export is inhibited [63].

Another scaffolding protein that recently caught attention in the field is β-arrestin, which was originally described as a mediator of receptor desensitization for agonist-bound GPCRs and a signaling scaffold in G protein-independent pathways [64]. Initial reports showed that β-arrestin1 promotes β-catenin-dependent transcriptional activity [65]. Further, β-arrestin is crucial for WNT-5A-induced FZD_4 internalization [66]. However, in contrast to the classical GPCR/β-arrestin model (where β-arrestin interacts with the agonist-bound and phosphorylated receptor), β-arrestin prefers FZD_4-bound DVL and mediates clathrin-dependent endocytosis of the receptor [66]. Thus, β-arrestin interaction with DVL could serve as a negative desensitizing signal, as positive signal-promoting mechanisms, and/or as a means of determining signal specification. Further studies illustrated that β-arrestin interacts and colocalizes with DVL, and that this interaction is a necessary component for WNT-induced β-catenin signaling *in vitro* and *in vivo* [44]. Furthermore, Han and Kim described the importance of β-arrestin for convergent extension (CE) movements in *Xenopus*, implicating a crucial role for the activation of small RHO-like GTPases [67] and that β-arrestin and CK1/2 define distinct branches of β-catenin-independent WNT signaling [68].

The interaction between DVL and β-arrestin has the potential to serve as a pathway specification scaffold, acting as a WNT/FZD signaling platform [69]. Still, it is unclear if β-arrestin, in the context of WNT signaling, is (1) capable of recruiting signaling modules such as other kinases, phosphodiesterases, and E3 ligases [64] to promote cross-talk with other pathways; (2) affects FZD desensitization; or (3) alters G protein coupling.

5.3.4. Evidence for G Protein Coupling of FZDs

There is accumulating evidence that FZDs indeed signal through the recruitment of heterotrimeric G proteins as recently reviewed by Egger-Adam and Katanaev [17]. The strongest evidence is based on genetic analysis in *Drosophila* [70], inhibition of signaling by the blockade of G proteins, and the use of chimeric FZDs [10]. Here, I would like to briefly summarize the facts from the literature in order to show that the evidence for G protein coupling of FZDs is rather indirect, being based on loss- or gain-of-function experiments rather than on a direct proof of a WNT-induced and FZD-mediated activation of guanine nucleotide exchange at a FZD-interacting G protein [9]. The reason we are forced to rely on circumstantial evidence is again the lack of tools such as purified WNTs, lack of information (e.g., pharmacological profiles of WNT–FZD interaction), as well as the peculiar characteristics of FZDs, which possibly represent an evolutionarily more modern form of seven transmembrane spanning receptors. As recently proposed by Egger-Adam and Katanaev [17], FZDs have strong responsibility during embryonic development that might have resulted in an evolutionary transformation of signaling modes adapting them from predominantly G protein-dependent to alternative signaling mechanisms. The striking difference between FZDs and "classical" GPCRs was recently emphasized by the separate classification of FZDs (and the related Smoothened) as a novel family of GPCRs by the International Union of Clinical and Experimental Pharmacology [12].

The first indication that FZDs could be GPCRs was based on structural analysis of the receptor by hydropathy plots revealing the presence of putative 7 transmembrane helices [4]. Further structural analysis indicated that the receptors might differentially couple to G proteins [71] and the first strong experimental indications resulted from experiments in zebrafish embryos overexpressing rat FZD_2 [72]. Coexpression of *Xenopus* WNT-5A induced pertussis toxin-sensitive Ca^{2+} transients, which could also be inhibited by βγ-sequestering transducin expression as well as GDP-β-S injection. The results implied a connection between Frizzleds, βγ release from $G_{i/o}$ family proteins and phospholipase C signaling. Subsequently, several reports showed that inhibition or downregulation of heterotrimeric G proteins affected both β-catenin-independent as well as β-catenin-dependent signaling pathways. Especially, work by Wang and collaborators led to the elucidation of a novel FZD-dependent Ca^{2+} pathway through cGMP-selective phosphodiesterases [73–76]. However, these reports have in common that the conclusions are based on downregulation/blockade of the G proteins rather than on a direct measure of WNT-induced and FZD-mediated activation of G proteins, for example, by a GTPγS binding assay. Thus, one explanatory model could still include the pure requirement of active G proteins for successful WNT signaling, which does not necessarily require direct FZD–G protein interaction. Thus far, the most compelling experimental evidence for a WNT-induced dissociation of a receptor–G protein complex was presented by Endo et al. [77]. They reported a rapid and prolonged WNT-3A-induced rearrangement of a FZD–DVL–G protein complex, which could be the consequence of a WNT-induced and FZD-mediated activation of a heterotrimeric G protein.

In addition, chimeric receptors consisting of FZD-intracellular loops and transmembrane and extracellular domains from adrenergic receptors [76, 78–81], designed with the aim of overcoming the lack of purified WNTs, provided important insight into FZD structure and function relationships as well as some pharmacological aspects. Even though it is difficult to interpret the results obtained with these chimeric receptors, with major parts from classical GPCRs, the experiments represent milestones in the understanding of FZDs as unconventional GPCRs. Most prominently, the agonist affinity shift in the presence or absence of guanine nucleotides, which was obtained in membrane preparation using isoproterenol competition binding to the FZD_1/β_2-adrenergic [78, 79] and FZD_2/β_2-adrenergic [76, 79] chimera, strongly suggests that the intracellular loops of FZDs could directly bind to and activate heterotrimeric G proteins.

5.3.5. Unconventional Signaling Modes

FZDs—similar to conventional GPCRs—undergo WNT-induced internalization and recycling [66, 82], which could be a means of rapid and transient desensitization and internalization-dependent signaling [83]. In the case of WNT signaling, endocytosis was also proposed as a means to establish morphogen gradients in the developing embryo [84]. As discussed recently [85], the role of endocytosis for WNT/FZD signaling remains rather unclear. Important players are again DVL and β-arrestin interacting with the clathrin adaptor protein-2 [86, 87]. Results from our group indicate that inhibition of endocytosis by hyperosmolaric sucrose and K^+ depletion leads to inhibition of WNT-3A-induced β-catenin signaling by rapid downregulation of DVL rather than by inhibition of receptor internalization [88]. The mechanism of the sucrose-induced degradation of DVL is not understood yet but may constitute a negative feedback loop involving the regulation of DVL stability, possibly in relation to agonist-dependent receptor dynamics.

Another layer of signaling complexity on the level of receptor dynamics was recently described in the *Drosophila* neuromuscular junction, where wingless activates FZD_2 resulting in internalization and C-terminal proteolysis of the receptor. The C-terminus is then translocated to the nucleus where it regulates gene transcription. This exciting signaling mode is awaiting confirmation in a mammalian setting. It will be important to delineate if this is a general signaling feature of other FZDs.

5.4. FRIZZLEDS—PHYSIOLOGY AND POSSIBLE THERAPY

5.4.1. Frizzleds in Physiology

It is well established that FZDs have a prominent role during embryonic development and in cancer. These topics exceed the scope of this chapter, and the reader is therefore referred to a series of excellent overviews [89–93]. One future challenge for the analysis of FZD function in physiology and pathophysiology will be to distinguish (dys-)function of FZDs in the adult organism from WNT/FZD-dependent defects arising during embryonic development. Understanding of basal signaling paradigms, FZD pharmacology, and their involvement in adult tissue homeostasis and pathophysiology will provide novel opportunities to treat diseases of the immune, nervous, and cardiovascular systems. Not unexpectedly, WNT/FZD signaling is important for the maintenance, proliferation, and differentiation of adult stem cells. Apart from that, very little is known as yet about the role of WNT/FZD signaling for physiology in the adult organism.

Novel physiological tasks for the WNT/FZD system are emerging from recent studies. Important roles in the brain include retrograde WNT signaling for axon guidance and synaptogenesis [94], the regulation of neurotransmitter release [95–97], and the modulation of N-methyl-D-aspartate (NMDA)-induced long-term potentiation [98]. Furthermore, as expected from their prominent role during organ development, signaling via WNT/FZD also serves some key but as yet poorly defined function in response to trauma, for tissue regeneration and tissue repair, such that WNT/FZD signaling might emerge as a hallmark of tissue trauma.

5.4.2. Therapeutic Potential

The WNT/FZD signaling system is broadly expressed in both the developing and the adult organism and its involvement in several disease states makes it a suitable target for therapy. Indeed, some forms of human disease are directly linked to mutations in FZD, as exemplified by the FZD_4-dependent familial exudative vitreoretinopathy [99]. Broad expression and a large number of receptor and ligand isoforms provide both possibilities and pitfalls for drug development. Despite the current lack of understanding of specificity of WNT–FZD interaction and the underlying signaling mechanisms, this receptor system has important therapeutic potential. Developing drugs that selectively attack FZDs and WNT-induced signaling positively and negatively could be very useful for the treatment of conditions such as cancer, neurological disorders, osteroporosis, and cardiovascular disease [93, 100–103], to name but a few. WNT/Frizzled signaling might also play a major role in the future use of stem cell-based therapies, particularly for the maintenance and priming of stem cells or precursors *in vitro*. Thus, basic research and molecular pharmacological approaches in combination with disease-orientated investigations will provide the possibility to employ Frizzleds and Frizzled-mediated signaling as targets for novel therapy [101].

5.4.3. Attacking WNT–FZD Interface?

To date, we are far from understanding the WNT–FZD interaction and attempts at screening for compounds that interfere with WNT–FZD binding are hampered by difficulties with WNT purification, the lack of suitable FZD expression systems, and the lack of selective radioligands. Recently, a WNT-5A-mimetic formylated peptide (formyl-Met-Asp-Gly-Cys-Glu-Leu) was described [104]. The peptide was shown to affect cell adhesion and migration in cancer cell lines through FZD_5, suggesting its use as an antimetastatic agent, for example, for human breast cancer patients. In addition to its promising potential as a therapeutic agent, this peptide could also be used as a tool to screen for small-molecule ligands that compete for binding sites. A similar short synthetic peptide was described for WNT-1, which promoted cell adhesion and hippocampal neural stem cell differentiation [105]. With regard to solubility, stability, administration, and drug distribution, a small-molecule ligand would be even more desirable than a peptide. Possible strategies are the development of orthosteric small-molecule agonists and antagonists, as well as positive or negative allosteric modulators. Based on the structure of FZDs and the above-described role of the CRD for ligand binding, it might be fruitful to develop an allosteric modulator that would bind to the receptor core and affect WNT binding with negative cooperativity. Similar approaches have led to important compounds in distantly related receptor systems, such as MPEP, a selective negative allosteric modulator for $mGLU_5$ receptors [106].

Another way to interfere with FZDs and WNT signaling is the use of FZD-specific antibodies, a strategy that has been put forward as a promising anticancer therapy, with several registered patents [107]. In addition, anti-FZD antibodies have been shown to have effects in an *in vitro* model for rheumatoid arthritis. Specifically, antibodies against the extracellular domain of FZD_5 were shown to inhibit cultured fibroblast-like synoviocytes from patients with rheumatoid arthritis [108].

5.4.4. Anti-DVL Drugs

DVL is a FZD pathway-specific signaling module and the molecular interface between DVL and FZD is dependent on specific PDZ domain interactions. Thus, the internal PDZ ligand domain of FZD is a potentially attractive target for drug design [109]. Indeed, a recent ligand screen identified an indole-3-carbinol compound (NSC668036 from the National Cancer Institute small-molecule library) that interfered with the DVL PDZ domain and disturbed binding and communication between FZD and DVL [110]. In *Xenopus laevis*-based assays, this compound was capable of inhibiting WNT-3A-induced signaling. Further development yielded FJ9 ((2-(1-hydroxypentyl)-3-(2-phenylethyl)-6-methyl)indole-5-carboxylic acid), a non-electrophilic indole-2-carbinol-based compound that inhibits the interaction between FZD and the PDZ domain of DVL and blocks the growth of tumor cells in a β-catenin-dependent manner [111]. The same structural approach was then further developed to optimize indole-2-carbinol-based compounds that selectively attack the DVL1-TCF pathway in cells and to implement a screening platform for developing compounds with higher potency, efficacy, and selectivity [112].

5.4.5. WNTs as Drugs

In view of the chemical characteristics of the WNT lipoglycoproteins, it appears unlikely that the WNTs will make it through drug development as full-length proteins for *in vivo* treatment. However, an important potential is the *in vitro* treatment of stem cells or precursors to prepare them for use in stem cell-based replacement therapy for Parkinson's disease, hematologic disorders (including malignancies), or organ injury and regeneration [107].

The first report of WNT-5A purification indicated that purified WNT-5A can be used to increase the number of dopaminergic neurons in rat ventral midbrain E14.5 precursor cultures [24], confirming previous results on the role of WNT-5A for the development of dopaminergic neurons in the midbrain [113]. More importantly, the proof of principle for WNT-5A-supported stem cell-based replacement therapy in parkinsonian mice [114] provided the basis to develop *ex-vitro* treatment methods of human stem cells for the treatment of Parkinson's disease [115].

A similar approach could be valuable for many different diseases, such as other neurodegenerative diseases, malignancies of the hematopoietic lineage, replacement of cardiomyocytes and hepatic cells, as well as for *ex vivo* generation of skin for transplants.

5.4.6. Future Directions

The understanding of the complexity of the WNT/FZD signaling system has made substantial progress during recent years. However, we are still far from grasping the array of signaling events initiated by WNT interaction with FZDs, the means by which the receptors selectively activate specific signaling pathways, and the physiological importance of the multitude of possible WNT/FZD combinations. The challenges for the coming years are the description of WNT–FZD interaction patterns in mammals, a more detailed understanding of the mechanisms determining pathway specificity, and the relevance of the different signaling pathways for physiology and pathophysiology in humans. Progress in these areas will enable us to employ this signaling cascade as a target for drug development and therapy in important areas such as the treatment of cancer, neurological disorders such as Parkinson's and Alzheimer's disease, cardiovascular and inflammatory diseases, and tissue injury. With increasing understanding of WNT/FZD signals in the adult organism, it will also be possible to identify even more disorders that may be amenable to treatment with WNT/FZD-related drugs.

ACKNOWLEDGMENTS

The research in the group was supported by the Swedish Research Council (vetenskapsrådet), the Swedish Cancer Society, STINT, The Foundations of the National Board of Health and Welfare of Sweden, Jeanssons, Tore Nilsons, Åhlén and Åke Wibergs Foundation, and Karolinska Institutet. I would also like to thank Dr. Vitezslav Bryja and Alexandra Schambony for their inspiration, support, and fruitful collaboration. Janet Holmén is kindly acknowledged for critical reading of this manuscript. I apologize to all authors whose relevant work could not be cited due to space restrictions.

REFERENCES

1. Rijsewijk, F., Schuermann, M., Wagenaar, E., Parren, P., Weigel, D., Nusse, R. (1987) The *Drosophila* homolog of the mouse mammary oncogene int-1 is identical to the segment polarity gene wingless. *Cell.* 50, 649–657.
2. Nusse, R., Brown, A., Papkoff, J., Scambler, P., Shackleford, G., McMahon, A., Moon, R., Varmus, H. (1991) A new nomenclature for int-1 and related genes: The Wnt gene family. *Cell.* 64, 231.
3. Nusse, R. (2003) Wnts and Hedgehogs: Lipid-modified proteins and similarities in signaling mechanisms at the cell surface. *Development.* 130, 5297–5305.
4. Vinson, C.R., Conover, S., Adler, P.N. (1989) A *Drosophila* tissue polarity locus encodes a protein containing seven potential transmembrane domains. *Nature.* 338, 263–264.

5. Guo, N., Hawkins, C., Nathans, J. (2004) Frizzled6 controls hair patterning in mice. *Proc Natl Acad Sci U S A.* 101, 9277–9281.
6. Krasnow, R.E., Wong, L.L., Adler, P.N. (1995) Dishevelled is a component of the frizzled signaling pathway in *Drosophila. Development.* 121, 4095–4102.
7. Bhanot, P., Brink, M., Samos, C.H., Hsieh, J.C., Wang, Y., Macke, J.P., Andrew, D., Nathans, J., Nusse, R. (1996) A new member of the frizzled family from *Drosophila* functions as a Wingless receptor. *Nature.* 382, 225–230.
8. Wang, Y., Macke, J.P., Abella, B.S., Andreasson, K., Worley, P., Gilbert, D.J., Copeland, N.G., Jenkins, N.A., Nathans, J. (1996) A large family of putative transmembrane receptors homologous to the product of the *Drosophila* tissue polarity gene frizzled. *J Biol Chem.* 271, 4468–4476.
9. Schulte, G., Bryja, V. (2007) The Frizzled family of unconventional G-protein-coupled receptors. *Trends Pharmacol Sci.* 28, 518–525.
10. Malbon, C.C. (2004) Frizzleds: New members of the superfamily of G-protein-coupled receptors. *Front Biosci.* 9, 1048–1058.
11. Ruiz-Gómez, A., Molnar, C., Holguín, H., Mayor, F., de Celis, J.F. (2007) The cell biology of Smo signalling and its relationships with GPCRs. *Biochim Biophys Acta.* 1768, 901–912.
12. Foord, S.M., Bonner, T.I., Neubig, R.R., Rosser, E.M., Pin, J.P., Davenport, A.P., Spedding, M., Harmar, A.J. (2005) International Union of Pharmacology. XLVI. G protein-coupled receptor list. *Pharmacol Rev.* 57, 279–288.
13. Barnes, M.R., Duckworth, D.M., Beeley, L.J. (1998) Frizzled proteins constitute a novel family of G protein-coupled receptors, most closely related to the secretin family. *Trends Pharmacol Sci.* 19, 399–400.
14. Fredriksson, R., Lagerström, M.C., Lundin, L.G., Schiöth, H.B. (2003) The G-protein-coupled receptors in the human genome form five main families. Phylogenetic analysis, paralogon groups, and fingerprints. *Mol Pharmacol.* 63, 1256–1272.
15. Palczewski, K. (2006) G protein-coupled receptor rhodopsin. *Annu Rev Biochem.* 75, 743–767.
16. Hawtin, S.R., Tobin, A.B., Patel, S., Wheatley, M. (2001) Palmitoylation of the vasopressin V1a receptor reveals different conformational requirements for signaling, agonist-induced receptor phosphorylation, and sequestration. *J Biol Chem.* 276, 38139–38146.
17. Egger-Adam, D., Katanaev, V.L. (2008) Trimeric G protein-dependent signaling by Frizzled receptors in animal development. *Front Biosci.* 13, 4740–4755.
18. Hendrickx, M., Leyns, L. (2008) Non-conventional Frizzled ligands and Wnt receptors. *Dev Growth Differ.* 50, 229–243.
19. Nam, J.S., Turcotte, T.J., Smith, P.F., Choi, S., Yoon, J.K. (2006) Mouse cristin/R-spondin family proteins are novel ligands for the Frizzled 8 and LRP6 receptors and activate beta-catenin-dependent gene expression. *J Biol Chem.* 281, 13247–13257.
20. Xu, Q., Wang, Y., Dabdoub, A., Smallwood, P.M., Williams, J., Woods, C., Kelley, M.W., Jiang, L., Tasman, W., Zhang, K., Nathans, J. (2004) Vascular development in the retina and inner ear: Control by Norrin and Frizzled-4, a high-affinity ligand-receptor pair. *Cell.* 116, 883–895.
21. Hsieh, J.C. (2004) Specificity of WNT-receptor interactions. *Front Biosci.* 9, 1333–1338.

22. Wu, C.H., Nusse, R. (2002) Ligand receptor interactions in the Wnt signaling pathway in *Drosophila*. *J Biol Chem.* 277, 41762–41769.
23. Willert, K., Brown, J.D., Danenberg, E., Duncan, A.W., Weissman, I.L., Reya, T., Yates, J.R., Nusse, R. (2003) Wnt proteins are lipid-modified and can act as stem cell growth factors. *Nature.* 423, 448–452.
24. Schulte, G., Bryja, V., Rawal, N., Castelo-Branco, G., So, K.M., Arenas, E. (2005) Purified Wnt-5a increases differentiation of midbrain dopaminergic cells and dishevelled phosphorylation. *J Neurochem.* 92, 1550–1553.
25. Hsieh, J.C., Rattner, A., Smallwood, P.M., Nathans, J. (1999) Biochemical characterization of Wnt-frizzled interactions using a soluble, biologically active vertebrate Wnt protein. *Proc Natl Acad Sci U S A.* 96, 3546–3551.
26. Rulifson, E.J., Wu, C.H., Nusse, R. (2000) Pathway specificity by the bifunctional receptor frizzled is determined by affinity for wingless. *Mol Cell.* 6, 117–126.
27. Chen, C.M., Strapps, W., Tomlinson, A., Struhl, G. (2004) Evidence that the cysteine-rich domain of *Drosophila* Frizzled family receptors is dispensable for transducing Wingless. *Proc Natl Acad Sci U S A.* 101, 15961–15966.
28. Povelones, M., Nusse, R. (2005) The role of the cysteine-rich domain of Frizzled in Wingless-Armadillo signaling. *EMBO J.* 24, 3493–3503.
29. Carron, C., Pascal, A., Djiane, A., Boucaut, J.C., Shi, D.L., Umbhauer, M. (2003) Frizzled receptor dimerization is sufficient to activate the Wnt/beta-catenin pathway. *J Cell Sci.* 116, 2541–2550.
30. Bryja, V., Schulte, G., Rawal, N., Grahn, A., Arenas, E. (2007) Wnt-5a induces dishevelled phosphorylation and dopaminergic differentiation via a CK1-dependent mechanism. *J Cell Sci.* 120, 586–595.
31. Bryja, V., Schulte, G., Arenas, E. (2007) Wnt-3a utilizes a novel low dose and rapid pathway that does not require casein kinase 1-mediated phosphorylation of Dvl to activate beta-catenin. *Cell Signal.* 19, 610–616.
32. Balla, S., Thapar, V., Verma, S., Luong, T., Faghri, T., Huang, C.H., Rajasekaran, S., del Campo, J.J., Shinn, J.H., Mohler, W.A., Maciejewski, M.W., Gryk, M.R., Piccirillo, B., Schiller, S.R., Schiller, M.R. (2006) Minimotif Miner: A tool for investigating protein function. *Nat Methods.* 3, 175–177.
33. Yanfeng, W.A., Tan, C., Fagan, R.J., Klein, P.S. (2006) Phosphorylation of frizzled-3. *J Biol Chem.* 281, 11603–11609.
34. Reiter, E., Lefkowitz, R.J. (2006) GRKs and beta-arrestins: Roles in receptor silencing, trafficking and signaling. *Trends Endocrinol Metab.* 17, 159–165.
35. Tobin, A.B. (2008) G-protein-coupled receptor phosphorylation: Where, when and by whom. *Br J Pharmacol.* 153 (Suppl. 1), S167–S176.
36. Macdonald, B.T., Semenov, M.V., He, X. (2007) SnapShot: Wnt/beta-catenin signaling. *Cell.* 131, 1204.
37. Semenov, M.V., Habas, R., Macdonald, B.T., He, X. (2007) SnapShot: Noncanonical Wnt signaling pathways. *Cell.* 131, 1378.
38. Mezhybovska, M., Wikström, K., Ohd, J.F., Sjölander, A. (2006) The inflammatory mediator leukotriene D4 induces beta-catenin signaling and its association with antiapoptotic Bcl-2 in intestinal epithelial cells. *J Biol Chem.* 281, 6776–6784.
39. Seifert, J.R., Mlodzik, M. (2007) Frizzled/PCP signalling: A conserved mechanism regulating cell polarity and directed motility. *Nat Rev Genet.* 8, 126–138.

40. Wallingford, J.B., Fraser, S.E., Harland, R.M. (2002) Convergent extension: The molecular control of polarized cell movement during embryonic development. *Dev Cell.* 2, 695–706.
41. Schambony, A., Wedlich, D. (2007) Wnt-5A/Ror2 regulate expression of XPAPC through an alternative noncanonical signaling pathway. *Dev Cell.* 12, 779–792.
42. McMahon, A.P., Moon, R.T. (1989) Ectopic expression of the proto-oncogene int-1 in Xenopus embryos leads to duplication of the embryonic axis. *Cell.* 58, 1075–1084.
43. Strutt, H., Price, M.A., Strutt, D. (2006) Planar polarity is positively regulated by casein kinase Iepsilon in *Drosophila*. *Curr Biol.* 16, 1329–1336.
44. Bryja, V., Gradl, D., Schambony, A., Arenas, E., Schulte, G. (2007) β-Arrestin is a necessary component of Wnt/beta-catenin signaling in vitro and in vivo. *Proc Natl Acad Sci U S A.* 104, 6690–6695.
45. Malbon, C.C., Wang, H.Y. (2006) Dishevelled: A mobile scaffold catalyzing development. *Curr Top Dev Biol.* 72, 153–166.
46. Cong, F., Schweizer, L., Varmus, H. (2004) Casein kinase Iepsilon modulates the signaling specificities of dishevelled. *Mol Cell Biol.* 24, 2000–2011.
47. Wolf, J., Palmby, T.R., Gavard, J., Williams, B.O., Gutkind, J.S. (2008) Multiple PPPS/TP motifs act in a combinatorial fashion to transduce Wnt signaling through LRP6. *FEBS Lett.* 582, 255–261.
48. MacDonald, B.T., Yokota, C., Tamai, K., Zeng, X., He, X. (2008) Wnt signal amplification via activity, cooperativity, and regulation of multiple intracellular PPPSP motifs in the Wnt co-receptor LRP6. *J Biol Chem.* 283, 16115–16123.
49. Dorsam, R.T., Gutkind, J.S. (2007) G-protein-coupled receptors and cancer. *Nat Rev Cancer.* 7, 79–94.
50. Gutkind, J.S. (2000) Regulation of mitogen-activated protein kinase signaling networks by G protein-coupled receptors. *Sci STKE.* 2000, RE1.
51. Schlange, T., Matsuda, Y., Lienhard, S., Huber, A., Hynes, N.E. (2007) Autocrine WNT signaling contributes to breast cancer cell proliferation via the canonical WNT pathway and EGFR transactivation. *Breast Cancer Res.* 9, R63.
52. Topol, L., Jiang, X., Choi, H., Garrett-Beal, L., Carolan, P.J., Yang, Y. (2003) Wnt-5a inhibits the canonical Wnt pathway by promoting GSK-3-independent beta-catenin degradation. *J Cell Biol.* 162, 899–908.
53. Umbhauer, M., Djiane, A., Goisset, C., Penzo-Méndez, A., Riou, J.F., Boucaut, J.C., Shi, D.L. (2000) The C-terminal cytoplasmic Lys-thr-X-X-X-Trp motif in frizzled receptors mediates Wnt/beta-catenin signalling. *EMBO J.* 19, 4944–4954.
54. Wharton, K.A. (2003) Runnin' with the Dvl: Proteins that associate with Dsh/Dvl and their significance to Wnt signal transduction. *Dev Biol.* 253, 1–17.
55. Wallingford, J.B., Habas, R. (2005) The developmental biology of Dishevelled: An enigmatic protein governing cell fate and cell polarity. *Development.* 132, 4421–4436.
56. Axelrod, J.D., Miller, J.R., Shulman, J.M., Moon, R.T., Perrimon, N. (1998) Differential recruitment of Dishevelled provides signaling specificity in the planar cell polarity and Wingless signaling pathways. *Genes Dev.* 12, 2610–2622.
57. Yang-Snyder, J., Miller, J.R., Brown, J.D., Lai, C.J., Moon, R.T. (1996) A frizzled homolog functions in a vertebrate Wnt signaling pathway. *Curr Biol.* 6, 1302–1306.

58. Rothbächer, U., Laurent, M.N., Deardorff, M.A., Klein, P.S., Cho, K.W., Fraser, S.E. (2000) Dishevelled phosphorylation, subcellular localization and multimerization regulate its role in early embryogenesis. *EMBO J.* 19, 1010–1022.
59. Yamanaka, H., Nishida, E. (2007) Wnt11 stimulation induces polarized accumulation of Dishevelled at apical adherens junctions through Frizzled7. *Genes Cells.* 12, 961–967.
60. Yanagawa, S., van Leeuwen, F., Wodarz, A., Klingensmith, J., Nusse, R. (1995) The dishevelled protein is modified by wingless signaling in *Drosophila*. *Genes Dev.* 9, 1087–1097.
61. Smalley, M.J., Signoret, N., Robertson, D., Tilley, A., Hann, A., Ewan, K., Ding, Y., Paterson, H., Dale, T.C. (2005) Dishevelled (Dvl-2) activates canonical Wnt signalling in the absence of cytoplasmic puncta. *J Cell Sci.* 118, 5279–5289.
62. Schwarz-Romond, T., Merrifield, C., Nichols, B.J., Bienz, M. (2005) The Wnt signalling effector Dishevelled forms dynamic protein assemblies rather than stable associations with cytoplasmic vesicles. *J Cell Sci.* 118, 5269–5277.
63. Itoh, K., Brott, B.K., Bae, G.U., Ratcliffe, M.J., Sokol, S.Y. (2005) Nuclear localization is required for Dishevelled function in Wnt/beta-catenin signaling. *J Biol.* 4, 3.
64. Dewire, S.M., Ahn, S., Lefkowitz, R.J., Shenoy, S.K. (2007) Beta-arrestins and cell signaling. *Annu Rev Physiol.* 69, 483–510.
65. Chen, W., Hu, L.A., Semenov, M.V., Yanagawa, S., Kikuchi, A., Lefkowitz, R.J., Miller, W.E. (2001) β-Arrestin1 modulates lymphoid enhancer factor transcriptional activity through interaction with phosphorylated dishevelled proteins. *Proc Natl Acad Sci U S A.* 98, 14889–14894.
66. Chen, W., ten Berge, D., Brown, J., Ahn, S., Hu, L.A., Miller, W.E., Caron, M.G., Barak, L.S., Nusse, R., Lefkowitz, R.J. (2003) Dishevelled 2 recruits β-arrestin 2 to mediate Wnt5A-stimulated endocytosis of Frizzled 4. *Science.* 301, 1391–1394.
67. Kim, G.H., Han, J.K. (2007) Essential role for beta-arrestin 2 in the regulation of *Xenopus* convergent extension movements. *EMBO J.* 26, 2513–2526.
68. Bryja, V., Schambony, A., Cajanek, L., Dominguez, I., Arenas, E., Schulte, G. (2008) β-arrestin and casein kinase 1/2 define distinct branches of non-canonical WBNT signaling pathways. *EMBO Rep.* 9, 1244–1250.
69. Force, T., Woulfe, K., Koch, W.J., Kerkelä, R. (2007) Molecular scaffolds regulate bidirectional crosstalk between Wnt and classical seven-transmembrane-domain receptor signaling pathways. *Sci STKE.* 2007, pe41.
70. Katanaev, V.L., Ponzielli, R., Sémériva, M., Tomlinson, A. (2005) Trimeric G protein-dependent frizzled signaling in *Drosophila*. *Cell.* 120, 111–122.
71. Wang, H.Y., Liu, T., Malbon, C.C. (2006) Structure-function analysis of Frizzleds. *Cell Signal.* 18, 934–941.
72. Slusarski, D.C., Corces, V.G., Moon, R.T. (1997) Interaction of Wnt and a Frizzled homologue triggers G-protein-linked phosphatidylinositol signalling. *Nature.* 390, 410–413.
73. Wang, H.Y., Malbon, C.C. (2003) Wnt signaling, Ca^{2+}, and cyclic GMP: Visualizing Frizzled functions. *Science.* 300, 1529–1530.
74. Wang, H.Y. (2004) WNT-frizzled signaling via cyclic GMP. *Front Biosci.* 9, 1043–1047.
75. Ma, L., Wang, H.Y. (2006) Suppression of cyclic GMP-dependent protein kinase is essential to the Wnt/cGMP/Ca^{2+} pathway. *J Biol Chem.* 281, 30990–31001.

76. Ahumada, A., Slusarski, D.C., Liu, X., Moon, R.T., Malbon, C.C., Wang, H.Y. (2002) Signaling of rat Frizzled-2 through phosphodiesterase and cyclic GMP. *Science.* 298, 2006–2010.
77. Endo, Y., Wolf, V., Muraiso, K., Kamijo, K., Soon, L., Uren, A., Barshishat-Küpper, M., Rubin, J.S. (2005) Wnt-3a-dependent cell motility involves RhoA activation and is specifically regulated by dishevelled-2. *J Biol Chem.* 280, 777–786.
78. Liu, T., DeCostanzo, A.J., Liu, X., Wang, H., Hallagan, S., Moon, R.T., Malbon, C.C. (2001) G protein signaling from activated rat frizzled-1 to the beta-catenin-Lef-Tcf pathway. *Science.* 292, 1718–1722.
79. DeCostanzo, A.J., Huang, X.P., Wang, H.Y., Malbon, C.C. (2002) The Frizzled-1/(β(2))-adrenergic receptor chimera: Pharmacological properties of a unique G protein-linked receptor. *Naunyn Schmiedebergs Arch Pharmacol.* 365, 341–348.
80. Liu, X., Liu, T., Slusarski, D.C., Yang-Snyder, J., Malbon, C.C., Moon, R.T., Wang, H. (1999) Activation of a frizzled-2/beta-adrenergic receptor chimera promotes Wnt signaling and differentiation of mouse F9 teratocarcinoma cells via $G\alpha o$ and $G\alpha t$. *Proc Natl Acad Sci U S A.* 96, 14383–14388.
81. Li, H., Malbon, C.C., Wang, H.Y. (2004) Gene profiling of Frizzled-1 and Frizzled-2 signaling: Expression of G-protein-coupled receptor chimeras in mouse F9 teratocarcinoma embryonal cells. *Mol Pharmacol.* 65, 45–55.
82. Yamamoto, H., Komekado, H., Kikuchi, A. (2006) Caveolin is necessary for Wnt-3a-dependent internalization of LRP6 and accumulation of beta-catenin. *Dev Cell.* 11, 213–223.
83. Kikuchi, A., Yamamoto, H. (2007) Regulation of Wnt signaling by receptor-mediated endocytosis. *J Biochem.* 141, 443–451.
84. Marois, E., Mahmoud, A., Eaton, S. (2006) The endocytic pathway and formation of the Wingless morphogen gradient. *Development.* 133, 307–317.
85. Gagliardi, M., Piddini, E., Vincent, J.P. (2008) Endocytosis: A positive or a negative influence on Wnt signalling? *Traffic.* 9, 1–9.
86. Yu, A., Rual, J.F., Tamai, K., Harada, Y., Vidal, M., He, X., Kirchhausen, T. (2007) Association of Dishevelled with the clathrin AP-2 adaptor is required for Frizzled endocytosis and planar cell polarity signaling. *Dev Cell.* 12, 129–141.
87. Laporte, S.A., Oakley, R.H., Holt, J.A., Barak, L.S., Caron, M.G. (2000) The interaction of beta-arrestin with the AP-2 adaptor is required for the clustering of beta 2-adrenergic receptor into clathrin-coated pits. *J Biol Chem.* 275, 23120–23126.
88. Bryja, V., Cajánek, L., Grahn, A., Schulte, G. (2007) Inhibition of endocytosis blocks Wnt signalling to beta-catenin by promoting dishevelled degradation. *Acta Physiol (Oxf).* 190, 55–61.
89. Wodarz, A., Nusse, R. (1998) Mechanisms of Wnt signaling in development. *Annu Rev Cell Dev Biol.* 14, 59–88.
90. Nusse, R. (2005) Wnt signaling in disease and in development. *Cell Res.* 15, 28–32.
91. Logan, C.Y., Nusse, R. (2004) The Wnt signaling pathway in development and disease. *Annu Rev Cell Dev Biol.* 20, 781–810.
92. Polakis, P. (2000) Wnt signaling and cancer. *Genes Dev.* 14, 1837–1851.
93. Malaterre, J., Ramsay, R.G., Mantamadiotis, T. (2007) Wnt-Frizzled signalling and the many paths to neural development and adult brain homeostasis. *Front Biosci.* 12, 492–506.

94. Salinas, P.C. (2005) Retrograde signalling at the synapse: A role for Wnt proteins. *Biochem Soc Trans.* 33, 1295–1298.
95. Cerpa, W., Godoy, J.A., Alfaro, I., Farías, G.G., Metcalfe, M.J., Fuentealba, R., Bonansco, C., Inestrosa, N.C. (2008) Wnt-7a modulates the synaptic vesicle cycle and synaptic transmission in hippocampal neurons. *J Biol Chem.* 283, 5918–5927.
96. Kishida, S., Hamao, K., Inoue, M., Hasegawa, M., Matsuura, Y., Mikoshiba, K., Fukuda, M., Kikuchi, A. (2007) Dvl regulates endo- and exocytotic processes through binding to synaptotagmin. *Genes Cells.* 12, 49–61.
97. Ahmad-Annuar, A., Ciani, L., Simeonidis, I., Herreros, J., Fredj, N.B., Rosso, S.B., Hall, A., Brickley, S., Salinas, P.C. (2006) Signaling across the synapse: A role for Wnt and Dishevelled in presynaptic assembly and neurotransmitter release. *J Cell Biol.* 174, 127–139.
98. Chen, J., Park, C.S., Tang, S.J. (2006) Activity-dependent synaptic Wnt release regulates hippocampal long-term potentiation. *J Biol Chem.* 281, 11910–11916.
99. Kaykas, A., Yang-Snyder, J., Héroux, M., Shah, K.V., Bouvier, M., Moon, R.T. (2004) Mutant Frizzled 4 associated with vitreoretinopathy traps wild-type Frizzled in the endoplasmic reticulum by oligomerization. *Nat Cell Biol.* 6, 52–58.
100. Cerpa, W., Dinamarca, M.C., Inestrosa, N.C. (2008) Structure-function implications in Alzheimer's disease: Effect of Abeta oligomers at central synapses. *Curr Alzheimer Res.* 5, 233–243.
101. Luo, J., Chen, J., Deng, Z.L., Luo, X., Song, W.X., Sharff, K.A., Tang, N., Haydon, R.C., Luu, H.H., He, T.C. (2007) Wnt signaling and human diseases: What are the therapeutic implications? *Lab Invest.* 87, 97–103.
102. Blankesteijn, W.M., van de Schans, V.A., ter Horst, P., Smits, J.F. (2008) The Wnt/frizzled/GSK-3 beta pathway: A novel therapeutic target for cardiac hypertrophy. *Trends Pharmacol Sci.* 29, 175–180.
103. Chan, A., van Bezooijen, R.L., Löwik, C.W. (2007) A new paradigm in the treatment of osteoporosis: Wnt pathway proteins and their antagonists. *Curr Opin Investig Drugs.* 8, 293–298.
104. Säfholm, A., Leandersson, K., Dejmek, J., Nielsen, C.K., Villoutreix, B.O., Andersson, T. (2006) A formylated hexapeptide ligand mimics the ability of Wnt-5a to impair migration of human breast epithelial cells. *J Biol Chem.* 281, 2740–2749.
105. Kajiwara, K., Kamamoto, M., Ogata, S.I., Tanihara, M. (2008) A synthetic peptide corresponding to residues 301–320 of human Wnt-1 promotes PC12 cell adhesion and hippocampal neural stem cell differentiation. *Peptides.* 29, 1479–1485.
106. Malherbe, P., Kratochwil, N., Zenner, M.T., Piussi, J., Diener, C., Kratzeisen, C., Fischer, C., Porter, R.H. (2003) Mutational analysis and molecular modeling of the binding pocket of the metabotropic glutamate 5 receptor negative modulator 2-methyl-6-(phenylethynyl)-pyridine. *Mol Pharmacol.* 64, 823–832.
107. Chien, A.J., Moon, R.T. (2007) WNTS and WNT receptors as therapeutic tools and targets in human disease processes. *Front Biosci.* 12, 448–457.
108. Sen, M., Chamorro, M., Reifert, J., Corr, M., Carson, D.A. (2001) Blockade of Wnt-5A/frizzled 5 signaling inhibits rheumatoid synoviocyte activation. *Arthritis Rheum.* 44, 772–781.
109. Wang, N.X., Lee, H.J., Zheng, J.J. (2008) Therapeutic use of PDZ protein–protein interaction antagonism. *Drug News Perspect.* 21, 137–141.

110. Shan, J., Shi, D.L., Wang, J., Zheng, J. (2005) Identification of a specific inhibitor of the dishevelled PDZ domain. *Biochemistry.* 44, 15495–15503.

111. Fujii, N., You, L., Xu, Z., Uematsu, K., Shan, J., He, B., Mikami, I., Edmondson, L.R., Neale, G., Zheng, J., Guy, R.K., Jablons, D.M. (2007) An antagonist of dishevelled protein–protein interaction suppresses beta-catenin-dependent tumor cell growth. *Cancer Res.* 67, 573–579.

112. You, L., Xu, Z., Punchihewa, C., Jablons, D.M., Fujii, N. (2008) Evaluation of a chemical library of small-molecule Dishevelled antagonists that suppress tumor growth by down-regulating T-cell factor-mediated transcription. *Mol Cancer Ther.* 7, 1633–1638.

113. Castelo-Branco, G., Wagner, J., Rodriguez, F.J., Kele, J., Sousa, K., Rawal, N., Pasolli, H.A., Fuchs, E., Kitajewski, J., Arenas, E. (2003) Differential regulation of midbrain dopaminergic neuron development by Wnt-1, Wnt-3a, and Wnt-5a. *Proc Natl Acad Sci U S A.* 100, 12747–12752.

114. Parish, C.L., Castelo-Branco, G., Rawal, N., Tonnesen, J., Sorensen, A.T., Salto, C., Kokaia, M., Lindvall, O., Arenas, E. (2008) Wnt5a-treated midbrain neural stem cells improve dopamine cell replacement therapy in parkinsonian mice. *J Clin Invest.* 118, 149–160.

115. Parish, C.L, Arenas, E. (2007) Stem-cell-based strategies for the treatment of Parkinson's disease. *Neurodegener Dis.* 4, 339–347.

116. Gao, Y., Wang, H.Y. (2007) Inositol pentakisphosphate mediates Wnt/beta-catenin signaling. *J Biol Chem.* 282, 26490–264502.

117. Tu, X., Joeng, K.S., Nakayama, K.I., Nakayama, K., Rajagopal, J., Carroll, T.J., McMahon, A.P., Long, F. (2007) Noncanonical Wnt signaling through G protein-linked PKCdelta activation promotes bone formation. *Dev Cell.* 12, 113–127.

118. Habas, R., Dawid, I.B., He, X. (2003) Coactivation of Rac and Rho by Wnt/Frizzled signaling is required for vertebrate gastrulation. *Genes Dev.* 17, 295–309.

119. Tsai, I.C., Amack, J.D., Gao, Z.H., Band, V., Yost, H.J., Virshup, D.M. (2007) A Wnt-CKIvarepsilon-Rap1 pathway regulates gastrulation by modulating SIPA1L1, a Rap GTPase activating protein. *Dev Cell.* 12, 335–347.

120. Chen, A.E., Ginty, D.D., Fan, C.M. (2005) Protein kinase A signalling via CREB controls myogenesis induced by Wnt proteins. *Nature.* 433, 317–322.

121. Inoki, K., Ouyang, H., Zhu, T., Lindvall, C., Wang, Y., Zhang, X., Yang, Q., Bennett, C., Harada, Y., Stankunas, K., Wang, C.Y., He, X., MacDougald, O.A., You, M., Williams, B.O., Guan, K.L. (2006) TSC2 integrates Wnt and energy signals via a coordinated phosphorylation by AMPK and GSK3 to regulate cell growth. *Cell.* 126, 955–968.

122. Wouda, R.R., Bansraj, M.R., de Jong, A.W., Noordermeer, J.N., Fradkin, L.G. (2008) Src family kinases are required for WNT5 signaling through the Derailed/RYK receptor in the *Drosophila* embryonic central nervous system. *Development.* 135, 2277–2287.

123. Lu, W., Yamamoto, V., Ortega, B., Baltimore, D. (2004) Mammalian Ryk is a Wnt coreceptor required for stimulation of neurite outgrowth. *Cell.* 119, 97–108.

CHAPTER 6

Secretin Receptor Dimerization: A Possible Functionally Important Paradigm for Family B G Protein-Coupled Receptors*

KALEECKAL G. HARIKUMAR, MAOQING DONG, and LAURENCE J. MILLER

Department of Molecular Pharmacology and Experimental Therapeutics, Mayo Clinic, Scottsdale, AZ

Protein–protein interactions represent a molecular mechanism for signal propagation and for regulation that is utilized extensively in cell biology and physiology. The interaction between single transmembrane tyrosine kinase receptors has long been recognized as a fundamental mechanism for the cross-phosphorylation involved in regulation of the activity of these important growth regulators [1]. After a long history of being considered as acting as single units, guanine nucleotide-binding protein (G protein)-coupled receptors (GPCRs) have recently also been reported to form dimers or higher-order oligomers [2]. However, the extent that this process occurs in normal physiology and its functional significance have been quite controversial [3, 4].

In this report, we will review the methodological approaches that have been utilized to demonstrate GPCR oligomerization, the structural themes for such association, and the functional effects that have been proposed. Family B GPCR dimerization may represent a unique and structurally specific process that can have clear functional implications. Within this chapter, we will carefully review the data relevant to the secretin receptor as a possible paradigm for how this process might affect Family B GPCRs, as well as the larger GPCR superfamily.

*This work was supported by grants from the National Institutes of Health, DK46577 (LJM), the Fiterman Foundation, and the Mayo Clinic.

GPCR Molecular Pharmacology and Drug Targeting: Shifting Paradigms and New Directions,
Edited by Annette Gilchrist

6.1. METHODOLOGICAL APPROACHES TO GPCR OLIGOMERIZATION

GPCR oligomerization has been explored a number of ways, including approaches dependent on structural associations and those involving functional reconstitution. It should be mentioned that procedures have often utilized the high-level expression of recombinant proteins, but such a level of expression may induce artificial interactions that might not be present in a physiological environment. Therefore, when possible, use of natural receptor-bearing cells or those expressing receptors at physiologic levels are preferred. The oldest approach to demonstrate physical association is the co-immunoprecipitation (co-IP) of two receptors, utilizing an antibody to one receptor to precipitate the partner receptor [5]. Since there is a paucity of high-quality, highly selective, high-affinity antibodies to natural receptors that are currently available, most of these studies have utilized epitope-tagged forms of receptors coexpressed in heterologous cell lines. This also permits the examination of receptor homodimerization, with the same receptor being prepared as two distinctly tagged forms. Controls to ensure adequate solubilization including the ultracentrifugation of the extracts or their passage through high-resolution filters prior to immunoprecipitation have been helpful, as have controls that involve the mixing of detergent extracts from two cell lines, each expressing only one receptor; however, this technique is still criticized and cannot be utilized as evidence of receptor oligomerization without the existence of complementary data utilizing another distinct technique. This relates to the tendency of these hydrophobic proteins to aggregate and to nonspecifically associate, as well as to the possibility that a detergent micelle might include multiple nonphysically associated membrane proteins.

The most common complementary technique involves resonance energy transfer, with nonradiative transfer of energy between a donor attached to one receptor and an acceptor attached to a second receptor (Fig. 6.1). This can utilize bioluminescence resonance energy transfer (BRET) [6] or fluorescence resonance energy transfer (FRET) [7]. The latter has even been utilized as a quantitative approach to measure distances between donor and acceptor [8]. BRET has a theoretical advantage over FRET, since donor effects on the acceptor signal are less and do not require correction. By choice of donor and acceptor with the appropriate overlap of donor emission and acceptor absorption, the resonance energy transfer signal can be limited to relatively short distances, but those that are adequate to identify complexes of physically associated receptors. Nevertheless, at high levels of engineered recombinant receptor expression, it becomes easy to visualize crowding of the lipid bilayer resulting in the generation of nonspecific signals in such studies. This phenomenon has been termed "nonspecific bystander effect" [9]. To overcome this potential problem, saturation resonance transfer techniques have been employed in which varied ratios of donor-tagged and acceptor-tagged receptors are coexpressed and evaluated for resonance transfer [9]. In such studies,

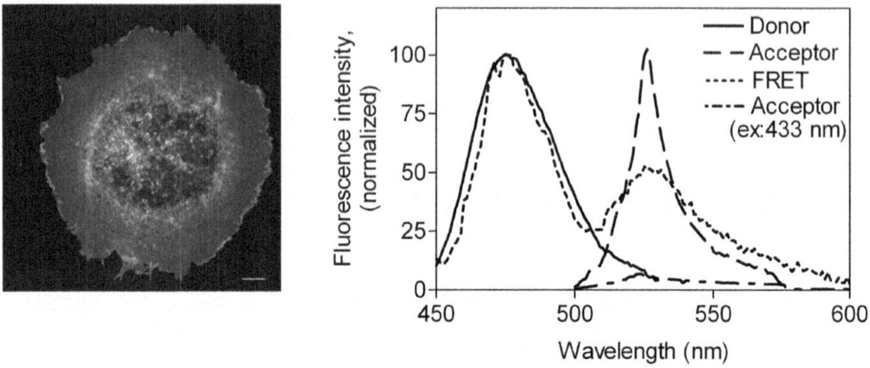

Figure 6.1 Study of secretin receptor oligomerization by fluorescence resonance energy transfer techniques. *Left* panel shows a representative confocal microscopic image of a COS (African green monkey kidney) cell expressing the YFP-tagged secretin receptor construct, demonstrating normal trafficking of the fluorescently tagged receptors to the cell surface. *Right* panel shows representative fluorescence emission spectra, including spectra of the donor (human secretin receptor–cyan fluorescent protein [HSecR-CFP]) excited at 433 nm, the acceptor (HSecR-YFP) excited at the donor excitation wavelength (433 nm) and excited at the wavelength optimal to elicit acceptor emission (480 nm), and a typical FRET emission profile from a system containing both donor and acceptor that was stimulated at 433 nm. Bar, 25 μm.

random collisions between bystanders are expected to increase in a linear manner with an increasing ratio of acceptor/donor, while significant structurally specific molecular associations would be expected to result in hyperbolic curves in which resonance energy transfer would approach a maximum level, as all available donors become complexed with the acceptors. Because of the requirement of fluorescent or luminescent tags that are not naturally present on receptors, these methods are not applicable to natural cellular systems in which there are normal levels of expression of receptors and relevant regulatory proteins. Receptor association might be suggested indirectly in such systems by demonstration of cooperativity of binding or action.

Other techniques have also been applied less commonly to explore receptor oligomerization. These include the stimulation of the internalization of one receptor by the treatment with an agonist of the second receptor [10], the trapping of one receptor in the biosynthetic cascade by the expression of a dominant negative partner receptor [11], rescuing an intracellularly trapped receptor by expression of another partner receptor [12], the modification of the binding properties of one receptor by the ligand occupation of the second receptor [13], and even by morphologic studies like high-resolution two-dimensional crystal structures [14] and atomic force microscopy [15]. Another novel approach called bimolecular fluorescence complementation (BiFC) has been utilized to demonstrate receptor dimerization by coexpressing the nonfluorescent halves of yellow fluorescent protein (YFP) attached to receptor constructs and monitoring the formation of functionally intact YFP [16].

Functional complementation is another unique and important methodological approach. In this, a structurally and functionally deficient form of a receptor is corrected by the expression of another functionally deficient receptor that has distinct structural defects, such that the association of the two receptors can contribute all the necessary structural domains to yield a fully functional unit [17]. This was first utilized to prove the possibility of "crossed-domain dimerization." While there is little doubt that this can occur under decidedly unique experimental conditions, with high levels of expression of complementary defective receptor constructs, sufficient data have not definitively established the relevance of this type of GPCR dimerization in normal physiology. Rather, most studies in recent years have focused on the association of intact receptors having complete heptahelical bundles with each other [18].

6.2. STRUCTURAL THEMES FOR GPCR OLIGOMERIZATION

There does not appear to be a consistent theme for the determinants of oligomeric complexes of GPCRs across this extensive superfamily. This should not be surprising, since the very large superfamily includes groups of receptors with quite divergent structures that have very distinct signature sequences representing their conserved patterns of residues. Although essentially all GPCRs are believed to have heptahelical architecture and to couple with heterotrimeric G proteins, even the structures of the helical bundle regions are predicted to be different for each major family, and the loop and tail regions may vary considerably among smaller groups within a given family.

As noted above, the concept of crossed-domain dimerization was developed based on the rescue of nonfunctional variants and mutants of receptors [19]. The most common structural theme for this was the association of the portion of one receptor extending from its amino terminus to the third intracellular loop, comprising transmembrane segments one through five, with a second receptor carboxyl terminus extending back to the same region, comprising transmembrane segments six and seven. This was possible with the very long third intracellular loop regions shown to be present in many GPCRs, which is particularly characteristic of the structures of Family A GPCRs. It is important to again emphasize that this type of dimerization has only been demonstrated in recombinant expression systems that often have involved high levels of expression of receptor constructs. Such dimerization has not been established to represent physiologically relevant structures.

The most common type of GPCR dimerization and oligomerization that has been explored involves the association of intact receptors with complete helical bundle domains. This provides the opportunity for the interaction and contribution of receptor ectodomains, cytosolic domains, or transmembrane domains. There are examples of at least two of these that have been well established with distinct receptors [20, 21]. A summary of the existing literature for GPCRs in each of the major families is listed in Table 6.1.

The strongest evidence for a dimeric GPCR complex of distinct structures comes from Family C [14, 22]. Here, crystal structures of amino-terminal ectodomains have revealed covalent disulfide-bond-linked dimers. The metabotropic glutamate receptors are among the best-defined structurally [14, 23]. Additionally, this family has multiple noncovalent associations between the dimeric molecules, including coiled–coil interactions in the intracellular carboxyl terminus. This has been demonstrated for γ-amino butyric acid (GABA) receptors [22]. Ligands bind to a venus flytrap motif in the extracellular amino terminus of these receptors. An interesting feature of these receptors is their constitutive heterodimerization, particularly noted for GABA and umami T1/T3 taste receptors, yielding distinct pharmacology [22, 24].

TABLE 6.1 Major Families of GPCRs

Family A Receptors			
Receptors	Oligomers		Constitutive (C), Ligand-Dependent (L+/L–)
	Homomers	Heteromers	
Adenosine			
A_1	N.D.	A_{2A}, P_2Y_1, P_2Y_2	C, L+
A_{2A}	+	D_2	C
Adrenergic			
α_{1A}	+	α_{1B}	C
α_{1B}	+		C
α_{2A}	+	α_{2C}, β_1	C
β_1	+	β_2	C
β_2	+	β_3, δ-OPR,κ-OPR	C
Angiotensin			
AT_1	+	AT_2, B_2	C
AT_2	+	B_2,β_2	C
Bradykinin			
B_1	+		C
B_2	+	AT_1, AT_2	C, L+
Cannabinoid			
CB1	+	D_2, μ-OPR, Orexin-1	C, L+
Chemokine			
CXCR1	+	CXCR2	C
CXCR2	+	δ-OPR	C
CCR5	+	C5aR, CCR2b, μ-OPR, δ-OPR, κ-OPR	C

The transmembrane helical bundle domain has been implicated as providing an important determinant for oligomerization in both Family A and Family B GPCRs; however, the details of these interactions have not been consistent across receptor families [18]. This region of the receptors has been demonstrated to be a major contributor by exclusion of other regions, for example, after truncating the associating receptors at their amino-terminal or carboxyl-terminal domains. The importance of the transmembrane domain has also been identified using bioinformatic computational approaches [25]. One experimental approach that has been quite powerful to gain additional insight has been the use of transmembrane segment peptides to compete for the resonance transfer signal indicating GPCR association. This approach was first introduced for the β2-adrenergic receptor [20]. In Family A GPCRs, there are examples of the importance of transmembrane segments one, two, four, five, six, and seven [26–31]. Family B GPCRs seem to be much more structurally consistent in the importance of a single helical segment (see below).

Techniques Employed	Functions Attributed	References
BRET, Co-IP, FRET	Binding	[56–59]
BRET, Co-IP, FRET	Trafficking	[60, 61]
BRET, Co-IP, FRET	Signaling, trafficking	[62–64]
FRET	Trafficking	[30]
BRET, Co-IP	Binding, signaling, trafficking	[65, 66]
BRET, Co-IP	Signaling, trafficking	[9, 67]
BRET, Co-IP	Signaling, trafficking	[68–70]
BRET, Co-IP, Cross linking	Binding, signaling, trafficking	[71–74]
Co-IP, FRET	Signaling	[75–77]
Co-IP	Signaling, trafficking	[78]
Co-IP, FRET Cross-linking	Signaling, trafficking	[71, 72]
BRET, Co-IP, FRET	Signaling, trafficking	[10, 79–81]
BRET, Co-IP, FRET	N.D.	[82]
BRET, Co-IP, FRET	Binding	[83]
BRET, Co-IP, FRET	Binding, signaling, trafficking	[84–87]

TABLE 6.1 (*Continued*)

Family A Receptors			
Receptors	Oligomers		Constitutive (C), Ligand-Dependent (L+/L−)
	Homomers	Heteromers	
Cholecystokinin			
CCK_1	+	CCK_2	C, L−
CCK_2	+		C
Dopamine			
D_1	+	D_2, D_3, NMDA	C
D_2	+	D_3, $SSTR_5$, CB_1	C, L+
D_3	+		C
Endothelin			
ET_A	+	ET_B	C
ET_B	+		C
Galanin			
Gal_1	+		C
Histamine			
H_1	+		C
H_3	+		C
H_4	+		C
Melanocortin			
MC_1	+		C
MC_3	+		C
MC_4	+		C
Melatonin			
MT_1	+	MT_2	C, L+
MT_2	+		C
Muscarinic mACh			
M1	+	M2, M3	C
M2	+		C
M3	+		C
Neurotensin			
NTS_1	+	NTS_2	C
Neuropeptide			
Y_1	+	Y_5	C
Y_2	+		C
Y_4	+		C, L−
Y_5	+		C
Opioid			
μ-OPR	+	δ-OPR, κ–OPR, α-AR, SSTR2A, NK1, CB_1	C

Techniques Employed	Functions Attributed	References
BRET, Co-IP, FRET	Signaling, trafficking	[27, 36, 88]
BRET	N.D.	[88]
BRET, Co-IP, FRET	Signaling, trafficking	[11, 12, 89–92]
Co-IP, FRET, ligand binding	Binding, signaling	[17, 79, 93, 94]
Co-IP	N.D.	[95]
Co-IP, FRET	Signaling	[96, 97]
Co-IP, FRET	Signaling, trafficking	[96]
FRET, Western blot	Trafficking	[98]
Co-IP, FRET	Binding	[99]
Co-IP	N. D.	[100]
BRET, Co-IP, FRET	N.D.	[101]
Co-IP	Trafficking	[102]
BRET	Binding	[103]
BRET	Binding	[104]
BRET, Co-IP	Binding	[35, 105]
BRET, Co-IP	N.D.	[35]
BRET	Binding	[106]
BRET	N.D.	[106]
BRET	N.D.	[106]
Co-IP, Ligand binding	Signaling, trafficking	[107, 108]
BRET, FRET	Binding, trafficking	[109, 110]
BRET, Co-IP	N.D.	[111]
FRET	N.D.	[109]
BRET	Binding, trafficking	[110]
BRET, Co-IP	Binding, signaling, trafficking	[69, 80, 112–118]

TABLE 6.1 (*Continued*)

Family A Receptors			
Receptors	Oligomers		Constitutive (C), Ligand-Dependent (L+/L–)
	Homomers	Heteromers	
δ-OPR	+	β2-AR	C, L–
κ-OPR	+	μ-OPR, δ-OPR, β2-AR	C
Opsin	+		C
Oxytocin			
OTR	+	V1a, V_2	C
Purinergic			
$P2Y_1$	+	$P2Y_{11}$	C, L+
Prostaglandin			
PR	+		C
EP_1		β-AR	C, L+
TPα		ProstacyclinR	C
Rhodopsin			
Rhodopsin	+		C
Serotonin			
$5\text{-}HT_{1A}$	+		C
$5\text{-}HT_{1B}$	+	$5\text{-}HT_{1D}$	C
$5\text{-}HT_{1D}$	+		C
$5\text{-}HT_{2C}$	+		C
$5\text{-}HT_4$	+		C
Somatostatin			
SSTR1	–	SSTR5	L+
SSTR2	+	SSTR3, μ-OPR	L+, L–
SSTR3	+	SSTR2	L+
SSTR5	+	SSTR1	L+
Thromboxane			
TPα	+	TPβ, ProstacyclinR	C
TPβ	+		C
Thyroid-stimulating hormone			
TSH	+	LH/CGR	C, L–
Thyrotropin releasing hormone			
TRHR1	+	TRHR2	C, L+
TRHR2	+		C, L+
Vasopressin			
V1a	+	V1a, V_2 Oxytocin	C
V_2	+		C

Techniques Employed	Functions Attributed	References
BRET, Co-IP, FRET, Cross linking	Signaling, trafficking	[5, 69]
BRET, Co-IP	Binding, signaling	[7, 119]
FRET, chemical cross-linking	N.D.	[120]
BRET, Co-IP	N.D.	[121, 122]
FRET	Binding, trafficking	[123, 124]
Co-IP	N.D.	[125]
BRET	Signaling	[126]
Western blot	Signaling, trafficking	[127, 128]
Atomic force and electron microscopy, FRET	N.D.	[15, 129, 130]
Co-IP, FLIM, FRET	No effect	[131]
Co-IP	N.D.	[132]
Co-IP	N.D.	[132]
BRET, Co-IP, FRET	Binding	[31, 32]
BRET, Co-IP	N.D.	[133]
Co-IP, FRET, Western blot	Binding, signaling, trafficking	[38, 134]
Co-IP, FRET, Western blot	Binding, signaling, trafficking	[117, 135, 136]
Co-IP	Binding, signaling, trafficking	[136]
Co-IP, FRET	Binding, signaling, trafficking	[38, 134]
Co-IP	Signaling, trafficking	[127, 128, 137, 138]
Western blot	Signaling, trafficking	[137, 138]
BRET, Co-IP, FRET	Binding	[139–141]
BRET, Co-IP	Signaling, trafficking	[142–144]
BRET	N.D.	[145]
BRET, Co-IP	Trafficking	[122, 146, 147]
BRET, Co-IP	No effect	[146]

TABLE 6.1 (*Continued*)

Family B Receptors			
Receptors	Oligomers		Constitutive (C), Ligand-Dependent (L+/L−)
	Homomers	Heteromers	
Calcitonin			
CTR	+		C
Calcitonin receptor-like receptor			
CRL	+		C
Corticotropin releasing factor			
CRF_1	+	Vasotocin, V_{1b}	C
Gastric inhibitory polypeptide receptor			
GIPR	+	Opsin, β_2-AR	C
Secretin			
Secretin	+	$VPAC_1$, $VPAC_2$, PTH_1, PTH_2, GLP_1, GLP_2, CRL, GHRH	C, L−
Vasoactive intestinal peptide receptor			
$VPAC_1$	+	$VPAC_2$	C, L−
$VPAC_2$	+		C, L−

Family C Receptors			
Receptors	Oligomers		Constitutive (C), Ligand-Dependent (L+/L−)
	Homomers	Heteromers	
Calcium-sensing			
CaR	+	mGluR1, mGluR5	C
GABA			
$GABA_B$	N.D.	$GABA_BR2$, mGluR	C
Metabotropic glutamate			
mGluR1	N.D.	A_1, CaR	C
mGluR5	+	A_{2A}, CaR	C
Taste			
T1Rs, T2Rs, T3Rs	N.D	T1R3	C

+, presence of homodimers; −, absence of homodimers; C, constitutive; L+, ligand-dependent association; L−, ligand-dependent dissociation; N.D., not determined; FLIM, fluorescence lifetime imaging.

Techniques Employed	Functions Attributed	References
Co-IP, FRET	Trafficking	[55]
BiFC, BRET, Co-IP	N.D.	[16, 148]
Co-IP, FRET	Signaling	[149, 150]
BRET	No effect	[151]
BiFC, BRET, FRET	Signaling	[28, 37, 50–52]
BRET, Co-IP, FRET	No effect	[37, 152]
BRET, Co-IP, FRET	No effect	[37, 152]

Techniques Employed	Functions Attributed	References
BRET, Co-IP	Signaling, trafficking	[153–156]
Co-IP	Binding, signaling	[21, 22, 157]
Co-IP	Signaling, trafficking	[154, 158, 159]
Co-IP, crystal structure	Signaling, trafficking	[14, 160]
Co-IP	Signaling	[24, 161]

The cytosolic region of GPCR dimeric complexes has also been implicated as playing an important functional role, such as the coiled–coil interaction demonstrated for some Family C GPCRs [22]. While far from proven and not currently representing a favored concept, the possibility of one heterotrimeric G protein interacting with a dimeric GPCR complex has been proposed [32]. The nature and stoichiometry of G protein interaction with one or more GPCR protomers in an oligomeric complex has not yet been critically examined. It is noteworthy that there is clear experimental evidence of the ability of a single, isolated GPCR unit to be fully functional [33, 34].

6.3. FUNCTIONAL EFFECTS OF GPCR OLIGOMERIZATION

There are numerous examples of natural ligands disrupting GPCR oligomeric complexes, in some cases inducing or stabilizing them, and in other instances having no effect on their status of association [35–38]. The absence of apparent consistent functional effects has been one of the features that critics of this phenomenon have focused on in supporting the concept of artifactual association of many of these receptors [3]. Nevertheless, there are adequate examples of clear and substantial functional effects to support the potential importance of GPCR association. Proposed functional roles for GPCR dimerization or oligomerization have spanned modifications of protein folding and intracellular trafficking, the selectivity of ligand binding and activation, the selectivity of G protein association, the efficiency of signal transduction, and the physiological functions observed.

One of the most interesting functional effects is the possibility of changing biological responsiveness of a receptor system. There is the possibility that a dimeric or oligomeric GPCR complex could bind and be activated by a distinct ligand that would not bind or activate a monomeric receptor. The best current example of this is 6′-guanidinonaltrindole, a delta and kappa opioid heterodimer-selective agonist [39]. Targeting of a broad variety of heterodimeric complexes of GPCRs has become an attractive goal for pharmaceutical companies to increase the specificity of drugs that are in development. There is also the possibility that dimeric or oligomeric GPCR complexes would couple to different effectors and result in distinct signaling events in the cell. Here, too, this is more theoretical than definitively established, but it may serve as an explanation for differences in signaling observed in distinct cellular backgrounds.

Another potentially important group of functional effects is the ability of binding to one protomer of a dimeric or oligomeric complex to result in functional effects on other receptor protomers in the complex. This has been termed asymmetry of action, with the occupation of one protomer resulting in G protein coupling to the associated protomer or to the internalization of the associated protomer. This might also manifest itself as positive or negative cooperativity, where the binding of a ligand molecule to a receptor changes the properties of binding the next ligand molecule to that receptor or to the other receptor protomer within a complex. Indeed, this phenomenon has been demonstrated for many members of this superfamily [40, 41]. The molecular basis of the cooperativity events will need to be sorted out, as it may be distinct for receptors in different families or for unique pairs of receptors in a given family.

6.4. SECRETIN RECEPTOR OLIGOMERIZATION

The secretin receptor was the first Family B GPCR that was identified [42]. It was soon followed by the identification of the receptors for calcitonin and parathyroid hormone [43, 44], with their sequence similarities defining this receptor family. All the receptors in Family B have moderately large natural peptide ligands that have diffused pharmacophoric domains. The amino-terminal region of these peptides is typically necessary for receptor activation, while the carboxyl-terminal region is typically necessary for high-affinity binding [45]. All of the Family B receptors have long amino-terminal tail regions in excess of 120 residues that incorporate six conserved cysteine residues that form three conserved intradomain disulfide bonds [46]. This domain of Family B GPCRs has been shown to be critical for natural ligand binding, providing a structural platform for the peptide interaction [47]. The sequences within the transmembrane segments that are typical of Family A GPCRs are absent in Family B GPCRs, although these receptors are also predicted to form a heptahelical transmembrane bundle [48]. Sequence analysis has predicted the structural features of the Family B GPCR helical bundle to be quite different from that of Family A GPCRs [49]. The secretin receptor is prototypic of this family in each of its structural and functional characteristics.

The secretin receptor has been shown to form homo-oligomeric complexes, utilizing BRET, FRET, saturation BRET, morphologic FRET, and BiFC [28, 37, 50–53]. These complexes have been shown to form constitutively [50] as they traverse the biosynthetic machinery [53]. Some typical data to support such complexes is illustrated in Fig. 6.2. When on the cell surface, these complexes are quite stable and are not affected by secretin binding and receptor activation [37]. Of note, the secretin receptor has also been shown to be able to form hetero-oligomeric complexes with almost all Family B GPCRs [52], yet has not been shown to form such complexes with GPCRs in Family A [53]. Clearly, there is structural specificity that determines which receptors are capable of associating with the secretin receptor.

In an effort to determine the portion of the secretin receptor that is most responsible for its oligomerization, a series of studies were performed [28, 53]. Truncation of the extracellular amino terminus of this receptor did not disrupt the BRET signal between receptor constructs [53]. Similarly, truncation of the intracellular carboxyl terminus of the secretin receptor did not disrupt the BRET signal between receptor constructs [53]. This focused attention on the helical bundle region. In an analogous approach to what had first been utilized

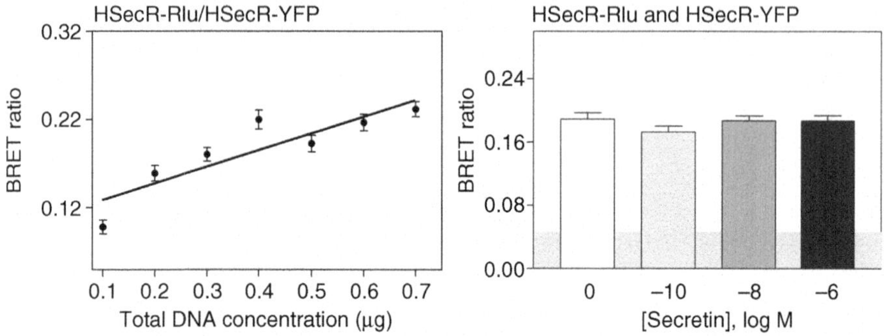

Figure 6.2 Study of secretin receptor oligomerization by BRET. *Left* panel shows that even at low levels of receptor expression, BRET ratios were significant, with these increasing in a linear manner with increasing amounts of donor (HSecR-Rlu) and acceptor (HSecR-YFP) expressed in COS cells. *Right* panel shows BRET ratios for COS cells expressing the donor (HSecR-Rlu) and acceptor (HSecR-YFP) in the presence of increasing of concentrations of the natural agonist ligand, secretin. The *shaded* area represents the nonspecific BRET signal that can be generated between Rlu-tagged receptor and soluble YFP protein or between YFP-tagged receptor and soluble Rlu, with BRET signals above this area considered to be significant. Secretin, at concentrations as high as 1 μM, had no effect on secretin receptor BRET ratios.

with the β2-adrenergic receptor [20], peptides corresponding to the predicted transmembrane segment peptides were utilized in an attempt to compete for the receptor oligomerization [28]. Of note, from all of the segments studied, only one of the transmembrane segment (TM4) successfully competed for the secretin receptor BRET signal, and this competition was absent even when all the other six transmembrane segment peptides were used in combination. This supported the importance of transmembrane segment four in secretin receptor dimerization. By mutating residues on the receptor that corresponded to predicted interhelical faces or the lipid-exposed face, it was suggested that the lipid face represented the relevant determinant of oligomerization [28]. This was then confirmed by introducing mutations into the lipid face of transmembrane helix four of the intact secretin receptor, and observing that this construct did not associate with itself, as reflected in the BRET assay [28]. Again, the specificity of this observation was carefully confirmed with saturation BRET experiments. This unique construct has also been quite useful for generating insights into the functional effects of dimerization of the secretin receptor, as there was a clear functional defect, with reduced cAMP stimulation, in the nondimerizing mutant [28].

It was noteworthy that only the transmembrane region of the secretin receptor appeared to be important for receptor oligomerization, and that only one of seven transmembrane segments appeared to be involved in this process. If the secretin receptor were involved in a high-order oligomerization complex, there would be the expectation that more than a single transmembrane segment interface would be critical. This led to the hypothesis that the secretin receptor was involved in a dimeric complex, rather than a higher-order complex. Additional studies were performed to experimentally test this hypothesis. One creative approach utilized BiFC, with receptors tagged with each of the component halves (YN and YC) of YFP [51]. When the non-fluorescent constructs were coexpressed in the same cells, the secretin receptor dimers formed and reconstituted an intact YFP that produced the characteristic fluorescence of that moiety. The YN/YC complemented secretin dimer provided a fluorescence acceptor for BRET studies with other secretin receptor constructs tagged at the carboxyl-terminal tail with *Renilla* luciferase (Rlu). However, coexpression of Rlu-tagged secretin receptor fluorescent donor with the reconstituted YN/YC secretin receptor acceptor dimer did not yield a significant resonance transfer signal. This experiment was repeated with secretin receptor constructs in which the Rlu or cyan fluorescent protein (CFP) was incorporated into each of the intracellular loop regions of the secretin receptor. These also failed to generate significant resonance transfer signals, suggesting that higher-order oligomers of the secretin receptor do not exist [51].

It will be important to extend these observations to further examine the molecular nature of the secretin receptor homodimeric complex, carefully mapping the interface and determining if this changes with activation. Such studies have recently been performed for Family A GPCRs, including the dopamine D2 receptor [26], the serotonin 5-HT4 receptor [54], and the serotonin 5-HT2C receptor [31]. Similarly, it will be critical to carefully examine the function of dimeric complexes of Family B GPCRs. Everything to date suggests that these studies will identify unique and important themes. It is remarkable that almost all other members of Family B GPCRs were shown to be capable of association with the secretin receptor [52]. Of the nine receptors tested, only the human calcitonin receptor did not yield a significant resonance transfer signal. Assuming that the fourth transmembrane segment might be the relevant determinant for this oligomerization, as it was for the secretin receptor, the human calcitonin receptor that was studied was found to have two differences in lipid-facing residues that were not present in other Family B GPCRs. Of interest, these residues were different in the rabbit calcitonin receptor, and, indeed, that receptor has been shown to form homo-oligomeric complexes [55]. While more work is clearly necessary to sort this out, it could provide further support for a common structural determinant for dimerization of receptors in this important family.

REFERENCES

1. Bouvier, M. (2001) Oligomerization of G-protein-coupled transmitter receptors. *Nat Rev Neurosci.* 2, 274–286.

2. Milligan, G., Ramsay, D., Pascal, G., Carrillo, J.J. (2003) GPCR dimerisation. *Life Sci.* 74, 181–188.
3. James, J.R., Oliveira, M.I., Carmo, A.M., Iaboni, A., Davis, S.J. (2006) A rigorous experimental framework for detecting protein oligomerization using bioluminescence resonance energy transfer. *Nat Methods*. 3, 1001–1006.
4. Bouvier, M., Heveker, N., Jockers, R., Marullo, S., Milligan, G. (2007) BRET analysis of GPCR oligomerization: Newer does not mean better. *Nat Methods*. 4, 3–4.
5. Cvejic, S., Devi, L.A. (1997) Dimerization of the delta opioid receptor: Implication for a role in receptor internalization. *J Biol Chem.* 272, 26959–26964.
6. Angers, S., Salahpour, A., Joly, E., Hilairet, S., Chelsky, D., Dennis, M., Bouvier, M. (2000) Detection of β2-adrenergic receptor dimerization in living cells using bioluminescence resonance energy transfer (BRET). *Proc Natl Acad Sci U S A*. 97, 3684–3689.
7. McVey, M., Ramsay, D., Kellett, E., Rees, S., Wilson, S., Pope, A.J., Milligan, G. (2001) Monitoring receptor oligomerization using time-resolved fluorescence resonance energy transfer and bioluminescence resonance energy transfer. The human delta-opioid receptor displays constitutive oligomerization at the cell surface, which is not regulated by receptor occupancy. *J Biol Chem.* 276, 14092–14099.
8. Stryer, L. (1978) Fluorescence energy transfer as a spectroscopic ruler. *Annu Rev Biochem*. 47, 819–846.
9. Mercier, J.F., Salahpour, A., Angers, S., Breit, A., Bouvier, M. (2002) Quantitative assessment of beta 1- and beta 2-adrenergic receptor homo- and heterodimerization by bioluminescence resonance energy transfer. *J Biol Chem.* 277, 44925–44931.
10. Ellis, J., Pediani, J.D., Canals, M., Milasta, S., Milligan, G. (2006) Orexin-1 receptor-cannabinoid CB1 receptor heterodimerization results in both ligand-dependent and -independent coordinated alterations of receptor localization and function. *J Biol Chem.* 281, 38812–38824.
11. Fiorentini, C., Busi, C., Gorruso, E., Gotti, C., Spano, P., Missale, C. (2008) Reciprocal regulation of dopamine D1 and D3 receptor function and trafficking by heterodimerization. *Mol Pharmacol*. 74, 59–69.
12. Kong, M.M., Fan, T., Varghese, G., O'Dowd, B.F., George, S.R. (2006) Agonist-induced cell surface trafficking of an intracellularly sequestered D1 dopamine receptor homo-oligomer. *Mol Pharmacol*. 70, 78–89.
13. Springael, J.Y., Urizar, E., Parmentier, M. (2005) Dimerization of chemokine receptors and its functional consequences. *Cytokine Growth Factor Rev.* 16, 611–623.
14. Kunishima, N., Shimada, Y., Tsuji, Y., Sato, T., Yamamoto, M., Kumasaka, T., Nakanishi, S., Jingami, H., Morikawa, K. (2000) Structural basis of glutamate recognition by a dimeric metabotropic glutamate receptor. *Nature*. 407, 971–977.
15. Fotiadis, D., Liang, Y., Filipek, S., Saperstein, D.A., Engel, A., Palczewski, K. (2003) Atomic-force microscopy: Rhodopsin dimers in native disc membranes. *Nature*. 421, 127–128.
16. Heroux, M., Hogue, M., Lemieux, S., Bouvier, M. (2007) Functional calcitonin gene-related peptide receptors are formed by the asymmetric assembly of a calcitonin receptor-like receptor homo-oligomer and a monomer of receptor activity-modifying protein-1. *J Biol Chem.* 282, 31610–31620.
17. Scarselli, M., Novi, F., Schallmach, E., Lin, R., Baragli, A., Colzi, A., Griffon, N., Corsini, G.U., Sokoloff, P., Levenson, R., Vogel, Z., Maggio, R. (2001) D2/D3

dopamine receptor heterodimers exhibit unique functional properties. *J Biol Chem.* 276, 30308–30314.

18. Milligan, G. (2008) A day in the life of a G protein-coupled receptor: The contribution to function of G protein-coupled receptor dimerization. *Br J Pharmacol.* 153 (Suppl. 1), S216–S229.
19. Gouldson, P.R., Reynolds, C.A. (1997) Simulations on dimeric peptides: Evidence for domain swapping in G-protein-coupled receptors? *Biochem Soc Trans.* 25, 1066–1071.
20. Hebert, T.E., Moffett, S., Morello, J.P., Loisel, T.P., Bichet, D.G., Barret, C., Bouvier, M. (1996) A peptide derived from a beta2-adrenergic receptor transmembrane domain inhibits both receptor dimerization and activation. *J Biol Chem.* 271, 16384–16392.
21. Jones, K.A., Borowsky, B., Tamm, J.A., Craig, D.A., Durkin, M.M., Dai, M., Yao, W.J., Johnson, M., Gunwaldsen, C., Huang, L.Y., Tang, C., Shen, Q., Salon, J.A., Morse, K., Laz, T., Smith, K.E., Nagarathnam, D., Noble, S.A., Branchek, T.A., Gerald, C. (1998) GABA(B) receptors function as a heteromeric assembly of the subunits GABA(B)R1 and GABA(B)R2. *Nature.* 396, 674–679.
22. White, J.H., Wise, A., Main, M.J., Green, A., Fraser, N.J., Disney, G.H., Barnes, A.A., Emson, P., Foord, S.M., Marshall, F.H. (1998) Heterodimerization is required for the formation of a functional GABA(B) receptor. *Nature.* 396, 679–682.
23. Romano, C., Miller, J.K., Hyrc, K., Dikranian, S., Mennerick, S., Takeuchi, Y., Goldberg, M.P., O'Malley, K.L. (2001) Covalent and noncovalent interactions mediate metabotropic glutamate receptor mGlu5 dimerization. *Mol Pharmacol.* 59, 46–53.
24. Zhao, G.Q., Zhang, Y., Hoon, M.A., Chandrashekar, J., Erlenbach, I., Ryba, N.J., Zuker, C.S. (2003) The receptors for mammalian sweet and umami taste. *Cell.* 115, 255–266.
25. Filizola, M., Weinstein, H. (2005) The study of G-protein coupled receptor oligomerization with computational modeling and bioinformatics. *FEBS J.* 272, 2926–2938.
26. Guo, W., Shi, L., Filizola, M., Weinstein, H., Javitch, J.A. (2005) Crosstalk in G protein-coupled receptors: Changes at the transmembrane homodimer interface determine activation. *Proc Natl Acad Sci U S A.* 102, 17495–17500.
27. Harikumar, K.G., Dong, M., Cheng, Z., Pinon, D.I., Lybrand, T.P., Miller, L.J. (2006) Transmembrane segment peptides can disrupt cholecystokinin receptor oligomerization without affecting receptor function. *Biochemistry.* 45, 14706–14716.
28. Harikumar, K.G., Pinon, D.I., Miller, L.J. (2007) Transmembrane segment IV contributes a functionally important interface for oligomerization of the Class II G protein-coupled secretin receptor. *J Biol Chem.* 282, 30363–30372.
29. Hernanz-Falcon, P., Rodriguez-Frade, J.M., Serrano, A., Juan, D., del Sol, A., Soriano, S.F., Roncal, F., Gomez, L., Valencia, A., Martinez, A.C., Mellado, M. (2004) Identification of amino acid residues crucial for chemokine receptor dimerization. *Nat Immunol.* 5, 216–223.
30. Lopez-Gimenez, J.F., Canals, M., Pediani, J.D., Milligan, G. (2007) The alpha1b-adrenoceptor exists as a higher-order oligomer: Effective oligomerization is required for receptor maturation, surface delivery, and function. *Mol Pharmacol.* 71, 1015–1029.

31. Mancia, F., Assur, Z., Herman, A.G., Siegel, R., Hendrickson, W.A. (2008) Ligand sensitivity in dimeric associations of the serotonin 5HT2c receptor. *EMBO Rep.* 9, 363–369.

32. Herrick-Davis, K., Grinde, E., Harrigan, T.J., Mazurkiewicz, J.E. (2005) Inhibition of serotonin 5-hydroxytryptamine2c receptor function through heterodimerization: Receptor dimers bind two molecules of ligand and one G-protein. *J Biol Chem.* 280, 40144–40151.

33. Meyer, B.H., Segura, J.M., Martinez, K.L., Hovius, R., George, N., Johnsson, K., Vogel, H. (2006) FRET imaging reveals that functional neurokinin-1 receptors are monomeric and reside in membrane microdomains of live cells. *Proc Natl Acad Sci U S A.* 103, 2138–2143.

34. Whorton, M.R., Bokoch, M.P., Rasmussen, S.G., Huang, B., Zare, R.N., Kobilka, B., Sunahara, R.K. (2007) A monomeric G protein-coupled receptor isolated in a high-density lipoprotein particle efficiently activates its G protein. *Proc Natl Acad Sci U S A.* 104, 7682–7687.

35. Ayoub, M.A., Couturier, C., Lucas-Meunier, E., Angers, S., Fossier, P., Bouvier, M., Jockers, R. (2002) Monitoring of ligand-independent dimerization and ligand-induced conformational changes of melatonin receptors in living cells by bioluminescence resonance energy transfer. *J Biol Chem.* 277, 21522–21528.

36. Cheng, Z.J., Miller, L.J. (2001) Agonist-dependent dissociation of oligomeric complexes of G protein-coupled cholecystokinin receptors demonstrated in living cells using bioluminescence resonance energy transfer. *J Biol Chem.* 276, 48040–48047.

37. Harikumar, K.G., Morfis, M.M., Lisenbee, C.S., Sexton, P.M., Miller, L.J. (2006) Constitutive formation of oligomeric complexes between family B G protein-coupled vasoactive intestinal polypeptide and secretin receptors. *Mol Pharmacol.* 69, 363–373.

38. Patel, R.C., Kumar, U., Lamb, D.C., Eid, J.S., Rocheville, M., Grant, M., Rani, A., Hazlett, T., Patel, S.C., Gratton, E., Patel, Y.C. (2002) Ligand binding to somatostatin receptors induces receptor-specific oligomer formation in live cells. *Proc Natl Acad Sci U S A.* 99, 3294–3299.

39. Waldhoer, M., Fong, J., Jones, R.M., Lunzer, M.M., Sharma, S.K., Kostenis, E., Portoghese, P.S., Whistler, J.L. (2005) A heterodimer-selective agonist shows in vivo relevance of G protein-coupled receptor dimers. *Proc Natl Acad Sci U S A.* 102, 9050–9055.

40. Albizu, L., Balestre, M.N., Breton, C., Pin, J.P., Manning, M., Mouillac, B., Barberis, C., Durroux, T. (2006) Probing the existence of G protein-coupled receptor dimers by positive and negative ligand-dependent cooperative binding. *Mol Pharmacol.* 70, 1783–1791.

41. Pin, J.P., Kniazeff, J., Liu, J., Binet, V., Goudet, C., Rondard, P., Prezeau, L. (2005) Allosteric functioning of dimeric class C G-protein-coupled receptors. *FEBS J.* 272, 2947–2955.

42. Ishihara, T., Nakamura, S., Kaziro, Y., Takahashi, T., Takahashi, K., Nagata, S. (1991) Molecular cloning and expression of a cDNA encoding the secretin receptor. *EMBO J.* 10, 1635–1641.

43. Lin, H.Y., Harris, T.L., Flannery, M.S., Aruffo, A., Kaji, E.H., Gorn, A., Kolakowski Jr., L.F., Lodish, H.F., Goldring, S.R. (1991) Expression cloning of an adenylate cyclase-coupled calcitonin receptor. *Science.* 254, 1022–1024.

44. McCuaig, K.A., Clarke, J.C., White, J.H. (1994) Molecular cloning of the gene encoding the mouse parathyroid hormone/parathyroid hormone-related peptide receptor. *Proc Natl Acad Sci U S A*. 91, 5051–5055.
45. Dong, M., Zang, M., Pinon, D.I., Li, Z., Lybrand, T.P., Miller, L.J. (2002) Interaction among four residues distributed through the secretin pharmacophore and a focused region of the secretin receptor amino terminus. *Mol Endocrinol.* 16, 2490–2501.
46. Lisenbee, C.S., Dong, M., Miller, L.J. (2005) Paired cysteine mutagenesis to establish the pattern of disulfide bonds in the functional intact secretin receptor. *J Biol Chem.* 280, 12330–12338.
47. Grace, C.R., Perrin, M.H., DiGruccio, M.R., Miller, C.L., Rivier, J.E., Vale, W.W., Riek, R. (2004) NMR structure and peptide hormone binding site of the first extracellular domain of a type B1 G protein-coupled receptor. *Proc Natl Acad Sci U S A*. 101, 12836–12841.
48. Harmar, A.J. (2001) Family-B G-protein-coupled receptors. *Genome Biol.* 2, 3013, 1–10.
49. Donnelly, D. (1997) The arrangement of the transmembrane helices in the secretin receptor family of G-protein-coupled receptors. *FEBS Lett.* 409, 431–436.
50. Ding, W.Q., Cheng, Z.J., McElhiney, J., Kuntz, S.M., Miller, L.J. (2002) Silencing of secretin receptor function by dimerization with a misspliced variant secretin receptor in ductal pancreatic adenocarcinoma. *Cancer Res.* 62, 5223–5229.
51. Harikumar, K.G., Happs, R.M., Miller, L.J. (2008) Dimerization in the absence of higher-order oligomerization of the G protein-coupled secretin receptor. *Biochim Biophys Acta*. 1778, 2555–2563.
52. Harikumar, K.G., Morfis, M.M., Sexton, P.M., Miller, L.J. (2008) Pattern of intra-family hetero-oligomerization involving the G protein-coupled secretin receptor. *J Mol Neurosci*. 36, 279–285.
53. Lisenbee, C.S., Miller, L.J. (2006) Secretin receptor oligomers form intracellularly during maturation through receptor core domains. *Biochemistry*. 45, 8216–8226.
54. Berthouze, M., Rivail, L., Lucas, A., Ayoub, M.A., Russo, O., Sicsic, S., Fischmeister, R., Berque-Bestel, I., Jockers, R., Lezoualc'h, F. (2007) Two transmembrane Cys residues are involved in 5-HT4 receptor dimerization. *Biochem Biophys Res Commun*. 356, 642–647.
55. Seck, T., Baron, R., Horne, W.C. (2003) The alternatively spliced deltae13 transcript of the rabbit calcitonin receptor dimerizes with the C1a isoform and inhibits its surface expression. *J Biol Chem.* 278, 23085–23093.
56. Ciruela, F., Casado, V., Rodrigues, R.J., Lujan, R., Burgueno, J., Canals, M., Borycz, J., Rebola, N., Goldberg, S.R., Mallol, J., Cortes, A., Canela, E.I., Lopez-Gimenez, J.F., Milligan, G., Lluis, C., Cunha, R.A., Ferre, S., Franco, R. (2006) Presynaptic control of striatal glutamatergic neurotransmission by adenosine A1-A2A receptor heteromers. *J Neurosci*. 26, 2080–2087.
57. Kamiya, T., Saitoh, O., Yoshioka, K., Nakata, H. (2003) Oligomerization of adenosine A2A and dopamine D2 receptors in living cells. *Biochem Biophys Res Commun*. 306, 544–549.
58. Suzuki, T., Namba, K., Tsuga, H., Nakata, H. (2006) Regulation of pharmacology by hetero-oligomerization between A1 adenosine receptor and P2Y2 receptor. *Biochem Biophys Res Commun*. 351, 559–565.

59. Yoshioka, K., Saitoh, O., Nakata, H. (2002) Agonist-promoted heteromeric oligomerization between adenosine A(1) and P2Y(1) receptors in living cells. *FEBS Lett.* 523, 147–151.
60. Canals, M., Burgueno, J., Marcellino, D., Cabello, N., Canela, E.I., Mallol, J., Agnati, L., Ferre, S., Bouvier, M., Fuxe, K., Ciruela, F., Lluis, C., Franco, R. (2004) Homodimerization of adenosine A2A receptors: Qualitative and quantitative assessment by fluorescence and bioluminescence energy transfer. *J Neurochem.* 88, 726–734.
61. Canals, M., Marcellino, D., Fanelli, F., Ciruela, F., de Benedetti, P., Goldberg, S.R., Neve, K., Fuxe, K., Agnati, L.F., Woods, A.S., Ferre, S., Lluis, C., Bouvier, M., Franco, R. (2003) Adenosine A2A-dopamine D2 receptor-receptor heteromerization: Qualitative and quantitative assessment by fluorescence and bioluminescence energy transfer. *J Biol Chem.* 278, 46741–46749.
62. Ramsay, D., Kellett, E., McVey, M., Rees, S., Milligan, G. (2002) Homo- and hetero-oligomeric interactions between G protein-coupled receptors in living cells monitored by two variants of bioluminescence resonance energy transfer (BRET): Hetero-oligomers between receptor subtypes form more efficiently than between less closely related sequences. *Biochem J.* 365, 429–440.
63. Stanasila, L., Perez, J.B., Vogel, H., Cotecchia, S. (2003) Oligomerization of the alpha 1a- and alpha 1b-adrenergic receptor subtypes. Potential implications in receptor internalization. *J Biol Chem.* 278, 40239–40251.
64. Uberti, M.A., Hall, R.A., Minneman, K.P. (2003) Subtype-specific dimerization of alpha 1-adrenoceptors: Effects on receptor expression and pharmacological properties. *Mol Pharmacol.* 64, 1379–1390.
65. Small, K.M., Schwarb, M.R., Glinka, C., Theiss, C.T., Brown, K.M., Seman, C.A., Liggett, S.B. (2006) Alpha2A- and alpha2C-adrenergic receptors form homo- and heterodimers: The heterodimeric state impairs agonist-promoted GRK phosphorylation and beta-arrestin recruitment. *Biochemistry.* 45, 4760–4767.
66. Xu, J., He, J., Castleberry, A.M., Balasubramanian, S., Lau, A.G., Hall, R.A. (2003) Heterodimerization of alpha 2A- and beta 1-adrenergic receptors. *J Biol Chem.* 278, 10770–10777.
67. Lavoie, C., Mercier, J.F., Salahpour, A., Umapathy, D., Breit, A., Villeneuve, L.R., Zhu, W.Z., Xiao, R.P., Lakatta, E.G., Bouvier, M., Hebert, T.E. (2002) Beta 1/beta 2-adrenergic receptor heterodimerization regulates beta 2-adrenergic receptor internalization and ERK signaling efficacy. *J Biol Chem.* 277, 35402–35410.
68. Breit, A., Lagace, M., Bouvier, M. (2004) Hetero-oligomerization between beta2- and beta3-adrenergic receptors generates a beta-adrenergic signaling unit with distinct functional properties. *J Biol Chem.* 279, 28756–28765.
69. Jordan, B.A., Trapaidze, N., Gomes, I., Nivarthi, R., Devi, L.A. (2001) Oligomerization of opioid receptors with beta 2-adrenergic receptors: A role in trafficking and mitogen-activated protein kinase activation. *Proc Natl Acad Sci U S A.* 98, 343–348.
70. Salahpour, A., Angers, S., Mercier, J.F., Lagace, M., Marullo, S., Bouvier, M. (2004) Homodimerization of the beta2-adrenergic receptor as a prerequisite for cell surface targeting. *J Biol Chem.* 279, 33390–33397.
71. AbdAlla, S., Lother, H., el Massiery, A., Quitterer, U. (2001) Increased AT(1) receptor heterodimers in preeclampsia mediate enhanced angiotensin II responsiveness. *Nat Med.* 7, 1003–1009.

72. AbdAlla, S., Lother, H., Quitterer, U. (2000) AT1-receptor heterodimers show enhanced G-protein activation and altered receptor sequestration. *Nature*. 407, 94–98.
73. AbdAlla, S., Zaki, E., Lother, H., Quitterer, U. (1999) Involvement of the amino terminus of the B(2) receptor in agonist-induced receptor dimerization. *J Biol Chem*. 274, 26079–26084.
74. Hansen, J.L., Theilade, J., Haunso, S., Sheikh, S.P. (2004) Oligomerization of wild type and nonfunctional mutant angiotensin II type I receptors inhibits $G_{\alpha q}$ protein signaling but not ERK activation. *J Biol Chem*. 279, 24108–24115.
75. Barki-Harrington, L., Luttrell, L.M., Rockman, H.A. (2003) Dual inhibition of beta-adrenergic and angiotensin II receptors by a single antagonist: A functional role for receptor-receptor interaction in vivo. *Circulation*. 108, 1611–1618.
76. Miura, S., Karnik, S.S., Saku, K. (2005) Constitutively active homo-oligomeric angiotensin II type 2 receptor induces cell signaling independent of receptor conformation and ligand stimulation. *J Biol Chem*. 280, 18237–18244.
77. Abadir, P.M., Periasamy, A., Carey, R.M., Siragy, H.M. (2006) Angiotensin II type 2 receptor-bradykinin B2 receptor functional heterodimerization. *Hypertension*. 48, 316–322.
78. Kang, D.S., Gustafsson, C., Morgelin, M., Leeb-Lundberg, L.M. (2005) B1 bradykinin receptor homo-oligomers in receptor cell surface expression and signaling: Effects of receptor fragments. *Mol Pharmacol*. 67, 309–318.
79. Kearn, C.S., Blake-Palmer, K., Daniel, E., Mackie, K., Glass, M. (2005) Concurrent stimulation of cannabinoid CB1 and dopamine D2 receptors enhances heterodimer formation: A mechanism for receptor cross-talk? *Mol Pharmacol*. 67, 1697–1704.
80. Rios, C., Gomes, I., Devi, L.A. (2006) mu opioid and CB1 cannabinoid receptor interactions: Reciprocal inhibition of receptor signaling and neuritogenesis. *Br J Pharmacol*. 148, 387–395.
81. Wager-Miller, J., Westenbroek, R., Mackie, K. (2002) Dimerization of G protein-coupled receptors: CB1 cannabinoid receptors as an example. *Chem Phys Lipids*. 121, 83–89.
82. Wilson, S., Wilkinson, G., Milligan, G. (2005) The CXCR1 and CXCR2 receptors form constitutive homo- and heterodimers selectively and with equal apparent affinities. *J Biol Chem*. 280, 28663–28674.
83. Parenty, G., Appelbe, S., Milligan, G. (2008) CXCR2 chemokine receptor antagonism enhances DOP opioid receptor function via allosteric regulation of the CXCR2-DOP receptor heterodimer. *Biochem J*. 412, 245–256.
84. El-Asmar, L., Springael, J.Y., Ballet, S., Andrieu, E.U., Vassart, G., Parmentier, M. (2005) Evidence for negative binding cooperativity within CCR5-CCR2b heterodimers. *Mol Pharmacol*. 67, 460–469.
85. Huttenrauch, F., Pollok-Kopp, B., Oppermann, M. (2005) G protein-coupled receptor kinases promote phosphorylation and beta-arrestin-mediated internalization of CCR5 homo- and hetero-oligomers. *J Biol Chem*. 280, 37503–37515.
86. Issafras, H., Angers, S., Bulenger, S., Blanpain, C., Parmentier, M., Labbe-Jullie, C., Bouvier, M., Marullo, S. (2002) Constitutive agonist-independent CCR5 oligomerization and antibody-mediated clustering occurring at physiological levels of receptors. *J Biol Chem*. 277, 34666–34673.

87. Suzuki, S., Chuang, L.F., Yau, P., Doi, R.H., Chuang, R.Y. (2002) Interactions of opioid and chemokine receptors: Oligomerization of mu, kappa, and delta with CCR5 on immune cells. *Exp Cell Res.* 280, 192–200.

88. Cheng, Z.J., Harikumar, K.G., Holicky, E.L., Miller, L.J. (2003) Heterodimerization of type A and B cholecystokinin receptors enhance signaling and promote cell growth. *J Biol Chem.* 278, 52972–52979.

89. Dziedzicka-Wasylewska, M., Faron-Gorecka, A., Andrecka, J., Polit, A., Kusmider, M., Wasylewski, Z. (2006) Fluorescence studies reveal heterodimerization of dopamine D1 and D2 receptors in the plasma membrane. *Biochemistry.* 45, 8751–8759.

90. Fiorentini, C., Gardoni, F., Spano, P., Di Luca, M., Missale C. (2003) Regulation of dopamine D1 receptor trafficking and desensitization by oligomerization with glutamate N-methyl-D-aspartate receptors. *J Biol Chem.* 278, 20196–20202.

91. O'Dowd, B.F., Ji, X., Alijaniaram, M., Rajaram, R.D., Kong, M.M., Rashid, A., Nguyen, T., George, S.R. (2005) Dopamine receptor oligomerization visualized in living cells. *J Biol Chem.* 280, 37225–37235.

92. So, C.H., Varghese, G., Curley, K.J., Kong, M.M., Alijaniaram, M., Ji, X., Nguyen, T., O'Dowd, B.F., George, S.R. (2005) D1 and D2 dopamine receptors form heterooligomers and cointernalize after selective activation of either receptor. *Mol Pharmacol.* 68, 568–578.

93. Armstrong, D., Strange, P.G. (2001) Dopamine D2 receptor dimer formation: Evidence from ligand binding. *J Biol Chem.* 276, 22621–22629.

94. Rocheville, M., Lange, D.C., Kumar, U., Patel, S.C., Patel, R.C., Patel, Y.C. (2000) Receptors for dopamine and somatostatin: Formation of hetero-oligomers with enhanced functional activity. *Science.* 288, 154–157.

95. Karpa, K.D., Lin, R., Kabbani, N., Levenson, R. (2000) The dopamine D3 receptor interacts with itself and the truncated D3 splice variant d3nf: D3-D3nf interaction causes mislocalization of D3 receptors. *Mol Pharmacol.* 58, 677–683.

96. Evans, N.J., Walker, J.W. (2008) Endothelin receptor dimers evaluated by FRET, ligand binding, and calcium mobilization. *Biophys J.* 95, 483–492.

97. Gregan, B., Jurgensen, J., Papsdorf, G., Furkert, J., Schaefer, M., Beyermann, M., Rosenthal, W., Oksche, A. (2004) Ligand-dependent differences in the internalization of endothelin A and endothelin B receptor heterodimers. *J Biol Chem.* 279, 27679–27687.

98. Wirz, S.A., Davis, C.N., Lu, X., Zal, T., Bartfai, T. (2005) Homodimerization and internalization of galanin type 1 receptor in living CHO cells. *Neuropeptides.* 39, 535–546.

99. Bakker, R.A., Dees, G., Carrillo, J.J., Booth, R.G., Lopez-Gimenez, J.F., Milligan, G., Strange, P.G., Leurs, R. (2004) Domain swapping in the human histamine H1 receptor. *J Pharmacol Exp Ther.* 311, 131–138.

100. Shenton, F.C., Hann, V., Chazot, P.L. (2005) Evidence for native and cloned H3 histamine receptor higher oligomers. *Inflamm Res.* 54 (Suppl. 1), S48–S49.

101. van Rijn, R.M., Chazot, P.L., Shenton, F.C., Sansuk, K., Bakker, R.A., Leurs, R. (2006) Oligomerization of recombinant and endogenously expressed human histamine H(4) receptors. *Mol Pharmacol.* 70, 604–615.

102. Sanchez-Laorden, B.L., Sanchez-Mas, J., Martinez-Alonso, E., Martinez-Menarguez, J.A., Garcia-Borron, J.C., Jimenez-Cervantes, C. (2006) Dimerization

of the human melanocortin 1 receptor: Functional consequences and dominant-negative effects. *J Invest Dermatol.* 126, 172–181.

103. Mandrika, I., Petrovska, R., Wikberg, J. (2005) Melanocortin receptors form constitutive homo- and heterodimers. *Biochem Biophys Res Commun.* 326, 349–354.
104. Kopanchuk, S., Veiksina, S., Mutulis, F., Mutule, I., Yahorava, S., Mandrika, I., Petrovska, R., Rinken, A., Wikberg, J.E. (2006) Kinetic evidence for tandemly arranged ligand binding sites in melanocortin 4 receptor complexes. *Neurochem Int.* 49, 533–542.
105. Ayoub, M.A., Levoye, A., Delagrange, P., Jockers, R. (2004) Preferential formation of MT1/MT2 melatonin receptor heterodimers with distinct ligand interaction properties compared with MT2 homodimers. *Mol Pharmacol.* 66, 312–321.
106. Goin, J.C., Nathanson, N.M. (2006) Quantitative analysis of muscarinic acetylcholine receptor homo- and heterodimerization in live cells: Regulation of receptor down-regulation by heterodimerization. *J Biol Chem.* 281, 5416–5425.
107. Perron, A., Sharif, N., Sarret, P., Stroh, T., Beaudet, A. (2007) NTS2 modulates the intracellular distribution and trafficking of NTS1 via heterodimerization. *Biochem Biophys Res Commun.* 353, 582–590.
108. White, J.F., Grodnitzky, J., Louis, J.M., Trinh, L.B., Shiloach, J., Gutierrez, J., Northup, J.K., Grisshammer, R. (2007) Dimerization of the class A G protein-coupled neurotensin receptor NTS1 alters G protein interaction. *Proc Natl Acad Sci U S A.* 104, 12199–12204.
109. Dinger, M.C., Bader, J.E., Kobor, A.D., Kretzschmar, A.K., Beck-Sickinger, A.G. (2003) Homodimerization of neuropeptide Y receptors investigated by fluorescence resonance energy transfer in living cells. *J Biol Chem.* 278, 10562–10571.
110. Gehlert, D.R., Schober, D.A., Morin, M., Berglund, M.M. (2007) Co-expression of neuropeptide Y Y1 and Y5 receptors results in heterodimerization and altered functional properties. *Biochem Pharmacol.* 74, 1652–1664.
111. Berglund, M.M., Schober, D.A., Esterman, M.A., Gehlert, D.R. (2003) Neuropeptide Y Y4 receptor homodimers dissociate upon agonist stimulation. *J Pharmacol Exp Ther.* 307, 1120–1126.
112. Angers, S., Bouvier, M. (2000) Reply: Beyond receptor dimerization. *Trends Pharmacol Sci.* 21, 326.
113. George, S.R., Fan, T., Xie, Z., Tse, R., Tam, V., Varghese, G., O'Dowd, B.F. (2000) Oligomerization of mu- and delta-opioid receptors. Generation of novel functional properties. *J Biol Chem.* 275, 26128–26135.
114. Gomes, I., Jordan, B.A., Gupta, A., Trapaidze, N., Nagy, V., Devi, L.A. (2000) Heterodimerization of mu and delta opioid receptors: A role in opiate synergy. *J Neurosci.* 20, RC110.
115. Jordan, B.A., Gomes, I., Rios, C., Filipovska, J., Devi, L.A. (2003) Functional interactions between mu opioid and alpha 2A-adrenergic receptors. *Mol Pharmacol.* 64, 1317–1324.
116. Pfeiffer, M., Kirscht, S., Stumm, R., Koch, T., Wu, D., Laugsch, M., Schroder, H., Hollt, V., Schulz, S. (2003) Heterodimerization of substance P and mu-opioid receptors regulates receptor trafficking and resensitization. *J Biol Chem.* 278, 51630–51637.
117. Pfeiffer, M., Koch, T., Schroder, H., Laugsch, M., Hollt, V., Schulz, S. (2002) Heterodimerization of somatostatin and opioid receptors cross-modulates phosphorylation, internalization, and desensitization. *J Biol Chem.* 277, 19762–19772.

118. Wang, D., Sun, X., Bohn, L.M., Sadee, W. (2005) Opioid receptor homo- and heterodimerization in living cells by quantitative bioluminescence resonance energy transfer. *Mol Pharmacol.* 67, 2173–2184.
119. Jordan, B.A., Devi, L.A. (1999) G-protein-coupled receptor heterodimerization modulates receptor function. *Nature.* 399, 697–700.
120. Kota, P., Reeves, P.J., Rajbhandary, U.L., Khorana, H.G. (2006) Opsin is present as dimers in COS1 cells: Identification of amino acids at the dimeric interface. *Proc Natl Acad Sci U S A.* 103, 3054–3059.
121. Devost, D., Zingg, H.H. (2003) Identification of dimeric and oligomeric complexes of the human oxytocin receptor by co-immunoprecipitation and bioluminescence resonance energy transfer. *J Mol Endocrinol.* 31, 461–471.
122. Terrillon, S., Durroux, T., Mouillac, B., Breit, A., Ayoub, M.A., Taulan, M., Jockers, R., Barberis, C., Bouvier, M. (2003) Oxytocin and vasopressin V1a and V2 receptors form constitutive homo- and heterodimers during biosynthesis. *Mol Endocrinol.* 17, 677–691.
123. Choi, R.C., Simon, J., Tsim, K.W., Barnard, E.A. (2008) Constitutive and agonist-induced dimerizations of the P2Y1 receptor: Relationship to internalization and scaffolding. *J Biol Chem.* 283, 11050–11063.
124. Ecke, D., Hanck, T., Tulapurkar, M.E., Schafer, R., Kassack, M., Stricker, R., Reiser, G. (2008) Hetero-oligomerization of the P2Y11 receptor with the P2Y1 receptor controls the internalization and ligand selectivity of the P2Y11 receptor. *Biochem J.* 409, 107–116.
125. Giguere, V., Gallant, M.A., de Brum-Fernandes, A.J., Parent, J.L. (2004) Role of extracellular cysteine residues in dimerization/oligomerization of the human prostacyclin receptor. *Eur J Pharmacol.* 494, 11–22.
126. McGraw, D.W., Mihlbachler, K.A., Schwarb, M.R., Rahman, F.F., Small, K.M., Almoosa, K.F., Liggett, S.B. (2006) Airway smooth muscle prostaglandin-EP1 receptors directly modulate beta2-adrenergic receptors within a unique heterodimeric complex. *J Clin Invest.* 116, 1400–1409.
127. Wilson, S.J., Dowling, J.K., Zhao, L., Carnish, E., Smyth, E.M. (2007) Regulation of thromboxane receptor trafficking through the prostacyclin receptor in vascular smooth muscle cells: Role of receptor heterodimerization. *Arterioscler Thromb Vasc Biol.* 27, 290–296.
128. Wilson, S.J., Roche, A.M., Kostetskaia, E., Smyth, E.M. (2004) Dimerization of the human receptors for prostacyclin and thromboxane facilitates thromboxane receptor-mediated cAMP generation. *J Biol Chem.* 279, 53036–53047.
129. Fotiadis, D., Liang, Y., Filipek, S., Saperstein, D.A., Engel, A., Palczewski, K. (2004) The G protein-coupled receptor rhodopsin in the native membrane. *FEBS Lett.* 564, 281–288.
130. Mansoor, S.E., Palczewski, K., Farrens, D.L. (2006) Rhodopsin self-associates in asolectin liposomes. *Proc Natl Acad Sci U S A.* 103, 3060–3065.
131. Kobe, F., Renner, U., Woehler, A., Wlodarczyk, J., Papusheva, E., Bao, G., Zeug, A., Richter, D.W., Neher, E., Ponimaskin, E. (2008) Stimulation- and palmitoylation-dependent changes in oligomeric conformation of serotonin 5-HT1A receptors. *Biochim Biophys Acta.* 1783, 1503–1516.
132. Xie, Z., Lee, S.P., O'Dowd, B.F., George, S.R. (1999) Serotonin 5-HT1B and 5-HT1D receptors form homodimers when expressed alone and heterodimers when co-expressed. *FEBS Lett.* 456, 63–67.

133. Berthouze, M., Ayoub, M., Russo, O., Rivail, L., Sicsic, S., Fischmeister, R., Berque-Bestel, I., Jockers, R., Lezoualc'h, F. (2005) Constitutive dimerization of human serotonin 5-HT4 receptors in living cells. *FEBS Lett.* 579, 2973–2980.
134. Rocheville, M., Lange, D.C., Kumar, U., Sasi, R., Patel, R.C., Patel, Y.C. (2000) Subtypes of the somatostatin receptor assemble as functional homo- and heterodimers. *J Biol Chem.* 275, 7862–7869.
135. Grant, M., Collier, B., Kumar, U. (2004) Agonist-dependent dissociation of human somatostatin receptor 2 dimers: A role in receptor trafficking. *J Biol Chem.* 279, 36179–36183.
136. Pfeiffer, M., Koch, T., Schroder, H., Klutzny, M., Kirscht, S., Kreienkamp, H.J., Hollt, V., Schulz, S. (2001) Homo- and heterodimerization of somatostatin receptor subtypes. Inactivation of sst(3) receptor function by heterodimerization with sst(2A). *J Biol Chem.* 276, 14027–14036.
137. Laroche, G., Lepine, M.C., Theriault, C., Giguere, P., Giguere, V., Gallant, M.A., de Brum-Fernandes, A., Parent, J.L. (2005) Oligomerization of the alpha and beta isoforms of the thromboxane A2 receptor: Relevance to receptor signaling and endocytosis. *Cell Signal.* 17, 1373–1383.
138. Sasaki, M., Miyosawa, K., Ohkubo, S., Nakahata, N. (2006) Physiological significance of thromboxane A(2) receptor dimerization. *J Pharmacol Sci.* 100, 263–270.
139. Latif, R., Graves, P., Davies, T.F. (2001) Oligomerization of the human thyrotropin receptor: Fluorescent protein-tagged hTSHR reveals post-translational complexes. *J Biol Chem.* 276, 45217–45224.
140. Latif, R., Graves, P., Davies, T.F. (2002) Ligand-dependent inhibition of oligomerization at the human thyrotropin receptor. *J Biol Chem.* 277, 45059–45067.
141. Urizar, E., Montanelli, L., Loy, T., Bonomi, M., Swillens, S., Gales, C., Bouvier, M., Smits, G., Vassart, G., Costagliola, S. (2005) Glycoprotein hormone receptors: Link between receptor homodimerization and negative cooperativity. *EMBO J.* 24, 1954–1964.
142. Kroeger, K.M., Hanyaloglu, A.C., Seeber, R.M., Miles, L.E., Eidne, K.A. (2001) Constitutive and agonist-dependent homo-oligomerization of the thyrotropin-releasing hormone receptor. Detection in living cells using bioluminescence resonance energy transfer. *J Biol Chem.* 276, 12736–12743.
143. Song, G.J., Hinkle, P.M. (2005) Regulated dimerization of the thyrotropin-releasing hormone receptor affects receptor trafficking but not signaling. *Mol Endocrinol.* 19, 2859–2870.
144. Song, G.J., Jones, B.W., Hinkle, P.M. (2007) Dimerization of the thyrotropin-releasing hormone receptor potentiates hormone-dependent receptor phosphorylation. *Proc Natl Acad Sci U S A.* 104, 18303–18308.
145. Hanyaloglu, A.C., Seeber, R.M., Kohout, T.A., Lefkowitz, R.J., Eidne, K.A. (2002) Homo- and hetero-oligomerization of thyrotropin-releasing hormone (TRH) receptor subtypes. Differential regulation of β-arrestins 1 and 2. *J Biol Chem.* 277, 50422–50430.
146. Schulz, A., Grosse, R., Schultz, G., Gudermann, T., Schoneberg, T. (2000) Structural implication for receptor oligomerization from functional reconstitution studies of mutant V2 vasopressin receptors. *J Biol Chem.* 275, 2381–2389.
147. Terrillon, S., Barberis, C., Bouvier, M. (2004) Heterodimerization of V1a and V2 vasopressin receptors determines the interaction with β-arrestin and their trafficking patterns. *Proc Natl Acad Sci U S A.* 101, 1548–1553.

148. Heroux, M., Breton, B., Hogue, M., Bouvier, M. (2007) Assembly and signaling of CRLR and RAMP1 complexes assessed by BRET. *Biochemistry*. 46, 7022–7033.

149. Kraetke, O., Wiesner, B., Eichhorst, J., Furkert, J., Bienert, M., Beyermann, M. (2005) Dimerization of corticotropin-releasing factor receptor type 1 is not coupled to ligand binding. *J Recept Signal Transduct Res.* 25, 251–276.

150. Mikhailova, M.V., Mayeux, P.R., Jurkevich, A., Kuenzel, W.J., Madison, F., Periasamy, A., Chen, Y., Cornett, L.E. (2007) Heterooligomerization between vasotocin and corticotropin-releasing hormone (CRH) receptors augments CRH-stimulated 3′,5′-cyclic adenosine monophosphate production. *Mol Endocrinol.* 21, 2178–2188.

151. Vrecl, M., Drinovec, L., Elling, C., Heding, A. (2006) Opsin oligomerization in a heterologous cell system. *J Recept Signal Transduct Res.* 26, 505–526.

152. Langer, I., Gaspard, N., Robberecht, P. (2006) Pharmacological properties of Chinese hamster ovary cells coexpressing two vasoactive intestinal peptide receptors (hVPAC1 and hVPAC2). *Br J Pharmacol.* 148, 1051–1059.

153. Bai, M., Trivedi, S., Kifor, O., Quinn, S.J., Brown, E.M. (1999) Intermolecular interactions between dimeric calcium-sensing receptor monomers are important for its normal function. *Proc Natl Acad Sci U S A.* 96, 2834–2839.

154. Gama, L., Wilt, S.G., Breitwieser, G.E. (2001) Heterodimerization of calcium-sensing receptors with metabotropic glutamate receptors in neurons. *J Biol Chem.* 276, 39053–39059.

155. Jensen, A.A., Hansen, J.L., Sheikh, S.P., Brauner-Osborne, H. (2002) Probing intermolecular protein-protein interactions in the calcium-sensing receptor homodimer using bioluminescence resonance energy transfer (BRET). *Eur J Biochem.* 269, 5076–5087.

156. Zhang, Z., Sun, S., Quinn, S.J., Brown, E.M., Bai, M. (2001) The extracellular calcium-sensing receptor dimerizes through multiple types of intermolecular interactions. *J Biol Chem.* 276, 5316–5322.

157. Tabata, T., Araishi, K., Hashimoto, K., Hashimotodani, Y., van der Putten, H., Bettler, B., Kano, M. (2004) Ca^{2+} activity at $GABA_B$ receptors constitutively promotes metabotropic glutamate signaling in the absence of GABA. *Proc Natl Acad Sci U S A.* 101, 16952–16957.

158. Ciruela, F., Escriche, M., Burgueno, J., Angulo, E., Casado, V., Soloviev, M.M., Canela, E.I., Mallol, J., Chan, W.Y., Lluis, C., McIlhinney, R.A., Franco, R. (2001) Metabotropic glutamate 1alpha and adenosine A1 receptors assemble into functionally interacting complexes. *J Biol Chem.* 276, 18345–18351.

159. Ferre, S., Karcz-Kubicha, M., Hope, B.T., Popoli, P., Burgueno, J., Gutierrez, M.A., Casado, V., Fuxe, K., Goldberg, S.R., Lluis, C., Franco, R., Ciruela, F. (2002) Synergistic interaction between adenosine A2A and glutamate mGlu5 receptors: Implications for striatal neuronal function. *Proc Natl Acad Sci U S A.* 99, 11940–11945.

160. Romano, C., Yang, W.L., O'Malley, K.L. (1996) Metabotropic glutamate receptor 5 is a disulfide-linked dimer. *J Biol Chem.* 271, 28612–28616.

161. Nelson, G., Hoon, M.A., Chandrashekar, J., Zhang, Y., Ryba, N.J., Zuker, C.S. (2001) Mammalian sweet taste receptors. *Cell.* 106, 381–390.

CHAPTER 7

Past and Future Strategies for GPCR Deorphanization

ANGÉLIQUE LEVOYE,[1,2,*] NATHALIE CLEMENT,[2,3,*] ELODIE TENCONI[2,3] and RALF JOCKERS[2,3]

[1]Unité de Pathogénie Virale, Department of Virology, Institut Pasteur, Paris, France

[2]Inserm, Paris, France

[3]Institut Cochin, Université Paris Descartes, Department of Cell Biology, Paris, France

7.1. INTRODUCTION

Seven transmembrane (7TM) domain G protein-coupled receptors (GPCRs) constitute the largest membrane receptor family. These proteins respond to a wide variety of extracellular molecules and play a crucial role in cell-to-cell communication by transmitting extracellular signals to cells [1]. Their involvement in a variety of physiological and pathophysiological processes makes this class of proteins the most common target of pharmaceutical drugs [2]. Recent genome sequencing projects indicated that approximately 400 sequences belong to the non-odorant GPCR family in the human genome [3–5]. Most of them have been matched with known ligands using different strategies. However, despite the vast and long-standing efforts of academic and industrial research to pair 7TM receptors to potential ligands, more than 100 GPCRs remain orphans, without an identified ligand [6]. These orphan 7TM proteins represent potential drug targets in analogy to the successful development of drugs targeting non-orphan GPCRs.

Historically, the characterization of the primary amino acid sequence and plausible topological structure of rhodopsin and the β2-adrenergic receptor

*These authors contributed equally.

GPCR Molecular Pharmacology and Drug Targeting: Shifting Paradigms and New Directions,
Edited by Annette Gilchrist

initiated the deorphanization process [7]. Indeed, comparison of these two sequences gave birth to the GPCR family concept and promoted homology-based screening approaches to identify new GPCR members by screening cDNA libraries at low stringency or by performing polymerase chain reactions using degenerated primers [8–10]. For example, the group of Parmentier and Vassart identified conserved sequence stretches in the third and sixth TM domain of many of the GPCRs cloned at that time, which allowed them to design DNA probes that could serve as degenerated primers under low stringency hybridization conditions [9]. At the same time, a strategy termed "reverse pharmacology" was born (Fig. 7.1) [6]. This approach is based on the exoge-

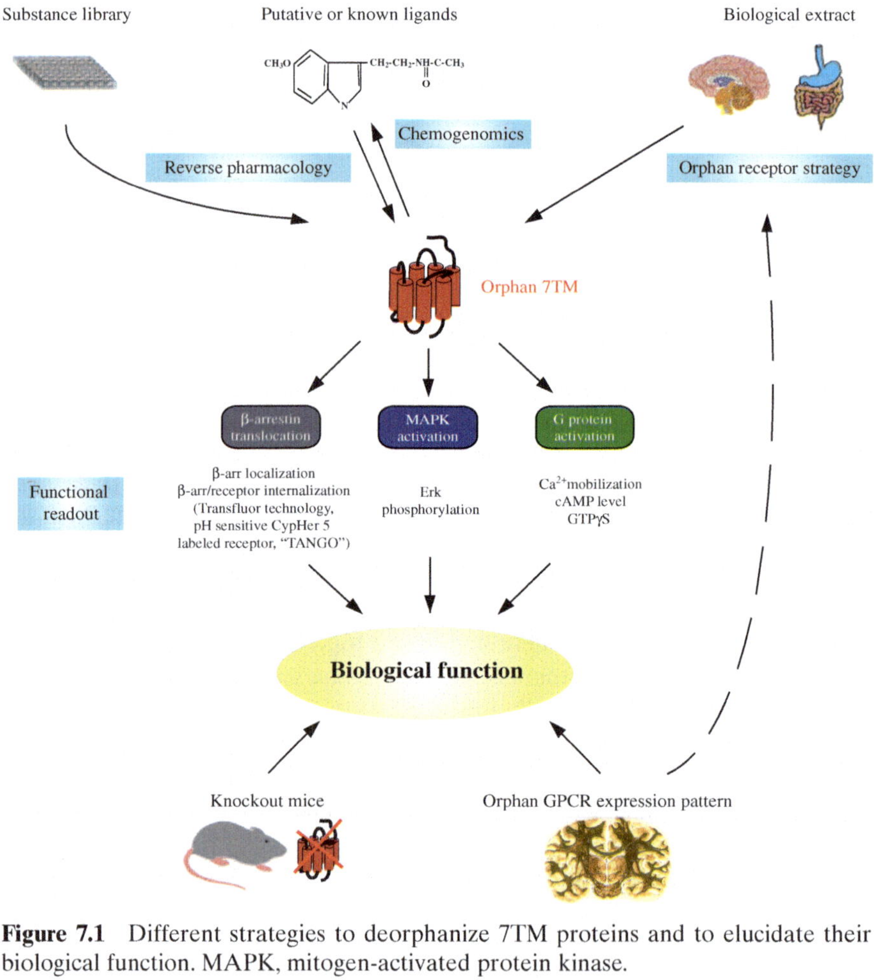

Figure 7.1 Different strategies to deorphanize 7TM proteins and to elucidate their biological function. MAPK, mitogen-activated protein kinase.

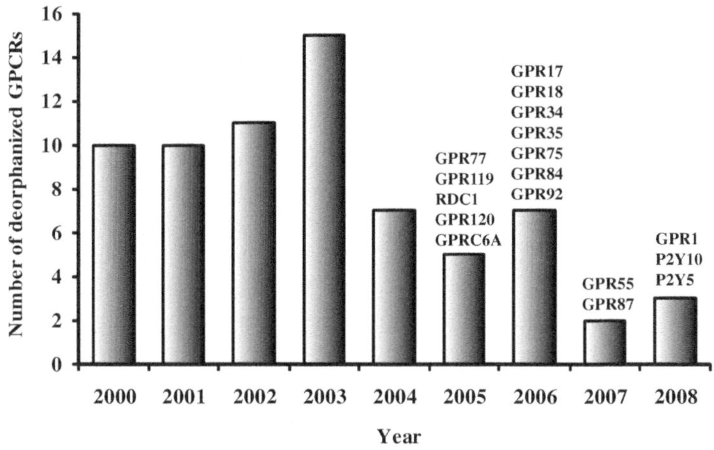

Figure 7.2 Deorphanization rate of 7TM proteins between 2000 and 2008.

nous expression of the orphan 7TM protein (a putative GPCR) in a suitable cell system. Activation of the receptor by exposure to a variety of potential ligands is monitored by changes in intracellular second messenger levels involving heterotrimeric G proteins. Early in the 1990s, the reverse pharmacology strategy led to the deorphanization of many orphan 7TM proteins and a subsequent outnumbering of the identified natural ligands. This led to the conclusion that the remaining orphans must bind ligands that have not been characterized. Thus, a variation of the reverse pharmacology approach called "orphan receptor strategy" was developed in 1995 (Fig. 7.1). Instead of screening potential transmitters, extracts of tissues were used as potential sources of new ligands [11, 12].

Today, these classical deorphanization strategies seem to have reached their limits with few new ligand–orphan GPCR pairings since 2004. Despite the fact that the number of deorphanized GPCRs has decreased since 2004 (Fig. 7.2), the identification of endogenous ligands for orphan 7TM proteins still remains a major goal as it often allows an understanding of their physiological role. Therefore, not only technical but also conceptual alternative approaches are required. The concept of GPCR dimerization and the discovery of G protein-independent pathways stimulated the development of new deorphanization strategies. In this chapter, we describe classical and recent strategies used to identify the natural ligands for orphan 7TM proteins and discuss the new concept suggesting that orphan 7TM proteins can have ligand-independent properties. Known examples supporting these concepts and some controversial cases illustrating the difficulty to deorphanize orphan 7TM proteins are also discussed.

7.2. CURRENT STRATEGIES TO IDENTIFY THE LIGAND AND FUNCTION OF ORPHAN 7TM PROTEINS

7.2.1. Reverse Pharmacology

From the early days of GPCR cDNA cloning in the late 1980s until now, the reverse pharmacology strategy participated in the deorphanization of 7TM proteins [6]. The dopamine D_2 and the serotonin 5-HT_{1A} receptor were the first GPCRs that were deorphanized by the reverse pharmacology approach [8, 13] with many others following over the years (Table 7.1) [14–19]. The approach is based on exogenous expression of orphan 7TM proteins in a suitable cell system. Receptors are typically stably overexpressed in model cell lines (i.e., Chinese hamster ovary [CHO] cells) and the activation of heterotrimeric G protein-dependent pathways are monitored (Fig. 7.1). Receptor overexpression lead also to the discovery that many 7TM proteins are constitutively active.

7.2.2. Orphan Receptor Strategy

As the number of identified natural ligands was rapidly outnumbered by the GPCRs discovered by DNA cloning, the orphan receptor strategy emerged in the mid-1990s [11]. This strategy, also referred to as "tissue-extract based approach," exposes orphan 7TM proteins to tissue extracts instead of purified ligands. Receptor activation is measured by changes in second messenger responses. Positive extracts are fractionated until the active component is isolated and characterized (Fig. 7.1). Despite low assay specificity, due to the complex composition of tissue extracts, the orphan receptor strategy proved to be useful for the identification of several natural ligands for orphan 7TM proteins. The first success of this strategy was the discovery of a novel transmitter for the ORL-1 receptor, which was cloned by homology to opioid receptors [20]. Based on the expression of ORL-1 in the central nervous system, brain tissue extracts were prepared and tested for their capacity to inhibit adenylyl cyclase activity in cells stably expressing ORL-1 receptors. A 17-residue long peptide sharing similarity with opioid peptides was ultimately isolated and named orphanin FQ or nociceptin (OFQ/N) [12, 21]. Subsequently, orphan 7TM proteins were screened randomly against large libraries of ligands [22]. These libraries typically contain all the ligands that have not been matched to any other receptor. Within a few years, more than 40 orphan 7TM proteins were deorphanized with this approach [23].

7.2.3. Use of Sequence Homology, Cross Genome Phylogenetic Analysis, and Chemogenomics to Predict Candidate Ligands

Sequence comparison between orphan 7TM proteins and GPCRs with known ligands has always been a straightforward strategy for candidate ligand prediction. Based on simple sequence comparisons, several orphan 7TM proteins

TABLE 7.1 Deorphanized 7TM Proteins since 2005

GPCRs	Ligands	Assay	Deorphanization Strategy	Reference
GPR1	Chemerin	Enzyme Fragment Complementation	β-arrestin recruitment	[70]
GPR17	Uracil nucleotides/ cycteinyl-leukotrienes	$[^{35}S]GTP\gamma S$	Phylogenetic analysis	[28]
GPR18	*N*-arachidonylglycine	Ca^{2+} (Fluo-3)	Reverse pharmacology	[107]
GPR34	Lysophophatidylserine	cAMP (EIA kit), $[^{35}S]GTP\gamma S$	Phylogenetic analysis	[108]
GPR35	Kynurenic acid	Ca^{2+} (aequorin)	Reverse pharmacology	[17]
GPR55	Canabinoid ligands	Ligand binding, $[^{35}S]GTP\gamma S$	Phylogenetic analysis	[109]
GPR75	RANTES	cAMP (luciferase reporter)	Reverse pharmacology	[19]
GPR77	Acylation-stimulating protein	β-arrestin translocation	Fluorescence-based β-arrestin translocation assay	[65]
GPR84	Medium-chain free fatty acids	Ca^{2+} (aequorin)	Reverse pharmacology	[18]
GPR87	Lysophosphatidic acid	GPR87-$G_{\alpha 16}$ fusion protein, Ca^{2+} (Fura-2)	Reverse pharmacology	[56]
GPR92	Lysophophatidic acid	Ligand binding, cAMP (EIA kit)	Phylogenetic analysis	[110]
		Ligand binding, internalization (IF)	Phylogenetic analysis	[111]
GPR119	Lysophophatildylcholine	cAMP (HTRF kit)	Reverse pharmacology	[112]
GPR120	Free fatty acids	GPR120-$G_{\alpha 16}$ fusion protein, internalization (flow cytometry and IF)	Reverse pharmacology	[57]
RDC1 (CXCR7)	CXCL12	Ligand binding, internalization (IF)	Phylogenetic analysis	[113]
GPRC6A	L-alpha-amino acids (L-Arg, L-Lys, L-omithine)	X-oocytes, CA^{2+} (Fluo-4)	Chemogenomics	[30]
P2Y5 receptor	Lysophosphatidic acid	cAMP (luciferase reporter)	Reverse pharmacology	[114]
P2Y10 receptor	Sphingosine-1-phosphate Lysophosphatidic acid	Ca^{2+} (Fura-2)	Phylogenetic analysis	[27]

IE, Immunofluorescence; EIA, enzyme-linked immunoassay; HTRF, homogenous time-resolved fluorscence; X-oocytes, *Xenopus* oocytes.

have been classified in the past into GPCR subfamilies and subsequently deorphanized [6, 24]. The complete sequencing of the human genome and the emergence of new bioinformatic tools further accelerated and refined this process. According to the most recent classification, some 7TM proteins segregate into subfamilies exclusively composed of orphans, and others into subfamilies with known ligands [5, 3]. Sequencing of the genome of many different species stimulated cross genome phylogenetic analysis that provided further insights in the correct classification of orphan 7TM proteins [25, 26].

Recent examples where phylogenetic analysis had a major impact on the deorphanization process are two P2Y-related receptors, the P2Y10 receptor and GPR17 [27, 28]. GPR17 constitutes an interesting example as this protein occupies an intermediate position between purinergic P2Y and cysteinyl-leukotrienes (CysLTs) receptors. Functional studies showed that GPR17 indeed seems to straddle both subfamilies and binds to both uracil nucleotides and CysLTs with high affinity (Table 7.1).

Further information came from a chemogenomic analysis of the ligand binding pocket of human non-odorant GPCRs [29]. Interestingly, clustering of 30 residues predicted or known to be important for ligand binding was similar but not identical compared to the phylogenetic tree derived from full-length GPCR cDNAs. This approach was used to deorphanize GPRC6A as the receptor for L-α-amino acids [30] and will certainly help to predict potential ligands for other 7TM proteins in the future (Fig. 7.1).

7.2.4. Determination of the Expression Pattern and the Phenotype of Knockout Mice of Orphan 7TM Proteins

Determination of the expression pattern of 7TM proteins may provide useful information for ligand–receptor pairing as exemplified by GPR109B, the receptor for nicotinic acid receptor [31]. GPR109B was deorphanized on the basis of its restricted expression pattern in adipose tissue and spleen, two main tissues of nicotinic acid action (Fig. 7.1). The generation of knockout animals is increasingly employed to elucidate the biological function of orphan 7TM proteins [32–34]) (Table 7.2; Fig. 7.1). For example, expression of the orphan Mas-related gene (Mrg) E receptor in the nervous system and the phenotype of mice lacking MrgE receptors revealed its involvement in pain [35]. Recently, a comparative study of GPR85 overexpressing transgenic mice and GPR85 knockout mice showed a role of GPR85 in the control of the brain size and in the pathophysiology of psychiatric disorders [36]. These functional studies constitute a first step toward the deorphanization of 7TM proteins.

7.3. FUNCTIONAL ASSAYS FOR DEORPHANIZATION

The nature of the functional assay used to deorphanize 7TM proteins is crucial for the success rate of ligand identification. A number of expression systems

TABLE 7.2 Identified Functions of Orphan 7TM Proteins Since 2006

GPCRs	Identified Functions	Assay	Expression System	Reference
GPR20	Constitutive inhibition of cAMP production through $G_{\alpha i}$	cAMP	HEK293 (stable)	[115]
GPR22	Constitutive inhibition of cAMP production through $G_{\alpha i}$	cAMP	HEK293 (stable)	[116]
	Protective role in response to hemodynamic stress	GPR22 deficient mice		
GPR26	Constitutive stimulation of cAMP production	cAMP	HEK293 (transient)	[117]
GPR78				
GPR37	Interacts with dopamine transporter and modulates its activity	Ligand binding, co-IP GPR37 deficient mice	HEK293 (transient)	[92]
GPR50	Heterodimerization with MT_1 inhibits melatonin binding and MT_1 signaling (G protein and β-arrestin recruitment)	BRET, co-IP, ligand binding, cAMP, GPR50 siRNA	HEK293 (transient), hCMEC/D3 (endogenous)	[82]
	Regulator of energy metabolism	GPR50 deficient mice	—	[34]
GPR56	Binds to tissue transglutaminase2 Inhibits tumor growth	Immunohistochemistry, radioimmunoprecipitation assay, proliferation	HEK293, MC-1 cells (transient)	[91]

TABLE 7.2 (*Continued*)

GPCRs	Identified Functions	Assay	Expression System	Reference
GPR81	Constitutively couples to G_i pathway	$G_{i/q}$ chimera, inositol phosphate assay GPR81 transgenic mice	HEK293 (transient)	[118]
GPR85 (SREB2)	Influences brain size, behavior and vulnerability to schizophrenia	GPR85 transgenic mice GPR85 deficient mice	—	[37]
GPR161	Controls neurulation and lens development	Vacuolated lens mouse model	—	[33]
GPRC5A	Constitutive inhibition of $G_{\alpha s}$ mRNA and cAMP production Increase the cell proliferation	cAMP, real-time PCR, proliferation assay, GPRC5A siRNA	Human thyroid follicular epithelial cells (endogenous)	[119]
EBI2	Constitutive inhibitions of cAMP production through $G_{\alpha i}$	cAMP, luciferase reporter	COS-7 (transient)	[120]
MrgE	Heterodimerization with MrgD decreases internalization and increases ERK activation and intracellular $[Ca^{2+}]$	Internalization, ERK 1/2 phosphorylation, Ca^{2+}	HEK293 (stable)	[83]
	Role in selective pain behavioural responses	MrgE deficient mice	—	[36]
LGR4	Regulation of postnatal epididymal morphogenesis via maintenance of extracellular matrix	*Lgr4* hypomorphic mutant mice	—	[121]

Co-IP, co-immunoprecipitation; hCMEC/D3, human endothelial cerebral cells.

including *Xenopus* oocytes, mammalian, and yeast cells combined with different functional readout systems contributed significantly to the deorphanization of orphan 7TM proteins in the past. These functional assays can be roughly divided into G protein-dependent assays, which have been extensively used in the past, and assays that rely on more recent strategies, namely receptor internalization and/or β-arrestin recruitment (Fig. 7.1). A recent review article provides an overview of the different assays with special emphasis on high-throughput screening formats [37].

7.3.1. Classical Assays of GPCR Deorphanization

Whenever high-affinity radioligands are available, these pharmacological tools are an integral part of the ligand screening process. Chemical libraries or tissue extracts are typically screened for compounds that compete with the radioligand for the same binding site.

Binding of agonists to GPCRs induces a conformational change and promotes the activation of heterotrimeric G proteins. Their activation can be most easily and directly detected by monitoring the binding of nonhydrolyzable guanosine 5′-O-(3-thiotriphosphate) (GTP) analogs such as [^{35}S]GTPγS to the G protein. These [^{35}S]GTPγS binding assays are widely used in pharmaceutical industry and 1536-well scintillation proximity assays have been developed [38].

Measuring second messengers such as cyclic adenosine monophosphate (cAMP) constitutes another widely used functional readout system (Table 7.1). Activation or inhibition of the cAMP pathway can be either measured directly on the second messenger level or in a reporter gene assay based on the activation of a cAMP response element located in a promoter that drives a luciferase reporter gene. These assays are specifically designed for those receptors that either activate (G_s-coupled) or inactivate (G_i-coupled) adenylyl cyclases. Natural cell systems such as *Xenopus laevis* melanophores offer additional options for ligand screening of orphan 7TM proteins [39]. The aggregation or dispersion of intracellular organelles termed melanosomes can be controlled by GPCR signaling mainly, but not exclusively, by the adenylyl cyclase pathway. This strategy has been recently used to deorphanize GPRC6A which binds several L-α-amino acids (Table 7.1) [30].

Many GPCRs physiologically regulate changes in intracellular Ca^{2+} either through an IP_3-mediated mechanism involving changes in stored Ca^{2+} or through the regulation of voltage-sensitive ion channels, which modulate Ca^{2+} influx. Indeed, this strategy has been applied to the ligand screening of about 40 orphan 7TM proteins (Table 7.1). Two of the most commonly employed instruments used to measure changes in intracellular calcium concentration are the Fluorescence Imaging Plate Reader (FLIPR™, Molecular Devices, Sunnyvale, CA) and the functional drug screening system (FDSS, Hamamatsu Corporation, Hamamatsu City, Japan) which use calcium indicators such as Fluo-3 or Fura-2 [40, 41]. Another Ca^{2+}-sensitive approach is based on aequorin, an induced Ca^{2+}-sensitive luciferase [42]. The assay was further optimized by the addition of promiscuous G_α proteins ($G_{\alpha15}$ and $G_{\alpha16}$) or chimeric G_α subunits (Gaq/i5) to facilitate the receptor–G protein coupling [43, 44].

7.3.2. Recent Assays in GPCR Deorphanization

Recent advances clearly show that not all receptor actions rely on the activation of heterotrimeric G proteins but rather can be promoted via activation of G protein-independent and/or β-arrestin-dependent pathways [45]. Two orphans that do not signal through G protein-dependent pathways are GPCR C5L2 (also known as GPR77) [46, 47] and the D6 chemokine receptor [48]. Whereas the first binds to C5a anaphylotoxin, the latter binds to chemokines. No evidence has been found for the activation of intracellular signaling pathways upon ligand binding to these receptors, which are also called decoy receptors as the ligands are captured and cointernalize with the receptor inside the cell.

GPCR–Gα Fusion Proteins Construction of fusion proteins between GPCRs and G_α subunits was reported more than 10 years ago [49]. The same strategy has more recently been applied to the deorphanization of orphan 7TM proteins [50, 51]. Typically, orphan 7TM proteins are fused to the promiscuous $G_{\alpha16}$ protein, and receptor activation is monitored by measuring [^{35}S] GTPγS binding in membranes prepared from Sf9 cells expressing the fusion protein [52]. Sf9 insect cells provide an excellent high-level expression system that is nearly devoid of mammalian endogenous G proteins. Using this approach, the 5-oxo-eicosatetraenoic acid was successfully identified as the ligand for GPCR48 [53, 54] followed by the deorphanization of GPR87 [55] and GPR120 [56].

Extracellular-Regulated Kinase (ERK) Activation Cell-Based Assay Whereas GPCR-G_α fusion protein assays are per definition restricted to G protein-dependent signaling events, activation of the ERKs involves multiple signaling pathways including G protein-dependent and -independent and β-arrestin-dependent pathways [45]. Most GPCRs are indeed able to activate the ERK pathway and thus, it may serve as a universal and broad-range readout of GPCR activation. A recently described cell-based assay that combines the homogeneous and nonradioactive properties of the alpha-screen technology to measure phosphorylation of ERK1 and ERK2, can be employed on endogenously expressed receptors and recombinant GPCRs and orphan 7TM proteins [57].

pH-Sensitive Probes: An Indicator of GPCR Internalization Agonist-promoted internalization is a regulatory phenomenon common to most GPCRs to limit receptor activation [58]. Activated GPCRs are rapidly phosphorylated by GRKs followed by the recruitment of β-arrestins and the internalization machinery. Several techniques have been developed to measure the disappearance of GPCRs from the cell surface and their accumulation in intracellular compartments. Recent strategies exploit the acidic nature of the endosomal compartments and the pH-sensitive cyanine dye CypHer 5, which is non-fluorescent at pH7.4 and fluoresces brightly in an acidic environment to measure receptor internalization [59, 60]. Surface-exposed receptors are labeled with CypHer 5-labeled antibodies, and ligand-dependent internalization into acidic endosomal compartments is monitored by the occurrence of intense intracellular fluorescence.

Assays Based on the Agonist-Promoted β-Arrestin Recruitment

Fluorescence and Bioluminescence-Based Protein Translocation Assays The translocation of β-arrestin to GPCRs is a hallmark of receptor activation that is increasingly used in ligand screening assay. Several assays have been developed, including a fluorescence-based assay that measures the cellular translocation of β-arrestin–green fluorescent protein (GFP) fusion proteins. Whereas β-arrestin-GFP is diffusely distributed in the cytoplasm under basal conditions, upon agonist stimulation, the fusion protein is first translocated to the plasma membrane and then internalized together with the activated receptor [61,62]. If the changes in β-ARR localization following agonist stimulation are small, data analysis, particularly in a high-throughput setting, may become difficult. A commercially available version of the assay is sold as "Transfluor technology". This assay has been used to identify novel ligands for *Drosophila* orphan 7TM proteins and to confirm ligand–receptor pairs [63]. The acylation stimulating protein (ASP) has been identified as a ligand for GPR77 (C5L2), using this assay (Table 7.1) [64]. Similar assays have been developed to monitor β-arrestin translocation by bioluminescence resonance energy transfer [65–67].

Protease-Mediated Transcriptional Reporter Gene Assay Another assay based on β-arrestin recruitment and activation of a transcriptional reporter gene has been developed recently [68]. In this assay, named TANGO™ (Invitrogen Corporation, Madison, WI), the receptor is fused at its C-terminal extremity to a protease cleavage site followed by the tetracycline-controlled transactivator (tTA). A second fusion protein of the corresponding protease fused to β-arrestin2 is coexpressed with the receptor fusion protein in a cell line expressing a tTA-dependent reporter gene. Ligand stimulation of the receptor leads to β-arrestin recruitment that induces the cleavage of the transcription factor. The released transcription factor translocates into the nucleus where it activates the transcription of the tTA-dependent luciferase reporter gene. This technology was used to identify the leukocyte chemoattractant chemerin as a ligand for the orphan GPR1 [69].

Enzyme Fragment Complementation This assay relies on the reconstitution of a functional β-galactosidase enzyme from two complementing β-galactosidase mutants, one fused to the receptor of interest, the other to β-arrestin [70]. The commercially available PathHunter technology from DiscoveRx (Fremont, CA) restores β-galactosidase activity by complementing a catalytically inactive form (enzyme acceptor) with a peptide containing the catalytic domain (enzyme donor). β-arrestin recruitment to the receptor and subsequent reconstitution of β-galactosidase activity is detected in lysed cells by measuring chemiluminescence. DiscoveRx is now extending this assay by generating PathHunter cell lines expressing orphan 7TM proteins either alone or in the presence of a non-orphan GPCR.

Collectively, these β-arrestin-dependent approaches are expected to be applied to a wide range of orphan 7TM proteins as the majority of GPCRs recruit β-arrestins [71, 72]. However, exceptions are likely to exist as some receptors may exhibit β-arrestin-independent internalization or interact constitutively (in the absence of ligand) with β-arrestin.

7.4. FUTURE DIRECTIONS AND NEW CONCEPTS

The inclusion of G protein-independent pathways in future ligand screening procedures will certainly help to deorphanize further orphan 7TM proteins. However, additional strategies may be necessary to deorphanize all remaining orphans. Accumulating evidence indicates that transfection of a single receptor cDNA into fibroblastic cell lines (typically CHO cells) may not always be sufficient for functional expression. The majority of GPCRs exist as dimers (or higher oligomers). This includes self-association into homodimers and association between different GPCRs into heterodimers. Heterodimer formation has been most extensively studied for non-orphan GPCRs demonstrating the reciprocal influence of heterodimerization on the functional properties of both protomers. Interestingly, several of the known heterodimers include also orphan 7TM proteins (Fig. 7.3) [73]. The first and most extensively studied heterodimer that is composed of an orphan and a non-orphan GPCR is the obligatory metabotropic γ-aminobutyric acid B ($GABA_B$) receptor heterodimer. A functional $GABA_B$ receptor is composed of two homologous subunits called $GABA_{B1}$ and $GABA_{B2}$ [74–76]. Whereas $GABA_{B1}$ provides ligand binding, the orphan $GABA_{B2}$ 7TM protein promotes efficient trafficking of $GABA_{B1}$ to the cell surface and G protein coupling [77]. An intracellular endoplasmic reticulum retention signal on $GABA_{B1}$, which prevents $GABA_{B1}$ from reaching the cell surface by itself, is masked in the $GABA_{B1}/GABA_{B2}$ heterodimer.

A similar configuration has been reported for the T1R taste receptors. $T1R_1$ and $T1R_2$, originally discovered as orphans, have subsequently been deorphanized by the discovery of $T1R_3$, their obligatory heterodimerization partner. Whereas $T1R_2/T1R_3$ heterodimers respond rather to sweet stimuli such as aspartame, the $T1R_1/T1R_3$ heterodimer responds to the umami taste of L-glutamate [78, 79]. Within these heterodimers, $T1R_1$ and $T1R_2$ bind their respective ligands with their large extracellular domain, whereas $T1R_3$ is devoid of any ligand binding capacity [78–80].

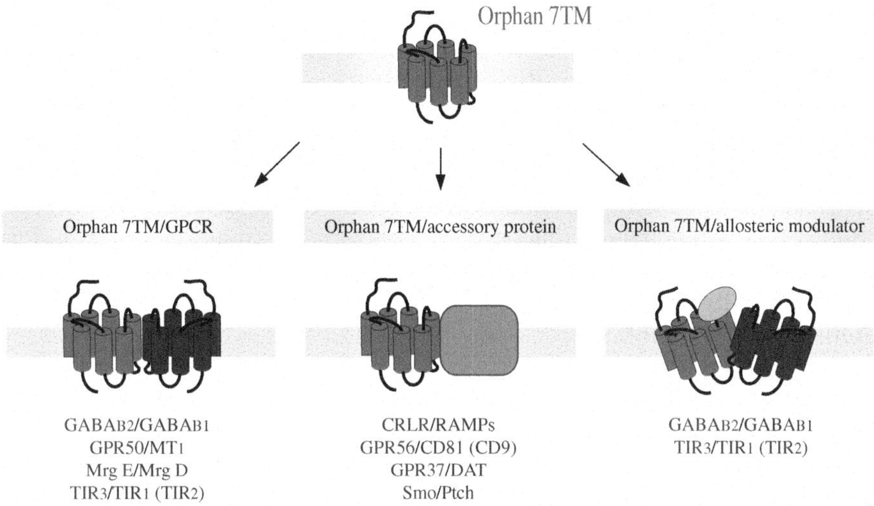

Figure 7.3 Novel functions of 7TM proteins. Association of 7TM proteins with deorphanized GPCRs or other accessory membrane proteins. Activation of orphan 7TM proteins by allosteric modulators. (See color insert.)

Three further cases illustrate the impact of orphan 7TM proteins on the signaling capacity of GPCRs. Engagement of the orphan GPR50 into heterodimers with MT_1, known to bind the circadian neurohormone melatonin, has profound consequences on MT_1 function, namely inhibition of high-affinity agonist binding, heterotrimeric G protein coupling, and β-arrestin binding [81]. A more detailed analysis provided a possible explanation pointing to the long carboxyl terminal tail of GPR50, which prevents recruitment of intracellular interaction partners such as G proteins and β-arrestins to MT_1 in the heterodimer. The second example concerns the Mrg family members MrgD and MrgE, which have been shown to form heterodimers. β-alanine binding to MrgD, when expressed alone, activates the ERK pathway and receptor internalization. Coexpression of MrgD with the orphan MrgE potentiates β-alanine-induced MrgD signaling and inhibits MrgD internalization in the MrgD/MrgE heterodimer [82]. The latest reported heterodimer concerns the serotonin 5-HT_{2A} receptor and the orphan trace amine associated receptor TAAR6, which are both associated with schizophrenia. In cells coexpressing both proteins, the maximal activation of the phospholipase C (PLC) pathway through 5-HT_{2A} receptors was significantly enhanced compared to cells expressing identical quantities of 5-HT_{2A} alone [83].

Taken together, the concept of heterodimer formation between GPCRs and orphan 7TM proteins clearly offers a new dimension for the potential function of orphan 7TM proteins, and ongoing work in several laboratories is likely to reveal more of these heterodimer couples (Fig. 7.3). Whereas the physiological importance of heterodimer formation has been clearly demonstrated in the case of $GABA_B$ and T1R receptors, the physiological relevance still remains to be firmly demonstrated for the other reported examples.

Heterodimers between DOR83b, a highly conserved odorant 7TM protein without any apparent affinity for odorants, and conventional ligand-binding odorant receptors, were initially classified as GPCR heterodimers. Similarly to the $GABA_{B2}$ subunit of the $GABA_B$ receptor, DOR83b promotes trafficking of its heterodimerization partners (i.e., DOR22a and DOR43a) to the cell surface. In addition to targeting odorant receptor heterodimers to sensory cilia membranes of olfactory sensory neurons, DOR43b was also shown to strongly increase the functional response of conventional odorant receptors [84, 85]. However, more recent reports indicate that DOR43b has indeed a 7TM structure, but with an inverted orientation in the plasma membrane compared to conventional GPCRs, with the N-terminus facing the cytoplasm and the C-terminus facing the inside of the cell [85]. Additional observations made by two independent groups indicate that coexpression of odorant receptors and DOR43b leads to inward cationic currents leading to the unexpected conclusion that this receptor complex is an ion channel [86, 87]. Whether DOR43b alone can form an ion channel or whether the entire heterodimeric complex is necessary is still a matter of debate.

Several other cases are known where coexpression of an orphan 7TM protein with an accessory protein is important for the functional expression of the orphan 7TM protein. A well-known example is the orphan calcitonin receptor-like receptor (CRLR) that requires single TM-spanning receptor activity-modifying proteins (RAMPs) for efficient transport to the cell surface. Importantly, the pharmacological phenotype can be modulated through differential interaction with one of the three RAMPs [88]. When associated with RAMP1, CRLR functions as calcitonin gene-related peptide (CPRP) receptor, whereas it acts as an adrenomedulin receptor when coexpressed with RAMP2 or RAMP3 (Fig. 7.3).

Two recent studies attempted to define the function of the orphan GPR56 7TM protein by characterizing associated proteins. Co-immunoprecipitation experiments revealed that GPR56 associates with $G_{q/11}$ proteins and CD81 and CD9 tetraspanins [89]. Interestingly, CD81 stabilized this ternary complex by directly binding to $G_{q/11}$ proteins. Although the signaling capacity of GPR56 remains currently unknown, one may speculate that CD81/CD9, by directly binding to GPR56 and $G_{q/11}$ proteins, provides the necessary scaffold for efficient coupling of GPR56 through $G_{q/11}$-mediated pathways. In a second study, the N-terminal extracellular domain of GPR56 has been shown to bind tissue transglutaminase TG2, an observation that has been correlated with the capacity of GPR56 to inhibit melanoma tumor growth [90].

The orphan GPR37, which is associated with Parkinson's disease, has recently been reported to associate with the dopamine transporter (DAT) and to modulate its activity. GPR37-null mice show enhanced dopamine uptake in striatal membranes associated with increased DAT cell surface expression [91]. A further example concerns Smoothened (Smo), a 7TM protein that mediates the effects of Hedgehog (Hh) in embryonic development. Interestingly, Hh does not bind directly to Smo but rather to the 12TM receptor Patched (Ptch), which inhibits the constitutive activity of Smo. Binding of Hh to Ptch relieves this repression of Smo. Consequently, Smo can be considered an orphan 7TM protein whose activity is regulated by its binding partner Ptch [92]. Taken together, these examples of heterodimers between orphan 7TM proteins and other membrane proteins offer new functional options for orphan 7TM proteins and novel perspectives for drug intervention [73].

In recent years, the concept of allosteric modulation of GPCRs has become progressively accepted and now represents an important aspect of the drug discovery process. Allosteric modulators act per definition outside of the orthosteric binding site of endogenous ligands. The action of allosteric ligands is generally considered to depend on the presence of the orthosteric ligand. However, several examples demonstrate that allosteric agonists may have also an effect, which is independent of the presence of the orthosteric ligand [93]. This notion is fully compatible with the definition for an allosteric agonist given by the International Union of Basis and Clinical Pharmacology (IUPHAR) committee: "... a ligand that is able to mediate receptor activation in its own right by binding to a recognition domain on the receptor macromolecules that is distinct from the primary (orthosteric) site." [94] (Fig. 7.3). This is nicely illustrated by two compounds acting on orphan 7TM proteins. The first compound, CGP7930, allosterically enhances GABA binding to the $GABA_B$ receptor heterodimer. Importantly, CGP7930 can activate the heptahelical domain of the orphan $GABA_{B2}$ subunit alone [95]. Similar observations have been made for taste T1R receptors. Lactisole and cyclamate have been proposed to bind to the 7TM domain of the orphan $T1R_3$ subunit and thus allosterically regulate ligand binding to the bilobate extracellular orthosteric sites of $T1R_1$ and $T1R_2$ in their respective heterodimers. Collectively, this indicates that orphan 7TM proteins can be targeted by allosteric agonists that may be interesting compounds even in the absence of orthosteric ligands for these proteins.

7.5. CONTROVERSIAL ISSUES

Deorphanization of 7TM proteins has become increasingly difficult. New screening strategies will help to deorphanize further orphans, and new concepts may highlight additional functions that may or may not depend on ligand binding to the predicted orthosteric binding site. However, deorphanization of the remaining orphan GPCRs is expected to remain a slow and tedious task.

From the early days of deorphanization on, researchers have been often confronted with high levels of apparent constitutive activity when expressing cDNAs of orphan 7TM proteins. Although many GPCRs exhibit a significant level of constitutive activity, it should be mentioned that in the case of orphan 7TM proteins, the possibility always remains that the apparent constitutive activity might actually reflect the presence of an endogenous ligand that either is difficult to remove or which is produced by the cell. This is illustrated by the adenosine A2 receptor whose apparent constitutive activity could be explained by endogenous adenosine [96]. A recent study on GPR40 shows that its apparent constitutive activity is due to the occupation of the receptor binding site by endogenous fatty acid ligands [97]. Screening programs for inverse agonists should take into account such potential complications.

The deorphanization process is not always linear. Reports claiming the successful deorphanization have to be double-checked by other groups. Initial results are indeed not always confirmed, leading to a controversy, which may persist for several years before finding a consensus. At least three groups of 7TM proteins are currently at that state. The first case concerns the GPCR subfamily comprising GPR4, GPR68 (OGR1), GPR65 (TDAG8), and GPR132 (G2A), which have been initially deorphanized as receptors of lipid messengers. However, these data could not be confirmed by others and were retracted several years later. More recently, GPR4, GPR65, and GPR68 were reported to be proton-sensing receptors that activate either phosphoinositol or cAMP pathways [98]. The nature of the physiologically relevant ligand of the more distantly related GPR132 is still not clear. The second debated orphan 7TM protein is GPR39. In 2005, GPR39 was reported to be the receptor for obestatin, a 23 amino acid peptide derived from the ghrelin precursor protein [99]. However, 2 years later, several other groups could not confirm these findings [100, 101]. The last case, which is extensively debated in the literature, is GPR30. Controversial features include its subcellular localization, at the plasma membrane versus the endoplasmic reticulum, its capacity to bind to estrogens, and the discrimination between effects mediated though GPR30 and the classical nuclear steroid receptors. Despite numerous studies and the development of specific tools such as GPR30-selective ligands and antibodies directed against GPR30, no clear consensus has been reached today on all these issues, demonstrating the difficulties that may arise in the deorphanization process [102, 103].

Systematic screening for putative heterodimerization partners of orphan 7TM proteins may represent a tremendous effort as more than 100 orphans have to be matched with approximately 260 GPCRs with known ligand. Biotechnology companies such as Dimerix Bioscience have already set up screening procedures for the identification of GPCR heterodimers. Candidate heterodimers could then be further characterized in established functional assays (binding, trafficking, and signaling assays). In respect to the other non-GPCR proteins that associate with orphan 7TM proteins, a systematic screening of potential interaction partners is more difficult in view of the large diversity of these proteins.

Finally, the question whether at least some of the remaining orphans are "true orphans" that do not bind to any endogenous ligand was suggested recently [104], will be difficult to answer even in the presence of compelling evidence for ligand-independent function such as the modification of the function of other proteins by heterodimerization. However, in some cases, evolutionary analysis may provide meaningful insights. Evolutionary trace analysis of the putative GABA binding site of the orphan $GABA_{B2}$ 7TM protein indicated that this binding site is not under evolutionary pressure strongly supporting the absence of any ligand binding capacity [105]. Furthermore, evolutionary analysis of GPR50, which belongs to the melatonin receptor subfamily but does not bind to melatonin or any other known ligand, showed that GPR50 is the mammalian ortholog of Mel1c, a high-affinity melatonin receptor in lower vertebrates [106]. This observation, together with the fact that the GPR50 gene evolved very rapidly in mammals, indicate that GPR50 has lost its capacity to bind melatonin during evolution. Although this finding does not completely rule out the possibility that GPR50 binds to another endogenous ligand, it strongly supports the idea that GPR50 is a true orphan receptor, whose expression during evolution is maintained due to essential ligand-independent functions.

In conclusion, new assays and concepts will certainly help to deorphanize many of the remaining orphan 7TM proteins; others might be true orphans with ligand-independent functions. Both categories, true orphans and deorphanized GPCRs, are likely to have great potential for drug development.

ACKNOWLEDGMENTS

This work was supported by grants from SERVIER, the Fondation Recherche Médicale ("Equipe FRM"), Fondation pour la Recherche sur le Cerveau (FRC) Neurodon, Institut National de la Santé et de la Recherche Médicale (INSERM), Centre National de la Recherche Scientifique (CNRS). We thank Jean-Luc Guillaume (Institut Cochin, Paris) for critical comments on the manuscript.

REFERENCES

1. Bockaert, J., Pin, J.P. (1999) Molecular tinkering of G protein-coupled receptors: An evolutionary success. *EMBO J.* 18, 1723–1729.
2. Drews, J. (2000) Drug discovery: A historical perspective. *Science.* 287, 1960–1964.
3. Fredriksson, R., Lagerstrom, M.C., Lundin, L.G., Schioth, H.B. (2003) The G-protein-coupled receptors in the human genome form five main families. Phylogenetic analysis, paralogon groups, and fingerprints. *Mol Pharmacol.* 63, 1256–1272.
4. Joost, P., Methner, A. (2002) Phylogenetic analysis of 277 human G-protein-coupled receptors as a tool for the prediction of orphan receptor ligands. *Genome Biol.* 3, RESEARCH0063.
5. Vassilatis, D.K., Hohmann, J.G., Zeng, H., Li, F., Ranchalis, J.E., Mortrud, M.T., Brown, A., Rodriguez, S.S., Weller, J.R., Wright, A.C., Bergmann, J.E., Gaitanaris,

G.A. (2003) The G protein-coupled receptor repertoires of human and mouse. *Proc Natl Acad Sci U S A*. 100, 4903–4908.

6. Civelli, O. (2005) GPCR deorphanizations: The novel, the known and the unexpected transmitters. *Trends Pharmacol Sci.* 26, 15–19.
7. Dixon, R.A.F., Kobilka, B.K., Strader, D.J., Benovic, J.L., Dohlman, H.G., Frielle, T., Bolanowski, M.A., Bennett, C.D., Rands, E., Diehl, R.E., Mumford, R.A., Slater, E.E., Sigal, I.S., Caron, M.G., Lefkowitz, R.J., Strader, C.D. (1986) Cloning of the gene and cDNA for mammalian β-adrenergic receptor and homology with rhodopsin. *Nature*. 321, 75–79.
8. Bunzow, J.R., Van Tol, H.H., Grandy, D.K., Albert, P., Salon, J., Christie, M., Machida, C.A., Neve, K.A., Civelli, O. (1988) Cloning and expression of a rat D2 dopamine receptor cDNA. *Nature*. 336, 783–787.
9. Libert, F., Parmentier, M., Lefort, A., Dinsart, C., Van Sande, J., Maenhaut, C., Simons, M.-J., Dumont, J.E., Vassart, G. (1989) Selective amplification and cloning of four new members of the G protein-coupled receptor family. *Science*. 244, 569–572.
10. O'Dowd, B.F., Nguyen, T., Jung, B.P., Marchese, A., Cheng, R., Heng, H.H., Kolakowski Jr., L.F., Lynch, K.R., George, S.R. (1997) Cloning and chromosomal mapping of four putative novel human G-protein-coupled receptor genes. *Gene*. 187, 75–81.
11. Civelli, O., Nothacker, H.P., Saito, Y., Wang, Z., Lin, S.H., Reinscheid, R.K. (2001) Novel neurotransmitters as natural ligands of orphan G-protein-coupled receptors. *Trends Neurosci*. 24, 230–237.
12. Reinscheid, R.K., Nothacker, H.P., Bourson, A., Ardati, A., Henningsen, R.A., Bunzow, J.R., Grandy, D.K., Langen, H., Monsma Jr., F.J., Civelli, O. (1995) Orphanin FQ: A neuropeptide that activates an opioidlike G protein-coupled receptor. *Science* 270, 792–794.
13. Fargin, A., Raymond, J.R., Lohse, M.J., Kobilka, B.K., Caron, M.G., Lefkowitz, R.J. (1988) The genomic clone G-21 which resembles a β-adrenergic receptor sequence encodes the 5HT1A receptor. *Nature*. 335, 358–360.
14. Libert, F., Schiffmann, S.N., Lefort, A., Parmentier, M., Gerard, C., Dumont, J.E., Vanderhaeghen, J.J., Vassart, G. (1991) The orphan receptor cDNA RDC7 encodes an A1 adenosine receptor. *EMBO J.* 10, 1677–82.
15. Libert, F., Vassart, G., Parmentier, M. (1991) Current developments in G-protein-coupled receptors. *Curr Opin Cell Biol.* 3, 218–223.
16. Mills, A., Duggan, M.J. (1994) Orphan seven transmembrane domain receptors: Reversing pharmacology. *Trends Biotechnol.* 12, 47–49.
17. Wang, J., Simonavicius, N., Wu, X., Swaminath, G., Reagan, J., Tian, H., Ling, L. (2006) Kynurenic acid as a ligand for orphan G protein-coupled receptor GPR35. *J Biol Chem.* 281, 22021–2208.
18. Wang, J., Wu, X., Simonavicius, N., Tian, H., Ling, L. (2006) Medium-chain fatty acids as ligands for orphan G protein-coupled receptor GPR84. *J Biol Chem.* 281, 34457–34464.
19. Ignatov, A., Robert, J., Gregory-Evans, C., Schaller, H.C. (2006) RANTES stimulates Ca2+ mobilization and inositol trisphosphate (IP3) formation in cells transfected with G protein-coupled receptor 75. *Br J Pharmacol.* 149, 490–497.

20. Henderson, G., McKnight, A.T. (1997) The orphan opioid receptor and its endogenous ligand—nociceptin/orphanin FQ. *Trends Pharmacol Sci.* 18, 293–300.
21. Meunier, J.C., Mollereau, C., Toll, L., Suaudeau, C., Moisand, C., Alvinerie, P., Butour, J.L., Guillemot, J.C., Ferrara, P., Monsarrat, B., Mazarguil, H., Vassart, G., Parmentier, M., Costentin, J. (1995) Isolation and structure of the endogenous agonist of opioid receptor-like ORL1 receptor. *Nature*. 377, 532–535.
22. Wise, A., Jupe, S.C., Rees, S. (2004) The identification of ligands at orphan G-protein coupled receptors. *Annu Rev Pharmacol Toxicol.* 44, 43–66.
23. Civelli, O., Saito, Y., Wang, Z., Nothacker, H.P., Reinscheid, R.K. (2006) Orphan GPCRs and their ligands. *Pharmacol Ther.* 110, 525–532.
24. Jacoby, E. (2006) Chemogenomics: Drug discovery's panacea? *Mol Biosyst.* 2, 218–220.
25. Metpally, R.P., Sowdhamini, R. (2005) Cross genome phylogenetic analysis of human and *Drosophila* G protein-coupled receptors: Application to functional annotation of orphan receptors. *BMC Genomics*. 6, 106.
26. Fredriksson, R., Schioth, H.B. (2005) The repertoire of G-protein-coupled receptors in fully sequenced genomes. *Mol Pharmacol.* 67, 1414–1425.
27. Murakami, M., Shiraishi, A., Tabata, K., Fujita, N. (2008) Identification of the orphan GPCR, P2Y(10) receptor as the sphingosine-1-phosphate and lysophosphatidic acid receptor. *Biochem Biophys Res Commun.* 371, 707–712.
28. Ciana, P., Fumagalli, M., Trincavelli, M.L., Verderio, C., Rosa, P., Lecca, D., Ferrario, S., Parravicini, C., Capra, V., Gelosa, P., Guerrini, U., Belcredito, S., Cimino, M., Sironi, L., Tremoli, E., Rovati, G.E., Martini, C., Abbracchio, M.P. (2006) The orphan receptor GPR17 identified as a new dual uracil nucleotides/cysteinyl-leukotrienes receptor. *EMBO J.* 25, 4615–4627.
29. Surgand, J.S., Rodrigo, J., Kellenberger, E., Rognan, D. (2006) A chemogenomic analysis of the transmembrane binding cavity of human G-protein-coupled receptors. *Proteins*. 62, 509–538.
30. Wellendorph, P., Hansen, K.B., Balsgaard, A., Greenwood, J.R., Egebjerg, J., Brauner-Osborne, H. (2005) Deorphanization of GPRC6A: A promiscuous L-alpha-amino acid receptor with preference for basic amino acids. *Mol Pharmacol.* 67, 589–597.
31. Tunaru, S., Kero, J., Schaub, A., Wufka, C., Blaukat, A., Pfeffer, K., Offermanns, S. (2003) PUMA-G and HM74 are receptors for nicotinic acid and mediate its antilipolytic effect. *Nat Med.* 9, 352–355.
32. Morita, H., Mazerbourg, S., Bouley, D.M., Luo, C.W., Kawamura, K., Kuwabara, Y., Baribault, H., Tian, H., Hsueh, A.J. (2004) Neonatal lethality of LGR5 null mice is associated with ankyloglossia and gastrointestinal distension. *Mol Cell Biol.* 24, 9736–9743.
33. Matteson, P.G., Desai, J., Korstanje, R., Lazar, G., Borsuk, T.E., Rollins, J., Kadambi, S., Joseph, J., Rahman, T., Wink, J., Benayed, R., Paigen, B., Millonig, J.H. (2008) The orphan G protein-coupled receptor, Gpr161, encodes the vacuolated lens locus and controls neurulation and lens development. *Proc Natl Acad Sci U S A.* 105, 2088–2093.
34. Ivanova, E.A., Bechtold, D., Dupre, S., Brennand, J., Barrett, P., Luckman, S., Loudon A.S. (2008) Altered metabolism in the melatonin-related receptor (GPR50) knock out mouse. *Am J Physiol Endocrinol Metab.* 294, E176–E182.

35. Cox, P.J., Pitcher, T., Trim, S.A., Bell, C.H., Qin, W., Kinloch, R.A. (2008) The effect of deletion of the orphan G-protein coupled receptor (GPCR) gene MrgE on pain-like behaviours in mice. *Mol Pain*. 4, 2.

36. Matsumoto, M., Straub, R.E., Marenco, S., Nicodemus, K.K., Matsumoto, S., Fujikawa, A., Miyoshi, S., Shobo, M., Takahashi, S., Yarimizu, J., Yuri, M., Hiramoto, M., Morita, S., Yokota, H., Sasayama, T., Terai, K., Yoshino, M., Miyake, A., Callicott, J.H., Egan, M.F., Meyer-Lindenberg, A., Kempf, L., Honea, R., Vakkalanka, R.K., Takasaki, J., Kamohara, M., Soga, T., Hiyama, H., Ishii, H., Matsuo, A., Nishimura, S., Matsuoka, N., Kobori, M., Matsushime, H., Katoh, M., Furuichi, K., Weinberger, D.R. (2008) The evolutionarily conserved G protein-coupled receptor SREB2/GPR85 influences brain size, behavior, and vulnerability to schizophrenia. *Proc Natl Acad Sci U S A*. 105, 6133–6138.

37. Eglen, R.M., Bosse, R., Reisine, T. (2007) Emerging concepts of guanine nucleotide-binding protein-coupled receptor (GPCR) function and implications for high throughput screening. *Assay Drug Dev Technol*. 5, 425–451.

38. Johnson, E.N., Shi, X., Cassaday, J., Ferrer, M., Strulovici, B., Kunapuli, P. (2008) A 1,536-well [(35)S]GTPgammaS scintillation proximity binding assay for ultra-high-throughput screening of an orphan galphai-coupled GPCR. *Assay Drug Dev Technol*. 6, 327–637.

39. Potenza, M.N., Graminski, G.F., Lerner, M.R. (1992) A method for evaluating the effects of ligands upon Gs protein-coupled receptors using a recombinant melanophore-based bioassay. *Anal Biochem*. 206, 315–322.

40. Ames, R.S., Li, Y., Sarau, H.M., Nuthulaganti, P., Foley, J.J., Ellis, C., Zeng, Z., Su, K., Jurewicz, A.J., Hertzberg, R.P., Bergsma, D.J., Kumar, C. (1996) Molecular cloning and characterization of the human anaphylatoxin C3a receptor. *J Biol Chem*. 271, 20231–20234.

41. Kojima, M., Hosoda, H., Date, Y., Nakazato, M., Matsuo, H., Kangawa, K. (1999) Ghrelin is a growth-hormone-releasing acylated peptide from stomach. *Nature*. 402, 656–660.

42. Schaeffer, M.T., Cully, D., Chou, M., Liu, J., Van der Ploeg, L.H., Fong, T.M. (1999) Use of bioluminescent aequorin for the pharmacological characterization of 5HT receptors. *J Recept Signal Transduct Res*. 19, 927–938.

43. Coward, P., Chan, S.D., Wada, H.G., Humphries, G.M., Conklin, B.R. (1999) Chimeric G proteins allow a high-throughput signaling assay of Gi-coupled receptors. *Anal Biochem*. 270, 242–248.

44. Offermanns, S., Simon, M.I. (1995) Ga15 and Ga16 couple a wide variety of receptors to phospholipase C. *J Biol Chem*. 270, 15175–15180.

45. Lefkowitz, R.J., Shenoy, S.K. (2005) Transduction of receptor signals by beta-arrestins. *Science*. 308, 512–517.

46. Cain, S.A., Monk, P.N. (2002) The orphan receptor C5L2 has high affinity binding sites for complement fragments C5a and C5a des-Arg(74). *J Biol Chem*. 277, 7165–7169.

47. Kalant, D., Cain, S.A., Maslowska, M., Sniderman, A.D., Cianflone, K., Monk, P.N. (2003) The chemoattractant receptor-like protein C5L2 binds the C3a des-Arg77/acylation-stimulating protein. *J Biol Chem*. 278, 11123–11129.

48. Mantovani, A., Locati, M., Vecchi, A., Sozzani, S., Allavena, P. (2001) Decoy receptors: A strategy to regulate inflammatory cytokines and chemokines. *Trends Immunol.* 22, 328–336.

49. Bertin, B., Freissmuth, M., Jockers, R., Strosberg, A.D., Marullo, S. (1994) Cellular signaling by an agonist-activated receptor/G(s)alpha fusion protein. *Proc Natl Acad Sci U S A.* 91, 8827–8831.

50. Guo, Z.D., Suga, H., Okamura, M., Takeda, S., Haga, T. (2001) Receptor-Galpha fusion proteins as a tool for ligand screening. *Life Sci.* 68, 2319–2327.

51. Suga, H., Haga, T. (2007) Ligand screening system using fusion proteins of G protein-coupled receptors with G protein alpha subunits. *Neurochem Int.* 51, 140–164.

52. Takeda, S., Yamamoto, A., Okada, T., Matsumura, E., Nose, E., Kogure, K., Kojima, S., Haga, T. (2003) Identification of surrogate ligands for orphan G protein-coupled receptors. *Life Sci.* 74, 367–377.

53. Takeda, S., Kadowaki, S., Haga, T., Takaesu, H., Mitaku, S. (2002) Identification of G protein-coupled receptor genes from the human genome sequence. *FEBS Lett.* 520, 97–101.

54. Hosoi, T., Koguchi, Y., Sugikawa, E., Chikada, A., Ogawa, K., Tsuda, N., Suto, N., Tsunoda, S., Taniguchi, T., Ohnuki, T. (2002) Identification of a novel human eicosanoid receptor coupled to G(i/o). *J Biol Chem.* 277, 31459–31465.

55. Tabata, K., Baba, K., Shiraishi, A., Ito, M., Fujita, N. (2007) The orphan GPCR GPR87 was deorphanized and shown to be a lysophosphatidic acid receptor. *Biochem Biophys Res Commun.* 363, 861–866.

56. Hirasawa, A., Tsumaya, K., Awaji, T., Katsuma, S., Adachi, T., Yamada, M., Sugimoto, Y., Miyazaki, S., Tsujimoto, G. (2005) Free fatty acids regulate gut incretin glucagon-like peptide-1 secretion through GPR120. *Nat Med.* 11, 90–94.

57. Crouch, M.F., Osmond, R.I. (2008) New strategies in drug discovery for GPCRs: High throughput detection of cellular ERK phosphorylation. *Comb Chem High Throughput Screen.* 11, 344–356.

58. Jockers, R., Da Silva, A., Strosberg, A.D., Bouvier, M., Marullo, S. (1996) New molecular and structural determinants involved in β2-adrenergic receptor desensitization and sequestration; delineation using chimeric β3/β2-adrenergic receptors. *J Biol Chem.* 271, 9355–9362.

59. Adie, E.J., Francis, M.J., Davies, J., Smith, L., Marenghi, A., Hather, C., Hadingham, K., Michael, N.P., Milligan, G., Game, S. (2003) CypHer 5: A generic approach for measuring the activation and trafficking of G protein-coupled receptors in live cells. *Assay Drug Dev Technol.* 1, 251–259.

60. Beletskii, A., Cooper, M., Sriraman, P., Chiriac, C., Zhao, L., Abbot, S., Yu, L. (2005) High-throughput phagocytosis assay utilizing a pH-sensitive fluorescent dye. *Biotechniques.* 39, 894–897.

61. Oakley, R.H., Hudson, C.C., Sjaastad, M.D., Loomis, C.R. (2006) The ligand-independent translocation assay: An enabling technology for screening orphan G protein-coupled receptors by arrestin recruitment. *Methods Enzymol.* 414, 50–63.

62. Fergusson, S.S., Caron, M.G. (2004) Green fluorescent protein-tagged β-arrestin translocation as a measure of G protein-coupled receptor activation. *Methods Mol Biol.* 237, 121–126.

63. Johnson, E.C., Bohn, L.M., Barak, L.S., Birse, R.T., Nassel, D.R., Caron, M.G., Taghert, P.H. (2003) Identification of *Drosophila* neuropeptide receptors by G protein-coupled receptors-beta-arrestin2 interactions. *J Biol Chem.* 278, 52172–52178.
64. Kalant, D., MacLaren, R., Cui, W., Samanta, R., Monk, P.N., Laporte, S.A., Cianflone, K. (2005) C5L2 is a functional receptor for acylation-stimulating protein. *J Biol Chem.* 280, 23936–23944.
65. Bertrand, L., Parent, S., Caron, M., Legault, M., Joly, E., Angers, S., Bouvier, M., Brown, M., Houle, B., Menard, L. (2002) The BRET2/arrestin assay in stable recombinant cells: A platform to screen for compounds that interact with G protein-coupled receptors (GPCRS). *J Recept Signal Transduct Res.* 22, 533–541.
66. Heding, A. (2004) Use of the BRET 7TM receptor/beta-arrestin assay in drug discovery and screening. *Expert Rev Mol Diagn.* 4, 403–411.
67. Vrecl, M., Jorgensen, R., Pogacnik, A., Heding, A. (2004) Development of a BRET2 screening assay using beta-arrestin 2 mutants. *J Biomol Screen.* 9, 322–333.
68. Verkaar, F., van Rosmalen, J.W., Blomenrohr, M., van Koppen, C.J., Blankesteijn, W.M., Smits, J.F., Zaman, G.J. (2008) G protein-independent cell-based assays for drug discovery on seven-transmembrane receptors. *Biotechnol Annu Rev.* 14, 253–274.
69. Barnea, G., Strapps, W., Herrada, G., Berman, Y., Ong, J., Kloss, B., Axel, R., Lee, K.J. (2008) The genetic design of signaling cascades to record receptor activation. *Proc Natl Acad Sci U S A.* 105, 64–69.
70. Olson, K.R., Eglen, R.M. (2007) Beta galactosidase complementation: A cell-based luminescent assay platform for drug discovery. *Assay Drug Dev Technol.* 5, 137–144.
71. Carman, C.V., Som, T., Kim, C.M., Benovic, J.L. (1998) Binding and phosphorylation of tubulin by G protein-coupled receptor kinases. *J Biol Chem.* 273, 20308–20316.
72. Pierce, K.L., Lefkowitz, R.J. (2001) Classical and new roles of beta-arrestins in the regulation of G-protein-coupled receptors. *Nat Rev Neurosci.* 2, 727–733.
73. Levoye, A., Jockers, R. (2008) Alternative drug discovery approaches for orphan GPCRs. *Drug Discov Today.* 13, 52–58.
74. Kaupmann, K., Malitschek, B., Schuler, V., Heid, J., Froestl, W., Beck, P., Mosbacher, J., Bischoff, S., Kulik, A., Shigemoto, R., Karschin, A., Bettler, B. (1998) GABA(B)-receptor subtypes assemble into functional heteromeric complexes. *Nature.* 396, 683–687.
75. White, J.H., Wise, A., Main, M.J., Green, A., Fraser, N.J., Disney, G.H., Barnes, A.A., Emson, P., Foord, S.M., Marshall, F.H. (1998) Heterodimerization is required for the formation of a functional GABA(B) receptor. *Nature.* 396, 679–682.
76. Jones, K.A., Borowsky, B., Tamm, J.A., Gerald, C. (1998) GABAb receptors function as a heterotrimeric assembly of the subunits GABAbR1 and GAGbR2. *Nature.* 396, 674–679.
77. Pin, J.P., Kniazeff, J., Liu, J., Binet, V., Goudet, C., Rondard, P., Prezeau, L. (2005) Allosteric functioning of dimeric class C G-protein-coupled receptors. *FEBS J.* 272, 2947–2955.
78. Nelson, G., Hoon, M.A., Chandrashekar, J., Zhang, Y., Ryba, N.J., Zuker, C.S. (2001) Mammalian sweet taste receptors. *Cell.* 106, 381–390.
79. Nelson, G., Chandrashekar, J., Hoon, M.A., Feng, L., Zhao, G., Ryba, N.J., Zuker, C.S. (2002) An amino-acid taste receptor. *Nature.* 416, 199–202.

80. Xu, H., Staszewski, L., Tang, H., Adler, E., Zoller, M., Li, X. (2004) Different functional roles of T1R subunits in the heteromeric taste receptors. *Proc Natl Acad Sci U S A.* 101, 14258–14263.

81. Levoye, A., Dam, J., Ayoub, M.A., Guillaume, J.L., Couturier, C., Delagrange, P., Jockers, R. (2006) The orphan GPR50 receptor specifically inhibits MT(1) melatonin receptor function through heterodimerization. *EMBO J.* 25, 3012–3023.

82. Milasta, S., Pediani, J., Appelbe, S., Trim, S., Wyatt, M., Cox, P., Fidock, M., Milligan, G. (2006) Interactions between the Mas-related receptors MrgD and MrgE alter signalling and trafficking of MrgD. *Mol Pharmacol.* 69, 479–491.

83. Dickson, L., Barclay, Z., Mitchell, C., Fortheringham, H., Robertson, D., Holland, P., Rosie, R., Johnson, M., Lutz, E., Mitchell, R. (2008) Interaction between 5-HT2AR and the orphan GPCR TAAR6. In *Keystone Symposium: G Protein-Coupled Receptors: New Insights in the Functiona Regulation and Clinical Application*, ed. H.E. Hamm, P.J. Conn, J.P. Pin, O. Civelli, pp. 60. Killarney, Ireland: INEC-Ireland's National Events & Conference Centre.

84. Neuhaus, E.M., Gisselmann, G., Zhang, W., Dooley, R., Stortkuhl, K., Hatt, H. (2005) Odorant receptor heterodimerization in the olfactory system of *Drosophila melanogaster*. *Nat Neurosci.* 8, 15–17.

85. Benton, R., Sachse, S., Michnick, S.W., Vosshall, L.B. (2006) Atypical membrane topology and heteromeric function of *Drosophila* odorant receptors in vivo. *PLoS Biol.* 4, e20.

86. Wicher, D., Schafer, R., Bauernfeind, R., Stensmyr, M.C., Heller, R., Heinemann, S.H., Hansson, B.S. (2008) *Drosophila* odorant receptors are both ligand-gated and cyclic-nucleotide-activated cation channels. *Nature.* 452, 1007–1011.

87. Sato, K., Pellegrino, M., Nakagawa, T., Nakagawa, T., Vosshall, L.B., Touhara, K. (2008) Insect olfactory receptors are heteromeric ligand-gated ion channels. *Nature.* 452, 1002–1006.

88. McLatchie, L.M., Fraser, N.J., Main, M.J., Wise, A., Brown, J., Thompson, N., Solari, R., Lee, M.G., Foord, S.M. (1998) RAMPs regulate the transport and ligand specificity of the calcitonin-receptor-like receptor. *Nature.* 393, 333–339.

89. Little, K.D., Hemler, M.E., Stipp, C.S. (2004) Dynamic regulation of a GPCR-tetraspanin-G protein complex on intact cells: Central role of CD81 in facilitating GPR56-Galpha q/11 association. *Mol Biol Cell.* 15, 2375–2387.

90. Xu, L., Begum, S., Hearn, J.D., Hynes, R.O. (2006) GPR56, an atypical G protein-coupled receptor, binds tissue transglutaminase, TG2, and inhibits melanoma tumor growth and metastasis. *Proc Natl Acad Sci U S A.* 103, 9023–9028.

91. Marazziti, D., Mandillo, S., Di Pietro, C., Golini, E., Matteoni, R., Tocchini-Valentini, G.P. (2007) GPR37 associates with the dopamine transporter to modulate dopamine uptake and behavioral responses to dopaminergic drugs. *Proc Natl Acad Sci U S A.* 104, 9846–9851.

92. Riobo, N.A., Lu, K., Emerson Jr., C.P. (2006) Hedgehog signal transduction: Signal integration and cross talk in development and cancer. *Cell Cycle.* 5, 1612–1615.

93. Schwartz, T.W., Holst, B. (2007) Allosteric enhancers, allosteric agonists and ago-allosteric modulators: Where do they bind and how do they act? *Trends Pharmacol Sci.* 28, 366–373.

94. Neubig, R.R., Spedding, M., Kenakin, T., Christopoulos, A. (2003) International Union of Pharmacology Committee on Receptor Nomenclature and Drug

Classification. XXXVIII. Update on terms and symbols in quantitative pharmacology. *Pharmacol Rev.* 55, 597–606.

95. Binet, V., Brajon, C., Le Corre, L., Acher, F., Pin, J.P., Prezeau, L. (2004) The heptahelical domain of GABA(B2) is activated directly by CGP7930, a positive allosteric modulator of the GABA(B) receptor. *J Biol Chem.* 279, 29085–29091.
96. Maenhaut, C., Van Sande, J., Libert, F., Abramowicz, M., Parmentier, M., Vanderhaegen, J.J., Dumont, J.E., Vassart, G., Schiffmann, S. (1990) RDC8 codes for an adenosine A2 receptor with physiological constitutive activity. *Biochem Biophys Res Commun.* 173, 1169–1178.
97. Stoddart, L.A., Brown, A.J., Milligan, G. (2007) Uncovering the pharmacology of the G protein-coupled receptor GPR40: High apparent constitutive activity in guanosine 5′-O-(3-[35S]thio)triphosphate binding studies reflects binding of an endogenous agonist. *Mol Pharmacol.* 71, 994–1005.
98. Seuwen, K., Ludwig, M.G., Wolf, R.M. (2006) Receptors for protons or lipid messengers or both? *J Recept Signal Transduct Res.* 26, 599–610.
99. Zhang, J.V., Ren, P.G., Avsian-Kretchmer, O., Luo, C.W., Rauch, R., Klein, C., Hsueh, A.J. (2005) Obestatin, a peptide encoded by the ghrelin gene, opposes ghrelin's effects on food intake. *Science.* 310, 996–999.
100. Chartrel, N., Alvear-Perez, R., Leprince, J., Iturrioz, X., Reaux-Le Goazigo, A., Audinot, V., Chomarat, P., Coge, F., Nosjean, O., Rodriguez, M., Galizzi, J.P., Boutin, J.A., Vaudry, H., Llorens-Cortes, C. (2007) Comment on "Obestatin, a peptide encoded by the ghrelin gene, opposes ghrelin's effects on food intake." *Science.* 315, 766.
101. Holst, B., Egerod, K.L., Schild, E., Vickers, S.P., Cheetham, S., Gerlach, L.O., Storjohann, L., Stidsen, C.E., Jones, R., Beck-Sickinger, A.G., Schwartz, T.W. (2007) GPR39 signaling is stimulated by zinc ions but not by obestatin. *Endocrinology.* 148, 13–20.
102. Otto, C., Rohde-Schulz, B., Schwarz, G., Fuchs, I., Klewer, M., Brittain, D., Langer, G., Bader, B., Prelle, K., Nubbemeyer, R., Fritzemeier, K.H. (2008) GPR30 localizes to the endoplasmic reticulum and is not activated by estradiol. *Endocrinology.* 149, 4846–4856.
103. Prossnitz, E.R., Oprea, T.I., Sklar, L.A., Arterburn, J.B. (2008) The ins and outs of GPR30: A transmembrane estrogen receptor. *J Steroid Biochem Mol Biol.* 109, 350–353.
104. Levoye, A., Dam, J., Ayoub, M.A., Guillaume, J.L., Jockers, R. (2006) Do orphan G-protein-coupled receptors have ligand-independent functions? New insights from receptor heterodimers. *EMBO Rep.* 7, 1094–1098.
105. Kniazeff, J., Galvez, T., Labesse, G., Pin, J.P. (2002) No ligand binding in the GB2 subunit of the GABA(B) receptor is required for activation and allosteric interaction between the subunits. *J Neurosci.* 22, 7352–7361.
106. Dufourny, L., Levasseur, A., Migaud, M., Callebaut, I., Pontarotti, P., Malpaux, B., Monget, P. (2008) GPR50 is the mammalian ortholog of Mel1c: Evidence of rapid evolution in mammals. *BMC Evol Biol.* 8, 105.
107. Kohno, M., Hasegawa, H., Inoue, A., Muraoka, M., Miyazaki, T., Oka, K., Yasukawa, M. (2006) Identification of N-arachidonylglycine as the endogenous ligand for

orphan G-protein-coupled receptor GPR18. *Biochem Biophys Res Commun*. 347, 827–832.

108. Sugo, T., Tachimoto, H., Chikatsu, T., Murakami, Y., Kikukawa, Y., Sato, S., Kikuchi, K., Nagi, T., Harada, M., Ogi, K., Ebisawa, M., Mori, M. (2006) Identification of a lysophosphatidylserine receptor on mast cells. *Biochem Biophys Res Commun*. 341, 1078–1087.

109. Ryberg, E., Larsson, N., Sjogren, S., Hjorth, S., Hermansson, N.O., Leonova, J., Elebring, T., Nilsson, K., Drmota, T., Greasley, P.J. (2007) The orphan receptor GPR55 is a novel cannabinoid receptor. *Br J Pharmacol*. 152, 1092–1101.

110. Kotarsky, K., Boketoft, A., Bristulf, J., Nilsson, N.E., Norberg, A., Hansson, S., Owman, C., Sillard, R., Leeb-Lundberg, L.M., Olde, B. (2006) Lysophosphatidic acid binds to and activates GPR92, a G protein-coupled receptor highly expressed in gastrointestinal lymphocytes. *J Pharmacol Exp Ther.* 318, 619–628.

111. Lee, C.W., Rivera, R., Gardell, S., Dubin, A.E., Chun, J. (2006) GPR92 as a new G12/13- and Gq-coupled lysophosphatidic acid receptor that increases cAMP, LPA5. *J Biol Chem.* 281, 23589–23597.

112. Soga, T., Ohishi, T., Matsui, T., Saito, T., Matsumoto, M., Takasaki, J., Matsumoto, S., Kamohara, M., Hiyama, H., Yoshida, S., Momose, K., Ueda, Y., Matsushime, H., Kobori, M., Furuichi, K. (2005) Lysophosphatidylcholine enhances glucose-dependent insulin secretion via an orphan G-protein-coupled receptor. *Biochem Biophys Res Commun*. 326, 744–751.

113. Balabanian, K., Lagane, B., Infantino, S., Chow, K.Y., Harriague, J., Moepps, B., Arenzana-Seisdedos, F., Thelen, M., Bachelerie, F. (2005) The chemokine SDF-1/CXCL12 binds to and signals through the orphan receptor RDC1 in T lymphocytes. *J Biol Chem.* 280, 35760–35766.

114. Pasternack, S.M., von Kugelgen, I., Aboud, K.A., Lee, Y.A., Ruschendorf, F., Voss, K., Hillmer, A.M., Molderings, G.J., Franz, T., Ramirez, A., Nurnberg, P., Nothen, M.M., Betz, R.C. (2008) G protein-coupled receptor P2Y5 and its ligand LPA are involved in maintenance of human hair growth. *Nat Genet*. 40, 329–334.

115. Hase, M., Yokomizo, T., Shimizu, T., Nakamura, M. (2008) Characterization of an orphan G protein-coupled receptor, GPR20, that constitutively activates Gi proteins. *J Biol Chem.* 283, 12747–12755.

116. Adams, J.W., Wang, J., Davis, J.R., Liaw, C., Gaidarov, I., Gatlin, J., Dalton, N.D., Gu, Y., Ross Jr., J.R., Behan, D., Chien, K.R., Connolly, D. (2008) Myocardial expression, signaling, and function of GPR22: A protective role for an orphan G-protein coupled receptor. *Am J Physiol Heart Circ Physiol*. 295, H509–H521.

117. Jones, P.G., Nawoschik, S.P., Sreekumar, K., Uveges, A.J., Tseng, E., Zhang, L., Johnson, J., He, L., Paulsen, J.E., Bates, B., Pausch, M.H. (2007) Tissue distribution and functional analyses of the constitutively active orphan G protein coupled receptors, GPR26 and GPR78. *Biochim Biophys Acta.* 1770, 890–901.

118. Ge, H., Weiszmann, J., Reagan, J.D., Gupte, J., Baribault, H., Gyuris, T., Chen, J.L., Tian, H., Li, Y. (2008) Elucidation of signaling and functional activities of an orphan GPCR, GPR81. *J Lipid Res.* 49, 797–803.

119. Hirano, M., Zang, L., Oka, T., Ito, Y., Shimada, Y., Nishimura, Y., Tanaka, T. (2006) Novel reciprocal regulation of cAMP signaling and apoptosis by orphan

G-protein-coupled receptor GPRC5A gene expression. *Biochem Biophys Res Commun*. 351, 185–191.

120. Rosenkilde, M.M., Benned-Jensen, T., Andersen, H., Holst, P.J., Kledal, T.N., Luttichau, H.R., Larsen, J.K., Christensen, J.P., Schwartz, T.W. (2006) Molecular pharmacological phenotyping of EBI2. An orphan seven-transmembrane receptor with constitutive activity. *J Biol Chem*. 281, 13199–13208.
121. Hoshii, T., Takeo, T., Nakagata, N., Takeya, M., Araki, K., Yamamura, K. (2007) LGR4 regulates the postnatal development and integrity of male reproductive tracts in mice. *Biol Reprod*. 76, 303–313.

CHAPTER 8

High-Throughput GPCR Screening Technologies and the Emerging Importance of the Cell Phenotype

TERRY REISINE* and RICHARD M. EGLEN

PerkinElmer Life and Analytical Sciences, Waltham, MA

8.1. INTRODUCTION

G protein-coupled receptors (GPCRs) are key factors in cell-to-cell communication in all living organisms. These surface proteins bind circulating neurotransmitters, hormones, and growth factors. This binding initiates a cascade of events in the target cell that modulates activity [1]. GPCR activation is necessary for normal physiology of all organisms while dysfunction of GPCR signaling is responsible for many of the diseases we encounter in our lives and as such is essential for survival.

Consequently, GPCRs have historically served a fundamental role in modern pharmacological research and have been targets for many drugs developed by the pharmaceutical industry to treat human disease. GPCR drug discovery has existed for decades, and the technologies used to identify such drugs are highly advanced and have been successful in the discovery of a large number of drugs that have been approved for marketing by the FDA. Due to the success in developing GPCR drugs and their pervasiveness in modern medicine, there is much interest in the pharmaceutical industry to produce improved and unique GPCR-based drugs. When one considers that the human genome expresses genes for between 800 and 1000 different GPCRs [2] and marketed drugs target less than 50 GPCRs, it is evident that the field of GPCR

*Terry Reisine, PhD, is an independent consultant.

GPCR Molecular Pharmacology and Drug Targeting: Shifting Paradigms and New Directions, Edited by Annette Gilchrist

drug discovery and development is still growing. Add to that the emergence of novel drugs such as inverse agonists or allosteric modulators and one can foresee an almost limitless horizon for GPCR targeted drug discovery [3].

8.2. HOW ARE GPCR DRUGS DISCOVERED?

Prior to the last several decades, most drug discovery was physiology-based [4, 5]. That is, one first identified a natural product that induced a desired pharmacological action. Then one identified bioassay systems in which the natural product produced a pronounced and selective action, and medicinal chemistry was then used to produce analogs that were easier to synthesize, more druggable, and induced desirable therapeutic effects. In fact, this approach was highly successful and has produced many of the drugs commonly employed in medical practice. For example, many opiate drugs were identified as natural products, their biological actions where assayed *in vitro* using the mouse vas deferens and guinea pig ileum assays; medicinal chemistry was employed to modify their pharmaceutical properties to obtained desired therapeutic effects, resulting in a large number of opiate receptor targeted drugs that are employed extensively in clinical practice nowadays [5].

Over the years, with the advance of biochemical and molecular biological technologies, drug discovery changed to a molecular target-based approach in which a protein was identified as having a critical function in a desired biological pathway, the protein was isolated, and drugs were developed that selectively interacted with the target to either activate or inhibit its function [3]. Due to the ease in defining the mechanism of action of a drug, this target-based approach is now the primary approach used by the pharmaceutical industry to discover new drugs.

The target-based approach employs a number of advanced technologies for drug discovery. For those proteins that can be isolated and purified in sufficient quantities in a manner which retains biological activity, crystallography can be used to identify the physical points of contact of a drug or natural regulator with the target. Using this structural information, rational medicinal chemistry can be used to design drugs to bind to the target in a desired manner and then those compounds are tested in biological systems for effectiveness.

The primary approach used to discover GPCR drugs involves the use of automated high-throughput screening (HTS) cell-based assays since approaches employing crystallography and rational medicinal chemistry cannot be easily employed to discover drugs against GPCRs. Over the last decade, HTS technologies to discover GPCR drugs have greatly expanded such that they can be readily adapted to automated fluid dispensing systems and microtiter plate detectors. They are also more sophisticated in terms of measuring the response of the GPCR in a cellular context. In fact, cell-based assays have become necessary as numerous GPCRs have no known endogenous ligand and are thus considered "orphan" in nature. Furthermore, the possibility of developing drugs targeting GPCRs yet acting through unique mechanisms, such as inverse agonists, allosteric modulators, and drugs interacting with GPCR heterodimers all require novel, cell-based HTS technologies.

HTS technologies for GPCR drugs require two critical factors. First, an assay must be identified to measure binding of drugs to the GPCR, and this may employ resultant functional consequences of that binding. Second, large libraries of compounds spanning various chemistries are needed to test in the assay. The general hypothesis of this approach is that by screening large enough numbers of compounds, a few will be identified that bind to the receptor with high affinity and selectivity, resulting in the modulation of its measured cellular activity. Then, one can employ medicinal chemistry to optimize the drug structures needed for binding to one with appropriate pharmaceutical properties while retaining target specificity and affinity. Cell-based HTS assays are the primary approach used by the pharmaceutical industry to discover new GPCR agonists, antagonists, allosteric modulators, and designer drugs.

8.3. GPCR DEPENDENCE ON G PROTEINS

Receptors are referred to as GPCRs because they couple to G proteins, and this association is essential for mediating the functions of these cell surface proteins. Importantly, G proteins produce the diversity of function of GPCRs by their ability to modulate activity of a large array of second messengers and cellular effector systems in cells [6]. The G protein superfamily consists of heterotrimeric complexes of distinct α, β, and γ subunits with at least 18 Gα, 5 Gβ, and 11 Gγ subunits, capable of creating a very large number of distinct heterotrimeric complexes [7]. The Gα subunits have been divided into subfamilies, Gα_s, Gα_i/Gα_o, Gα_q/Gα_{11}, and Gα_{12}/Gα_{13} based on similarities of function. Importantly, these different G protein heterotrimers have significant specificity both with regard to the GPCRs they interact with and the cellular effector systems that they regulate [6–9].

G proteins serve as a switchboard in creating the diversity of GPCR function [3, 10, 11]. Thus, each GPCR only interacts, in a natural manner, with a set array of G proteins (GPCRs can also be forced to interact with individual G proteins by overexpression), and the G proteins have specificity in the second messenger systems they regulate. For example Gα_s is known to couple GPCRs to adenylyl cyclase to stimulate formation of the second messenger cAMP. In contrast, Gα_i mediates the inhibition of adenylyl cyclase by activation of a distinct and, in some cases, overlapping set of GPCRs. Furthermore, Gα_i is critical for other functions of GPCRs including modulating inward rectifying K^+ channels (GIRKs) [12, 13]. Therefore, stimulation of the same receptor/Gα_i complex can lead to turning off the cAMP pathway and simultaneously inhibiting cell firing by stimulating GIRKs.

Similarly, $G\alpha_o$ has been shown to link GPCRs to Ca^{++} conductance channels to regulate the influx of Ca^{++} to cells [14]. $G\alpha_o$ provides further diversity in function because it can also link GPCRs to phosphoinositol phospholipase Cβ, which hydrolyzes phosphatidylinositol 4,5 bisphosphate (PIP_2) forming *sn* 1,2 diacylgycerol (DG) and inositol 1,4,5 trisphosphate (IP_3). IP_3 binds and opens endoplasmic IP_3-gated calcium channels, causing release of bound calcium into the cytosol. Thus, the same GPCR coupled to $G\alpha_o$ can modulate Ca^{++} influx from extracellular and intracellular sites to maintain Ca^{++} homeostasis.

In addition to $G\alpha_o$, GPCRs also couple to another subfamily of G protein subunits, $G\alpha_q/G\alpha_{11}$, to regulate phospholipase C to increase Ins P_3 and intracellular Ca^{++} release to active downstream regulators such as protein kinase C. Protein kinase C activation can then lead to modulation of the MAP kinase pathway [15]. This involves stimulation of Raf, Mek, and then MAP kinase ERK. MAP kinase ERK can then phosphorylate and activate transcription factors to change gene expression and produce long-term alterations in cell activity. In fact, ERKs may be a convergent target for activation of most GPCRs since cAMP-dependent protein kinase is also involved in regulating ERK, and therefore, those GPCRs that regulate the cAMP signal transduction pathway also modulate ERK activity. Furthermore, GPCRs acting via $G\alpha_q$ can also modulate activity of growth factor receptor pathways in cells including stimulation of the tyrosine kinase Pyk2 which can activate Src to cause phosphorylation and activation of the epidermal growth factor (EGF) receptor [16]. This intracellular pathway provides a mechanism for cross-talk of GPCRs and growth factor receptors.

The fourth G protein subfamily $G\alpha_{12}/G\alpha_{13}$ has also been found to be important for GPCR signaling [17]. These Gα subunits are activated by GPCR stimulation and have been found to regulate the activity of RhoA GTPase. Through this downstream effector, $G\alpha_{12}/G\alpha_{13}$ can influence cell morphology, movement, and proliferation. Specifically, they have been found to affect neurite and axonal morphology and to be critical in regulation of migration of neurons in brain during development. These Gα subunits have also been implicated in the modulation of mitochrondrial function and cell apoptosis and their regulation, or dysregulation of p53 is likely to be involved in their control of cell proliferation, induction of metastasis and cancer, as well as other disorders.

In addition to the alpha subunit, βγ subunits of the G proteins are also critical for GPCR signaling [9]. Following α dissociation from βγ in response to GPCR stimulation, the βγ subunits are also able to directly interact with K^+ and Ca^{++} channels to regulate ionic conductance [9]. βγ subunits can also modulate the activity of phosphoinositol phospholipase Cβ [3]. Thus, following GPCR activation, both the α and βγ are released to modulate both overlapping as well as distinct enzymes and ionic conductance channels. This creates both the diversity of function of individual GPCRs as well as serving as an amplification process in cellular signaling. Importantly, because G proteins and second messenger systems are critical for the functions of GPCRs, they become critical components of assays used to study GPCRs and for GPCR drug discovery.

8.4. TECHNOLOGIES FOR GPCR COMPOUND SCREENING AND DRUG DISCOVERY

8.4.1. Cell-Free Assays

Drug discovery for GPCRs has involved either the use of cell free or cell-based HTS assays. Cell-free assays include ligand binding and GTPase assays. Ligand binding technologies have been one of the linchpins in the GPCR drug discovery process over the years, and most FDA approved GPCR drugs were discovered using this approach. However, ligand binding assays are not employed as much for drug discovery nowadays because they require the use of high-affinity, selective ligands that can be chemically radiolabeled, and for many receptors, such ligands are either not available or are cost prohibitive.

An alternative approach to ligand binding assays focuses on the interaction of GPCRs with G proteins and the fact that stimulation of GPCRs by an agonist results in an increase in GTPase catalytic activity of the G protein. Thus, one assay that has been employed for identifying GPCR agonists measures the ability of a compound to promote the binding of radiolabeled nonhydrolyzable GTP analogs, such as ^{35}S-GTPγS, to G proteins which are coupled to a GPCR [18]. Such binding can be detected using a scintillation proximity assay (SPA) format and is easily adaptable for HTS. Furthermore, nonradioactive alternatives exist to measure GTPγS binding, such as the dissociation-enhanced lanthanide fluorescence immunoassay (DELFIA) Eu-GTPγS. These non-isotopic ligands have also been adapted for HTS with the advantage of reduced cost due to a lack of radioactive waste. Importantly, novel technologies using GPCR-Gα fusion proteins eliminate the problem of whether the target cells expressing the recombinant GPCR under study have the appropriate endogenous G proteins that couple with the GPCR or whether the ratio of expression of GPCR to G protein is unnatural since by nature the stoichiometry of GPCR to G proteins in the fusion protein is 1:1 [19].

8.4.2. Cell-Based Assays

Although assays measuring binding of ligands to GPCRs have been used historically to identify GPCR drugs, most current technologies used for GPCR drug discovery are cell-based [20]. The primary readout of such assays is accumulation of second messenger levels in response to GPCR activation [21–24]. Assays measuring cellular levels of cAMP are dependent on the activity of adenylyl cyclase and detect GPCRs coupled to the G proteins, $G\alpha_s$ and $G\alpha_i$. Many commercial assays are available to measure this cAMP response. These include radioimmunoassay, time-resolved fluorescence resonance energy transfer (TR-FRET) and bioluminescence resonance energy transfer (BRET) assays, and those employing β-galactosidase enzyme fragment complementation (EFC) technology [3, 10, 11].

Activation of GPCRs coupled to G_q and G_o is generally measured by detecting cellular levels of IP3, the end product of activation of phosphoinositide phospholipase. IP3 levels can be detected using radioisotopes and biochemical approaches employing column chromatography. To adapt the assays for HTS, non-isotopic assays have been developed that measure IP3 levels such as PerkinElmer's ALPHAscreen (PerkinElmer, Waltham, MA) [3].

8.4.3. Ca^{++} Transients for GPCR HTS

Because activation of the IP3 pathways leads to changes in stored Ca^{++}, one can measure intracellular Ca^{++} changes as a response to GPCR stimulation. Classically, the principal technology to measure cytosolic Ca^{++} levels employs probes, either dyes or photoproteins, whose fluorescent or luminescent response changes in accordance with changes in intracellular Ca^{++} levels. In fact, the most commonly employed assay used in GPCR drug discovery is the fluorometric imaging plate reader (FLIPR) or functional drug screening system (FDSS) using the dyes Fluo-3, Fluo-4, Calcium 3 [25–27]. In these assays, cells are treated with the acetoxymethyl ester forms of these dyes which are cell permeable and thus freely enter the cell. Once in the cell, the dyes are de-esterified to a form that is less permeable and are thus retained for the period of the experiment. The dyes can chelate Ca^{++}, and in the presence of Ca^{++}, the emission wavelength of the dye changes, providing a means to detect changes in intracellular Ca^{++} levels. Thus, a light source is used to excite the dye in cells, and the difference in fluorescent emission when the dye is bound or not bound to Ca^{++} reflects changes in intracellular Ca^{++} levels. This simple procedure to measure Ca^{++} transients has improved with both the development of newer dyes, such as Fluo-4 and Calcium 3 as well as Fluo-8 automated "inject and read" detection systems [25]. Finally, the implementation of procedures that obviate extensive cellular washing to remove extracellular dye has increased both the viability of the dye-loaded cells and diminished cellular stress, both of which induce an artifactual excitation of the cell [25].

A major advance, in the last decade or so, involves use of photoprotein biosensors to detect free intracellular Ca^{++} ions [26–30]. By using recombinant approaches, photoproteins can be transiently or stably expressed into a range of cell phenotypes. The photoproteins emit an intense luminescent signal in the presence of free Ca^{++} in the absence of light stimulation which reduces background signals, a problem found when using dyes.

Several photoproteins have been isolated and cloned, although the one most widely used for measuring GPCR-induced Ca^{++} transients is aequorin, originally isolated from jellyfish [26, 29, 31]. This protein has a central hydrophobic core which binds the imidazopyrazinone chromophore, coelenterazine [32, 33], and Ca^{++} binding to aequorin induces oxidation of coelenterazine to the excited form, coelenteramide. Relaxation from the excited state results in blue light emission at 465 nm which can be detected using either a photomultiplier tube or a charge-coupled device (CCD) camera. There is generally low background luminescent emission from most cells used for drug discovery, and in those cells expressing aequorin, there is little background bioluminescence unless aequorin has access to increased Ca^{++}. Consequently, the flash luminescence associated with aequorin provides a very high signal to noise response ratio.

An important advantage of aequorin over fluorescent dyes is that its cloning allows the protein to be expressed stably in cell lines, thereby providing a consistent source of cells expressing the biosensor. One then can optimize GPCR signaling detected by aequorin bioluminescence by recombinantly manipulating the appropriate levels of receptor, G protein, and aequorin expressed in the cell. Another advantage of bioluminescent responses is that they have a high quantal efficiency such that only a few photons need to be emitted in order to be detected by CCD, which is now a standard automated detection system routinely used in GPCR drug discovery. Due to the greatly sensitive response, relatively few bioluminescent photoprotein emitters are needed to be expressed per cell for detection. In fact, photoproteins only need to be expressed at attomolar levels in cells to detect intracellular Ca^{++} changes [29]. This provides major advantages over fluorescent dyes since fewer cells need to be used in the assay, and miniaturization to very small fluid volumes (as would be used in 1536 microtiter plate assays) becomes feasible. Furthermore, because of the high sensitivity of these readout systems, as few as 100 cells per well can be used to detect Ca^{++} transients using photoproteins [30], making cells expressing these probes easily adaptable for ultra-HTS of drugs. In addition, the large signal to background ratios of the luminescent assays allows for robust detection of weak partial agonists or positive allosteric modulators of the GPCR in a primary HTS mode, both of which cannot be easily seen with responses having high background and low dynamic range of response.

Also important is that by using targeting sequences, aequorin can be selectively expressed in specific organelles to measure microdomains of Ca^{++} signaling, something which cannot be done easily with dyes [28]. Targeting aequorin to the mitochondria provides a more accurate and robust measure of GPCR-stimulated Ca^{++} transients than expression throughout the cytoplasm given that the mitochondria are spatially close to Ca^{++} channels and release sites from the endoplasmic reticulum [28]. Thus, aequorin targeted to those sites is better at monitoring the rapid increases in Ca^{++} levels following GPCR/Gq/IP3 signaling than if the photoprotein is diffusely expressed throughout the cytoplasm where its detection of Ca^{++} may be diluted because the site of detection is distal from the site of generation of Ca^{++} transients. Furthermore, targeting aequorin or a newly improved photoprotein Photina [30] to other cellular sites, such as the plasma membrane, allows for highly localized detection of Ca^{++} transients in cellular microdomains. This may provide the means to distinguish Ca^{++} transients resulting for increased Ca^{++} influx due to ionic conductance channels versus release for storage sites in the endoplasmic reticulum. This may allow for discrimination of the signaling of the same GPCR via distinct G proteins coupled to phospholipase C/endoplasmic reticulum/PI linked Ca^{++} transients versus plasma membrane ionic conductance channel Ca^{++} influx.

8.4.4. Reporter Assays for GPCR HTS

In addition to direct measurements of second messengers, reporter gene assays can be used to detect the consequence of changes in second messengers [21–23]. These HTS assays employ constructs consisting of second messenger response elements such as the cAMP response element (CRE) or the calcium-sensitive activator protein 1 (AP1) or nuclear factor of activated T cell (NFAT) elements linked to genes that encode for enzymes, such as luciferase or β lactamase, which act to catalyze the formation of luminescent or fluorescent products. Like the photoprotein assays, this technology is highly sensitive to changes in GPCR-mediated changes in second messengers and is easily adapted to HTS format.

8.4.5. Universal HTS Assays for GPCRs?

In general, the set of G proteins normally coupled to a target GPCR determines whether one employs the cAMP versus the IP3/Ca^{++} assays to measure activation. However, more generalized assays which can be used as universal GPCR screening technologies have also been developed which can be used for most, if not all receptors. One such universal screening approach for GPCR activity measures mitogen activated protein (MAP) kinase stimulation, since the MAP kinases are believed to be a point of convergence of the vast majority of GPCRs in regulating cell function. A commercially available assay to measure the activity of one of the MAP kinases, ERK, is the AlphaScreen SureFire ERK marketed by Perkin Elmer. This is a cell-based, homogenous, nonradioactive assay that measures phosphorylated ERK1 and ERK2 [34].

Another universal GPCR assay measures β-arrestin translocation in response to agonist stimulation [35]. This technology is based on the hypothesis that activation of many if not most GPCRs results in the translocation of a family of protein kinases referred to as G protein receptor kinases (GRKs) to the receptor which phosphorylate the GPCRs. This promotes the translocation of β-arrestin to the receptor that interrupts GPCR/G protein signaling. While other protein kinases such as protein kinase C can phosphorylate GPCRs, it is believed that activation of only the GRK/β-arrestin system selectively leads to phosphorylation and modulation of the agonist-bound receptor whereas other protein kinases are more pervasive in their actions, and their phosphorylation of a receptor is not dependent on agonist binding to that receptor. For example, stimulation of the β-adrenergic receptor with agonist can lead to activation of a specific GRK, leading to phosphorylation of the β-adrenergic receptor and association of specific β-arrestins. However, this process (GRK and β-arrestin activation) does not lead to specific phosphorylation of other GPCRs which are free of agonist. In fact, activation of protein kinase C can cause phosphorylation of multiple GPCRs at a time.

The technology was originally designed to measure β-arrestin translocation using confocal microscopy but has been modified to be detected by BRET [36, 37]. More recently, Wehrman et al. [38] have developed an enzyme fragment complementation assay to measure β-arrestin translocation in a format adapted for HTS by DiscoveRx. Here, two fragments of the enzyme β-galactosidase are employed, a small N-terminal fragment (Prolink) and a large truncated form of β-galactosidase (EA). Each fragment alone is inactive but when added together, they recombine to form an active β-galactosidase enzyme that can produce thousand of luminescent molecules in a short time. In the GPCR assay format, Prolink is tagged to the GPCR C-terminus and EA is tagged to β-arrestin. When agonist stimulates the GPCR, the β-arrestin-EA associates with the receptor-Prolink and complementation occurs, generating a highly luminescent response. Like the MAP kinase assays, β-arrestin-based assays such as the EFC-GPCR are employed for HTS of both known and orphan GPCRs, and as they are not dependent on G protein-mediated signaling events, they can also be considered a universal approach for identifying agonists, inverse agonists, and antagonists.

8.5. IMPORTANCE OF TARGET CELLS IN GPCR HTS ASSAYS

Generally, cell-based assays involve the expression of recombinant GPCRs in modified tumor cell lines. By expressing the molecular target in such cells, large quantities of the target are generated for testing in HTS assays. A major advantage of these approaches is that the immortalized recombinant cells provide a relatively naive background for target expression and a relatively homogenous target expression system that facilitates consistency in screening. Furthermore, not only can the GPCR be stably expressed at physiologically relevant levels, but other proteins can be engineered into these cells that can provide reporter readouts of drug–target interactions. Additionally, unique molecular targets can also be expressed in tumor cells in order to identify specific types of drugs, including small molecules selectively interacting with GPCR oligomers as well as allosteric regulators [3, 11, 39]. Collectively, due to their versatility, wide applicability, naïve background, and ease of use, a limited number of tumor cell lines expressing recombinant molecular targets provide the main systems for drug screening against GPCRs.

Despite their extensive use, there are limitations in screening compounds using tumor cell lines, particularly with regard to the applicability of the data generated on the action of the compound to the human physiological setting. A major limitation is related to the levels of expression of the transfected target protein. In the case of GPCRs, the expression levels in tumor cell lines employed in HTS are frequently much higher than occurs with endogenous levels found physiologically or even patho-physiologically. Such high expression levels significantly change the ratio of GPCR to G protein and, consequently, the inherent efficiency in receptor activation [40–42]. This may cause misleading interpretation of drug actions on GPCRs such that compounds can act as full agonists on the recombinant receptor while being only a partial agonist or antagonist at the receptor endogenously expressed in cells and *in vivo*.

Another important issue with regard to GPCR overexpression in tumor cells concerns the creation of constitutively active receptors, that is, receptors producing a functional response in the absence of activating ligand [3, 11, 20, 39, 40]. Since constitutively active receptors exhibit an intrinsically high basal activity, antagonists that reduce basal activity are designated as inverse agonists. Kenakin [41] suggests that many antagonists in clinical use today act as inverse agonists, and it has been suggested [43] that the use of constitutively active GPCRs can provide the means to identify an important new class of drugs.

These postulates are based on pharmacological studies using recombinant receptors, without concern for potential physiological relevance. Importantly, it is not clear whether there are GPCRs that are constitutively active in natural tissues and whether inverse agonists identified in HTS assays act as inverse agonists on endogenous receptors *in vivo*. If in fact, the drugs identified on recombinant GPCRs overexpressed in tumor cell lines act differently on GPCRs *in vivo*, it raises questions as to whether most drugs identified using immortalized tumor cells are clinically relevant.

In addition to the concerns of GPCR overexpression, a number of studies have shown that the overall cellular environment of the receptor profoundly influences the properties of GPCRs and the drugs targeted to those receptors. Thus, the rank order of agonist potencies as well as efficacy at a given GPCR can differ depending on the cell line in which the receptor is expressed. This can be due in part to differences in the G protein expression patterns in the different cell lines used in GPCR drug discovery, and Kenakin [41] suggested that cell:cell variations in G protein association affects receptor reserve and thus the efficacy of agonists and other ligands to modulate signaling pathways. Such issues create a dilemma in determining whether the cell line choice employed for HTS does or does not express the "correct" portfolio of G proteins that couple endogenously to the GPCR under examination. Thus, the cell line selected can significantly affect the pharmacological profile leading to identification of leads that will possess different activities *in vivo* when compared to their effects in clonal cell lines.

These potential disparities between the physiological environment of screening systems using recombinant tumor cell lines compared to natural tissues has led to a growing interest in the use of primary mammalian cells for drug screening and discovery [44–46]. In primary cells, the endogenous molecular targets are tacitly assumed to be expressed in an environment that more closely resembles that found in the patient, and with levels that reproduce those found naturally and endogenously. Consequently, novel drugs characterized using these primary cell systems are presumed to act in a more predictable fashion in clinical evaluation than those characterized in tumor cells in which recombinant targets are overexpressed.

Primary cell lines may consist of cells derived from embryonic tissues, including neuronal cultures, as well as those from adult tissues such as pituitary cells or hepatocytes. Primary cells can be used to evaluate endogenous targets for drug discovery or be generated to express recombinant targets using viral vector systems as well as employing tissues from transgenic animals. As with recombinant cells, assays using primary cells may employ responses such as transient changes in intracellular calcium using fluorometric dyes, detection of second messenger accumulation or other proteins reflective of cell function, including reporter enzymes via response elements recombinantly engineered into cells derived from transgenic animals as well as the use of electrophysiological approaches that measure changes in membrane potential.

A major limitation of the use of standard primary mammalian cells in HTS is the limited availability. To some extent, this is compensated for by using highly sensitive assay techniques, and miniaturized detection systems, collectively allowing the use of very few cells per assay. Alternatively, the broader availability of embryonic stem cells (ESC) or adult pluripotent stem cells (APSC) provides for cells that can be grown in relatively high abundance in a similar manner to tumor cells, yet which retain several phenotypic characteristics of the "natural" cells [45, 47, 48]. Furthermore, ESCs and APSCs can be induced to differentiate into distinct cell types, each reflective of specific organs and tissues, such as hepatocytes, cardiomyocytes, kidney cells, and neurons [49]. In some cases, these differentiated cells develop the characteristics of the mature cell, such as cell:cell networks, and in many ways, return compound screening to classical pharmacological approaches used 30 or so years ago, in which organs and tissues were used to assess drug actions and interactions.

While primary cells are rarely, if ever, used in HTS for drug discovery, they have been employed extensively for drug toxicity screening. Indeed, HTS approaches using microfluidic technologies to examine large numbers of compounds for toxic effects on liver cells can now be routinely assessed either as release of lactate dehydrogenase (LDH) from injured cells or scanned as changes in respiratory cellular activity. Such assays are employed as a secondary screen to test potential "hits" for potential toxicity and can also be employed as a primary screen to exclude compounds.

Toxicity screening is now employed using primary neuronal cultures to predict side effects of potential novel CNS drugs. Peripheral neurons, such as motor neurons or central neurons derived from embryonic tissues can be cultured using microtiter plate format. Here, the toxicity of novel compounds is assessed in terms of cell viability. For example, the effect of drugs on these neurons can be quantified by counting the living fluorescent cells in automated fashion using systems such as a FlashCytometer (Trophos, France) which image and quantify the number of luminescent cells per well. Many other dyes are now used to measure oxidative stress (ThermoFisher/Cellomics, Waltham, MA), changes in redox potential, and mitochondrial function as initial measures of neuronal toxicity.

Although this approach is employed to study neuronal drug toxicity, a similar approach is used for drug discovery, notably to identify novel neuroprotective agents [50]. Such drugs have the potential to treat degenerative central or peripheral nervous system diseases, such as Alzheimer's disease, Parkinson's disease, and amyotrophic lateral sclerosis (ALS). Essentially, the approach involves stressing neurons, either by removal of growth factors, or addition of toxic agents such as glutamate, to induce neuronal death. Consequently, the stressed neurons are used to screen small molecule libraries for compounds that can reverse neuronal loss. This technique has been most useful in screening for compounds that protect motor neurons from necrosis.

Importantly, Rogers et al. [51] have recently generated transgenic mice expressing a green fluorescent protein (GFP)-aequorin fusion protein targeted to the mitochondria to measure intracellular Ca^{++}. These authors employed this approach to measure Ca^{++} transients *in vivo* by detecting bioluminescent images using either a "photon imager" or the "video imager" consisting of intensified charge-coupled device (ICCD) cameras. Their studies suggested that the aequorin probe was expressed in most if not all cells and that Ca^{++} signaling in many different cells could be detected both under basal physiological conditions and in response to drugs. This is critical because primary cells, including ESC derived from such animals, could be employed for drug screening in much the same manner that immortalized tumor cells are used for GPCR drug discovery. This is particularly important since aequorin is such a highly sensitive detector of Ca^{++} transients that relatively few cells per well (500 or less) can be used for drug screening, possibly making the system adaptable for HTS format.

Furthermore, changes in intracellular signaling can be readily measured in primary cells using calcium-sensitive dyes such as Fluo3 or Fluo4 just as in immortalized recombinant cell lines. Viero et al. [52] has now developed an imaging system to measure calcium ion signaling in single adult ventricular myocytes—measuring not only Ca^{++} ion transients, but also myocyte contractility. These cells can be maintained in culture for extended periods with only minimal changes over time in calcium ion signaling or electrical "pacing" activity or contractility. These parameters are unaltered even when the cells are engineered to express calcium ion sensors that allow for rapid, continuous calcium ion measurements in HTS. Consequently, these systems, particularly when linked to HTS imaging systems, allows for high-throughput drug screening against individual, physiologically active, cardiomyocytes.

Classical microplate reader assay systems that detect calcium ion transients are also used in HTS campaigns with cultured brain neurons [53]. Moreover, imaging-based systems using FRET technologies to detect membrane potential changes in an HTS format have also been employed. Indeed, most, if not all, pharmaceutical companies have now developed automated microscopic HTS systems that measure discrete changes in protein translocation, and many of these assays have been adapted for primary cells. Therefore, despite the limited abundance of primary cells, the recent growth in available sensitive imaging technologies facilitates HTS against a range of molecular targets expressed in primary neuronal or cardiac myocyte cultures.

Thus, primary cells may provide major advantages over immortalized tumor cell lines for GPCR drug screening. In particular, the use of primary neuronal cells provides advantages because they can maintain an environment that replicates the complex interplay of endogenously expressed ion channels, second messengers, and other cell signaling proteins to which central nervous system GPCRs are normally exposed. The result is that the primary cells with the endogenous receptors better recapitulate the physiological milieu than when the GPCR is recombinantly expressed in a tumor cell line. Furthermore, using primary cells from transgenic animals expressing disease phenotypes may provide the basis for developing drugs targeting the disease condition rather than the normal conditions providing the means to develop more effective drugs with fewer side effects. Therefore, in the future, HTS campaigns using primary cells, and perhaps even cellular networks, could serve as the basis for identification of novel compounds and become an essential component of future GPCR drug discovery programs.

8.6. SUMMARY

The central role of GPCRs in human physiology and their diversity of function have made the GPCR family one of the prime targets for drug discovery. Many GPCRs have discrete biological roles indicating that drugs targeting those proteins are likely to produce specific therapeutic effects. Importantly, the pharmaceutical industry has been highly successful over the years in developing GPCR-directed drugs that provide good therapeutics able to treat diseases and disorders for which there are few alternatives.

Advances made in the GPCR research field in the last decade have heightened interest in discovering new GPCR drugs. Specifically, the ability to deorphanize some GPCRs having powerful biological functions has led to the emergence of an entire subfield devoted to better understanding the roles of the remaining orphan GPCRs and using HTS assays to develop novel therapeutics targeting those receptors. These receptors represent a relatively unexplored field of drug discovery that may have high value for the pharmaceutical industry because of the unique functions they mediate.

The identification of drugs targeting orphan GPCRs, as well as inverse agonists, allosteric regulators, pharmacological chaperones, and hetero-oligomers, while providing important new directions for drug discovery, has also directed the industry to develop novel technologies for HTS. This is necessary since classical drug discovery approaches cannot be easily used to identify drugs targeting orphan GPCRs nor hetero-oligomer-selective drugs. This has led to attempts to develop universal GPCR assays that are independent of the classical second messenger approaches such as assays that measure protein–protein interactions as well as assays that measure subtle conformational changes in GPCRs such as physical techniques [54, 55] including dielectric spectroscopy, optical biosensors, isothermal titration calorimetry, second harmonic generation, and differential scanning calorimetry. Indeed, several "label free" technologies are now being rapidly optimized for HTS purposes [56] such that assays can be run in 1536-well microtiter plate formats, similar to capabilities in more standard assays. Such technologies offer new opportunities to discover novel drugs because they provide alternative methods to measure interaction of the drug with the GPCR. One anticipates that when such techniques are used in conjunction with "classical" approaches, the identification of new generations of drugs, possibly with therapeutic properties beyond those GPCR drugs presently available, will become a reality.

REFERENCES

1. Limbird, L. (1995) Identification of receptors using direct ligand binding techniques. In *Cell Surface Receptors: A Short Course on Theory and Method*, 2nd ed., ed. M. Nijhoff, pp. 61–119. Boston: Kluwer Adacemic.
2. Venter, J.C., Adams, M.D., Myers, E.W., Li, P.W., Mural, R.J., Sutton, G.G., Smith, H.O., Yandell, M., Evans, C.A., Holt, R.A., Gocayne, J.D., Amanatides, P., Ballew, R.M., Huson, D.H., Wortman, J.R., Zhang, Q., Kodira, C.D., Zheng, X.H., Chen, L., Skupski, M., Subramanian, G., Thomas, P.D., Zhang, J., Gabor Miklos, G.L., Nelson, C., Broder, S., Clark, A.G., Nadeau, J., McKusick, V.A., Zinder, N., Levine, A.J., Roberts, R.J., Simon, M., Slayman, C., Hunkapiller, M., Bolanos, R., Delcher, A., Dew, I., Fasulo, D., Flanigan, M., Florea, L., Halpern, A., Hannenhalli, S., Kravitz, S., Levy, S., Mobarry, C., Reinert, K., Remington, K., Abu-Threideh, J., Beasley, E., Biddick, K., Bonazzi, V., Brandon, R., Cargill, M., Chandramouliswaran, I., Charlab, R., Chaturvedi, K., Deng, Z., Di Francesco, V., Dunn, P., Eilbeck, K., Evangelista, C., Gabrielian, A.E., Gan, W., Ge, W., Gong, F., Gu, Z., Guan, P., Heiman, T.J., Higgins, M.E., Ji, R.R., Ke, Z., Ketchum, K.A., Lai, Z., Lei, Y., Li, Z., Li, J., Liang, Y., Lin, X., Lu, F., Merkulov, G.V., Milshina, N., Moore, H.M., Naik, A.K., Narayan, V.A., Neelam, B., Nusskern, D., Rusch, D.B., Salzberg, S., Shao, W., Shue, B., Sun, J., Wang, Z., Wang, A., Wang, X., Wang, J., Wei, M., Wides, R., Xiao, C., Yan, C., Yao, A., Ye, J., Zhan, M., Zhang, W., Zhang, H., Zhao, Q., Zheng, L., Zhong, F., Zhong, W., Zhu, S., Zhao, S., Gilbert, D., Baumhueter, S., Spier, G., Carter, C., Cravchik, A., Woodage, T., Ali, F., An, H., Awe, A., Baldwin, D., Baden, H., Barnstead, M., Barrow, I., Beeson, K., Busam, D., Carver, A., Center, A., Cheng, M.L., Curry, L., Danaher, S., Davenport, L., Desilets, R., Dietz, S., Dodson, K., Doup, L., Ferriera, S., Garg, N., Gluecksmann, A., Hart, B., Haynes, J., Haynes, C., Heiner, C., Hladun, S., Hostin, D., Houck, J., Howland, T., Ibegwam, C., Johnson, J., Kalush, F., Kline, L., Koduru, S., Love, A., Mann, F., May, D., McCawley, S., McIntosh, T., McMullen, I., Moy, M., Moy, L., Murphy, B., Nelson, K., Pfannkoch, C., Pratts, E., Puri, V., Qureshi, H., Reardon, M., Rodriguez, R., Rogers, Y.H., Romblad, D., Ruhfel, B., Scott, R., Sitter,

C., Smallwood, M., Stewart, E., Strong, R., Suh, E., Thomas, R., Tint, N.N., Tse, S., Vech, C., Wang, G., Wetter, J., Williams, S., Williams, M., Windsor, S., Winn-Deen, E., Wolfe, K., Zaveri, J., Zaveri, K., Abril, J.F., Guigó, R., Campbell, M.J., Sjolander, K.V., Karlak, B., Kejariwal, A., Mi, H., Lazareva, B., Hatton, T., Narechania, A., Diemer, K., Muruganujan, A., Guo, N., Sato, S., Bafna, V., Istrail, S., Lippert, R., Schwartz, R., Walenz, B., Yooseph, S., Allen, D., Basu, A., Baxendale, J., Blick, L., Caminha, M., Carnes-Stine, J., Caulk, P., Chiang, Y.H., Coyne, M., Dahlke, C., Mays, A., Dombroski, M., Donnelly, M., Ely, D., Esparham, S., Fosler, C., Gire, H., Glanowski, S., Glasser, K., Glodek, A., Gorokhov, M., Graham, K., Gropman, B., Harris, M., Heil, J., Henderson, S., Hoover, J., Jennings, D., Jordan, C., Jordan, J., Kasha, J., Kagan, L., Kraft, C., Levitsky, A., Lewis, M., Liu, X., Lopez, J., Ma, D., Majoros, W., McDaniel, J., Murphy, S., Newman, M., Nguyen, T., Nguyen, N., Nodell, M., Pan, S., Peck, J., Peterson, M., Rowe, W., Sanders, R., Scott, J., Simpson, M., Smith, T., Sprague, A., Stockwell, T., Turner, R., Venter, E., Wang, M., Wen, M., Wu, D., Wu, M., Xia, A., Zandieh, A., Zhu, X. (2001) The sequence of the human genome. *Science.* 291, 1304–1351.

3. Eglen, R., Bosse, R., Reisine, T. (2007) Review: Emerging concepts of guanine nucleotide binding protein coupled receptor function and implications for high throughput screening. *Assay Drug Dev Technol.* 5, 425–451.
4. Furchgott, R.F. (1984) The role of endothelium in the responses of vascular smooth muscle to drugs. *Annu Rev Pharmacol Toxicol.* 24, 175–197.
5. Brown, J.H., Taylor, P. (1996) Muscarinic receptor agonists and antagonists. In *Goodman & Gilman's The Pharmacological Basis of Therapeutics*, 9th ed., ed. J.G. Hardman, L.L. Limbird, pp. 141–160. New York: McGraw-Hill.
6. Ross EM. (1996) Pharmacodynamics. *Goodman & Gilman's The Pharmacological Basis of Therapeutics*, 9th ed., ed. J.G. Hardman, L.L. Limbird, pp. 29–41. New York: McGraw-Hill.
7. Simon, M., Strathman, M., Gautam, N. (1991) Diversity of G proteins in signal transduction. *Science.* 252, 802–808.
8. Dessauer, C.W., Posner, B.A., Gilman, A.G. (1996) Visualizing signal transduction: receptors, G proteins and adenylate cyclases. *Clin Sci.* 91, 527–37.
9. Clapham, D.E., Neer, E. (1993) New roles for G protein βγ-dimers in transmembrane signaling. *Nature.* 365, 403–406.
10. Eglen, R.M., Gilchrist, A., Reisine, T. (2008) An overview of high throughput screening using primary and embryonic stem cells. *Comb Chem High Throughput Screen.* 11, 566–572.
11. Eglen, R.M., Gilchrist, A., Reisine, T. (2008) The use of immortalized cell lines in GPCR screening: The good, bad and ugly. *Comb Chem High Throughput Screen.* 11, 560–565.
12. Brown, A., Birnbaumer, L. (1988) Direct G protein gating of ion channels. *Am J Physiol.* 254, H401–H410.
13. Hille, B. (1992) *Ionic Channels of Excitable Membranes*. Sunderland, MA: Sinauer.
14. Kleuss, C., Hescheler, J., Ewel, C., Rosenthal, W., Schultz, G., Wittig, B. (1991) Assignment of G protein subtypes to specific receptors inducing inhibition of calcium currents. *Nature.* 353, 43–48.
15. Lefkowitz, R.J., Shenoy, S.K. (2005) Transduction of receptor signals by β-arrestins. *Science.* 308, 512–517.

16. Carpenter, G. (1999) Mini-Review: Employment of the epidermal growth factor receptor in growth factor-independent signaling pathways. *J Cell Biol.* 146, 697–702.
17. Worzfeld, T., Wettschureck, N., Offermanns, S. (2008) G(12)/G(13)-mediated signalling in mammalian physiology and disease. *Trends Pharmacol Sci.* 11, 582–589.
18. De Lapp, N.W. (2004) The antibody-capture [(35)S]GTPgammaS scintillation proximity assay: A powerful emerging technique for analysis of GPCR pharmacology. *Trends Pharmacol Sci.* 25, 400–401.
19. Pascal, G., Milligan, G. (2005) Functional complementation and the analysis of opioid receptor homodimerization. *Mol Pharmacol.* 68, 905–915.
20. Rees, S. (2002) Functional assay systems for drug discovery at G-protein coupled receptors and ion channels. *Receptors Channels.* 8, 257–259.
21. Kunapuli, P., Ransom, R., Murphy, K.L., Pettibone, D., Kerby, J., Grimwood, S., Zuck, P., Hodder, P., Lacson, R., Hoffman, I., Inglese, J., Strulovici, B. (2003) Development of an intact cell reporter gene beta-lactamase assay for G protein-coupled receptors for high-throughput screening. *Anal Biochem.* 314, 16–29.
22. Kornienko, O., Lacson, P., Kunapuli, P., Schneewels, J., Hoffman, I., Smith, T., Alberts, M., Inglese, J., Strulovici, B. (2004) Miniaturization of whole live cell-based GPCR assays using microdispensing and detection systems. *J Biomol Screen.* 9, 186–195.
23. Williams, C. (2004) cAMP detection methods in HTS: Selecting the best from the rest. *Nature Rev Drug Discov.* 3, 125–135.
24. Goetz, A.S., Liacos, J., Yingling, J., Ignar, D.M. (1999) A combination assay for simultaneous assessment of multiple signaling pathways. *J Pharmacol Toxicol Methods.* 42, 225–235.
25. Zhang, Y., Kowal, D., Kramer, A., Dunlop, J. (2003) Evaluation of FLIPR Calcium 3 Assay Kit—A new no-wash fluorescence calcium indicator reagent. *J Biomol Screen.* 8, 571–577.
26. Dupriez, V., Maes, K., Le Poul, E., Burgeon, E., Detheux, M. (2002) Aequorin-based functional assays for G-protein-coupled receptors, ion channels, and tyrosine kinase receptors. *Receptors Channels.* 8, 319–330.
27. Le Poul, E., Hisada, S., Miziguchi, Y., Dupriez, V.J., Burgeon, E., Detheux, M. (2002) Adaptation of aequorin functional assay to high throughput screening. *J Biomol Screen.* 7, 57–65.
28. Rizzuto, R., Pozzan, T. (2006) Microdomains of intracellular Ca^{+2}: Molecular determinants and functional consequences. *Physiol Rev.* 86, 369–408.
29. Roda, A., Pasini, P., Mirasoli, M., Michelini, E., Massimo, G. (2004) Biotechnological applications of bioluminescence and chemiluminescence. *Trends Biotechnol.* 22, 295–303.
30. Bovolenta, S., Foti, M., Lohmer, S., Corazza, S. (2007) Development of a Ca2+-activated photoprotein, photina, and its application to high-throughput screening. *J Biomol Screen.* 12, 694–704.
31. Ungrin, M.D., Singh, L., Stocco, R., Sas, D., Abramovitz, M. (1999) An automated aequorin luminescence-based functional calcium assay for G-protein-coupled receptors. *Anal Biochem.* 272, 34–42.
32. Prendergast, F. (2000) Bioluminescence illuminated. *Nature.* 405, 291–293.
33. Vysotski, E., Lee, J. (2005) Ca2+-regulated photoproteins: Structural insight into the bioluminescence mechanism. *Acc Chem Res.* 37, 405–415.

34. Leroy, D., Missotten, M., Waltzinger, C., Matrin, T., Scheer, A. (2007) G protein-coupled receptor-mediated ERK 1/2 phosphorylation: Towards a generic sensor of GPCR activation. *J Recept Signal Transduct Res.* 27, 83–97.
35. Barak, L.S., Ferguson, S.S., Zhang, J., Caron, M.G. (1997) A beta-arrestin/GFP biosensor for detecting GPCR activation. *J Biol Chem.* 272, 27497–27500.
36. Heding, A. (2004) Use of BRET 7TM receptor/beta arrestin assay in drug discovery and screening. *Expert Rev Mol Diagn.* 3, 403–411.
37. Vrecl, M., Jorgensen, R., Pogacnik, A., Heding, A. (2004) Development of a BRET2 screening assay using beta-arrestin 2 mutants. *J Biomol Screen.* 4, 322–333.
38. Wehrman, T.S., Casipit, C.L., Gewertz, N.M., Blau, H.M. (2005) Enzymatic detection of protein translocation. *Nat Methods.* 2, 521–527.
39. Gilchrist, A. (2007) Modulating G-protein-coupled receptors: From traditional pharmacology to allosterics. *Trends Pharmacol Sci.* 28, 431–437.
40. Kenakin, T. (2005) New concepts in drug discovery: Collateral efficacy and permissive antagonism. *Nat Rev Drug Discov.* 4, 919–927.
41. Kenakin, T. (2004) Efficacy as a vector: The relative prevalence and paucity of inverse agonism. *Mol Pharmacol.* 65, 2–11.
42. Kenakin, T. (1997) Differences between natural and recombinant G protein-coupled receptor systems with varying receptor/G protein stoichiometry. *Trends Pharmacol Sci.* 18, 456–464.
43. Chen, G., Way, J., Armour, S., Watson, C., Queen, K., Jayawickreme, C.K., Chen, W.J., Kenakin, T. (2000) Use of constitutive G protein-coupled receptor activity for drug discovery. *Mol Pharm.* 57, 125–134.
44. McNeish, J. (2004) Embryonic stem cells in drug discovery. *Nat Rev Drug Discov.* 3, 70–80.
45. Pouton, C.W., Haynes, J.M. (2007) Embryonic stem cells as a source of models for drug discovery. *Nature Rev Drug Discov*. 6, 605–616.
46. Nolan, G.P. (2007) What's wrong with drug screening today. *Nature Chem Biol.* 3, 187–191.
47. Gorba, T., Allsopp, T.E. (2003) Pharmacological potential of embryonic stem cells. *Pharmacol Res.* 47, 269–278.
48. Davila, J.C., Cezar, G.G., Thiede, M., Strom, S., Miki, T., Trosko, J. (2004) Use and application of stem cells in toxicology. *Toxicol Sci.* 79, 214–223.
49. Keller, G., Snodgrass, H.R. (1999) Human embryonic stem cells: The future is now. *Nat Med.* 5, 151–152.
50. Bordet, T., Buisson, B., Michaud, M., Drouot, C., Galéa, P., Delaage, P., Akentieva, N.P., Evers, A.S., Covey, D.F., Ostuni, M.A., Lacapère, J.J., Massaad, C., Schumacher, M., Steidl, E.M., Maux, D., Delaage, M., Henderson, C.E., Pruss, R.M. (2007) Identification and characterization of cholest-4-en-3-one, oxime (TRO19622), a novel drug candidate for amyotrophic lateral sclerosis. *J Pharmacol Exp Ther.* 322, 709–720.
51. Rogers, K.L., Picaud, S., Roncali, E., Boisgard, R., Colasante, C., Stinnakre, J., Tavitian, B., Brûlet, P. (2007) Non-invasive in vivo imaging of calcium signaling in mice. *PLoS ONE.* 2, e974.
52. Viero, C., Kraushaar, U., Ruppenthal, S., Kaestner, L., Lipp, P. (2008) A primary culture system for sustained expression of a calcium sensor in preserved adult rat ventricular myocytes. *Cell Calcium.* 43, 59–71.

53. Hemstapat, K., Smith, M.T., Monteith, G.R. (2004) Measurement of intracellular Ca2+ in cultured rat embryonic hippocampal neurons using a fluorescence microplate reader: Potential application to biomolecular screening. *J Pharmacol Toxicol Methods.* 49, 81–87.

54. Cooper, M.A. (2006) Non-optical screening platforms: The next wave in label-free screening? *Drug Discov Today.* 11, 1068–1074.

55. Cooper, M.A. (2006) Optical biosensors: Where next and how soon? *Drug Discov Today.* 11, 1061–1067.

56. Fang, Y. (2006) Label-Free Cell-Based Assays with Optical Biosensors in Drug Discovery. *Assay Drug Develop Technol.* 4, 583–595.

CHAPTER 9

Are "Traditional" Biochemical Techniques Out of Fashion in the New Era of GPCR Pharmacology?

MARIA TERESA DELL'ANNO and MARIA ROSA MAZZONI

Department of Psychiatry, Neurobiology, Pharmacology and Biotechnologies, University of Pisa, Pisa, Italy

9.1. OVERVIEW

Today, G protein-coupled receptors (GPCRs) represent the most important class of drug targets with ~50% of current drugs targeting them and ~20% of the top 50 best-selling drugs acting through this receptor superfamily. GPCRs are involved in many major human diseases, including pain, asthma, inflammation, obesity, cancer, as well as cardiovascular, metabolic, gastrointestinal, and CNS diseases. The action of a drug at the target receptor depends on two events: the binding of the drug to the receptor and the response triggered by the drug on receptor signaling in the cell and associated tissue. The binding of the drug is reflected in the affinity with which it binds to the receptor. The affinity can be described in molecular terms and is physicochemical in nature; it is the reciprocal of the equilibrium dissociation constant for the ligand receptor interaction. The ability of the drug to alter the activity of signaling systems linked to the receptor is indicated as "efficacy" and is reflected in differences in the extent and potency of the response. Considering an efficacy scale, full agonists will be found at the extreme positive side followed by partial agonists while inverse agonists will be located at the negative side. In this hypothetical scale, neutral antagonists will be in the middle having zero efficacy.

Whereas binding affinity can be estimated by equilibrium binding studies using the radiolabeled ligand and either membrane preparations or intact cells

GPCR Molecular Pharmacology and Drug Targeting: Shifting Paradigms and New Directions, Edited by Annette Gilchrist

expressing the appropriate GPCR, the evaluation of drug "efficacy" requires suitable functional assays that allow a change in the biological state to be measured. Classical functional assays that measure variations of second messenger levels such as calcium and adenosine-3′,5′ cyclic monophosphate (cAMP) have been extensively employed. New assays that take under consideration other or alternative molecules in the downstream signaling cascades have been also introduced, including inositol-1,4,5-trisphosphate (IP_3) production assay, extracellular signal-regulated kinase (ERK) activation assay, and regulation of gene expression studied by reporter gene assays [1]. Alternatively, the level of G protein activation following agonist occupation of a GPCR can be measured by determining the binding of a radiolabeled nonhydrolyzable analog of GTP, [^{35}S]guanosine-5′-O-(3-thio)-triphosphate ([^{35}S]GTPγS), to Gα subunits. This assay measures the functional consequence of receptor occupancy at one of the earliest receptor-mediated events (reviewed in References 2–4).

Quantifying both biochemical interactions and functional actions of ligands with their target GPCRs is fundamental not only to the understanding of receptor pharmacology, but it is also central in the development of new selective drugs to be used in the treatment of diseases. In recent years, the increased pace of GPCR identification together with the expansion of compound libraries has presented the compelling need to develop technologies to screen multiple GPCRs simultaneously. Thus, new assay systems have been developed or adapted to screen GPCR binding to ligands and partner proteins, such as GPCR microarrays [5] and flow cytometric assays [6]. Although these and other innovative techniques have an important impact into the drug discovery process, traditional radioligand binding and functional assays still represent easily available methodologies for pharmacological characterization or the identification of natural or surrogate GPCR ligands. In this chapter we intend to give a general overview of two traditional techniques to characterize GPCR ligand interactions and measure the functional consequence of receptor activation: radioligand binding and cAMP production assays.

9.2. RECEPTOR BINDING ASSAYS

The biological identification and pharmacological characterization of receptors have been aided immensely by the advent of radiolabeled agents that retained their biological activity after iodination or tritiation, allowing the development of a method known as the radioligand binding or receptor binding technique. Due to the relative easy way with which radioligand binding assays can be performed, they are still widely used in both academic and industrial laboratories. The quality of data obtained allows the determination of drug affinity, allosteric interactions, the existence of receptor subtypes, and estimates of the receptor numbers. However, the ease of the method does not imply that rigorous criteria are not required for studying binding to functional relevant receptors. These criteria and some practical guidelines are given in the present section while a more detailed information on radioligand binding methodologies, including theoretical aspects, can be found in a variety of other sources [7–9].

The most frequently used assay based on the radioligand binding technique is the membrane filtration receptor assay [9]. A variation of the membrane assay is the intact cell radioligand binding assay which has some advantages in special circumstances such as in studies of receptor internalization [9]. To analyze individual receptor pharmacological properties in the setting of intact tissues, tissue segment binding assays have been also proposed [10]. Other additional approaches that are useful in studying receptors in relation to their physiological functions and diseases are receptor autoradiography and *in vivo* labeling of receptors with positron emission tomography (PET).

The basics of radioligand binding assays with membrane preparations are rather simple: membranes containing the receptor of interest are incubated with a suitable radioligand for an appropriate period of time and then receptor-bound radioactivity is measured. The experiments can be performed according to three major types: saturation, kinetic, and inhibition. A saturation experiment is carried out by holding constant the receptor amount and varying the radioligand concentration, thus allowing to generate a saturation curve. By definition, the radioligand binding must be saturable because there is a finite number of receptors. From this type of experiment, both the receptor density (B_{max}) and dissociation constant for the radioligand (K_D) can be estimated. If the amount of radioligand and receptor is maintained constant and the time varied, then kinetic data are obtained from which association (k_{on} or k_{+1}) and dissociation (k_{off} or k_{-1}) rate constants can be estimated. If the amount of a competing nonradioactive drug included in the incubation is varied while both radioligand and receptor are held constant, then the inhibition constant (K_i) of the drug for the receptor labeled by the radioligand can be calculated.

Membrane fraction is prepared by standard procedures including homogenization of tissue or cells in a hypotonic buffer followed by differential centrifugation. The need of more than two centrifugation steps on washing purpose is required to remove unwanted substances as endogenous ligands and guanine nucleotides. Details of buffer composition can be found in many papers and together with laboratory experience aid in the choice of the appropriate procedure to prepare the membrane fraction.

Generally for a given receptor, there are now several radiolabeled agonists and antagonists that are commercially available. Some characteristics of the radioligand to be considered include: radioisotope (^{3}H or ^{125}I); the extent of nonspecific binding; the selectivity and affinity of the radioligand for the receptor; whether the radioligand is an agonist or an antagonist. Tritiated ligands have the advantage that are chemically unaltered and are thus biologically indistinguishable from the unlabeled compound and possess a long half-life (12 years). On the other hand, iodinated radioligands have short half-lives (60 days), but they possess the advantage of having a high specific activity

(~2,000 Ci/mmol). In addition, iodinated radioligands do not need scintillation fluid to count radioactivity, thus eliminating purchasing and disposal costs.

The use of selective radioligands for a given receptor or receptor subtype is required for labeling them and their pharmacological characterization. Usually, high-affinity radioligands (K_D, 10 pM–10 nM) are preferable since lower concentrations need to be used in the assay, and this results in lower levels of nonspecific binding. In order to be useful in binding assays, high-affinity radioligands must also possess high specific activity. Thus, iodinated radioligands have K_D values for binding to the receptor in the pmolar range while tritiated radioligands have K_D values in the 0.1–10 nM range. When needed, it is always possible to reduce the specific activity by diluting the ^{125}I-radioligand with the unlabeled ligand in order to limit the amount of radioactivity in each assay tube (<10^6 cpm). Another important issue to be considered is whether we are using an agonist or antagonist radioligand to selectively label the GPCR under investigation. Agonist radioligands preferentially label the portion of the total receptor population which is in the high-affinity state, namely those GPCRs tightly coupled to heterotimeric G proteins with no guanine nucleotide bound to the Gα subunit. Antagonist radioligands generally label all available GPCRs and this is certainly the case of neutral antagonists.

Considering the binding assay conditions, the choice of the buffer type, the addition of cations or protease inhibitors depends on both the radioligand and the GPCR under investigation. Generally, the pH should be in the physiological range (pH 7–8) while tris(hydroxymethyl)amino-methane (Tris) buffer is often used as buffer for binding reactions. Some GPCRs such as the A_{2a} adenosine receptor [11] have the absolute requirement of Mg^{2+} ions for agonist binding. The addition to the incubation buffer of GTP or a nonhydrolyzable analog such as GTPγS decreases the affinity of agonists for the GPCR and is able to convert a biphasic or shallow inhibition curve of an unlabeled agonist for radiolabeled antagonist binding into a monophasic inhibition curve (Fig. 9.1). Protease inhibitors may be added to the incubation buffer to prevent degradation of peptide radioligands.

Since for saturation and inhibition experiments it is required to work at equilibrium, the incubation time needs to be sufficient to ensure reaction equilibrium or at least steady state. The time to reach steady state is dependent on the radioligand concentration, but at radioligand concentrations near to their K_D value, most of them reach steady state at room temperature within 20–60 min. The demonstration that the amount of specific binding is constant over a period of time indicates that a steady state has been reached. Indeed, most assays are performed at steady state rather than at equilibrium.

The radioligand concentration in the assay is dependent on the type of experiment being performed. In kinetic experiments, the concentration should be reasonably low but high enough to obtain a reasonable level of specific binding. In inhibition assays, the general rule is to use the K_D concentration or lower. In saturation experiments, the range of radioligand concentrations

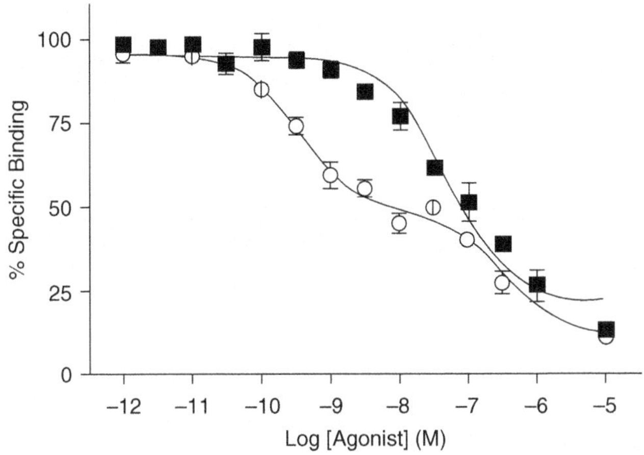

Figure 9.1 Competition curve of a radiolabeled neutral antagonist bound to a GPCR by an unlabeled agonist. In the absence of GTP, the curve is clearly biphasic and better fitted by a two-site competion model of the nonlinear regression analysis (GraphPad Prism Version 4.0). The addition of 100 μM GTP in the reaction mixture causes a right shift of the curve, which is now monophasic, and it is better fitted by one-site model. This is a typical example of GPCR two affinity states for the agonist ligand. The agonist has higher affinity for GPCRs tightly coupled to G proteins than for those receptors uncoupled to G proteins. On the other hand, the neutral antagonist binds with the same affinity to coupled and uncoupled receptors.

should be from approximately 0.1 × to 10 × K_D value. Of course, this is in ideal conditions that are not always possible, but some alternative options are available to study binding saturability, such as homologous competition experiments.

Usually, the higher the membrane protein concentration, the better is the binding. This is especially true for membranes prepared from tissues and untransfected cells. In this case, a membrane protein concentration range of 100–500 μg/mL is adequate. For transfected cells overexpressing the receptor, the protein concentration required in the assay is much lower. Increasing receptor concentration causes an increase of the ratio between specific binding to nonspecific binding. The amount of specific binding should be linearly related to the membrane protein concentration. However, the rule is that less than 10% of the added radioligand must be bound.

In inhibition experiments, it is important to consider the concentration range of the inhibiting or competing drug. When the inhibition of radioligand binding by a drug follows a single site model, then about 10 inhibitor concentrations spanning the range of at least 100-fold on both sides of the 50% inhibitory concentration (IC_{50}) are adequate. In the case of multiple binding sites or affinity states, at least 20 concentrations of the inhibitor over a larger range are required.

In any radioligand binding assay, the most important consideration is the definition of specific binding. By definition, specific binding is the binding to the receptor of interest while nonspecific binding is any other binding. Nonspecific binding is measured in the presence of an appropriate excess of an unlabeled drug (e.g., 100-fold its IC_{50} value) to completely block the receptor of interest. Nonspecific binding includes radioligand binding to other receptor sites, to glass fiber filters, adsorption, and dissolution in the membrane lipids. Specific binding is calculated as the difference between total and nonspecific binding. Nonspecific binding reaches steady state more rapidly than specific binding, but it does not saturate in the same ways as total and specific binding do. To determine nonspecific binding, it is better to use a drug that is chemically dissimilar from the radioligand. For an adequate binding assay, specific binding must be at least 50% of total binding, but the best conditions are at 70–90% of total binding. Peptide radioligands frequently give problems of high nonspecific binding. In this case, some strategies can be adopted to reduce nonspecific binding [9], but unfortunately, they are not always successful.

A crucial step in receptor binding assays is the separation of bound radioligand from free radioligand. During separation, it is important to prevent dissociation of receptor radioligand complex. This is obtained by reducing the temperature and performing separation as rapidly as possible. On this purpose, in membrane binding assays, the most frequently used technique is vacuum filtration through glass fiber filters which retain membranes. Membranes and filters are also washed with large amount of cold buffer which reduce nonspecific binding. In addition, to decrease nonspecific binding, filters can be presoaked with a 0.1% aqueous solution of polyethylenimine or bovine serum albumin (0.2% BSA). Filter presoaking with a BSA solution is used in the case of peptide or protein radioligands. Filtration under reduced pressure can be used only when the radioligand K_D for binding to the receptor is lower than 10 nM. In the case of K_D values in the 10 nM–1 μM range bound from free radioligand is separated by centrifugation which is also able to reduce nonspecific binding. An aliquot of the radioligand as that added to each assay tube should be also counted.

Analysis of data derived from radioligand binding experiments is crucial for studying receptor pharmacology and identification of new ligands. The best way to fit saturation data and thus derive K_D and B_{max} values is to use nonlinear regression analyses such as that provided by GraphPad Prism (GraphPad Software Inc., San Diego, CA) or a variety of other software packages. Data are visualized as hyperbolic (bound vs. free radioligand concentration) or sigmoidal (bound vs. logarithm of free radioligand concentration) curves. The GraphPad Prism software also possesses excellent graphic capabilities which aid in data visualization. It is frequently desirable to present data transformed into a linear form, such as bound/free vs. bound or Rosenthal plot (also known as Scatchard plot) [12]. In this case, the intercept on the x-axis gives the B_{max}

value while K_D corresponds to the negative reciprocal of the slope. To identify the presence of two binding sites or affinity states, nonlinear regression analysis of bound vs. free radioligand or eventually the Rosenthal plot is preferable over the semilogarithmic plot that can mask the complexity of the data. However, a nonlinear regression analysis of saturation data is always the best choice since linear transformation distorts experimental errors.

Data from inhibition experiments are also better analyzed using nonlinear regression techniques. Data are visualized as sigmoidal inhibition curves of bound radioligand expressed as percent of maximum vs. the logarithm of the inhibitor concentrations. The concentration of the inhibitor that reduces bound radioligand by half is the IC_{50} (inhibitory concentration 50%) or EC_{50} (effective concentration 50%) and can be estimated by the inhibition curve. The GraphPad Prism software fits inhibition curves using a nonlinear regression analysis, directly calculates Log EC_{50}, and derives K_i using the equation of Cheng and Prusoff [13]. If the radioligand and the inhibitor compete for a single class of binding sites, the curve is really sigmoidal descending from 90% to 10% bound over an 81-fold increase in inhibitor concentration and the slope factor (also called Hill slope) is −1. If the inhibition binding curve is shallow with the slope factor less than −1, we may consider the existence of a heterogeneous receptor population, more than one receptor affinity state (Fig. 9.1) or negative cooperativity. The GraphPad Prism software allows to fit data to two equations according to one-site or two-site competitive binding models and compare the two fits [14] (more detailed information are available at: http://www.curvefit.com). If the existence of multiple binding sites is suspected, data can also be visualized using a plot of bound vs. bound × inhibitor concentration [15]. The plot is linear for a single class of binding sites but is markedly nonlinear for two or more classes of binding sites. The intercept on the *y*-axis is the amount of binding in the absence of the inhibitor while the slope is the negative reciprocal of the IC_{50}.

A particular case of competition binding experiments is homologous competition. Sometimes, this type of experiment is performed to study binding saturability without varying the radioligand concentration. However, to obtain appropriate data, four conditions must be respected: (1) receptors must have identical affinity for the labeled and unlabeled ligand; (2) there is no cooperativity; (3) there is no ligand depletion; and (4) nonspecific binding is proportional to the concentration of the labeled ligand. Analysis of homologous competition binding data can be tricky, but the use of the nonlinear regression technique is quite useful [14] (detailed information are at: http://www.curvefit.com).

Kinetic binding data can be also analyzed using a nonlinear regression analysis (GraphPad Prism). From dissociation binding data, k_{off} (min^{-1}) and $t_{1/2}$ (min) are obtained while K_{on} ($molar^{-1}$ min^{-1}) is derived from association binding data. Thus, from kinetic experiments, the calculation of K_D is possible since it is equal to the ratio k_{off}/k_{on}.

Radioligand binding assays have been the mainstay of drug discovery and development. Even in the era of high-throughout screening (HTS), this versatile technique has retained its fundamental role. Indeed, incorporation of scintillation proximity technology together with automation and ratiometric counting instrumentation have served to maintain receptor binding assays as one of the primer tools of drug discovery (reviewed in Reference 16).

9.3. METHODS FOR MEASUREMENT OF cAMP

Many drugs and hormones interact with plasma membrane receptors to induce changes in the production of intracellular second messengers. The first such messenger to be identified was cAMP discovered by Sutherland and Rall in the late 1950s [17]. As for other signaling molecules, the levels of intracellular cAMP are tightly regulated. Production of cAMP is controlled through the adenylyl cyclase family of enzymes, which convert adenosine triphosphate (ATP) to cAMP and inorganic pyrophosphate. These enzymes are activated or inhibited via direct interaction with Gα subunits and, for some isoforms, with calcium and calmodulin. Following G_s-coupled receptor activation, GTP-bound Gα_s exerts a positive effect on adenylyl cyclase catalysis. Cyclic AMP is produced and is able to bind to and activate cAMP-dependent protein kinase A (PKA) within the cell, initiating phosphorylation events that regulate target enzymes and transcription factors (Fig. 9.2). Following G_i-coupled receptor activation, GTP-bound Gα_i exerts a negative effect on enzyme catalytic activity. Degradation of cAMP is controlled by cAMP phosphodiesterase (PDE) enzymes, which catalyze the hydrolysis of the 3′ ester bond of the cAMP to form 5′ adenosine monophosphate (AMP). There are several PDE isoforms located in different subcellular compartments and various tissue types; the isoforms are activated by a number of mechanisms. The most widespread mechanism is phosphorylation by cAMP-dependent protein kinases. Thus, cAMP-dependent PDE enzymes act as an important negative feedback system on the receptor-mediated signaling cascade, regulating the extent of changes in intracellular cAMP concentrations [18].

Measurement of adenylyl cyclase activity is carried out by quantitative determination of cAMP produced from ATP as a substrate. However, in the plethora of available methods to study variations in the cAMP levels occurring in cells, tissues, or organic fluids, a first distinction has to be made. Indeed, some assays start from labeled ATP as a substrate while others employ nonlabeled ATP. In the following paragraphs, this criterion will be adopted.

9.3.1. Assessments of Adenylyl Cyclase Activity: Methods Using Labeled ATP

These assays have been widely described by Salomon et al. [19, 20]. The first method utilizes [α^{32}P]ATP, which is converted by adenylyl cyclase to [^{32}P]cAMP to assess enzyme activity in cell and tissue homogenates or membranes. The second one involves incubating cells or tissues with [^{3}H]adenine to label

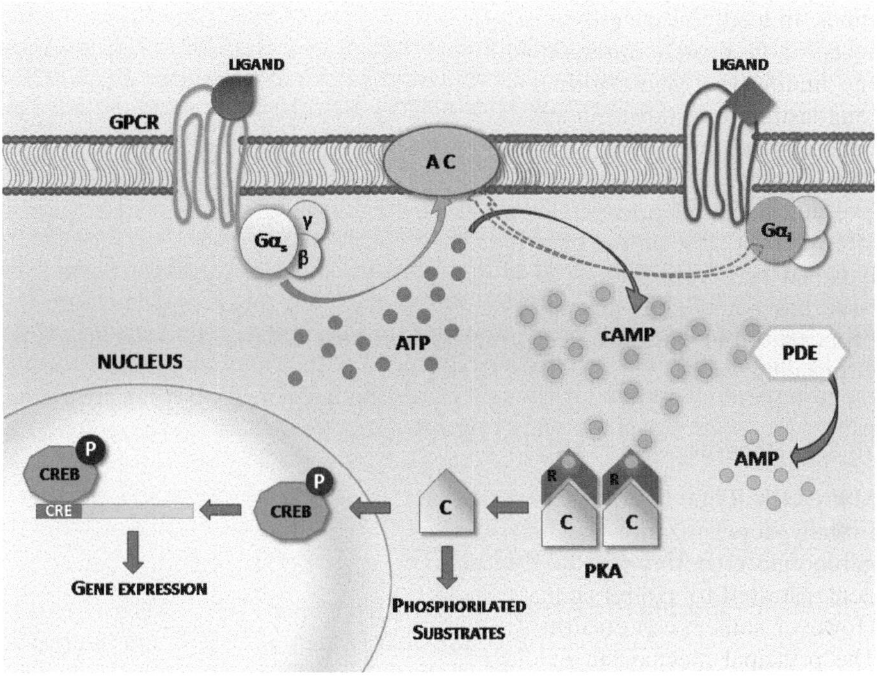

Figure 9.2 This cartoon represents the signaling pathways of G_s- and G_i-coupled receptors. Adenylyl cyclase (AC) is at the center of $G\alpha_s$- and $G\alpha_i$-mediated regulation. Arrows represented by continous lines indicate activation while the arrow represented by dotted line indicates inhibition. PDE, phosphodiesterase; PKA, cAMP-dependent protein kinase; C, PKA catalytic subunit; R, PKA regulatory subunit; CREB, cAMP response element-binding protein; CRE, cAMP response element.

intracellular pools of adenine nucleotides [21]. Accumulation of [^{3}H]cAMP is then used as an index of adenylyl cyclase activity. Both of these methods require the separation of radioactively labeled cAMP from other components of the reaction mixture and particularly from adenine nucleotides other than cAMP. This is efficiently accomplished utilizing sequential chromatography on Dowex cation exchange and alumina columns.

When performing these assays, it is important to note that adenylyl cyclase activity can be influenced by several factors. First, degradation of ATP by enzymes (nucleotidases and hydrolases) present in cell homogenates can result in depletion of substrate. To prevent this problem, incubations are performed in an ATP regeneration system typically consisting in creatinine phosphate and creatinine phosphokinase (or phosphoenolpyruvate and pyruvate kinase). Second, adenylyl cyclase activity is dependent on the presence of divalent cation (Mg^{2+} or Mn^{2+}) which must be present during incubation. Third, phosphodiesterases present in cell membranes can hydrolyze cAMP to AMP.

Therefore, incubation buffers typically contain one or more phosphodiesterase inhibitors (e.g., isobutylmethylxantine, rolipram, and RO 20-1724) to prevent cAMP breakdown. Fourth, the regulation of adenylyl cyclase by GPCRs requires GTP (or a nonhydrolyzable GTP analog) be included in the incubations. Fifth, it is important that cAMP generation is linear with respect to time and protein content. This may require modification of the amount of cellular protein, length of incubation, or incubation temperature. A factor that contributes to nonlinear (and loss) enzymatic activity is the relative instability of adenylyl cyclase owing to denaturation, especially at temperatures typically used for enzyme assays [22]. Furthermore, when studying a G_i-coupled receptor signaling pathway, in order to better appreciate the reduction in cAMP levels, the enzyme must be firstly activated, and this is usually done by using the diterpene forskolin. Although these methods are very sensitive and still widely used nowadays, they rely upon dangerous and costly radioactive compounds.

9.3.2. Methods Using Nonlabeled ATP

Enzymatic Fluorimetric Assay In order to eliminate risks and costs associated with the use of radioactive compounds, Weign et al. [23] and Sugiyama et al. [24,25] introduced a novel enzymatic fluorimetric assay to assess adenylyl cyclase activity. This method, firstly applied on ventricular membrane preparations and later extended also to tissues, is based upon previous observations by Lowry and Passonneau [26]. These authors developed a number of sensitive assays which can measure small amounts of biological compounds based on the fluorescence of reduced pyridine nucleotides. These methods employ one or more of a series of enzymatic reactions which ultimately lead to the production of either β-nicotinamide-adenine dinucleotide phosphate ($NADP^+$) or β-nicotinamide-adenine dinucleotide (NAD^+) or the reduced forms, NADPH and NADH. Indeed NADPH is a fluorescent molecule (340 nm excitation wavelength/460 emission wavelength) which can be easily detected using a fluorescence spectrophotometer. The enzymatic fluorimetric assay consists of two parts, namely: the production of cAMP by the adenylyl cyclase in samples (membranes or tissues), and the measurement of NADPH which is generated in proportion to newly formed cAMP. The second part of the procedure comprises several steps:

1. Enzymatic destruction of the adenine nucleotides other than cAMP in the samples with a mixture of apyrase, 5′nucleotidase and adenosine deaminase;
2. Conversion of cAMP to AMP with phosphodiesterase;
3. Quantification of the resulting AMP (Fig. 9.3). This can be used to stimulate the activity of added glycogen phosphorylase A, which converts added glycogen and inorganic phosphate into glucose-1-

(a)

$$ATP \xrightarrow{\text{apyrase}} AMP + 2P_i$$

$$ATP + H_2O \xrightarrow{\text{5' nucleotidase}} \text{adenosine} + P_i$$

$$\text{Adenosine} + H_2O \xrightarrow{\text{adenosine deaminase}} \text{inosine} + NH_3$$

(b)

$$cAMP \xrightarrow{\text{phosphodiesterase}} AMP$$

(c)

$$\text{Glycogen} + P_i \xrightarrow[\text{AMP}]{\text{glycogen phosphorilase A}} \text{glucose-1-phosphate}$$

$$\text{Glucose-1-phosphate} \xrightarrow{\text{phosphoglucomutase}} \text{glucose-6-phosphate}$$

$$\text{Glucose-6-phosphate} + NADP^+ \xrightarrow{\text{glucose-6-phosphate dehydrogenase}} \text{6-phosphogluconolactone} + NADPH + H^+$$

Figure 9.3 Enzymatic fluorimetric assay of cAMP. Schematic representation of the steps required for detection of cAMP using coupled enzymatic reactions to glycogen phosphorylase A.

phosphate. Ultimately, glucose-1-phosphate is enzymatically converted into 6-phosphogluconolactone, NADPH and H^+. The NADPH concentration is then determined fluorimetrically and can be correlated with the concentration of adenylyl cyclase, cAMP or AMP in the sample.

Alternatively, AMP can be converted to ADP by combining it with ATP in the presence of myokinase (Fig. 9.4, panel a). The resulting ADP is then converted to ATP and pyruvate by combining it with 2-phospho(enol)pyruvate (PEP) and pyruvate kinase. ATP is finally involved in a reaction with fructose and esokinase to produce 6-phosphogluconolactone and NADPH.

This enzymatic assay has a number of potential advantages over the classical radioactive methods: (1) the risks and costs associated with radioactivity are eliminated; (2) no overtime incubation or counting of radioactivity is necessary, so the time required for the assay is significantly shorter; and (3) finally, the standard curve is linear over a wide range of cAMP concentration.

Bioluminescent Enzymatic Assay The bioluminescent assay of adenylyl cyclase activity shares benefits of previously described fluorimetric assays including high degree of sensitivity, less expense, more versatility, and no radioactivity. In addition, it has two additional advantages: (1) it can be performed in fewer steps and, under similar experimental conditions, (2) it is more

(a)

$$\text{AMP + ATP (trace)} \xrightarrow{\text{myokinase}} \text{2 ADP}$$

$$\text{ADP + phospho(enol)pyruvate} \xrightarrow{\text{pyruvate kinase}} \text{ATP + pyruvate}$$

$$\text{ATP + fructose} \xrightarrow{\text{esokinase}} \text{fructose-6-phosphate + ADP}$$

$$\text{Fructose-6-phosphate} \xrightarrow{\text{phosphoglucose isomerase}} \text{glucose-6-phosphate}$$

$$\text{Glucose-6-phosphate + NADP}^{+} \xrightarrow{\text{glucose-6-phosphate dehydrogenase}} \text{6-phosphogluconolactone + NADPH + H}^{+}$$

(b)

$$\text{Luciferin + ATP + O}_2 \xrightarrow{\text{luciferase}} \text{dehydroluciferin + AMP + CO}_2 \text{ + PP}_i \text{ + hv}$$

Figure 9.4 Enzymatic fluorimetric and bioluminescent assays of cAMP. (a) Schematic representation of the steps required for fluorimetric detection of cAMP using coupled enzymatic reactions to myokinase; (b) schematic representation of cAMP detection using luciferase.

sensitive than the fluorimetric techniques. The first steps are identical to those previously described: there is a cleaning reaction to degrade adenine nucleotides other than cAMP, the conversion of cAMP to AMP with phosphodiesterase, and the conversion of AMP to ATP using myokinase and pyruvate kinase. What is different is just the final reaction to measure ATP. In this case, luciferin is used, which is enzymatically converted to dehydroluciferin by luciferase in an ATP-consuming reaction and production of light (Fig. 9.4, panel b). Resulting luminescence is proportional to initial cAMP concentration [27].

Another assay employing bioluminescence, also feasible for wide compound library screening (HTS) has been recently introduced by Kumar et al. [28]. The assay is based on the principle that cAMP modulates PKA holoenzyme activity, decreasing available ATP and leading to decreased light production in a coupled luciferase reaction. Since the amount of relative luminescence units (RLU) generated is a measure of the remaining ATP, a reciprocal relationship between RLU and both PKA activity and cAMP intracellular concentration is observed. Thus, the functional activity of agents that modulate G_s- or G_i-coupled GPCRs can be measured by changes in the amount of RLU readout.

Immunoassays: RIA and EIA The immunoassays' approach for determining cAMP concentration relies on the highly specific antigen–antibody interaction. The RIA, introduced by Steiner et al. [29] uses an antibody generated against 2′-*O*-monosuccinyl cAMP and ^{125}I labeled cAMP. The concept behind this method is competition between radiolabeled cAMP and cAMP from samples and standards, for binding to the antibody. Whereas the first RIA protocol was sensitive to 1–2 pmoles of cAMP, the method has been subsequently improved by means of 2′-*O*-acetylation of cAMP in samples and standards. This step allows to reach a higher sensitivity to readily detect fmole amounts of cAMP in tissue extracts [30, 31]. RIA has also been adapted to investigate adenylyl cyclase activity in cell membranes with several advantages over the conventional method utilizing [α-^{32}P]ATP. First, rather than using up to a million cpm of ^{32}P to perform a single assay, only about 10,000 cpm of ^{125}I are required, thus increasing safety. Second, no column chromatography or purification of any kind are necessary. Moreover, even if the protocol is easier in comparison with the one elaborated by Salomon's methods [19–21], sensitivity is comparable [31].

Immunoassays, both RIA and EIA, require separation of antibody-bound cAMP fraction from free cAMP; the efficiency with which this is done is crucial to the overall assay performance and to the simplicity and convenience of the protocol. The most common methods for separation are adsorption and precipitation. The first one utilizes a suspension of particles that could absorb free cAMP onto their surface (e.g., charcoal); following centrifugation, an aliquot of the supernatant is transferred to a scintillation vial to enable determination of the antibody-bound fraction. Alternatively, addition of a precipitating agent (a second antibody, or, less selective, ammonium sulfate or polyethylene glycol) can be useful to isolate immunoglobulins from the reaction mixture. After incubation with the precipitating agent, the antibody-bound cAMP is separated from the unbound fraction by centrifugation. Supernatant is decanted away and the activity (radioactivity or enzyme activity) is determined in the pellet.

RIA protocol has been adapted also to a solid-phase procedure in order to simplify bound and unbound cAMP separation. This assay, useful for tissues, body fluids, and cultured cells, includes microtiter wells or strips coated with polyclonal anti-cAMP antibody. In this solid-phase procedure, separation is achieved by pouring the content of the wells away or washing strips, respectively, leaving the bound fraction physically attached [32].

RIA has been subsequently improved in order to completely skip the separation step, thus obtaining a new simpler protocol called scintillation proximity assay (SPA). The Flashplate technology, introduced by Perkin Elmer, uses SPA method: microtiter plates coated with scintillant enable the detection of specific binding of radiolabeled molecules. These plates are coated with an anti-cAMP antibody and the assay uses ^{125}I-cAMP as a tracer. In the absence of cellular cAMP, the antibody sequesters ^{125}I-cAMP, bringing it in close enough proximity to the scintillant on the plate, such that light is produced. In the presence of cellular cAMP, the unlabelled cAMP competes off the iodinated molecule and thereby reduces the signal [18]. Since any unbound radioligand remains too distant to activate scintillant, the need for physical separation process is eliminated.

Even if RIA is surely a highly sensitive and precise method to detect cAMP in different biological samples, given the safety and environmental concerns, the use of radioactive materials should be avoided. On these bases, the introduction of EIA has been very useful. In this case, cAMP is not radioiodinated but it is labeled with an enzyme, for instance β-D-galactosidase [33] or acetylcholinesterase [34]. The assessment of cAMP concentration is determined evaluating enzymatic activity usually by means of a colorimetric or a fluorimetric assay. Like RIA, EIA too needs a separation step of bound from unbound cAMP. Therefore, starting from the early 1990s, several solid-phase strategies have been introduced [35, 36]. At present, there are many commercially available kits for solid-phase EIA like the one distributed by Sigma-Aldrich (St. Louis, MO). The kit uses a polyclonal antibody to cAMP to bind, in a competitive manner, the cAMP in the sample or an alkaline phosphatase molecule that has cAMP covalently bound. Samples or standard, alkaline phosphatase conjugate, and antibody are simultaneously incubated in a secondary antibody-coated multiwell plate. After a short incubation time with alkaline phosphatase substrate, the reaction is stopped and the yellow color generated is read on a multiwell plate reader. The intensity of the yellow color is inversely proportional to the concentration of cAMP in either the standards or the samples. Even if these kits, either for RIA or EIA, are very simple to use, they are usually expensive.

HTS Methods The concept behind immunoassays, that is competition between labeled or nonlabeled cAMP for binding immunoglobulins, has been widely extended and modified, especially to develop new assays suitable for HTS campaigns aimed at seeking novel receptor modulators. Both fluorescence polarization (FP) and time-resolved fluorescence resonance energy transfer (TR-FRET) technologies have been applied to measurement of cAMP for HTS. A description of these and other methods for HTS is not in the scope of this chapter, but it can be found in some recent reviews [18].

Protein Binding Assay This protocol was firstly described by Gilman [37] and Brown et al. [38] in the early 1970s. According to their works, the concentration of cAMP can be assessed by means of competition between the cAMP present in samples or standards and [^{3}H]cAMP for association with a cAMP-dependent protein kinase. In this case, a separation step is also required to separate bound from free cAMP. The original method has been improved with the introduction of faster separation techniques such as ammonium sulfate precipitation [39], or filtration by polyethylenimine-treated glass filters [40]. This protocol, used firstly to determine cAMP concentration in tissues, was adapted also to cell culture supernatants and body fluid [41]. This procedure can be easily performed for samples containing ≥3 nM concentration of cAMP [22]. However, the efficacy of protein binding assay is surely lower than immunoassays (RIA and EIA) since antibodies have clear advantages over cAMP binding proteins in terms of affinity and stability.

Protein Kinase Activation Together with the introduction of RIA by Steiner et al. [29], another method was suggested to assess tissue cAMP concentration: the protein kinase activation assay. This method is based upon the ability of the cyclic nucleotide to activate cAMP-dependent protein kinase, which in turn phosphorylates a substrate (such as caseine), using ATP. Thus, by means of exposure of tissue extracts to γ^{32}P-ATP, a certain amount of phosphorylated casein is obtained in relation with cAMP tissue concentration. The phosphorylated product can be isolated through filtration [42, 43]. However, this technique has the disadvantage of requiring radiolabeled ATP.

9.4. CONCLUSIONS

We can summarize that over the years, both receptor binding and cAMP production assays have maintained their roles in GPCR pharmacology and drug discovery. Indeed, in the era of HTS, improvement and automation of these techniques have allowed them to retain their importance as primary tools of drug discovery.

REFERENCES

1. Hill, S.J., Baker, J.G., Rees, S. (2001) Reporter-gene systems for the study of G-protein- coupled receptors. *Curr Opin Pharmacol.* 1, 526–532.
2. Windh, R.T., Manning, D.R. (2002) Analysis of G protein activation in Sf9 and mammalian cells by agonist-promoted [^{35}S]GTPγS binding. *Methods Enzymol.* 344, 3–14.
3. Milligan, G. (2003) Principles: Extending the utility of [^{35}S]GTPγS binding assays. *Trends Pharmacol Sci.* 24, 87–90.
4. Harrison, C., Traynor, J.R. (2003) The [^{35}S]GTPγS binding assay: Approaches and applications in pharmacology. *Life Sci.* 74, 489–508.
5. Fang, Y., Lahiri, J., Picard, L. (2003) G protein-coupled receptor microarrays for drug discovery. *Drug Discov Today.* 8, 755–761.
6. Waller, A., Simons, P.C., Biggs, S.M., Edwards, B.S., Prossnitz, E.R., Sklar, L.A. (2004) Techniques: GPCR assembly, pharmacology and screening by flow cytometry. *Trends Pharmacol Sci.* 25, 663–669.
7. Kenakin, T. (ed) (1993) *Pharmacological Analysis of Drug-Receptor Interactions.* New York: Raven.
8. Limberd, L.E. (1996) *Cell surface receptors: A Short Course on Theory and Methods*, 2nd ed. Boston: Martinus Nijhoff.
9. Bylund, D.B., Toews, M.L. (1993) Radioligand binding methods: Practical guide and tips. *Am J Physiol.* 265, L421–L429.
10. Muramatsu, I., Tanaka, T., Suzuki, F., Li, Z., Hiraizumi-Hiraoka, Y., Anisuzzaman, A.S., Yamamoto, H., Horinouchi, T., Morishima, S. (2005) Quantifying receptor properties: The tissue segment binding method—A powerful tool for the pharmacome analysis of native receptors. *J Pharmacol Sci.* 98, 331–339.

11. Mazzoni, M.R., Martini, C., Lucacchini, A. (1993) Regulation of agonist binding to A_{2A} adenosine receptors: Effects of guanine nucleotides (GDPβS and GTPγS) and Mg^{2+} ion. *Biochim Biophys Acta.* 1220, 76–84.

12. Rosenthal, H.E. (1967) A graphic method for the determination and presentation of binding parameters in a complex system. *Anal Biochem.* 20, 525–532.

13. Cheng, Y., Prusoff, W.H. (1973) Relationship between the inhibition constant (K_i) and the concentration of inhibitor which causes 50 per cent inhibition (IC_{50}) of an enzymatic reaction. *Biochem Pharmacol.* 22, 3099–3108.

14. Motulsky, H.J. (1999) The complete guide to nonlinear regression. Published online by GraphPad Software. http://www.curvefit.com.

15. Bylund, D.B. (1986) Graphic presentation and analysis of inhibition data from ligand-binding experiments. *Anal Biochem.* 159, 50–57.

16. Carpenter, J.W., Laethem, C., Hubbard, F.R., Eckols, T.K., Baez, M., McClure, D., Nelson, D.L., Johnston, P.A. (2002) Configuring radioligand receptor binding assays for HTS using scintillation proximity assay technology. *Methods Mol Biol.* 190, 31–49.

17. Sutherland, E.W., Rall, T.W. (1958) Fractionation and characterization of a cyclic adenine ribonucleotide formed by tissue particles. *J Biol Chem.* 232, 1077–1091.

18. Williams, C. (2004) cAMP detection methods in HTS: Selecting the best from the rest. *Nat Rev Drug Discov.* 3, 125–135.

19. Salomon, Y. (1979) Adenylate cyclase assay. *Adv Cyclic Nucleotide Res.* 10, 35–55.

20. Salomon, Y., Londos, C., Rodbell, M. (1974) A highly sensitive adenylate cyclase assay. *Anal Biochem.* 58, 541–548.

21. Johnson, R.A., Salomon, Y. (1991) Assay of adenylyl cyclase enzymatic activity. *Methods Enzymol.* 195, 3–21.

22. Post, S.R., Ostrom, R.S., Insel, P.A. (2000) Biochemical methods for detection and measurement of cyclic cAMP and adenylyl cyclase activity. *Methods Mol Biol.* 126, 363–374.

23. Wiegn, P., Dutton, J., Lurie, K.G. (1993) An enzymatic fluorimetric assay for adenylate cyclase activity. *Anal Biochem.* 208, 217–222.

24. Sugiyama, A., Lurie, K.G. (1994) An enzymatic fluorimetric assay for adenosine 3′:5′-monophosphate. *Anal Biochem.* 218, 20–25.

25. Sugiyama, A., McKnite, S., Lurie, K.G. (1995) Measurement of adenylylcyclase activity with an enzymatic fluorimetric assay. *Anal Biochem.* 225, 368–371.

26. Lowry, O.H., Passonneau, J.V. (1972) *A Flexible System of Enzymatic Analysis*. New York: Accademic.

27. McKnite, S., Evingson, M., Pennington, J., Adkisson, W., Sugiyama, A., Lurie, K.G. (1996) A bioluminescent enzymatic assay for adenylylcyclase activity. *Anal Biochem.* 235, 103–106.

28. Kumar, M., Hsiao, K., Vidugiriene, J., Goueli, S.A. (2007) A bioluminescent-based, HTS-compatible assay to monitor G-protein-coupled receptor modulation of cellular cAMP. *Assay Drug Dev Technol.* 5, 237–245.

29. Steiner, A.L., Kipnis, D.M., Utiger, R., Parker, C. (1969) Radioimmunoassay for the measurement of adenosine 3′,5′-cyclic monophosphate. *Proc Natl Acad Sci U S A.* 64, 367–373.

30. Harper, J.F., Brooker, G. (1975) Femtomole sensitive radioimmunoassay for cyclic AMP and cyclic GMP after 2'0 acetylation by acetic anhydride in aqueous solution. *J Cyclic Nucleotide Res.* 1, 207–218.

31. Brooker, G., Harper, J.F., Terasaki, W.L., Moylan, R.D. (1979) Radioimmunoassay of cyclic AMP and cyclic GMP. *Adv Cyclic Nucleotide Res.* 10, 1–33.

32. Daniels, C.K., Zhang, L., Musser, B., Vestal, R.E. (1994) A solid-phase radioimmunoassay for cyclic AMP. *J Pharmacol Toxicol Methods.* 31, 41–46.

33. Yamamoto, I., Tsuji, J. (1981) Enzyme immunoassay of cyclic adenosine 5′, 5′-monophosphate (AMP) using β-D-galactosidase as label. *Immunopharmacology.* 3, 53–59.

34. Pradelles, P., Grassi, J., Chabardes, D., Guiso, N. (1989) Enzyme immunoassays of adenosine cyclic 3′,5′-monophosphate and guanosine cyclic 3′,5′-monophosphate using acetylcholinesterase. *Anal Chem.* 61, 447–453.

35. Linden, J., Vandenhoff, G.E., Taylor, D., Finkelstein, G.L. (1992) Solid phase enzyme immunoassay of cyclic adenosine 3′,5′-monophosphate. Effect of coating strategy upon assay performance in comparison with radioimmunoassay. *J Immunol Methods.* 151, 209–216.

36. Tsugawa, M., Iida, S., Fujii, H., Moriwaki, K., Tarui, S., Sugi, M., Yamane, R., Fujimoto, M. (1990) An enzyme-linked immunosorbent assay (ELISA) for adenosine 3′,5′-cyclic monophosphate (cAMP) in human plasma and urine using monoclonal antibody. *J Immunoassay.* 11, 49–61.

37. Gilman, A.G. (1970) A protein binding assay for adenosine 3′:5′-cyclic monophosphate. *Proc Natl Acad Sci U S A.* 67, 305–312.

38. Brown, B.L., Albano, J.D.M., Ekins, R.P., Sgherzi, A.M. (1971) A simple and sensitive saturation assay method for the measurement of adenosine 3′:5′-cyclic monophosphate. *Biochem J.* 121, 561–562.

39. Santa Coloma, T.A., Bley, M.A., Charreau, E.H. (1987) Improvement on the competitive binding assay for the measurement of cyclic cAMP by using ammonium sulphate precipitation. *Biochem J.* 245, 923–924.

40. Takeda, T., Kuno, T., Shuntoh, H., Tanaka, C. (1989) A rapid filtration assay for cAMP. *J Biochem.* 105, 327–329.

41. Nordstedt, C., Fredholm, B.B. (1990) A modification of protein binding method for rapid quantification of cAMP in cell-culture supernatants and body fluid. *Anal Biochem.* 189, 231–234.

42. Kuo, J.F., Greengard, P. (1970) Cyclic nucleotide-dependent protein kinases. An assay method for the measurement of adenosine 3′,5′-monophosphate in various tissues and a study of agents influencing its level in adipose cells. *J Biol Chem.* 245, 4067–4073.

43. Wastila, W.B., Stull, J.T., Mayer, S.E., Walsh, D.A. (1971) Measurement of cyclic 3′,5′-adenosine monophosphate by the activation of skeletal muscle protein kinase. *J Biol Chem.* 246, 1996–2003.

CHAPTER 10

Fluorescence and Resonance Energy Transfer Shine New Light on GPCR Function

CARSTEN HOFFMANN and MORITZ BÜNEMANN

Department of Pharmacology and Toxicology, University of Würzburg, Würzburg, Germany

10.1. OVERVIEW

Recent advancements in site-directed labeling of G protein-coupled receptors (GPCRs) and interacting proteins with fluorescent probes such as variants of the green fluorescent proteins (GFP) and small fluorescent molecules enabled detailed studies of receptor function. In the context of developments in fluorescence spectroscopy and life cell imaging, a considerable amount of new insights into the dynamics and extent of ligand-induced conformational changes of receptors have been achieved. This chapter discusses details of how to label a GPCR with different fluorescent dyes, it highlights advantages and disadvantages of fluorescence resonance energy transfer (FRET) and bioluminescence resonance energy transfer (BRET) in order to study protein conformations as well as protein–protein interactions, and also contains some advice of when and how to use these techniques in order to study receptor function.

10.2. INTRODUCTION

More than three decades of intensive research on GPCRs have brought about insights into some aspects of structure and function or membrane receptors, whereas many fundamental questions remain unresolved. Most of the unre-

*Present address: Pharmacology and Clinical Pharmacy, Philipps-Universitaet Marburg, Marburg, Germany

GPCR Molecular Pharmacology and Drug Targeting: Shifting Paradigms and New Directions, Edited by Annette Gilchrist

solved questions are related to the dynamics of GPCR function, assembly, and signaling. Until 15 years ago, receptors could only be directly detected by means of radioligand binding assays and antibody binding. The function of receptors was determined either by measuring cellular responses resulting from the respective signaling pathway, or by means of radioactively labeled GTPγS binding to heterotrimeric G proteins.

The technical advancement which allowed labeling of receptors and proteins that interact with them using fluorescent tags revolutionized GPCR research as it enabled to image subcellular localization and to track internalization processes. One can now study the interaction of GPCRs with G proteins as well as with arrestins and can do so in a temporally resolved fashion, either indirectly by means of G protein activation or by monitoring conformational changes underlying receptor activation. A key to these developments was the discovery of fluorescent proteins [1, 2], which were critically important for monitoring localization of receptors in living cells in real time. The ability to fuse fluorescent proteins to GPCRs or their interacting proteins accelerated the understanding of receptor trafficking and subcellular localization (see Section 10.3., Labeling GPCRs with Fluorescent Tags). The recent introduction of smaller fluorophores suitable for site-directed labeling of proteins in intact cells (see Section 10.3.2., Labeling of GPCRs with Fluorescent Dyes) accelerated functional studies on GPCRs [3–5]. A major breakthrough for investigation of receptor assemblies, as well as dynamics of receptor function in living cells, came with the introduction of variants of GFP suitable for FRET [6] and, similarly, the establishment of luciferase/GFP variants useful for BRET [7], which allowed addition of optical "rulers" with which to measure receptors and their interacting binding partners (see Section 10.6., Resonance Energy Transfer, a Tool to Get New Insight into GPCR Function). FRET and BRET assays are extremely sensitive and can be used to measure changes in distance close to the Förster radius (typically between 4 nm and 6 nm). This distance is optimal for studying the interaction between small- to medium-size proteins and also allows detection of conformational changes within a protein that lead to significant intramolecular movements. In addition to the different labeling approaches in this chapter, we discuss BRET and FRET as methods to study interactions of receptors and their function in living cells. A special emphasis will be the potential pitfalls and problems in the practical use of these methods.

10.3. LABELING GPCRs WITH FLUORESCENT TAGS

10.3.1. Tagging GPCRs with Fluorescent Proteins

The introduction of fluorescent proteins has accelerated research on GPCRs similar to that seen for many other topics in the life science field. The first fluorescent protein cloned was GFP [2], and it has since been widely used. The generation of color variants and variants that fluoresce as monomeric entities have been very useful in many studies and are reviewed elsewhere [8]. Recently, additional red shifted fluorescent proteins have become available, and GFP variants useful for studying signal transduction and trafficking such as photo-activatable fluorescent proteins have been generated [9]. Furthermore, a tremendous amount of work has been done to improve fluorescent properties and photo stability of these fluorophores [10]. To visualize protein–protein interactions, GFP variants have been split into two nonfluorescent proteins that upon colocalization can complement each other, resulting in a fluorescent protein complex [11]. Fusion of the two parts of these GFP variants to two potentially interacting proteins has been used to localize protein complexes by means of bimolecular fluorescence complementation (BiFC). The advantage of this technique is a low background since only protein complexes will fluoresce [11]. In the context of GPCR research, BiFC has been successfully used to study G proteins ensuring a subtype-specific combination of G protein subunits [12]. To detect and localize stable protein complexes, BiFC offers great sensitivity and specificity. Still, it is important to control the extent of specific BiFC signals with suitable control proteins, to exclude complex formation in the absence of specific binding. The BiFC approach reaches its limits when it comes into studying rapid dynamics of unstable protein–protein interactions due to the possible introduction of additional binding affinities. To address these questions, resonance energy transfer (RET)-based methods are superior; however, BiFC can complement a RET approach and can even be integrated into it, for instance, by function as a RET acceptor [13].

Even though most fluorescent proteins are inferior to many small synthetic fluorophores in terms of quantum efficiency, quantum yield, and photo stability, as well as spectral properties, there are striking advantages that led to their enormous popularity. One important feature is the specificity of the fluorescence signal due to the possibility of generating fusion proteins with the protein of interest. In such cases, no unspecific labeling is expected unless the fluorescent protein is cleaved from its fused protein. No other labeling method is available that reaches such degree of specificity. The fact that by means of molecular biology the fluorescent protein can potentially be fused at any position within the protein of interest makes this method unique. Since these proteins are genetically encoded, it is also possible to generate transgenic organisms that express the fluorescent receptors as demonstrated for dopamine receptors [14]. Detailed analysis of receptor trafficking in cell culture models by means of GFP-labeled receptors [15] have demonstrated that receptors undergo agonist-induced sequestration and internalization and subsequently either recycle back to the membrane or target to the lysosomal pathway, leading to their degradation.

10.3.2. Labeling of GPCRs with Fluorescent Dyes

Fluorescent tagging of proteins has become a broad applicable technique in GPCR research with respect to dynamic protein studies. However, the use of

GFP and its color variants is just one of the many ways in which to achieve this goal. Depending on the questions to be answered, a variety of different labeling approaches can be taken. In this section, we will briefly introduce several of the different approaches.

One of the earliest examples was chemical labeling of GPCRs using cysteine-reactive fluorescent probes on purified β_2-adrenoceptors. The system employed purified and reconstituted β_2-adrenoceptors that were modified on all but the essential cysteines [16]. The modified receptors can then be reacted with small cysteine-reactive fluorescent probes, which can be monitored for several different properties, like side-chain mobility, fluorescence intensity, or fluorescence lifetime [17, 18]. This system has been employed with great success to investigate conformational changes induced by ligands with varying degrees of efficacy and has helped to resolve some major questions of GPCR activation [18].

The use of cysteine-reactive fluorescent probes has not only been limited to the β_2-adrenoceptor but has also been used to study GPCRs that bind peptide ligands [19–21]. The studies used fluorescently modified ligands combined with receptors that were selectively modified on the extracellular domains to study the ligand binding mode by FRET. This technique was used successfully for the cholecystokinin receptor [19, 20] and the secretin receptor [21]. The approach uses a special methanethiosulfonate reagent that permits labeling of receptors in whole living cells, rather than purified receptors. A combination of several fluorophore attachment points on the ligand and receptor allowed measuring a number of distance constraints which were then used to generate a model of the ligand–receptor complex [21]. However, since steady-state fluorescence was used as the readout, no dynamic information of ligand binding could be obtained. Dynamic binding information was reported studying an N-terminally GFP-labeled parathyroid hormone (PTH) receptor construct and fluorescently modified PTH or PTH derivatives [22] in living cells. The studies revealed a two-step binding mode of the ligand to the receptor. The initial contact was made between the ligand and the receptor N-terminus, and a second binding phase involved contacts within the transmembrane domains of the receptor.

Successful chemical labeling of peptide ligands with fluorescent probes is more readily achieved compared to small biogenic amine ligands, as with biogenic amines the functional groups that could be used for chemical labeling are often also involved in receptor binding [23], and chemical modifications of these groups can result in reduced binding affinities. These problems have been widely recognized and currently, more suitable fluorescent ligands are being developed [24, 25].

An alternative labeling strategy for proteins utilizing small fluorescent dyes is the tetracystein labeling technology which employs a modified fluorescein derivative called FlAsH. FlAsH stands for fluorescein arsenical hairpin binder [26] and was originally introduced in 1998 [27]. This labeling strategy uses a genetically encoded sequence of minimally six amino acids with the sequence

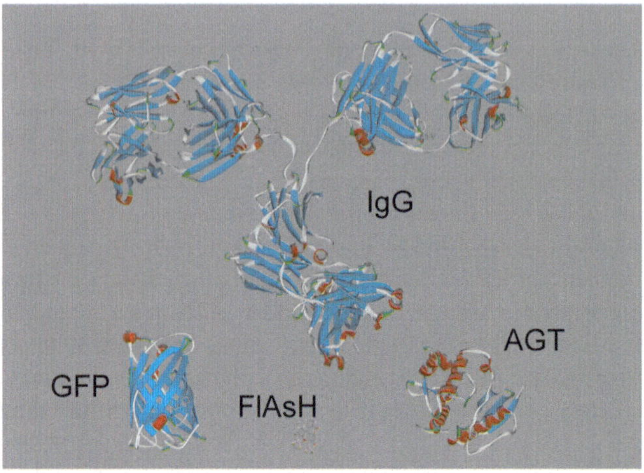

Figure 10.1 Depicted are structures of moieties that are suitable to label proteins with fluorescent tags. The relative size of the depicted structures of an antibody (IgG, pdb 1IGT), green fluorescent protein (GFP, pdb 1emb), fluorescein arsenical hairpin binder (FlAsH), and O^6-alkylguanine-DNA alkyltransferease (AGT, pdb 1EHG) are shown for comparison. The color code depicts alpha-helical structures in red and beta strands in silver blue.

CCPGCC [26, 28]. The advantages of this technique are the small size of the tag (see Fig. 10.1 for comparison of the different sizes of labeling tags) and that the sequence can be fused or inserted into the protein of interest [29]. The corresponding protein construct can be expressed and selectively labeled with the membrane permeable FlAsH in living cells [74, 75]. Thus, this technique can be viewed as a bridge, in between genetically encoded fluorescent proteins and fluorescent labeling of cysteines. This approach allowed the replacement of yellow fluorescent protein (YFP) as a FRET partner for cyan fluorescent protein (CFP) [29] and has been applied to several GPCRs to investigate receptor activation and signaling properties in living cells and in real time [30–34]. Currently, the fluorescein-based FlAsH is the only variant that has been applied to label GPCRs; however, a red resuferin-based color variant has been developed [26] and is available as resorufin arsenical hairpin binder (ReAsH). Additionally, more photo stable variants of FlAsH have been reported but are not yet commercially available for general use [35].

If one needs to achieve greater color variability and does not want to use genetically encoded proteins based on GFP [8, 10], an alternative approach is the use of an enzyme-based system like O^6-alkylguanine-DNA alkyltransferase (AGT) [4] or the acyl carrier protein (ACP) [5]. These enzymes are genetically encoded and can be fused to the protein of interest. The expressed protein is then labeled by activated small organic dyes, which get transferred to the fused protein by enzymatic transfer. However, the size of the AGT-derived SNAP-tag is about 20 kDa, and is thus closer to the size of GFP and significantly larger than the above-mentioned tetracystein technology (see Fig. 10.1). The SNAP-tag can be used with greater color variability since a selection of small dyes is available [36]. Furthermore, unlike for fluorescent proteins, the same construct can be used for labeling with different colors and thus, no additional cloning is needed to achieve different colors. SNAP-tag labeling has been applied in living cells to track the neuropeptide Y receptor [37] and gave similar results to those obtained with a GFP variant. When the ACP tag was employed to label the neurokinin-1 (NK1) receptor [38, 39], it provided a means to label surface-expressed receptors with very little background. Using two different colors and an elaborate labeling protocol, it was possible to investigate the NK1 receptor in single cells by FRET microscopy and to demonstrate that the receptors are monomeric and reside in microdomains in intact cells [39]. Receptor dimerization was further investigated using a time-resolved FRET approach involving the SNAP-tag technology [40], and evidence was found that family A and family C GPCRs could not only dimerize, but could also occur in even higher oligomeric states. Recently, a further engineered variant of the SNAP-tag, named CLIP-tag, was reported. This CLIP-tag can be labeled with chemically different compounds compared to the SNAP-tag, and Gautier et al. reported four color labeling of two different proteins with two different colors at different time points [41]. Although this labeling was not done with GPCRs, it highlights the potential of this technology.

Any fluorescent label can potentially interfere with the function and fate of the labeled receptor. It is actually never possible to prove that a labeled receptor behaves in all aspect exactly as the nonlabeled receptors. Therefore, it is very important to carefully control for the function of the labeled receptor in respect to the specific aspect that is under investigation. In many cases, the trafficking and signaling of a given receptor is the focus of the research, and labeled receptor should be analyzed in detail regarding its signaling properties. There are many examples where C-terminal tags on receptors have had no detectable influence on the G protein-activating properties of the receptors; however, the C-terminal tags may influence desensitization mechanisms and arrestin-mediated signaling properties, and these properties have not always been investigated. In addition, for many receptors such as the PTH receptor [42], the distal C-terminus serves either as an anchoring point for adaptors such as NHERF1, or to exert other important functions such as targeting, which in turn may influence signaling [42, 43]. Careful consideration of appropriate functional controls is critical when fluorescent dyes have been used to label GPCRs.

10.4. DETECTION OF FLUORESCENCE AND BIOLUMINESCENCE

Photons derived either from fluorescence or bioluminescence need to be detected and quantified. For imaging purposes, the light derived from labeled proteins contains not only information about the extent of expression, but in the case of fluorescence imaging, also provides where this expression takes place. There are many detectors that can be used to measure light with high accuracy and sensitivity. The following paragraph briefly summarizes the advantages and disadvantages of fluorescence detection methods suitable for studies of GPCR function and trafficking. Photodiodes have the advantage that they exhibit linear responses upon illumination with intensities varying over a large range. In general, they are robust and reliable, and the speed of responses can be quite fast. Avalanche photodiodes for instance are fast enough to measure the lifetime of fluorescence in the femtosecond range. Photodetection by photodiodes is used for a broad spectrum of applications. Photometric detection of FRET or ratiometric detection of Ca^{2+} by fluorescence indicators are typical examples of their application. Photomultipliers (PMT) have an advantage when detection of light at very low intensities is necessary. Based on their ability to detect light at high frequencies and with excellent quantum yields, PMTs are used in laser scanning microscopes as well as for photometry systems. Charge-coupled devices (CCD) function similarly to an array of photodiodes, and due to their small pixel sizes, they are popular for many imaging devices such as cameras. Technology improvements gave rise to ultrafast and sensitive electron-modifying charge-coupled device (EM-CCD) chips which can take images within a few milliseconds. The sensitivity of these chips has increased over the years, which has helped to increase speed and sensitivity of fluorescence imaging. The application of CCD cameras is very broad and reaches from regular epifluorescence imaging, to single particle tracking, to total internal reflection fluorescence (TIRF) microscopy, to spinning disc confocal microscopy.

10.5. FLUORESCENCE-BASED ASSAYS TO STUDY RECEPTOR LOCALIZATION, TRAFFICKING AND RECEPTOR FUNCTION

Localization of fluorescent receptors and its adaptor proteins within subcellular compartments has been made possible with fluorescence microscopy. The trafficking of receptors was initially studied by means of immunofluorescence. This typically requires fixation of cells, thus excluding the possibility of being able to track receptors on their way to different intracellular compartments. However, this approach helped pave the way for understanding basic mechanisms of receptor trafficking, particular internalization, and recycling [44]. Fusion of GFP variants to the C-terminus of receptors in many cases did not measurably alter their function, and they have served as useful tools to study the fate of GPCRs. For this purpose, high spatial resolution is important, and therefore, either laser scanning confocal microscopes or fluorescence microscopes equipped with high-resolution CCD cameras with or without spinning disks are frequently used. The tools to study targeting of receptors to subdomains of cell membranes are increasing more and more: the very recent introduction of novel microscopy techniques achieving sub diffraction-limited resolution such as stimulated emission depletion (STED) [45] will boost our capabilities in visualizing receptor localization and function. If trafficking events in and out of the plasma membrane are of special interest, total internal reflection fluorescence (TIRF) microscopy provides the advantage of fast measurements with 100–300 nm resolution in Z-direction. For instance, the group of Mark von Zastrow very elegantly used TIRF microscopy to study β-receptor sequestration and recycling [46]. Their work demonstrated that the receptor stimulated internalization through a subset of clathrin-coated pits and uncovered that the residence time of a clathrin-coated pit on the cell surface is dependent on its cargo [46].

10.5.1. How to Monitor Receptor Function by Means of Fluorescence Microscopy

Receptors get activated due to binding agonists in their respective binding pocket. During this activation process, receptors will undergo conformational changes which allow them to productively interact with G proteins, arrestins, and receptor kinases in order to regulate downstream events. Recently, two different fluorescence-based assays have allowed direct monitoring of conformational changes that underlie ligand-induced receptor activation. The first assay was based on tetramethylrhodamine (TMR) maleimide labeling of purified β_2-adrenergic receptors at cysteines introduced at different sites as described above [17]. The basic premise was that ligand-induced changes in the receptor conformation will alter the hydrophobicity close to the TMR maleimide label. This can be monitored spectroscopically as a change in fluorescence. This method principally provides information on the kinetics of ligand-induced conformational change, and has been used successfully to characterize conformational changes of the β_2 adrenergic receptor that occurred upon binding of different ligands [18]. Due to the small size of the fluorophores utilized for site-specific labeling of the receptor, the approach allowed mapping hydrophobicity changes within small regions of the receptor. In regard to precise structural information, the approach may provide more information of where the fluorophore actually moves. Critical limitations of this technique include the requirement to purify receptors, to functionally reconstitute them, and to remove any reactive cysteines in the protein. Even when these limitations have been overcome, such as was demonstrated for β_2-adrenergic receptors, the kinetics of receptor activation seem to be unrealistically slow compared to situations in intact cells. An alternative approach for detecting receptor activation on the level of the receptor itself is based on FRET and is discussed in detail later in this chapter (*detection of receptor function by means of FRET*).

10.6. RESONANCE ENERGY TRANSFER, A TOOL TO GET NEW INSIGHTS INTO GPCR FUNCTION

10.6.1. BRET

Renilla luciferase is able to oxidize substrates such as h-coelenterazines or DeepBlueC, a reaction that generates bioluminescence with peaks in the emission spectra either at 475 nm (h-coelenterazine) or 395 nm (DeepBlueC) [47, 48]. In order to use bioluminescence as a donor for RET, acceptor fluorophores with suitable excitation and emission spectra need to be used. Typically, enhanced yellow fluorescent protein (eYFP) is used as an acceptor for h-coelenterazines (BRET1), whereas GFP^2 is used as an acceptor for DeepBlueC (BRET2) [48]. BRET2 has the advantage that the emission of DeepBlueC does not overlap with the emission spectra of GFP^2, resulting in reduced bleed-through of the donor into the acceptor channel. Thus, for accurate measurement of the absolute BRET efficiency, BRET2 is preferable. However, a disadvantage of BRET2 is the lower quantum yield and therefore the requirement to use higher amounts of enzyme, either by increasing the sample size, or more typically, by increasing the expression [51]. For the latter case, overexpression artifacts have to be considered with BRET2, even more than when BRET1 is utilized. For luciferase assays such as BRET, the time point of substrate application prior to the actual measurement is very critical for the amount of bioluminescence measured, due to the depletion of substrate over time. This has been problematic with respect to repeated long-term measurements, and the difficulty has led to the development of coelenterazine as a precursor, which is taken up by cells and then later enzymatically releases h-coelenterazine.

BRET is most often measured by means of ratiometric detection of YFP or GFP^2 emission over donor emission derived from the bioluminescence emitting cleavage of suitable substrates by the luciferase. The detection occurs via a sensitive luminometer, typically based on photomultiplier tubes. As the bioluminescence intensity is orders of magnitudes lower than that of fluorescence derived from direct excitation of fluorescence by high power excitation light sources, typically, luminescence is detected in multicellular preparations [49].

10.6.2. FRET

Energy from excited fluorophores will be transmitted to nearby electron ring systems by means of nonradiant dipole–dipole resonance phenomena, a process called FRET (also known as Förster resonance energy transfer, named after the German scientist Theodor Förster). The efficiency of the energy transfer depends not only on the emission and excitation spectra of donor and acceptor fluorophores, but is very strongly coupled to the distance between fluorophores. In fact, the FRET efficiency is inversely proportional, $1/(1 + R^6/R_0^6)$, where R is the distance between donor and acceptor chromophores and

R_0 is the Förster radius, the distance at which transfer is 50% efficient (50% of the maximal possible energy transfer from the donor to the acceptor has occurred). The great sensitivity toward changes in distance between the fluorophores is one reason FRET serves as a useful method to measure distance between fluorophores. For fusion proteins of GFP variants and signaling molecules such as GPCRs, the Förster radius is typically within the range of protein diameters. Therefore, interaction between the proteins can be measured by FRET. Moreover, if proteins are labeled with two different fluorophores, distance changes between the fluorophore can be detected by FRET and can be used to monitor conformational changes of the carrier protein. For FRET studies in intact cells, popular GFP-based FRET pairs are enhanced cyan fluorescent proteins (eCFPs and cerulean) and eYFP and venus. Cerulean maturates faster and gives rise to higher fluorescence emission, which is an advantage if the donor emission is noise limiting for ratiometric FRET recordings [50]. eCFP offers the advantage over cerulean to be a better energy donor toward the YFP. Currently, many labs are trying to optimize pairs of donor- and acceptor-fluorescent proteins to increase the efficiency of resonance energy transfer, which in the future should improve signal-to-noise ratios. The use of more red-shifted fluorophores for future FRET studies will allow expanding FRET measurements to simultaneous multicolor detection of more than one signaling event.

Time-resolved FRET represents a specialized FRET method that utilizes lanthanide donors such as europium (as a Eu^{3+}-cryptate complex) and suitable acceptor fluorophores such as XL665. The use of lanthanides as donors offers two advantages: first, the donor emission is very weak in the spectral range of the acceptor emission [51]. Second, lanthanides exhibit very long fluorescence lifetimes which allows measurement of emission temporally separated from excitation. These advantages give rise to very low background signals. A key feature of the use of these fluorophores is the requirement of antibodies for labeling. Therefore, the specificity of antibodies and the efficiency of antibody/receptor interactions largely determine the applicability of this assay. This assay has been successfully applied to study oligomerization of GPCRs [52]. For these studies, the use of polyclonal antibodies should be avoided because of the possibility to cross-link receptors by means of antibody binding. The Förster radius for these FRET pairs is about 9 nm, which is important to consider when interpreting data.

10.6.3. Comparison of BRET and FRET

Both methods, BRET and FRET, have been successfully applied to the field of GPCR research. The methods have unique advantages and disadvantages, which will be discussed only briefly in this context. Due to strong signal strength, FRET clearly has the advantage in terms of imaging where things happen. For certain fluorophores, FRET can be detected even between single molecules [53]; though this has not yet been applied to GFP-based receptor studies, in the future it will likely be an important tool. FRET has demonstrated to be useful when analyzing receptor conformational changes as well as interaction characteristics with associating proteins. FRET can be measured by means of at least three reliable and quantitative techniques: Donor dequenching by acceptor photo bleaching (a method preferable to measure FRET efficiency), ratiometric detection of sensitized emission (preferable for resolving fast kinetics) and by fluorescence lifetime imaging (FLIM; a method ideally suited to image steady state FRET signals with subcellular resolution) [54]. As all three methods detect FRET by different means, the combination of all three methods can deliver accurate results regarding the occurrence of FRET, including its subcellular localization. In its current form, BRET relies only on ratiometric detection of light derived from acceptor and donor, and these measurements can be done with high sampling frequency. However, given that BRET measurements have to be done with typically at least 1000 cells per well, the kinetics of BRET responses are often limited by mixing time and always represent the average of a cell population. For very weak RET, BRET2 is probably more sensitive compared to FRET, due to the separated emission spectra from donor and acceptor. BRET assays are quite straightforward to perform, rely on automated luminescence readers, and offer the advantage of being able to measure many different experimental conditions within a short period of time [48]. It should come as no surprise that BRET has become a popular and powerful method by which receptor–receptor interactions can be studied, with a particular strength on generation of unbiased results (due to averaging and automation).

10.7. ANALYSIS OF STEADY-STATE PROTEIN–PROTEIN INTERACTION BY MEANS OF RET

Many studies are designed to identify steady-state interactions between proteins. RET-techniques, if performed appropriately, offer great sensitivity toward identification of whether RET between two tagged proteins can be detected. However, if the aim of a study is to test for specificity of the observed RET, as well as quantify these interactions, the process is far more complicated. In the following paragraph, the major problems will be discussed in greater detail in order to help to set guidelines for future studies.

How does one determine whether a RET signal between two proteins is specific? It is critically important to use appropriate controls for determining the amount of unspecific RET. However, in many cases, including those of protein interactions with GPCRs, appropriate controls are difficult to establish: We propose the following three criteria be met for a RET-acceptor-control for it to qualify for the term "appropriate control": (1) Comparable expression levels of the RET-acceptor-control and the RET acceptor of interest need to be achieved. (2) For membrane proteins, the attachment points of the RET-acceptor-fluorophore on control proteins and proteins of interest should be comparable in terms of parameters such as flexibility of the linker, as well as distance to the membrane surface. (3) The RET-acceptor-control should localize to the identical microdomains of the membrane but should not interact with the protein of interest. If one carefully considers these points, it will be extremely difficult to find controls that qualify as appropriate. For example, with our studies on receptor/G protein interactions, none of the acceptor controls that were used satisfactorily qualified for appropriate when the three criteria are applied.

The second problem is the quantification of protein–protein interactions by RET. RET is proportional to the extent of interaction, consequently for *in vitro* studies with purified RET acceptor—and in RET-donor-proteins, the quantification of interaction by means of RET is straightforward: specific RET signals in the absence of RET-acceptor molecules (RET_{min}—*should be 0*) as well as in the presence of excess acceptor molecules (RET_{max}) will be measured. The extent of interaction between donor and acceptor molecule is linearly dependent on the RET signal, with (RET_{max}—RET_{min})/2 corresponding to the half-maximal interaction. When using intact cells, in many cases, this issue is even more difficult. The availability of an appropriate acceptor control is needed to determine the amount of unspecific RET (corresponding to RET_{min}). Next, RET_{max} needs to be determined in living cells. This requires titration of the expression levels of acceptor molecules relative to those of the donor molecules. Moderate expression levels of the donor need to be achieved, whereas acceptor molecules should be expressed in great excess over donor molecules. Both high expression of acceptors and low expression of donors is not always easy to achieve. In addition, correct localization of donor and acceptor molecules needs to be controlled. The above-mentioned pitfalls are sometimes difficult to circumvent. Additional methods to determine protein–protein interactions are still needed in order to ascertain the specificity and extent of steady-state interactions.

10.8. KINETIC ANALYSIS OF PROTEIN–PROTEIN INTERACTIONS BY MEANS OF FRET

Receptors are activated by agonists, and this event subsequently alters their interactions with G proteins and other downstream effectors such as arrestins. The possibility of studying these conformational changes by agonist application (or withdrawal) and monitoring receptors switching from one state to another and accordingly altering their interaction pattern makes GPCRs a preferred object for kinetically resolved FRET and BRET studies. In most cases, for ligand-induced alterations in BRET and FRET signals, the absolute RET efficiency is of minor interest, and much less care needs to be taken for controlling basal and maximal FRET levels.

Current approaches in GPCR research measure in many cases ligand-induced changes indirectly by means of assessing downstream signaling events at various points along the signaling pathway. For instance, popular assays detect cAMP level, inositol-phosphates or Ca^{2+} signals. These assays are typically robust and easy to perform; however, they measure events relatively far downstream of the receptor. That such events occur indirectly, for example, as a result of receptor cross-talk driven through other signaling pathways needs to be considered. Furthermore, when the observed signal is downstream of receptor activation, it has been amplified manifold, and the quantitative correlation is very difficult. Similarly, kinetics of second messenger production do not adequately represent kinetics of upstream events. Receptor-induced binding of radioactive GTPγS to G proteins allows for better quantification of receptor activity; however, fast kinetics are difficult to measure and the assay cannot be performed in intact cells.

10.8.1. G Protein Activity Measured by FRET

To try and measure G protein activation directly and in intact cells, many labs have explored using RET-based approaches. Janetopoulos et al. [55] were the first to succeed in resolving kinetics of G protein activation by means of FRET in *Dictyostilium discoideum* by attaching CFP and YFP variants to Gα and Gβγ subunits, respectively. In cells coexpressing both fluorescent G protein subunits, FRET signals decreased during activation, suggesting an increase in distance between CFP and YFP, which is in accordance with the classical concept of activation-induced subunit dissociation [56]. The first publication on FRET-based detection of mammalian G protein activation demonstrated a counterintuitive increase in FRET between Gαi and Gβγ subunits upon activation, not compatible with complete separation of G protein subunits (see Fig. 10.2) [57]. Rather, the results supported the concept that G protein subunits did not completely separate prior to functionally regulating their effectors, an idea originally proposed by Levitzki and coworkers [58], which has gained more acceptance [59]. BRET technology has also been successfully applied to study complex formation within the G protein signaling pathway [60]. The results from these studies also support Levitzki's concept. It should be noted however, that the published data so far do not predict that Gα and Gβγ subunits, once formed, will stay together as a lifelong trimer. A recent study that uses a specialized fluorescence recovery after photobleaching (FRAP) approach to study stability of protein–protein interactions actually demonstrated a decrease in stability of Gα/Gβγ interaction upon receptor-mediated interaction in intact cells [61]. In contrast to Gi proteins, FRAP measurements indicated activated Gαs and Gβγ subunits continue to form stable complexes [61]. RET-based detection of G protein activity has been successfully applied to Gi/o-, Gz-, Gq-, and Gs-proteins [31, 57, 62]. The ability to study receptor-dependent activation of G proteins in a subtype-specific manner makes the FRET approach unique.

The possibility to study receptor activation, receptor–G protein interaction, as well as G protein activation in intact cells by means of FRET, uncovered speed-limiting steps in this initial part of GPCR signaling. Whereas receptor

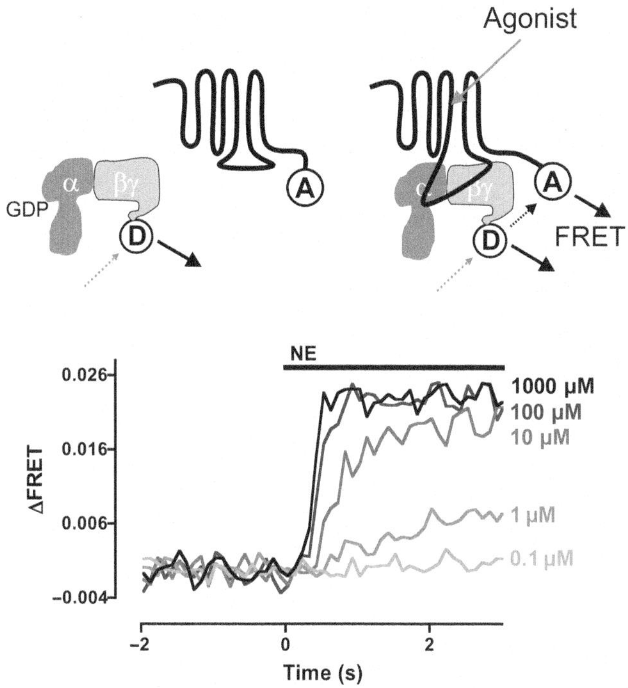

Figure 10.2 Interaction between heterotrimeric G proteins and receptors determined by FRET. Receptors labeled at the C-terminus with YFP (Y) and G proteins containing a CFP moiety at the N-terminus (C) will allow for FRET upon interaction as indicated in the upper scheme. The time course of agonist-induced alterations in FRET are illustrated in dependence of agonist concentration.

activation for small molecule neurotransmitter receptors typically exhibited time constants of less than 100 ms [29, 31], subsequent G protein activation required several hundreds of millisecond [31, 57]. At high expression level of receptors as well as G protein kinetics of the actual coupling between receptors and G proteins as detected by FRET was not different compared to receptor activation [30, 31]. However, lowering G protein expression levels to endogenous levels of a HEK cell did result in a substantial slowing of the interaction kinetics [30], in line with collision-coupling models. Considering that expression levels of endogenous receptors in many cases are considerably lower compared to those of the above-mentioned studies, we have to expect that receptor–G protein interaction might as well be a speed-limiting step in the initial steps of G protein signaling.

10.8.2. Receptor–G Protein Interaction Studied by RET

FRET- and BRET-based assays have been developed to study interactions between G proteins and receptors. To do so, receptors were labeled at the

C-terminus with either CFP or YFP variants (for FRET studies) or luciferase (for BRET studies), and one of the three G protein subunits (α, β, γ) was labeled with the corresponding RET acceptor or donor. The detection of a specific RET signal in the absence of receptor stimulation would argue for the existence of precoupling between receptor and G protein.

Several RET-based studies reported that precoupling between certain GPCRs and heterotrimeric G proteins exists in the absence of ligand-induced receptor activation. The most rigorously tested receptor in this respect is the α_{2A}-adrenergic receptor which couples to pertussis toxin-sensitive G proteins and was the object of three independent RET-based studies [30, 63, 64]. Unfortunately, the conclusions drawn from these studies are somewhat contradictory: the formation of complexes between receptors and G proteins in the absence of agonists was proposed from two independent studies [65, 66], whereas a third study found support for interaction only in the presence of agonist [30]. These contradictory results could only partially be explained by the hypothesis that receptor–G protein precoupling might be receptor specific. In these studies, the receptor G protein precoupling is reduced to the question of whether or not a specific RET signal between labeled receptors and G protein subunits can be detected. All three studies detected RET between receptors and G proteins in the absence of agonists; however, the question of whether this RET signal is specific was answered differently. This case can be valued as an example of the weakness of RET approaches to study steady-state interactions. The choice of control proteins to distinguish between bystander RET and specific FRET is often very difficult, and all three studies used quite different control proteins, none of which qualified for the term "appropriate" as discussed above. A recent study used mobility measurements by means of FRAP [65] in order to study receptor–G protein interaction. This study found no restriction in mobility of fluorescent Gi proteins was found upon experimental immobilization of α_{2A}-adrenergic receptors in living cells. The result supports the results and conclusions of Hein et al. [30] that these receptors interact only in the presence of agonist with G proteins, and that this interaction is short-lived relative to the lifetime of an active G protein, suggesting the occurrence of catalytic collision coupling. The controversy over precoupling of GPCRs and G proteins is far from being resolved. Although there is solid evidence indicating an interaction between GPCRs and G proteins prior to activation for a number of receptors suggesting that precoupling exists for some GPCRs, more studies are needed to clarify this issue, including those that utilize fluorescence-based approaches. Such assays will help address the important functional consequences of precoupling between inactive receptors and G proteins.

10.8.3. Kinetics of Receptor–G Protein Interactions

Upon receptor activation, a small but significant increase in RET between receptors and G proteins has been observed (see Fig. 10.3; [30, 63]). There are

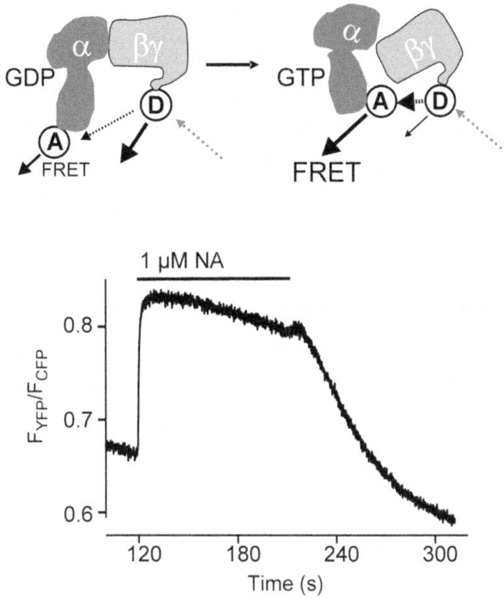

Figure 10.3 G protein activation measured by means of FRET microscopy. G protein activity can be visualized in intact cells by means of FRET between fluorescent Gα and Gβγ subunits as indicated in the upper scheme. A single cell ratiometric FRET recording of cells expressing $G\alpha i_1$ containing a YFP insertion between position 91 and 92 and $G\beta_1\gamma_2$ bearing a cerulean moiety at the N-terminus of the Gβ-subunit is illustrated. Upon stimulation of α_{2A}-adrenergic receptors with 10 μM noradrenaline, a fast increase in FRET was observed, which recovered upon withdrawal of the agonist much more slowly. The increase in FRET upon G protein activation is not compatible with complete separation of Gαi- and Gβγ-subunits, but rather might reflect a rearrangement of the subunits.

three striking features associated with this increase in RET: (1) the amplitude in RET was very small; (2) the kinetics were indistinguishable from receptor activation, and were much faster than G protein activation (30, 57; also compare Figs. 10.2 and 10.3), and (3) the FRET increase quickly equilibrated to maximal steady-state amplitude. Why is the amplitude of agonist-induced FRET between G proteins and receptors so small? The amplitude of agonist-induced FRET was increased severalfold with G protein mutants exhibiting lower affinity for GTP [30]. Accordingly, Hein et al. concluded that even in the presence of agonist, only a very minor fraction of wt-G proteins will interact at any given time with receptors. Resolving the kinetics of receptor–G protein interactions has demonstrated that agonist-induced G protein interaction occurs on a similar timescale as receptor activation. The apparent absence of a lag time between receptor activation and receptor–G protein interaction could be interpreted as evidence for a strict precoupling between receptors and G proteins in the absence of agonist. However, it needs to be considered that both fluorescent G proteins and receptors were exogenously expressed with typical robust expression levels, leading to potentially very short diffusion times. Indeed, lowering expression levels of G proteins led to a slowing of receptor–G protein interaction [30]. Based on theoretical considerations, the finding that agonist-induced receptor–G protein interaction reaches a long-lasting plateau within less than a second was surprising. Considering the generally accepted view is that active receptors will interact predominantly with inactive G proteins, and accepting that the equilibrium between inactive and active G proteins is strongly shifted toward active within seconds after agonist application, one would expect a corresponding transient FRET between receptors and G proteins after agonist stimulation. So far, this issue has not been resolved; however, the results obtained have led to the proposal of the existence of interactions between active receptors and active G proteins.

10.8.4. Receptor–β-arrestin Interaction Detected by RET

Many GPCRs undergo desensitization upon prolonged exposure to agonist. Initially, desensitization mechanisms were worked out for rhodopsin and β_2-adrenergic receptors [66, 67]. Subsequently, numerous other GPCRs were demonstrated to exhibit similar patterns and mechanisms of desensitization. However, some receptors utilize different mechanisms for desensitization [68]. The principal mechanism of desensitization can be described briefly: upon agonist binding, receptors are recognized and phosphorylated by kinases such as G protein-coupled receptor kinases (GRKs) and protein kinase A (PKA) on intracellular serine and threonine residues. Subsequently, β-arrestins recognize the phosphorylated receptors and bind to their intracellular surface. This, in turn, blocks interactions between receptors and G proteins due to steric hindrance. Kinetic analysis of arrestin recruitment by receptors has been made possible by means of fluorescence imaging of GFP-tagged arrestins [69]. However, membrane translocation of fluorescent arrestins represents only a semiquantitative approach for studying GPCR–arrestin interactions, particularly because arrestins may interact with other membrane proteins subsequent to receptor binding and therefore exhibit prolonged membrane localization. In other words, arrestins may stay at the membrane even after dissociation from the GPCR. A more direct method to study receptor–arrestin interactions is the use of RET. Both BRET and FRET have been successfully employed to study arrestin binding to receptors [70, 71]. These studies have demonstrated that for Class A receptors such as β_2-adrenergic receptors, phosphorylation of the receptors is not the only prerequisite for GPCR–arrestin interaction. Phosphorylated receptors need to be in an agonist-activated state in order to bind arrestins, and withdrawal of agonist leads to a fast dissociation of arrestins from receptors. These studies uncovered that phosphorylation of the receptors represents a rate-limiting step in receptor–arrestin interactions, and phosphorylated receptors actually exhibit an agonist-dependent interaction with arrestins. The onset of arrestin interaction with phosphorylated receptors is fast (within a few seconds), and is only slightly slower than the interaction of receptors with G proteins [30, 71, 72]. The results from RET-based studies highlight that GPCR–arrestin interactions represent a kind of coincidence detector, requiring both the presence of agonist and phosphorylation of the receptor.

10.9. DETECTION OF RECEPTOR FUNCTION BY FLUORESCENCE RESONANCE ENERGY

Direct detection of the receptor function has been established by means of intramolecular FRET. For receptor sensors, both a FRET donor and acceptor needed to be inserted into conformationally sensitive positions of the receptor. For optimal receptor sensors, a distance between the fluorophores close to the Förster radius (~5 nm) is required. Based on recent structural models on how transmembrane helices move during receptor activation, the third intracellular loop and the (truncated) C-terminus of the receptor were chosen as promising positions for fluorophore insertion. More precisely, Vilardaga et al. mapped suitable fluorophore insertion sites to regions distal of the fifth transmembrane helix and to the C-terminus 20–30 residues distal to the 7th transmembrane helix [73]. This approach was initially validated for the PTH receptor, as well as for the α_{2A}-adrenergic receptor (see Fig. 10.4). Later, it was transferred to several other GPCRs including β-adrenergic [32] and adenosine receptors [31]. Insertion of CFP and YFP into the receptor gave rise to receptors that exhibited functionality toward G protein activation, albeit at a much reduced coupling efficiency. The replacement of YFP by flash within the third intracellular loop allowed measurement of receptor activation in receptors fully functional in terms of G protein signaling [29]. Structural activation models for receptors predicted that the distance between the fifth helix and the C-terminal part of helix 7 will increase upon activation. This prediction was supported by two recent crystal structures of opsin, which most likely display different activation states [74, 75]. Therefore, a decrease in FRET upon activation of receptors was anticipated and indeed could be experimentally demonstrated (see Reference 73 and Fig. 10.4). However, distance changes within the compact receptor molecule were not expected to be very large, and consequently, agonist-induced changes in FRET were expected to be rather small. Thus, a key to the success of the experiments was a robust and sensitive measurement technique. FRET between CFP and YFP can be easily detected by many methods. However, subtle changes in FRET are much more difficult to measure reliably. A major problem is the inherent property of these fluorophores to bleach quite quickly. Even worse, in the case of strong FRET, the donor is somewhat protected from bleaching, leaving the acceptor exposed to photobleaching. In other words, FRET is particularly prone to photobleaching.

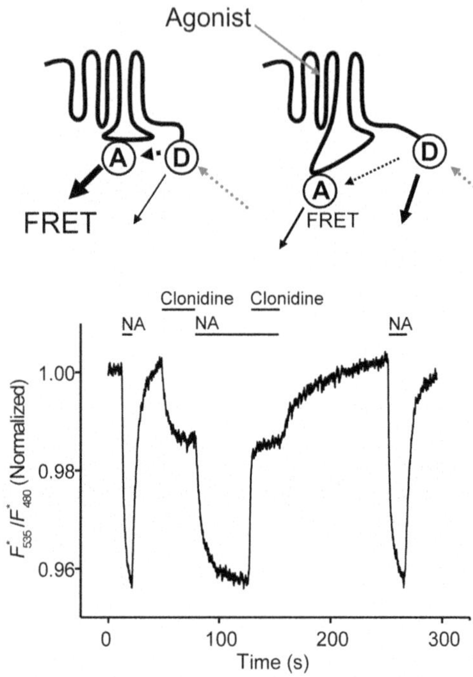

Figure 10.4 Detection of the conformational change associated with agonist-induced receptor activation by means of intramolecular FRET. Yellow and cyan fluorescent moieties were tagged to conformational sensitive domains at the C-terminus and the third intracellular loop of GPCRs. Upon ligand-induced activation of the receptor, conformational changes of the receptor were translated into changes in distance and/or orientation of the inserted fluorophores, resulting in an attenuation of intramolecular FRET. The representative single cell FRET recording of a cell expressing the α_{2A}-FRET receptor demonstrates partial FRET responses to saturating concentrations of clonidine (10 µM) compared to responses evoked by noradrenaline (100 µM).

Therefore, it requires caution to determine whether the agonist or the photobleaching effect decreased the apparent FRET signal. Assays that allow measurement of FRET with high temporal resolution and fast application devices for agonists are very useful for inducing and detecting FRET changes fast enough to distinguish them from photobleaching effects. Obviously, more sensitive light detectors will also help to minimize excitation intensities, and therefore reduce photobleaching.

By using a single cell microscopic assay allowing for rapid exchange of agonist in the context of a sensitive photometric detection unit, Vilardaga et al. were the first to directly measure receptor activation in single living cells [73]. The extent of FRET change due to receptor activation was in the range of a few percent of the initial FRET ratio. For both PTH receptor and α_{2A}-adrenergic receptor, the addition of agonist resulted in a decrease in FRET, reflecting relative movements at the intracellular attachment points of the fluorophores [78]. The kinetics of the decay of the FRET signal was dependent on agonist concentration. At agonist concentrations more than 200-fold higher than the dissociation constant, the speed of activation was maximal. These measurements demonstrated that receptor activation can occur within tens of milliseconds in the case of receptors activated by small ligands. Even though receptor activation can be much faster than previously reported for purified receptors, it is considerably slower than those of ligand-gated ion channels which can operate within 1–5 ms [76]. The GPCR rhodopsin is known to be activated within a timescale similar to that found for ligand-gated ion channels. Since rhodopsin is activated by light-induced isomerization of *cis*-retinal prebound in the ligand binding pocket, the difference in activation kinetics of rhodopsin and ligand-operated receptors could be related to the binding event itself. Given that the binding pocket for adrenaline and adenosine is known to be buried deep within the receptor, one might speculate that the actual rate-limiting step in receptor activation is the actual binding process of the ligand. For PTH receptors, FRET-based detection of the binding of fluorescent ligands to fluorescent receptors revealed a two-step binding process. Fast binding was observed to the N-terminal part of the receptor, whereas binding to the "body" of the receptor was slower and coincided with the actual conformational switch corresponding to receptor activation [22]. Whether the binding of the complex of PTH and N-terminal part of the receptor [77] to the binding site on the receptor is comparable to small ligand binding such as neurotransmitters is questionable.

10.9.1. Partial Agonism Detected on the Level of the Receptor

Many clinically relevant drugs actually act not as full agonists but rather as partial agonists on GPCRs. However, functional assays to determine the degree of agonism are very limited. In the past, only downstream cellular assays existed to study receptor activity, and such assays are dependent on the cellular context, as well as the expression level of the receptor. A significant problem with respect to GPCR expression levels is the existence of spare receptors (receptor reserve); if neglected, a spare receptor phenomenon may produce a seemingly full functional response for ligands that in fact are weak partial agonists. FRET offers the ability to study the degree of GPCR activity directly at the level of the receptor, and the approach may ultimately provide a means to more accurately determine the efficacy of ligands (see clonidine-induced FRET response in Fig. 10.4). In FRET assays, distance changes and/or rotation of fluorophores due to ligand-induced conformational changes of the receptor is measured. How strong is the correlation between agonist-induced FRET change and change in activity of the receptor? If temporally resolved FRET-based receptor activation assays will eventually be robust enough to be introduced to high-throughput screening, a tremendous gain of functional information as well as kinetic information is possible.

How do we know that the change in FRET directly correlates with receptor activation? The agonist-induced FRET change between fluorophores attached to certain sites on the receptor reports changes in distance between these fluorophores and/or their orientation toward each other. That agonists induce these changes in FRET, whereas antagonists do not indicate that activation of receptors is somehow related to the observed change in FRET. However, it is possible that events subsequent to receptor activation, such as interactions with G proteins or arrestins or other unknown intracellular interaction partners, lead to bending of the fluorophores which then result in alterations of the FRET signal. A straightforward approach to strip associated proteins off the membranes is treatment with a 6-M urea containing buffer. For α_{2A}-adrenergic receptor, the agonist-induced reduction in FRET was not altered in urea-stripped membranes [73]. Furthermore, the concentration dependence for agonist-induced FRET changes was sensitive to the presence of nucleotide-free G proteins, and an excess of nucleotide-free G proteins increased the sensitivity of the FRET response toward agonist [73], a phenomena also observed with ligand binding studies. Interestingly, a reasonable correlation between the degree of agonism of partial agonist and the maximal achieved FRET change was also observed [33].

10.9.2. Inverse Agonism Detected at the Level of the Receptor

FRET-based detection of receptor conformational changes which are directly related to activation proved to be a useful tool to determine the action of inverse agonists on GPCRs. Both inverse agonists for α_{2A}-adrenergic receptors, as well as those acting on β_1-adrenergic receptors have been studied using this approach [32, 78]. All inverse agonists tested induced a FRET change in the opposite direction compared to agonists. In the case of α_{2A}-adrenergic receptor, the introduction of a point mutation which substantially increases the constitutive activity of the receptor allowed a detailed analysis of receptor activation [78]. As expected, inverse agonists induce FRET changes in the opposite direction compared to agonists. However, two features were found when using FRET that were not expected. The first surprise was the extent of inverse agonist-induced changes were much larger than expected, based on the degree of inhibition of basal receptor activity [78]. This finding suggests that inverse agonists not only change the equilibrium toward inactive receptor conformations, but also induce larger conformational changes of the receptor. Unlike partial agonists, for which the degree of partial agonism correlated roughly with the detected FRET amplitude, the amplitude of FRET changes induced by inverse agonists did not match differences in receptor activity. For example, with β_1-adrenergic receptor, the clinically relevant inverse agonist carvedilol exerted a much more dramatic increase in FRET compared to the equally potent inverse agonist bisoprolol [32]. Notably, a frequent occurring polymorphism of β_1-adrenergic receptor (Arg389) exhibited a substantially larger FRET increase specifically for carvedilol but not bisoprolol [32]. Some indications exist that the population which carries this polymorphism also respond particularly well therapeutically to carvedilol. The FRET study may be a hint at how even clinical pharmacology can benefit from FRET-based studies of receptor activation. The second surprise was that the kinetics of inverse agonist action on receptors were much slower than the kinetics of agonist-induced receptor activation [73, 78]. So far, we do not understand the underlying molecular basis for this observation. This might change in the future, as more structural information on the receptors is becoming available, and computational approaches to model dynamics of conformational movements are forthcoming.

REFERENCES

1. Tsien, R.Y. (1998) The green fluorescent protein. *Annu Rev Biochem.* 67, 509–544.
2. Shimomura, O., Johnson, F.H., Saiga, Y. (1962) Extraction, purification and properties of aequorin, a bioluminescent protein from the luminous hydromedusan, *Aequorea. J Cell Comp Physiol.* 59, 223–239.
3. Griffin, B.A., Adams, S.R., Tsien, R.Y. (1998) Specific covalent labeling of recombinant protein molecules inside live cells. *Science.* 281, 269–272.
4. Keppler, A., Pick, H., Arrivoli, C., Vogel, H., Johnsson, K. (2004) Labeling of fusion proteins with synthetic fluorophores in live cells. *Proc Natl Acad Sci U S A.* 101, 9955–9959.
5. George, N., Pick, H., Vogel, H., Johnsson, N., Johnsson, K. (2004). Specific labeling of cell surface proteins with chemically diverse compounds. *J Am Chem Soc.* 126, 8896–8897.
6. Miyawaki, A., Llopis, J., Heim, R., McCaffery, J.M., Adams, J.A., Ikura, M., Tsien, R.Y. (1997) Fluorescent indicators for Ca2+ based on green fluorescent proteins and calmodulin. *Nature.* 388, 882–887.
7. Xu, Y., Piston, D.W., Johnson, C.H. (1999) A bioluminescence resonance energy transfer (BRET) system: Application to interacting circadian clock proteins. *Proc Natl Acad Sci U S A.* 96, 151–156.
8. Giepmans, B.N., Adams, S.R., Ellisman, M.H., Tsien, R.Y. (2006) The fluorescent toolbox for assessing protein location and function. *Science.* 312, 217–224.
9. Patterson, G.H., Lippincott-Schwartz, J. (2002) A photoactivatable GFP for selective photolabeling of proteins and cells. *Science.* 297, 1873–1877.
10. Shaner, N.C., Lin, M.Z., McKeown, M.R., Steinbach, P.A., Hazelwood, K.L., Davidson, M.W., Tsien, R.Y. (2008) Improving the photostability of bright monomeric orange and red fluorescent proteins. *Nat Methods.* 5, 545–551.
11. Ghosh, I., Hamilton, A.D., Regan, L. (2000) Antiparallel leucine zipper-directed protein reassembly: Application to the green fluorescent protein. *J Am Chem Soc.* 122, 5658–5659.
12. Neu-Hynes, T.R., Yost, E., Mervine, S., Berlot, C.H. (2008) Multicolor BiFC analysis of competition among G protein beta and gamma subunit interactions. *Methods.* 45, 207–213.

13. Gandia, J., Galino, J., Amaral, O.B., Soriano, A., Lluís, C., Franco, R., Ciruela, F. (2008) Detection of higher-order G protein-coupled receptor oligomers by a combined BRET-BiFC technique. *FEBS Lett.* 582, 2979–2984.
14. Scherrer, G., Tryoen-Tóth, P., Filliol, D., Matifas, A., Laustriat, D., Cao, Y.Q., Basbaum, A.I., Dierich, A., Vonesh, J.L., Gavériaux-Ruff, C., Kieffer, B.L. (2006) Knockin mice expressing fluorescent delta-opioid receptors uncover G protein-coupled receptor dynamics in vivo. *Proc Natl Acad Sci U S A.* 103, 9691–9696.
15. Kallal, L., Benovic, J.L. (2000) Using green fluorescent proteins to study G-protein-coupled receptor localization and trafficking. *Trends Pharmacol Sci.* 21, 175–180.
16. Gether, U., Lin, S., Kobilka, B.K. (1995) Fluorescent labeling of purified beta 2 adrenergic receptor. Evidence for ligand-specific conformational changes. *J Biol Chem.* 270, 28268–28275.
17. Kobilka, B.K., Deupi, X. (2007) Conformational complexity of G-protein-coupled receptors. *Trends Pharmacol Sci.* 28, 397–406.
18. Kobilka, B.K. (2007) G protein coupled receptor structure and activation. *Biochim Biophys Acta.* 1768, 794–807.
19. Harikumar, K.G., Pinon, D.I., Wessels, W.S., Dawson, E.S., Lybrand, T.P., Prendergast, F.G., Miller, L.J. (2004) Measurement of intermolecular distances for the natural agonist peptide docked at the cholecystokinin receptor expressed in situ using fluorescence resonance energy transfer. *Mol Pharmacol.* 65, 28–35.
20. Harikumar, K.G., Miller, L.J. (2005) Fluorescence resonance energy transfer analysis of the antagonist- and partial agonist-occupied states of the cholecystokinin receptor. *J Biol Chem.* 280, 18631–18635.
21. Harikumar, K.G., Lam, P.C., Dong, M., Sexton, P.M., Abagyan, R., Miller, L.J. (2007) Fluorescence resonance energy transfer analysis of secretin docking to its receptor: Mapping distances between residues distributed throughout the ligand pharmacophore and distinct receptor residues. *J Biol Chem.* 282, 32834–32843.
22. Castro, M., Nikolaev, V.O., Palm, D., Lohse, M.J., Vilardaga, J.P. (2005) Turn-on switch in parathyroid hormone receptor by a two-step PTH binding mechanism. *Proc Natl Acad Sci U S A.* 102, 16084–16089.
23. Shi, L., Javitch, J.A. (2002) The binding site of aminergic G protein-coupled receptors: The transmembrane segments and second extracellular loop. *Annu Rev Pharmacol Toxicol.* 42, 437–467.
24. Daly, C.J., McGrath, J.C. (2003) Fluorescent ligands, antibodies, and proteins for the study of receptors. *Pharmacol Ther.* 100, 101–118.
25. Middleton, R.J., Kellam, B. (2005) Fluorophore-tagged GPCR ligands. *Curr Opin Chem Biol.* 9, 517–525.
26. Adams, S.R., Campbell, R.E., Gross, L.A., Martin, B.R., Walkup, G.K., Yao, Y., Llopis, J., Tsien, R.Y. (2002) New biarsenical ligands and tetracysteine motifs for protein labeling in vitro and in vivo: Synthesis and biological applications. *J Am Chem Soc.* 124, 6063–6076.
27. Griffin, B.A., Adams, S.R., Tsien, R.Y. (1998) Specific covalent labeling of recombinant protein molecules inside live cells. *Science.* 281, 269–272.
28. Martin, B.R., Giepmans, B.N., Adams, S.R., Tsien, R.Y. (2005) Mammalian cell-based optimization of the biarsenical-binding tetracysteine motif for improved fluorescence and affinity. *Nat Biotechnol.* 23, 1308–1314.

29. Hoffmann, C., Gaietta, G., Bünemann, M., Adams, S.R., Oberdorff-Maass, S., Behr, B., Vilardaga, J.P., Tsien, R.Y., Ellisman, M.H., Lohse, M.J. (2005) A FlAsH-based FRET approach to determine G protein-coupled receptor activation in living cells. *Nat Methods.* 2, 171–176.

30. Hein, P., Frank, M., Hoffmann, C., Lohse, M.J., Bünemann, M. (2005) Dynamics of receptor/G protein coupling in living cells. *EMBO J.* 24, 4106–4114.

31. Hein, P., Rochais, F., Hoffmann, C., Dorsch, S., Nikolaev, V.O., Engelhardt, S., Berlot, C.H., Lohse, M.J., Bünemann, M. (2006) Gs activation is time-limiting in initiating receptor-mediated signaling. *J Biol Chem.* 281, 33345–33351.

32. Rochais, F., Vilardaga, J.P., Nikolaev, V.O., Bünemann, M., Lohse, M.J., Engelhardt, S. (2007) Real-time optical recording of beta1-adrenergic receptor activation reveals supersensitivity of the Arg389 variant to carvedilol. *J Clin Invest.* 117, 229–235.

33. Nikolaev, V.O., Hoffmann, C., Bünemann, M., Lohse, M.J., Vilardaga, J.P. (2006) Molecular basis of partial agonism at the neurotransmitter {alpha}2A-adrenergic receptor and Gi-protein heterotrimer. *J Biol Chem.* 281, 24506–24511.

34. Lohse, M.J., Nikolaev, V.O., Hein, P., Hoffmann, C., Vilardaga, J.P., Bünemann, M. (2008) Optical techniques to analyze real-time activation and signaling of G-protein-coupled receptors. *Trends Pharmacol Sci.* 29, 159–165.

35. Spagnuolo, C.C., Vermeij, R.J., Jares-Erijman, E.A. (2006) Improved photostable FRET-competent biarsenical-tetracysteine probes based on fluorinated fluoresceins. *J Am Chem Soc.* 128, 12040–12041.

36. Keppler, A., Arrivoli, C., Sironi, L., Ellenberg, J. (2006) Fluorophores for live cell imaging of AGT fusion proteins across the visible spectrum. *Biotechniques.* 41, 167–170.

37. Böhme, I., Morl, K., Bamming, D., Meyer, C., Beck-Sickinger, A.G. (2007) Tracking of human Y receptors in living cells—A fluorescence approach. *Peptides.* 28, 226–234.

38. Meyer, B.H., Martinez, K.L., Segura, J.M., Pascoal, P., Hovius, R., George, N., Johnsson, K., Vogel, H. (2006) Covalent labeling of cell-surface proteins for in-vivo FRET studies. *FEBS Lett.* 580, 1654–1658.

39. Meyer, B.H., Segura, J.M., Martinez, K.L., Hovius, R., George, N., Johnsson, K., Vogel, H. (2006) FRET imaging reveals that functional neurokinin-1 receptors are monomeric and reside in membrane microdomains of live cells. *Proc Natl Acad Sci U S A.* 103, 2138–2143.

40. Maurel, D., Comps-Agrar, L., Brock, C., Rives, M.L., Bourrier, E., Ayoub, M.A., Bazin, H., Tinel, N., Durroux, T., Prézeau, L., Trinquet, E., Pin, J.P. (2008) Cell-surface protein-protein interaction analysis with time-resolved FRET and snap-tag technologies: Application to GPCR oligomerization. *Nat Methods.* 5, 561–567.

41. Gautier, A., Juillerat, A., Heinis, C., Corrêa Jr., I.R., Kindermann, M., Beaufils, F., Johnsson, K. (2008). An engineered protein tag for multiprotein labeling in living cells. *Chem Biol.* 15, 128–136.

42. Wang, B., Bisello, A., Yang, Y., Romero, G.G., Friedman, P.A. (2007) NHERF1 regulates parathyroid hormone receptor membrane retention without affecting recycling. *J Biol Chem.* 282, 36214–36222.

43. Bockaert, J., Fagni, L., Dumuis, A., Marin, P. (2004) GPCR interacting proteins (GIP) *Pharmacol Ther.* 103, 201–221.

44. Drake, M.T., Shenoy, S.K., Lefkowitz, R.J. (2007) Trafficking of G protein-coupled receptors. *Circ Res.* 99, 570–582.
45. Willig, K.I., Rizzoli, S.O., Westphal, V., Jahn, R., Hell, S.W. (2006) STED microscopy reveals that synaptotagmin remains clustered after synaptic vesicle exocytosis. *Nature.* 440, 935–939.
46. Puthenveedu, M.A., Yudowski, G.A., von Zastrow, M. (2007) Endocytosis of neurotransmitter receptors: Location matters. *Cell.* 130, 988–989.
47. Eidne, K.A., Kroeger, K.M., Hanyaloglu, A.C. (2002) Applications of novel resonance energy transfer techniques to study dynamic hormone receptor interactions in living cells. *Trends Endocrinol Metabol.* 13, 415–421.
48. Ramsay, D., Kellett, E., McVey, M., Rees, S., Milligan, G. (2002) Homo and hetero-oligomeric interactions between G protein-coupled receptors in living cells monitored by two variants of bioluminescence resonance energy transfer (BRET): Hetero-oligomers between receptor subtypes form more efficiently than between less closely related sequences. *Biochem J.* 365, 429–440.
49. Milligan, G. (2004) Applications of bioluminescence- and fluorescence resonance energy transfer to drug discovery at G protein-coupled receptors. *Eur J Pharm Sci.* 21, 397–405.
50. Kremers, G.J., Goedhart, J., van Munster, E.B., Gadella Jr., T.W. (2006) Cyan and yellow super fluorescent proteins with improved brightness, protein folding, and FRET Förster radius. *Biochem.* 45, 6570–6580.
51. Bazin, H., Trinquet, E., Mathis, G. (2002) Time resolved amplification of cryptate emission: A versatile technology to trace biomolecular interactions. *J Biotechnol.* 82, 233–250.
52. van Rijn, R.M., Chazot, P.L., Shenton, F.C., Sansuk, K., Bakker, R.A., Leurs, R. (2006) Oligomerization of recombinant and endogenously expressed human histamine H(4) receptors. *Mol Pharmacol.* 70, 604–615.
53. Roy, R., Hohng, S., Ha, T. (2008) A practical guide to single-molecule FRET. *Nat Methods.* 5, 507–516.
54. van Munster, E.B., Gadella, T.W. (2005) Fluorescence lifetime imaging microscopy (FLIM). *Adv Biochem Eng Biotechnol.* 95, 143–175.
55. Janetopoulos, C., Jin, T., Devreotes, P. (2001) Receptor-mediated activation of heterotrimeric G-proteins in living cells. *Science.* 291, 2408–2411.
56. Gilman, A.G. (1987) G proteins: Transducers of receptor-generated signals. *Annu Rev Biochem.* 56, 615–649.
57. Bünemann, M., Frank, M., Lohse, M.J. (2003) Gi protein activation in intact cells involves subunit rearrangement rather than dissociation. *Proc Natl Acad Sci U S A.* 100, 16077–16082.
58. Levitzki, A., Klein, S., Reuveni, H. (2002) G-protein subunit dissociation is not an integral part of G-protein action signal transduction by a nondissociable heterotrimeric yeast G protein. *Chembiochem.* 3, 815–818.
59. Robishaw, J.D., Berlot, C.H. (2004) Translating G protein subunit diversity into functional specificity. *Curr Opin Cell Biol.* 16, 206–209.
60. Galés, C., Van Durm, J.J., Schaak, S., Pontier, S., Percherancier, Y., Audet, M., Paris, H., Bouvier, M. (2006) Probing the activation-promoted structural rearrangements in preassembled receptor-G protein complexes. *Nat Struct Mol Biol.* 13, 778–786.

61. Digby, G.J., Lober, R.M., Sethi, P.R., Lambert, N.A. (2006) Some G protein heterotrimers physically dissociate in living cells. *Proc Natl Acad Sci U S A.* 103, 17789–17794.
62. Frank, M., Thümer, L., Lohse, M.J., Bünemann, M. (2005) G protein activation without subunit dissociation depends on a G{alpha}(i)-specific region. *J Biol Chem.* 280, 24584–24590.
63. Nobles, M., Benians, A., Tinker, A. (2005) Heterotrimeric G proteins precouple with G protein-coupled receptors in living cells. *Proc Natl Acad Sci U S A.* 102, 18706–18711.
64. Galés, C., Rebois, R.V., Hogue, M., Trieu, P., Breit, A., Hébert, T.E., Bouvier, M. (2005) Real-time monitoring of receptor and G-protein interactions in living cells. *Nat Methods.* 2, 177–184.
65. Qin, K., Sethi, P.R., Lambert, N.A. (2008) Abundance and stability of complexes containing inactive G protein-coupled receptors and G proteins. *FASEB J.* 22, 2920–2927.
66. Palczewski, K., Saari, J.C. (1997) Activation and inactivation steps in the visual transduction pathway. *Curr Opin Neurobiol.* 7, 500–504.
67. Luttrell, L.M., Lefkowitz, R.J. (2002) The role of beta-arrestins in the termination and transduction of G-protein-coupled receptor signals. *J Cell Sci.* 115, 455–465.
68. Bünemann, M., Hosey, M.M. (1999) G-protein coupled receptor kinases as modulators of G-protein signalling. *J Physiol.* 517, 5–23.
69. Barak, L.S., Ferguson, S.S., Zhang, J., Caron, M.G. (1997) A beta-arrestin/green fluorescent protein biosensor for detecting G protein-coupled receptor activation. *J Biol Chem.* 272, 27497–27500.
70. Pfleger, K.D., Dalrymple, M.B., Dromey, J.R., Eidne, K.A. (2007) Monitoring interactions between G-protein-coupled receptors and beta-arrestins. *Biochem Soc Trans.* 35, 764–766.
71. Krasel, C., Bünemann, M., Lorenz, K., Lohse, M.J. (2005) β-Arrestin binding to the β_2-adrenergic receptor requires both receptor phosphorylation and receptor activation. *J Biol Chem.* 280, 9528–9535.
72. Violin, J.D., Ren, X.R., Lefkowitz, R.J. (2006) G-protein-coupled receptor kinase specificity for beta-arrestin recruitment to the beta2-adrenergic receptor revealed by fluorescence resonance energy transfer. *J Biol Chem.* 281, 20577–20588.
73. Vilardaga, J.P., Bünemann, M., Krasel, C., Castro, M., Lohse, M.J. (2003) Measurement of millisecond activation switch of G protein-coupled receptors in living cells. *Nat Biotech.* 7, 807–812.
74. Scheerer, P., Park, J.H., Hildebrand, P.W., Kim, Y.J., Krauss, N., Choe, H.W., Hofmann, K.P., Ernst, O.P. (2008) Crystal structure of opsin in its G-protein-interacting conformation. *Nature.* 455, 497–502.
75. Park, J.H., Scheerer, P., Hofmann, K.P., Choe, H.W., Ernst, O.P. (2008) Crystal structure of the ligand-free G-protein-coupled receptor opsin. *Nature.* 454, 183–187.
76. Hille, B. (2001) *Ionic Channels of Excitable Membranes*, 3rd ed. Sunderland, MA: Sinauer.
77. Pioszak, A.A., Xu, H.E. (2008) Molecular recognition of parathyroid hormone by its G protein-coupled receptor. *Proc Natl Acad Sci U S A.* 105, 5034–5039.
78. Vilardaga, J.P., Steinmeyer, R., Harms, G.S., Lohse, M.J. (2005) Molecular basis of inverse agonism in a G protein-coupled receptor. *Nat Chem Biol.* 1, 25–28.

CHAPTER 11

Integration of Label-Free Detection Methods in GPCR Drug Discovery

OLIVER NAYLER, MAGDALENA BIRKER-ROBACZEWSKA, and JOHN GATFIELD

Actelion Pharmaceuticals Ltd., Allschwil, Switzerland

11.1. OVERVIEW

G protein-coupled receptors (GPCRs) represent one of the most important classes of drug targets in the pharmaceutical industry. Due to a limited number of well-characterized downstream signaling pathways, GPCR signaling can be monitored with a relatively small standard set of assay formats, most of which have been adapted for high-throughput screening. These formats include, among others, real-time Ca^{2+} release assays for Gαq coupling, cAMP assays for Gαs and Gαi coupling, and GTPγS assays, mainly for Gαi coupling. New assay technologies are continuously developed, thereby creating a variety of possibilities to choose the appropriate assay platform for every specific need or budget. Nevertheless, most of them are variations of well-known signaling pathways, and the majority of these assays are based on the use of dyes, antibodies, or reporter gene constructs. These conventional assay technologies often rely on the use of highly expressed recombinant receptors or signal transducers in a cellular background that is compatible with the assay but, in many instances, nonphysiological. In addition, the majority of assays are end point assays, thereby preventing a continued monitoring of cell signaling. It is therefore a challenge to develop label-free, real-time detection assays to characterize GPCR signaling and to circumvent some of the drawbacks of traditional assay technologies. Recent developments, such as the introduction of automated microscopy, impedance, or resonant waveguide grating tech-

GPCR Molecular Pharmacology and Drug Targeting: Shifting Paradigms and New Directions, Edited by Annette Gilchrist

nologies, allow high-throughput-compatible, label-free, real-time GPCR assays with the required accuracy and robustness to make them amenable to drug discovery. In this review, these novel assay principles are discussed in the context of GPCR receptor pharmacology, especially information content and comparability with conventional assay methods.

11.2. INTRODUCTION

The members of the G protein-coupled receptor (GPCR) superfamily are important contributors to many processes of development and disease [1, 2]. They consist of at least 300 nonsensory human family members that are known or predicted to be activated by endogenous ligands. GPCRs are an important drug target class for the pharmaceutical industry, and more than 20% of currently top selling drugs are GPCR related [3, 4]. However, only a small proportion of the GPCR family has been fully exploited, and a significant number of orphan GPCRs are considered potential drug targets [5].

With the implementation of high-throughput screening (HTS) and the exponential growth of compound libraries over the past two decades, it has become increasingly important to consider assay time, robustness, and costs in the design of screening campaigns for any new GPCR target. This has led to the development of a relatively small set of HTS-compatible GPCR assays that are currently used in drug discovery. These formats include, among others, real-time Ca^{2+} release assays for $G\alpha q$ coupling, cAMP assays for $G\alpha s$ and $G\alpha i$ coupling, $GTP\gamma S$ assays, mainly for $G\alpha i$ coupling, or reporter gene assays that can be easily adapted for almost any known G protein coupling pathway [6]. This apparent simplicity of GPCR biology has led to wide implementation of drug discovery platforms that are specialized in the development of GPCR assays adapted for HTS.

The process of GPCR drug discovery has thus became fully industrialized, allowing the screening of dozens of GPCR targets per year. However, as a typical HTS campaign often leads to hit rates in excess of 1% of the total library, which may contain up to 2 million compounds, secondary assays have become increasingly important to validate hits emerging from primary screens. Nevertheless, the industrialization process of the GPCR drug discovery workflow has not fully translated into the expected flood of novel drugs, and there is much debate on the reasons of this apparent "failure" and how it can be remedied [7–9].

Besides nonmechanism-based toxicological or tolerability problems of compounds in preclinical or clinical development, there may be an additional issue with respect to the target GPCR biology and the pharmacological behavior of synthetic GPCR modulators. It is evident from the current literature that clinically successful GPCR-targeting compounds, exemplified by dopamine receptor agonists and β-adrenergic agonists, do not necessarily have a straightforward pharmacological profile when they are carefully characterized

TABLE 11.1 Summary of Label-Free Technologies and Equipment

Equipment	Provider	Assay Principle	Provider Home Page
Label-Free Systems with Potential Applications in GPCR Cell Signaling Pathway Analysis			
Cytosensor microphysiometer	Molecular Devices Inc.	Extracellular pH	http://www.moleculardevices.com
XF analyzer	Seahorse Bioscience	O_2 consumption/ extracellular pH	http://www.seahorsebio.com
EPIC System	Corning	RWG	http://www.corning.com
BIND Reader	SRU Biosystems	RWG	http://www.srubiosystems.com
Owls	Microvacuum	Optical grating-coupled waveguide	http://www.owls-sensors.com
xCELLigence system	Roche Applied Science	Impedance	http://www.roche-applied-science.com
CellKey	MDS Sciex	Impedance	http://www.moleculardevices.com
ECIS	Applied Biophysics	Impedance	http://www.biophysics.com
Bionas 2500 analyzer	Bionas	Extracellular pH, O_2, and impedance	http://www.bionas.de
Other Label-Free Systems			
IncuCyte	Essen	Automated microscopy	http://www.essen-instruments.com
Cell-IQ	Chip Man Technologies	Automated microscopy	http://www.chipmantech.com
Q-Sense E1 or E4	Q-Sense	Microbalance	http://www.q-sense.com

in a panel of different GPCR signaling assays [10–13]. Matters become even more complex, when considering that the pharmacology of GPCR targeting compounds may significantly depend on the cellular background or disease condition [14]. Based on these recent insights, which are translated into the setting of modern GPCR drug discovery, it is currently considered essential to investigate novel synthetic modulators in multiple signaling pathways for any given GPCR and to fully characterize the pharmacological behavior of these compounds, especially in primary cells [15]. It is thus assumed that a detailed pharmacological understanding of the compound *in vitro* will eventually help to design innovative compounds with potentially fewer side effects *in vivo*.

Today, it is common practice to consider the full spectrum of potential synthetic GPCR modulators that could emerge from a HTS campaign, including full or partial agonists or antagonists, pathway-selective (biased) agonists, inverse agonists, allosteric modulators, or compounds with the potential to specifically modulate GPCR dimers [16–18]. Consequently, the assay development and compound profiling strategy must be adequately adapted to identify suitable lead molecules and to allow a careful characterization of the target GPCR pharmacology in physiological or disease-relevant cells.

In this context, it is questionable whether the traditional screening strategy, which includes the widespread use of recombinant systems, often combined with forced coupling using chimeric Gα proteins or promiscuous Gα_{16} proteins, is sufficient to identify and characterize molecules with complex GPCR pharmacology. Furthermore, by using highly artificial screening assays, there is an inherent risk to find molecules that display nonphysiological pharmacology in disease-relevant target systems. It is for these reasons that the pharmaceutical industry has developed an increasing interest in label-free systems, avoiding the use of dyes and other molecular interventions to measure compound activity and to incorporate real-time kinetics. Label-free systems thus simplify assay development and offer novel opportunities to measure GPCR pharmacology in nonrecombinant or primary cells even in high-throughput screens. This review looks at the evolution of label-free detection methods over the past years and describes current technologies (Table 11.1) and their use in the context of GPCR pharmacology, especially in drug discovery.

11.3. LABEL-FREE TECHNOLOGIES—PAST AND PRESENT

GPCR stimulation by agonists activates a variety of intracellular pathways via associated heterotrimeric G proteins and β-arrestin signaling. These molecular events are mediated through specific interactions with downstream effector molecules, which ultimately lead to changes in cell physiology, such as cell adhesion, cell migration, cell proliferation, or even neurite outgrowth. A variety of technologies are used to measure either direct cell behavior or surrogate end points in response to GPCR stimulation in a quantitative and, occasionally, semiquantitative way. For example, cell migration can be measured in modified Boyden chambers in low-throughput to medium-throughput formats, or cell proliferation can be quantified using end point assays and surrogate markers such as DNA synthesis (^{3}H-thymidine, BrdU, DNA dye staining) or mitochondrial activity (MTT). Unfortunately, most of these traditional assay methods are laborious and frequently involve the use of radioactivity or dyes. It is therefore a challenge to develop label-free assay systems that allow precise monitoring of physiological changes of cells in response to ligand stimulation.

During the past years, several assay platforms and instruments were introduced to allow label-free, quantitative, and even real-time measurements of complex cell responses. The majority of these novel technologies find general and widespread applications in multiple areas of cell biology, but some of them, including microphysiometry, impedance, and resonant waveguide grating (RWG), are of special interest to the GPCR research field as they potentially offer novel information on GPCR pathway signatures and their consequences in cell physiology. For this particular reason, we elaborate here in particular on novel GPCR-relevant assay technologies.

11.3.1. Automated Microscopes and Microbalances

Automated microscopes, such as the very recently launched Cell-IQ® (Chipman Technologies, Tampere, Finland) or IncuCyte™ instruments (Essen Instruments, Ann Arbor, MI), offer unattended, label-free, and real-time kinetic analysis of cell proliferation, cell migration, or neurite outgrowth in a quantitative way. Both instruments are the result of continued development of classical phase contrast microscopy combined with sophisticated software for image analysis. Both instruments offer a technical solution to automatically find or display the required focal plane. Both are also compatible with classical 96-well and various other assay plate formats, and they are thus able to acquire data in low to mid-throughput. Besides their obvious advantage of measuring microscopic changes in cell behavior, automated microscopes find increasing use in cell quality control to improve the reliability of downstream cellular assays especially when using primary cells.

The IncuCyte instrument collects phase contrast images of live cells in microplates or other cell culture vessels, and is placed in a conventional cell culture incubator. Image acquisition is accomplished by taking autofocused images at user-defined times and locations within the sample drawer. Once the digital images are collected, custom image processing software calculates image metrics, such as monolayer confluence, providing a quantitative and consistent measure of cellular proliferation. A novelty of the IncuCyte instrument is the possibility to perform so-called scratch wound assays of cell monolayers using a special plate design and a Woundmaker™ (Essen Instruments, Ann Arbor, MI), which is essentially a mechanical device that produces well-defined scratch wounds using sterile tips. Scratch wound assays are commonly used to assess the involvement of GPCRs in cellular proliferation and/or migration associated with wound closing [19]. This instrument offers a very convenient way to quantify these effects in a label-free format.

In contrast to the IncuCyte, which is housed in a conventional cell culture chamber, the Cell-IQ device is a fully integrated, self-contained solution offering continuous live cell imaging and analysis. The instrument contains a cell growth environmental chamber, where microplates and other cell culture flasks can be maintained under optimal growth conditions, and features a built-in microscope, light source, and camera. The Cell-IQ system scans through the cells (z-stack), taking images at regular intervals, for example, 1 μM distance. These images are then combined to produce an all-in-focus single resultant image.

Another recently introduced label-free technology (Q-Sense E series) is based on the principle of thin quartz discs that are placed between two gold layers (Q-Sense, Västra Frölunda, Sweden). The device is primarily designed to measure cell adhesion or cell proliferation. Application of an alternating current (AC) leads to an excitation and oscillation of the quartz crystal, which can be influenced by associating cells. Measuring the resonance, decaying, and dissipative frequencies can accurately quantify this influence. So far, only limited information is available to demonstrate wide applicability in cell biological experiments, and there are currently no published records demonstrating GPCR-related applications.

Despite the obvious potential of quartz sensors or automated microscopes as a means to measure GPCR activation or inhibition, and the consequences at the microscopic level, they carry the disadvantage that changes in proliferation or cell migration are not universal readouts in GPCR biology, and furthermore, in absence of pharmacological intervention (e.g., pertussis toxin), such readouts have so far not been indicative of the activation of specific G proteins, effectors, or pathways.

11.3.2. Microphysiometry

One of the first label-free, real-time detection principles offering information on pathway-specific activation of GPCRs was introduced with the Cytosensor™ microphysiometer (Molecular Devices, Inc., Sunnyvale, CA). The Cytosensor uses silicon semiconductors to measure extracellular acidification as a consequence of increased cell metabolism and secretion of acid metabolites due to GPCR activation. Importantly, the microphysiometer was successfully used to monitor activation of many different GPCR protein family members, and the assay principle appears to be independent of a particular signaling pathway. Remarkably, in some cases, the extracellular acidification kinetics provided some information on the particular G protein pathway involved. Thus, for a given cellular background, the GPCR-induced responses differed in terms of onset and persistence of the extracellular acidification depending on the different signaling pathways. Gαq coupling induced a rapid, transient, and sharp acidification response in Chinese hamster ovary (CHO) cells, exemplified in a study that used the neurokinin receptor subtype 3 (NK_3)-selective agonist senktide in cells expressing recombinant NK_3 receptors. In this study, the senktide-induced effects were inhibited by the protein kinase C (PKC) inhibitor staurosporine and by depletion of Ca^{2+} stores with thapsigargin. However, the agonist-induced effects were insensitive to both cholera toxin and pertussis toxin treatment, suggesting a phospholipase C (PLC)-mediated Ca^{2+} response pathway as a consequence of Gαq protein coupling [20]. In contrast, CHO cells expressing the recombinant human corticotropin-releasing factor receptor subtype 2 (CRF_2) receptor, which is a prototypical Gαs protein-coupled GPCR, responded to CRF stimulation with an increase in acidification rate that was characterized by a prolonged monophasic response that was broader than a typical Gαq trace, reaching a maximum at 10–12 min after ligand stimulation and returning slowly to baseline [21]. Gαi coupling induced an extracellular acidification rate change in CHO cells that was reminiscent of a Gαs coupling reaching a maximum between 1 and 6 min after ligand stimulation. However, the time to reach the response peak varied between receptor types and cell clones, and therefore, pertussis toxin treatment was used to distinguish Gαs-mediated from Gαi-mediated coupling [22].

With the Cytosensor, a significant number of GPCR ligands were functionally characterized, including structure–function relationships, dissection of signaling pathways using pathway-specific inhibitors and receptor desensitization, and successful identification of partial agonists and even inverse agonists [23]. Although the potencies of GPCR agonists did not always correlate between other GPCR assay methods and microphysiometer assays, the rank order of compound potencies was usually preserved. It is also important to remember that the GPCR-induced extracellular acidification rate may be modulated or influenced by the expression levels of effector proteins in a particular cellular background, and, if synthetic molecules are investigated, there might be an influence on the cell metabolism due to physicochemical properties of the compound and through unknown molecular interactions. Furthermore, the G protein-specific signatures can be influenced by the desensitization kinetics of the GPCR upon ligand stimulation, and this must be considered in structure–activity studies using either peptidic or synthetic small molecules that display partial agonism. In these cases, the signature may not be very indicative of the coupling mechanism or potency, but it may instead indicate the different propensities of the molecules to desensitize a target GPCR upon stimulation [24].

Despite autosampling capabilities, the Cytosensor is a very low-throughput machine (either 4 or 8 wells), and it features a complex mechanical design involving peristaltic pumps and plungers. Furthermore, the cell growth medium has to be exchanged for a low-buffering-capacity medium to allow accurate measurements, which might explain the scarce information on the use of non-recombinant or primary cells with the Cytosensor. Together, these issues may have contributed to the discontinuation of this product by the manufacturer. The recently introduced XF24 analyzer (Seahorse Biosciences, North Billerica, MA) now offers improved throughput (24-well and 96-well) and combines acidification and oxygen consumption measurements in a single assay well. Oxygen levels and pH are measured by optical biosensors that are placed approximately 300 µm above the cells. Measurements are made by monitoring changes in the surrounding media in as little as 1.5 min and can be extended over a period of days. As oxygen consumption is a direct consequence of mitochondrial activity, whereas extracellular acidification is predominantly a consequence of increased glycolysis, there is a potential advantage of using both parameters in one assay, as this may lead to novel and interesting insights of GPCR activity on cell metabolism, and this could help to unravel new mechanisms of GPCR signaling.

Until now, only one published study demonstrates the applicability of the XF24 analyzer to GPCR pharmacology in primary cells [25]. Feline cardiomyocytes, which naturally express muscarinic receptors, were stimulated with 1 µM of carbachol to induce a time-dependent rise in acidification. It remains to be seen if the XF24 analyzer will revive the microphysiometry assay principle in the GPCR field, although this methodology will clearly help to identify potentially cytotoxic molecules and may reveal metabolic side effects of an exploratory compound. Furthermore, as a significant number of targets, including GPCRs, are currently explored to discover novel drugs in metabolic disease, this technology may help to optimize certain compounds in disease-relevant cells.

11.3.3. Impedance/RWG

Two very recently introduced technologies, based on impedance and RWG, have received considerable attention from cell biologists and, in particular, from researchers interested in GPCR signaling. Both platforms carry translational potential and open the stage for new insights into GPCR-induced signaling pathways using a combination of label-free detection and real-time kinetics with high accuracy and throughput. In addition, these technologies offer enhanced information content over other assay technologies. Both platforms will be described here in more detail with respect to the underlying biological processes, assay principles, and their potential in GPCR research.

Eukaryotic cells rely on subcellular structures, such as the nucleus, mitochondria, endoplasmic reticulum (ER), and transport vesicles. They continuously adapt to changing environmental conditions or extracellular ligand stimulation through coordinated molecular and mechanical responses. These require a cytoskeleton-guided transport of molecules through intracellular regions with differing viscosity or elasticity. This leads to minute changes in cell morphology or induces adaptations of cell–cell and cell–substrate interactions. In the case of GPCR activation, dynamic molecular interactions at the plasma membrane initiate intense trafficking of subcellular structures, which includes the internalization of receptors in vesicles as part of the desensitization process. Microrheology experiments, for example, demonstrated that lysophosphatidic acid receptor (LPA) activation leads to transient changes in cytoplasmic stiffness and viscosity in serum-starved Swiss 3T3 cells [26, 27].

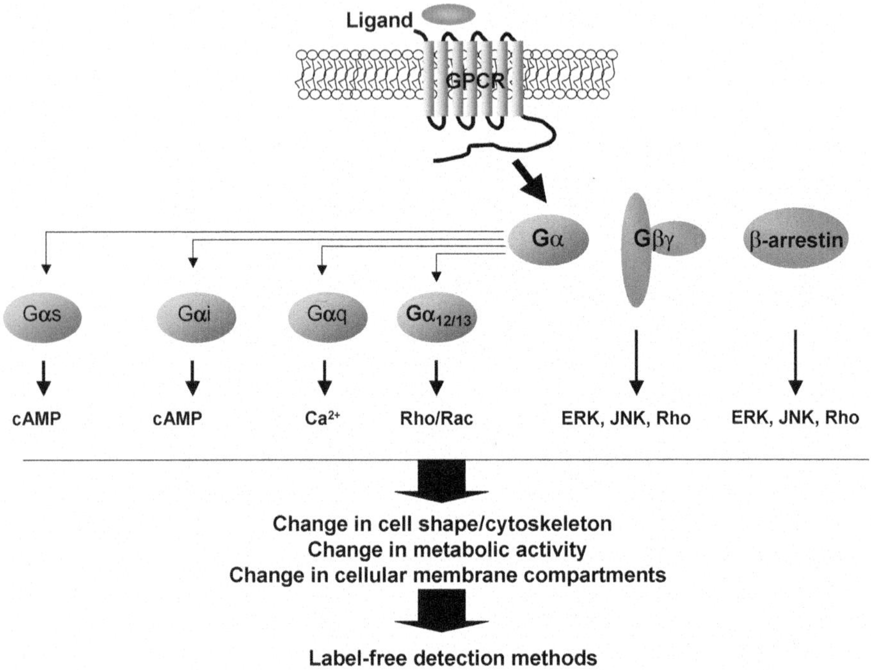

Figure 11.1 GPCR-induced signaling pathways measured with label-free detection technologies. Stimulation of GPCRs leads to activation of one or several downstream effectors, which induce alterations in subcellular compartmentalization and cytoskeletal architecture. Label-free detection methods measure the integrated cellular response of a ligand-stimulated/agonist-stimulated GPCR.

Cytoplasmic subcellular structures are highly organized by a dynamic cytoskeletal scaffold that is controlled by effector molecules, which transmit extracellular signals received at the plasma membranes to induce changes in cellular structure (Fig. 11.1). It has been reported that heterologously expressed endothelin A receptor (ETA) or endogenously expressed LPA receptor stimulation induces stress fiber formation in Swiss 3T3 fibroblasts [28], and it was shown that sphingosine-1-phosphate (S1P) stimulation of $S1P_2$ receptor-expressing 3T3 fibroblasts leads to inhibition of insulin-like growth factor-1-mediated chemotaxis and membrane ruffling, whereas stress fiber formation is stimulated [29]. The angiotensin II (Ang II)-induced morphological effects were investigated in recombinant Ang II receptor subtype 1 (AT_1R) expressing HEK293 cells. Intense remodeling of the actin cytoskeleton and formation of membrane protrusions and ruffles were observed 10–15 min after Ang II treatment [30]. These examples clearly demonstrate that GPCR stimulation or inhibition alters cell morphology and behavior of cells on a microscopic and also submicroscopic level.

On a molecular level, it is known that GPCR activation translates into changes in the actin cytoskeleton involving different Gα protein-induced pathways and downstream GTPases, such as Rho, Rac, or CDC42 [31–33]. Furthermore, GPCR-induced changes in second messenger levels (cAMP and Ca^{2+}) modulate the cytoskeleton via protein kinase A (PKA) activation or by changing calmodulin activity. It was also shown that stimulation of some GPCRs leads to activation of focal adhesion kinase (FAK) and other well-known mediators of cytoskeletal rearrangement [34, 35]. More recently, it was demonstrated that dissociating Gβγ subunits can also influence cytoskeletal architecture via direct molecular interactions involving pleckstrin homology (PH) domains. The influence of GPCR-induced cytoskeletal modulation in pathophysiological conditions such as cardiac hypertrophy, hypertension, asthma, inflammatory disease, cancer, and neurological disease is well documented and highlights the physiological significance of such measurements in the context of GPCR drug discovery [36].

Although these effects are well described in the literature, it is technically very difficult to obtain a highly accurate quantitative measurement using morphological changes and conventional microscopy. Both the impedance and RWG technologies make use of subtle ligand-induced cellular rearrangements and now provide a technical solution to measure changes in cytoskeletal architecture in a highly accurate and reproducible way. As the input of multiple signals or the activation of parallel signaling pathways induced by a single receptor converges at the cytoskeletal level to induce changes in morphology, an accurate readout provides information on an integrated response and thereby offers the potential to investigate the pharmacological effects of ligands or synthetic GPCR modulators in a physiological context (Fig. 11.1). The following describes findings in the GPCR field using either impedance or RWG, with particular emphasis on the correlation of data with results using established GPCR assay tools, including IC_{50} or EC_{50} values, and on partial or biased agonism and inverse agonism. We also examine the information content that is obtained by both technologies in comparison to other assay platforms.

Electric cell-substrate impedance sensing technology (ECIS™, Applied BioPhysics Inc., New York, NY) was originally invented by Giaever and Keese, who presented their first label-free biosensor to monitor cell morphology in 1984 [37, 38]. These morphological changes include cell locomotion and other behaviors directed by the cell's cytoskeleton. In 1991, Giaever and Keese founded Applied BioPhysics Inc., a company that still commercializes and markets ECIS instruments. The ECIS system makes use of small circular gold electrodes (250 μm diameter) to which the cells adhere and a larger counter electrode. Both electrodes are deposited on the cell culture vessel bottom by a lithographic process. When a small AC (1 μA) is applied, the media electrolytes create an ionic environment that is influenced by the cells adhering to the smaller electrode. The ECIS instruments measure the impedance of the cell-covered electrode, which is determined by the capacitance of the cell membrane, the resistance between adjacent cells caused by junction

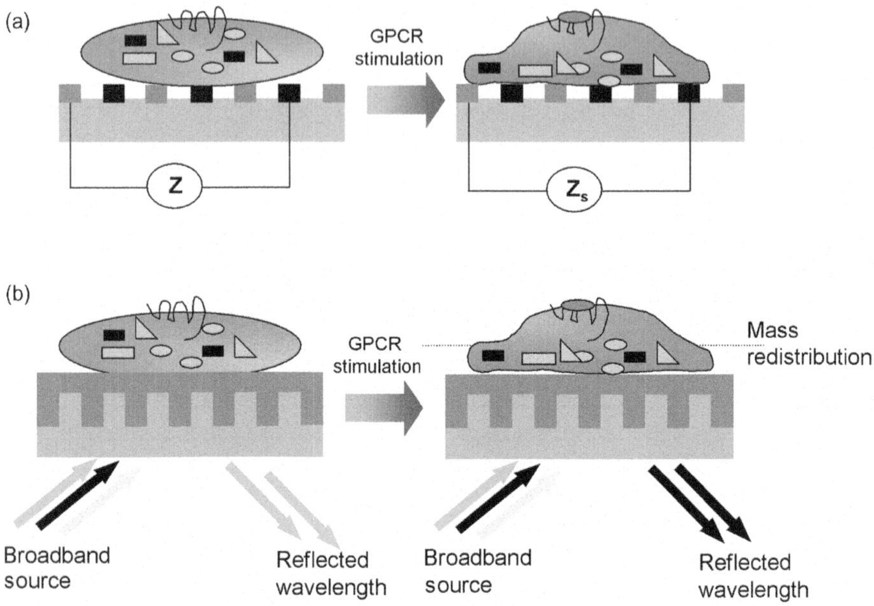

Figure 11.2 Principles of impedance and resonance waveguide grating (RWG). (a) Impedance is determined by the ratio of voltage to current as defined by Ohm's law ($Z = V/I$). Cells are seeded onto a cell culture vessel that contains electrodes at the bottom of the plate, and low voltage at various frequencies is applied. This causes a current to flow either through or around cells. Impedance assay technology is based on the complex impedance (Z) changes ($Z_s - Z$) resulting from stimulation-induced changes in cell shape, cell–cell contact, and cell–substrate interactions, which influence the extracellular and transcellular currents. (b) RWG is the underlying principle that is used to measure changes of dynamic mass redistribution (DMR). Cells are seeded onto a cell culture vessel that contains an optical biosensor (waveguide) at the bottom of the plate. Stimulation-induced changes in subcellular structure localization, referred to as DMR, influence the refractive index of the measured reflected resonant wavelength in an area up to 150 nm distance from the biosensor.

formation, and the distance between the cell surface and the electrode. Ligand-induced alterations in cell morphology, cell–cell contacts, or cell–substrate interaction affect the flow of extracellular and transcellular current, and this is quantitatively measured as changes in impedance (Fig. 11.2a).

Increased sophistication can be added by using ECIS instruments that are capable of monitoring both the voltage and the phase of the voltage relative to the current. Combining these parameters, the impedance can be broken down into two parts—one due to pure resistance and the other to the reactance of the system resistance and capacitance. These additional features are most useful for cell layer integrity assays. ECIS instruments can be used with a cell wounding module, which allows the creation of a well-defined, circular cell layer wound by a strong electrical pulse and the monitoring of healing through changes in impedance as cells migrate over the gold electrode to occupy the empty space [39]. More recently, other companies, such as ACEA Biosciences/Roche, Molecular Devices, and Bionas, have introduced their own impedance-based technology products. The various instruments work on the same principle and differ mainly in the geometry and surface size of the electrodes, the magnitude and frequency of the applied AC signal, and additional features such as on-board pipetting capability. Table 11.1 provides more detailed information on this technology platform.

The impedance readout is influenced by changes of cell number, changes in cell–substrate interactions, changes of cell–cell contacts, and subtle changes in individual cell morphology. Therefore, a significant number of different assay designs can be applied to this platform. The sensitivity of impedance measurements can reach the single-cell level, although most experiments are performed with 1000–50,000 cells. The original ECIS instruments have been most widely used to measure extracellular ligand-induced changes in endothelial or epithelial layer integrity in immortalized and primary cells [40–42]. Despite the very attractive assay platform that was offered by the ECIS instrument and the ensuing appearance of numerous applications in many areas of cell biology, impedance has not been considered a breakthrough in GPCR research and has not drawn the attention of drug discovery programs in the pharmaceutical industry. One reason for this lack of attention was the unavailability of 96-well assays, which were only recently introduced, and the lack of G protein-specific pathway information. Furthermore, the surface area offered by the ECIS electrodes was apparently not sufficient to generate the required accuracy to perform quantitative pharmacology in the context of drug discovery. A new generation of impedance technology-based instruments, such as CellKey™ (Molecular Devices) and xCELLigence™ (Roche Applied Sciences, Indianapolis, IN), formerly known as RT-CES® (Acea Biosciences, San Diego, CA), provide improved electrode designs, on-board pipetting capabilities (CellKey), standard 96-well or 384-well formats, and improved sensitivity, reproducibility, and throughput of the assay system. However, these features alone have not been sufficient to significantly raise the attention for impedance technology, especially among GPCR-related drug discovery programs.

This has changed with the remarkable finding that impedance measurements can be used to discover G protein-specific patterns of ligand-stimulated GPCRs [43]. The distinct signatures were observed during the first period of data acquisition, that is, during the first few minutes and within a time frame that is known to involve changes in Ca^{2+} release or cAMP levels. In a given cellular background, such as CHO-K1 cells, characteristic patterns of known $G\alpha q$-, $G\alpha s$-, and $G\alpha i$-coupled receptors were clearly shown, although it must be pointed out that the observed pattern for a given receptor is not necessarily conserved across cell lines. Figure 11.3 illustrates the three prototypic response patterns in CHO-K1 cells that can be observed for recombinant GPCRs, which

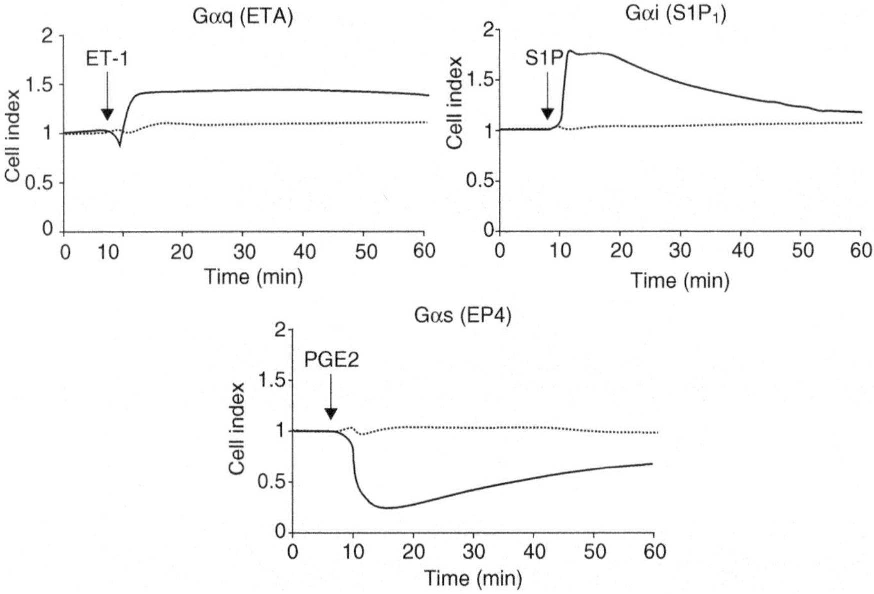

Figure 11.3 Prototypic Gα subunit signatures are detected in impedance measurements (RT-CES/xCELLigence). Stable CHO-K1 cells expressing either recombinant ETA, $S1P_1$, or prostanoid EP4 receptor (EP4) were used to characterize pathway-specific signatures. Cells were stimulated with their cognate ligands (solid lines) or vehicle (dotted line), and impedance changes were monitored during 60 min. Characteristic signatures were identified for Gαq (ETA), Gαi (S1P), and Gαs (EP4) protein-mediated signaling.

are known to signal exclusively through Gαq-, Gαs-, or Gαi-mediated pathways. It is immediately obvious from this figure that Gαq-coupled responses are characterized by an immediate and transient reduction in impedance, which is followed by a larger increase in impedance and a gradual decline during a period of recovery. In contrast, a prototypic Gαi response is characterized by an immediate rise in impedance, which decreases over time, suggesting that it lacks the initial decrease in impedance that is seen in a prototypic response in Gαq-mediated pathways. Gαs-induced responses are characterized by a gradual decrease of impedance upon ligand stimulation, which is long lasting and requires several hours for full recovery to baseline. It is noteworthy that such patterns can be observed using CellKey or xCELLigence (RT-CES) equipment, provided that measurements are started immediately after ligand addition and that the interval of data collection in each well is maximally reduced. These findings also imply that the assay temperature, usually room temperature or 37°C, does not critically influence the response pattern and that the number of frequencies that is used to apply the AC current can be reduced without significant loss of information.

Most published protocols describe a change from cell growth medium to buffer followed by an equilibration period of 15 min just prior to ligand or compound addition. However, recent experiments show that this may not be required in all cases, and comparable dose response curves can be established without any medium change before the onset of the experiment (J. Gatfield, unpublished observation). This observation suggests that these systems can be used in conditions that closely resemble standard cell growth conditions. This can be considered another significant advantage of this technology compared to other assay platforms, especially when using primary cells.

Subsequent reports further substantiated these initial findings and demonstrated pathway-specific signatures in various cellular backgrounds using recombinant systems [44]. However, it should be again emphasized that signatures for a given receptor coupling do not necessarily translate from one cell type to another, although they likely remain constant in a given cell background. Based on the initial data using GPCR signatures, impedance technology was applied to identify endogenously expressed receptors in several cell lines using receptor-panning experiments, and pathway analysis was performed to successfully identify particular receptor subtypes that were known to have different G protein coupling.

In the U2OS cell line, receptor-panning experiments showed a variety of functionally expressed GPCRs, including muscarinic, prostanoid, thrombin, and histamine receptors. A careful analysis of the receptor signature of the histamine-stimulated response then identified a Gαq-coupled process, and this correctly pointed to the histamine receptor subtype 1, as it is the only known Gαq-coupled histamine receptor subtype [45]. This study not only demonstrated the successful use of impedance technology to perform GPCR receptor subtype identification based on pathway signatures, but also demonstrated the use of this technology to perform extensive GPCR receptor panning in a multitude of cell lines in parallel. Previously, such data were laboriously acquired using quantitative polymerase chain reaction (Q-PCR) analysis combined with functional analysis. It is therefore conceivable that impedance technology will not only help to profile cell lines of interest for the functional expression of GPCRs, but could also be integrated in approaches to deorphanize GPCRs, which often rely on the use of promiscuous or chimeric G proteins in addition to the recombinant GPCR of interest to make the cells amenable for conventional high-throughput assay formats. The significant advantage of measuring pathway-specific signatures of GPCRs in recombinant cells or in cells expressing the endogenous receptor opens the field to perform pharmacological studies in a noninvasive fashion and with limited assay development time.

As many GPCRs are known to mediate signals through activation of multiple G proteins simultaneously, impedance technology is also predisposed as a tool to dissect the individual coupling pathways. As an example, Fig. 11.4 shows the activation profile of the endogenous bradykinin B2 (BKB2) receptor in the human epithelial carcinoma cell line A431 treated with 10 nM BK

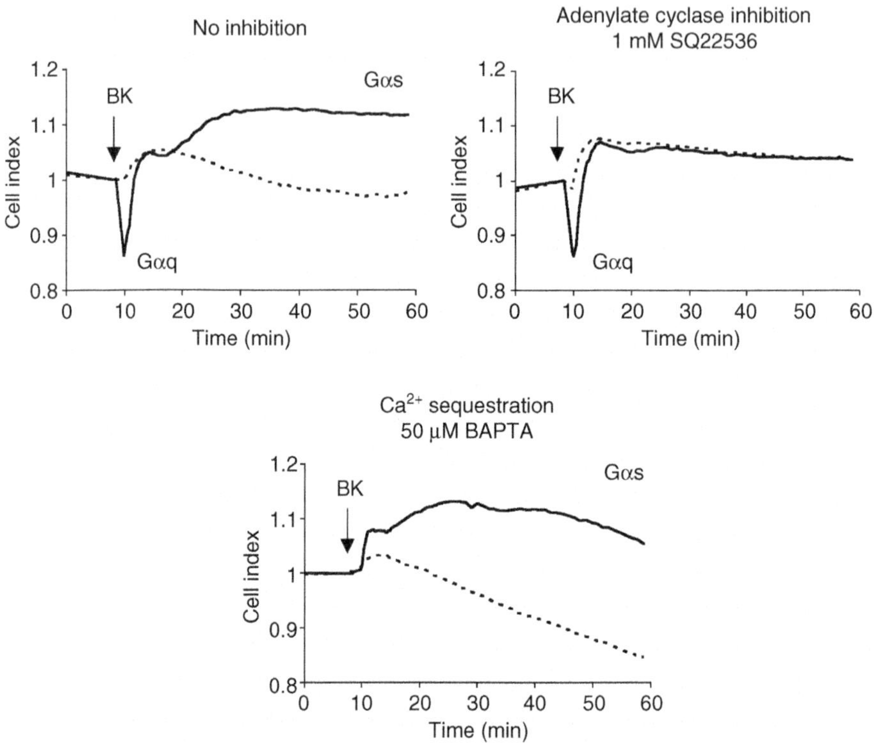

Figure 11.4 Impedance technology can be used to dissect coupling to multiple pathways (RT-CES/xCELLigence). A431 cells endogenously expressing the BKB2 receptor were triggered at the indicated time point (arrow) with 10 nM BK (solid line) or vehicle (dotted line), giving rise to a specific biphasic response (top left panel). Pretreatment with 1 mM SQ22536 (adenylate cycle inhibitor, top right panel) or 50 μM BAPTA-AM (Ca^{2+} chelator, bottom panel) specifically blocked either the second phase or the first phase of the response, allowing a clear assignment of Gαq and Gαs coupling to the two response phases.

(solid line) or vehicle (dotted line). Using conventional assays, it was shown that the activated BKB2 receptor couples through Gαq and Gαs proteins in A431 cells [46]. In the impedance assay, a biphasic response to BK is observed, consisting of a rapid but transient decrease in cell index, followed by an increase of the cell index to a stable plateau. Pretreatment of the cells with either the adenylate cyclase inhibitor SQ22536, to block the Gαs pathway, or the calcium chelator BAPTA-AM, to block the Gαq pathway, specifically abolishes either the plateau phase of the response or the initial transient decrease. Impedance technology, in combination with selective pathway inhibitors, is therefore a useful tool to separate and clearly assign the Gαs- and Gαq-mediated coupling of the endogenous BKB2 receptor in A431 cells. It is also interesting to note that the Gαs signature induced by the endogenous

BKB2 receptor in A431 cells differs from the prototypical Gαs pattern observed in CHO-K1 cells (see above), which again highlights the cell background dependence for specific GPCR signaling pathway signatures. In extrapolation, the unique feature of an integrated readout of multiplexed coupling allows a characterization of GPCR agonists and antagonists on recombinant and nonrecombinant cells with respect to pathway bias, a beneficial feature that is presently of large interest to the pharmaceutical industry.

Since impedance technology provides information on an integrated cell response upon ligand stimulation, it is of critical importance to compare the pharmacological parameters of cognate ligands and synthetic GPCR modulators measured by impedance assays with measurements using existing *in vitro* assays. A recent report describes three examples of Gαi-coupled GPCRs to evaluate impedance measurements for intra-experimental and inter-experimental precision, and the reproducibility of a known structure—activity relationships [47]. Agonists and antagonists spanning a range of potencies and efficacies were used, and compounds displaying positive allosteric modulator activity were included in this study. The investigators chose Gαi-coupled receptors because they pose a particular problem in drug discovery, as the relevant functional cellular assay that measures cAMP levels involves the combined use of forskolin and test compound, which often leads to lack of assay precision. Using the dopamine D2 receptor expressed in CHO-K1 cells, they found a generally good agreement between impedance measurements and preexisting cAMP values of synthetic agonists, including efficacy and potency values. The M4 muscarinic receptor was used to investigate three known positive allosteric modulators, and they found a conserved rank order between these compounds in impedance and GTPγS assays with a greater dynamic range in impedance. Other studies on a wider range of GPCR coupling types, including Gαq- and Gαs-coupled receptors, suggest that this technology delivers values for natural ligands and synthetic compounds that are highly comparable to existing technologies and highlight the possibility to measure inverse agonism in a real-time and in a quantitative manner [48]. The described z' values for impedance measurements are usually above 0.5, which is indicative of a robust assay with the potential to be used in screening campaigns.

Peters et al. also investigated the pharmacological responses of an endogenous GPCR with low expression levels in CHO-K1 cells [47]. The expression level was insufficient to establish a robust assay using traditional assays. Even under these difficult conditions, impedance technology was successfully used to establish a robust assay and to measure compound potency. Thus, impedance technology was clearly superior in measuring endogenous receptor activity under a situation of low receptor abundance. It should be noted, however, that an accurate evaluation of experimental compounds requires well-characterized tools (agonist or antagonists) to fully attribute the response to the receptor of interest. However, provided that these tools are available, impedance technology has a significant potential to be used in compound characterization in physiologically relevant cells, even with low GPCR expression levels, not only in immortalized cell lines but also in primary cells from healthy or diseased conditions. There are also ongoing efforts to combine impedance electrodes with additional pH and oxygen sensors in a single well (Bionas, Rostock, Germany); however, this is still done in a low throughput format, and a change of buffer prior to the experiment is mandatory. Nevertheless, it will be interesting to see if these combined measurements allow an even more detailed pharmacological characterization of synthetic compounds, especially in the context of partial or biased agonists.

Summarizing, there is sufficient evidence to show that impedance measurements are a useful addition to the repertoire of existing GPCR assay tools, with the possibility to investigate receptor pharmacology and compound characteristics in physiologically relevant cell systems. Based on the current literature, impedance technology does not suffer from the known problems using microphysiometry, thus offering higher throughput and increased reproducibility. Clearly, it is a significant advantage that impedance technology does not involve prelabeling or postlabeling of cells, that it can be performed without changes of the cell growth medium, and that it provides real-time data to allow acquisition of highly reproducible and accurate quantitative data. In this context, impedance technology can be applied in drug discovery settings that are aimed at compound characterization in secondary assays using physiologically relevant target cells.

One of the problems that must be addressed to fully exploit this technology in high-throughput screens is the theoretical difficulty to detect false positives due to activation of nontarget-related receptors, which is a particular concern if agonists are sought. However, the technology will be very useful to identify cytotoxic compounds in a high-throughput screen without the need for additional assays. The coming years will show how this new technology will fully establish itself, although it is already evident today that impedance technology became a valuable tool in many research facilities, especially those with a particular focus on GPCR pharmacology.

Almost concomitant with the appearance of impedance technology in GPCR signaling research, RWG became known as a novel and interesting technology to examine cell signaling in a noninvasive, real-time, label-free, and high-throughput format [49, 50]. Similar to impedance technology, RWG makes use of ligand-induced changes in cellular composition; however, instead of measuring changes in cell shape, cell–substrate, or cell–cell interaction, RWG makes use of the process of dynamic redistribution of cellular content, also referred to as dynamic mass redistribution (DMR). DMR is measured with an optical biosensor that is integrated into each well of the cell culture vessel, and it uses a microplate reader. Cells are grown on the biosensor, and the microplate is illuminated with broadband light at 830 nm. Measurements are made of the reflected light, which is sensitive to changes in refractive index caused by the cells and their content at a distance of up to 150 nm from the biosensor surface. DMR is determined as a result of GPCR stimulation by its ligand or synthetic compounds (Fig. 11.2b). Based on this principle, RWG measures integrated and converging signaling events, similar to impedance technology. It is therefore not surprising that RWG shares many characteristics with impedance technology. Currently, the best known system that is based on the RWG principle is the EPIC® System (Corning, Lowell, MA), which features a 384-well format, and because it is compatible with standard pipetting robotics equipment, it is suitable for large-scale HTS campaigns. Similar instruments, such as BIND® Reader (SRU Biosystems, Woburn, MA) or Owls (Microvacuum, Budapest, Hungary), are available from other providers.

Using the DMR readout, GPCR activation revealed a G protein-dependent response pattern in a given cell background [50], and analysis of the prototypical response curves showed a rapid response onset with a peak at 10 min after stimulation. The kinetics are therefore slower than those measured in impedance assays, and the shape of the characteristic response curves for Gαq-, Gαi-, or Gαs-mediated responses is quite different from impedance curves. Based on a most recent publication, it was, however, not possible to distinguish a Gαq- from a Gαi-mediated response in CHO-K1 cells that expressed two known recombinant GPCRs (dopamine D3 or muscarinic M1 receptor). To resolve the issue, forskolin was used to correctly identify the Gαi protein-mediated coupling [51]. In the same publication, a small collection of compounds was used to examine the pharmacological behavior of known GPCR modulators, and the measured IC_{50} and EC_{50} values were compared with data derived from conventional assay platforms. In most cases, especially those using antagonists, the measured DMR responses were of comparable magnitude and potency. However, there were some exceptions that highlight the potential problems and opportunities in using an integrated readout. The agonist potencies measured in DMR assays in the dopamine D3 receptor-expressing cells were slightly lower than in cAMP assays, with the exception of the selective synthetic agonist PD 128907, which showed a more than 1000-fold shift to lower potency in the DMR assays compared to the cAMP assay. It was speculated that the discrepancy could be caused by unknown activation pathways that desensitize or inhibit the D3 receptor-mediated response induced by PD 128907. This clearly shows that results from DMR measurements, or results from any other label-free technology using integrated end point signals, need careful analysis to avoid misleading conclusions.

The potential of DMR assays to measure GPCR responses in cells expressing endogenous receptors was explored in A431 cells using a selection of well-known β2-adrenergic receptor modulators [52]. In this study, a multiparameter analysis revealed distinct receptor activation patterns that were induced by different agonists, demonstrating the sensitivity and utility of DMR assays to investigate pharmacological characteristics of compounds. In this context, DMR will be particularly useful in screens that are designed to identify pathway-biased compounds, provided that the required signature is well known and sufficiently correlates to a desired pharmacological output. The results of a full-scale high-throughput screen using the EPIC System recently became available [53], and they clearly demonstrate the applicability of this technology in this specialized area. The reported data indicate good z' values >0.6 and an acceptable false positive rate.

11.4. DISCUSSION

In recent years, it has become clear that a thorough characterization of GPCR-modulating compounds in terms of *in vitro* pharmacological profile is required, especially in the context of selective coupling to various signaling pathways and their biological consequences. There is also a continued need to increase our knowledge in terms of GPCR confirmation and compound binding modes. Furthermore, the past years have highlighted the increasing complexity in drug discovery designing GPCR-targeting molecules with optimal efficacy and tolerability. In this context, it was proposed that biased agonists or antagonists might have the potential to be superior drugs compared to the majority of currently known GPCR-targeting molecules. Three principle requirements are essential to reach those goals: (1) to increase the use of target-relevant primary cells, possibly even from disease conditions, (2) to develop assays that require minimal intervention prior to the addition of ligand or compounds, and (3) to increase the information content of preexisting assay technologies using real-time readout modes and to use integrated signals as outputs to measure compound characteristics.

The recent addition of label-free, noninvasive, real-time technologies to supplement the existing GPCR assay repertoire provides an opportunity to approach most of the requirements. We describe here a selection of novel technologies that make use of different signaling endpoints to measure integrated cell behavior in a quantitative and, in some cases, even high-throughput format. It is also evident that some of these novel tools have not yet been fully explored, and further experimentation is required. The implementation of fully automated microscopy combined with sophisticated and user-friendly image analysis software will dramatically increase efficiency and allows real-time quality control of cells to determine the optimal timepoint for further experiments. These technologies will soon find their way to a majority of cell biology laboratories. Other platform technologies that are based on extracellular acidification rate or oxygen consumption rates may soon become standard equipment in specialized laboratories dedicated to toxicology, and provided that the assays are robust enough to obtain G protein-specific pathway signatures, there will be an increasing interest in the GPCR research community.

In this review, we describe two technology platforms, impedance technology and RWG, in more detail, as these technologies are of particular interest to the GPCR community and have been most widely discussed during the past few years. It is remarkable that the acquired EC_{50} values of natural GPCR ligands corresponded very well with values acquired with established technologies, and in most cases, the IC_{50} values of GPCR antagonists agreed with values generated in conventional methods. However, interesting discrepancies were reported using some synthetic agonists, and it is tempting to speculate that these novel technologies detected pharmacological features that would have otherwise remained unnoticed. More detailed studies will be required to clarify these differences on a molecular level.

Both technology platforms initially claimed to detect G protein coupling-specific signatures. While most of the data have been confirmed, it became clear that a "prototypic" pattern is not conserved across cell lines and may strongly depend on the particular set of downstream effector molecules and their cell type-specific activity. Nevertheless, once a set of signatures has been established for a particular cell type, it is likely that this pattern is maintained for an extended number of GPCRs, either recombinant or endogenously expressed, provided that the expression levels of downstream effectors, including G proteins, remain unchanged. This knowledge will clearly help to identify the coupling preferences of GPCRs in various cell lines and primary cells expressing the endogenous receptor.

Furthermore, as it was shown that combined G protein signals can be dissected using pharmacological tools, we can expect a more detailed understanding of a particular GPCR coupling pattern. As these label-free technologies measure converging and integrated responses, we may also have versatile tools at hand to investigate biased agonists and their physiological consequences in endogenous settings, and we may be able to investigate pharmacological consequences of GPCR heterodimerization or homodimerization or GPCR-RTK crosstalk in native cells.

Another interesting area, especially in the pharmaceutical industry, is the potential application of label-free technology in HTS. This would clearly facilitate screening of primary cells, which often express low levels of a target GPCR, and it would decrease assay development time/costs compared to traditional assays. The use of impedance technology in high-throughput applications has not been explored well enough to judge its general applicability, whereas RWG technology has successfully demonstrated its potential in standard HTS settings. Nevertheless, as with other screening platforms, a panel of supporting secondary assays will be required to distinguish false positives from true positive hits.

GPCR research is striving to reach a comprehensive understanding of GPCR function and signaling, and in drug discovery, the race to search for compounds with increasing pharmacological sophistication has just begun. The availability of label-free technologies will supplement existing technologies to increase the reliability of cellular assays, and it provides new capabilities to look at GPCR function in physiological and disease-relevant cell backgrounds.

ACKNOWLEDGMENTS

We thank Nico Angliker, Lucile Monnier, Sylvie Poirey, Bénédicte Haenig, and Rolf Studer for technical assistance and Susan Flores for assistance with the manuscript. We especially thank Walter Fischli for his support.

REFERENCES

1. Schöneberg, T., Schulz, A., Gudermann, T. (2002) The structural basis of G-protein-coupled receptor function and dysfunction in human diseases. *Rev Physiol Biochem Pharmacol*. 144, 143–227.
2. Coughlin, S.R. (1994) Expanding horizons for receptors coupled to G proteins: Diversity and disease. *Curr Opin Cell Biol*. 6, 191–197.
3. Drews, J. (2000) Drug discovery: A historical perspective. *Science*. 287, 1960–1964.
4. Wilson, S., Bergsma, D. (2000) Orphan G-protein coupled receptors: Novel drug targets for the pharmaceutical industry. *Drug Des Discov*. 17, 105–114.
5. Wilson, S., Bergsma, D.J., Chambers, J.K., Muir, A.I., Fantom, K.G., Ellis, C., Murdock, P.R., Herrity, N.C., Stadel, J.M. (1998) Orphan G-protein-coupled receptors: The next generation of drug targets? *Br J Pharmacol*. 125, 1387–1392.
6. Jacoby, E., Bouhelal, R., Gerspacher, M., Seuwen, K. (2006) The 7 TM G-protein-coupled receptor target family. *ChemMedChem*. 1, 761–782.
7. Butcher, E.C. (2005) Can cell systems biology rescue drug discovery? *Nat Rev Drug Discov*. 4, 461–467.
8. Overington, J.P., Al-Lazikani, B., Hopkins, A.L. (2006) How many drug targets are there? *Nat Rev Drug Discov*. 5, 993–996.
9. van der Greef, J., McBurney, R.N. (2005) Innovation: Rescuing drug discovery: In vivo systems pathology and systems pharmacology. *Nat Rev Drug Discov*. 4, 961–967.
10. DeLeon, A., Patel, N.C., Crismon, M.L. (2006) Aripiprazole: A comprehensive review of its pharmacology, clinical efficacy, and tolerability. *Clin Ther*. 26, 649–666.
11. Horacek, J., Bubenikova-Valesova, V., Kopecek, M., Palenicek, T., Dockery, C., Mohr, P., Höschl, C. (2006) Mechanism of action of atypical antipsychotic drugs and the neurobiology of schizophrenia. *CNS Drugs*. 20, 389–409.
12. Wisler, J.W., DeWire, S.M., Whalen, E.J., Violin, J.D., Drake, M.T., Ahn, S., Shenoy, S.K., Lefkowitz, R.J. (2007) A unique mechanism of beta-blocker action: Carvedilol stimulates beta-arrestin signaling. *Proc Natl Acad Sci U S A*. 104, 16657–16662.
13. Urban, J.D., Clarke, W.P., von Zastrow, M., Nichols, D.E., Kobilka, B., Weinstein, H., Javitch, J.A., Roth, B.L., Christopoulos, A., Sexton, P.M., Miller, K.J., Spedding, M., Mailman, R.B. (2007) Functional selectivity and classical concepts of quantitative pharmacology. *J Pharmacol Exp Ther*. 320, 1–13.
14. Kenakin, T. (2002) Recombinant roulette versus the apparent virtues of "natural" cell receptor systems: Receptor genotypes versus phenotypes. *Trends Pharmacol Sci*. 23, 403–404.
15. Goldbard, S. (2006) Bringing primary cells to mainstream drug development and drug testing. *Curr Opin Drug Discov Devel*. 9, 110–116.

16. Bosier, B., Hermans, E. (2007) Versatility of GPCR recognition by drugs: From biological implications to therapeutic relevance. *Trends Pharmacol Sci.* 28, 438–446.
17. Gilchrist, A. (2007) Modulating G-protein-coupled receptors: From traditional pharmacology to allosterics. *Trends Pharmacol Sci.* 28, 431–437.
18. Kenakin, T. (2007) Collateral efficacy in drug discovery: Taking advantage of the good (allosteric) nature of 7TM receptors. *Trends Pharmacol Sci.* 28, 407–415.
19. Baroni, A., Perfetto, B., Canozo, N., Braca, A., Farina, E., Melito, A., De Maria, S., Cartenì, M. (2008) Bombesin: A possible role in wound repair. *Peptides.* 29, 1157–1166.
20. Jordan, R.E., Smart, D., Grimson, P., Suman-Chauhan, N., McKnight, A.T. (1998) Activation of the cloned human NK3 receptor in Chinese Hamster Ovary cells characterized by the cellular acidification response using the Cytosensor microphysiometer. *Br J Pharmacol.* 125, 761–766.
21. Smart, D., Coppell, A., Rossant, C., Hall, M., McKnight, A.T. (1999) Characterisation using microphysiometry of CRF receptor pharmacology. *Eur J Pharmacol.* 379, 229–235.
22. Coldwell, M.C., Boyfield, I., Brown, A.M., Stemp, G., Middlemiss, D.N. (1999) Pharmacological characterization of extracellular acidification rate responses in human D2(long), D3 and D4.4 receptors expressed in Chinese hamster ovary cells. *Br J Pharmacol.* 127, 1135–1144.
23. Smart, D., Wood, M.D. (2000) Cytosensor techniques for examining signal transduction of neurohormones. *Biochem Cell Biol.* 78, 281–288.
24. Engström, M., Savola, J.M., Wurster, S. (2006) Differential efficacies of somatostatin receptor agonists for G-protein activation and desensitization of somatostatin receptor subtype 4-mediated responses. *J Pharmacol Exp Ther.* 316, 1262–1268.
25. Ferrick, D.A., Neilson, A., Beeson, C. (2008) Advances in measuring cellular bioenergetics using extracellular flux. *Drug Discov Today.* 13, 268–274.
26. Kole, T.P., Tseng, Y., Wirtz, D. (2004) Intracellular microrheology as a tool for the measurement of the local mechanical properties of live cells. *Methods Cell Biol.* 78, 45–64.
27. Kole, T.P., Tseng, Y., Jiang, I., Katz, J.L., Wirtz, D. (2005) Intracellular mechanics of migrating fibroblasts. *Mol Biol Cell.* 16, 328–338.
28. Gohla, A., Offermanns, S., Wilkie, T.M., Schultz, G. (1999) Differential involvement of Galpha12 and Galpha13 in receptor-mediated stress fiber formation. *J Biol Chem.* 274, 17901–17907.
29. Sugimoto, N., Takuwa, N., Okamoto, H., Sakurada, S., Takuwa, Y. (2003) Inhibitory and stimulatory regulation of Rac and cell motility by the $G_{12/13}$-Rho and G_i pathways integrated downstream of a single G protein-coupled sphingosine-1-phosphate receptor isoform. *Mol Cell Biol.* 23, 1534–1545.
30. Cotton, M., Boulay, P.L., Houndolo, T., Vitale, N., Pitcher, J.A., Claing, A. (2007) Endogenous ARF6 interacts with Rac1 upon angiotensin II stimulation to regulate membrane ruffling and cell migration. *Mol Biol Cell.* 18, 501–511.
31. Seasholtz, T.M., Majumdar, M., Brown, J.H. (1999) Rho as a mediator of G protein-coupled receptor signaling. *Mol Pharmacol.* 55, 949–956.
32. Bhattacharya, M., Babwah, A.V., Ferguson, S.S. (2004) Small GTP-binding protein-coupled receptors. *Biochem Soc Trans.* 32, 1040–1044.

33. Charest, P.G., Firtel, R.A. (2007) Big roles for small GTPases in the control of directed cell movement. *Biochem J.* 401, 377–390.

34. Jiang, X., Sinnett-Smith, J., Rozengurt, E. (2006) Differential FAK phosphorylation at Ser-910, Ser-843 and Tyr-397 induced by angiotensin II, LPA and EGF in intestinal epithelial cells. *Cell Signal.* 19, 1000–1010.

35. Turner, C.E. (2000) Paxillin interactions. *J Cell Sci.* 113, 4139–4140.

36. Siehler, S. (2007) $G_{12/13}$-dependent signalling of G-protein-coupled receptors: Disease context and impact on drug discovery. *Expert Opin Drug Discov.* 2, 1591–1604.

37. Giaever, I., Keese, C.R. (1984) Monitoring fibroblast behavior in tissue culture with an applied electric field. *Proc Natl Acad Sci U S A.* 81, 3761–3764.

38. Giaever, I., Keese, C.R. (1993) A morphological biosensor for mammalian cells. *Nature.* 366, 591–592.

39. Keese, C.R., Wegener, J., Walker, S.R., Giaever, I. (2004) Electrical wound-healing assay for cells in vitro. *Proc Natl Acad Sci U S A.* 101, 1554–1559.

40. Garcia, J.G., Liu, F., Verin, A.D., Birukova, A., Dechert, M.A., Gerthoffer, W.T., Bamberg, J.R., English, D. (2001) Sphingosine 1-phosphate promotes endothelial cell barrier integrity by Edg-dependent cytoskeletal rearrangement. *J Clin Invest.* 108, 689–701.

41. Yin, F., Watsky, M.A. (2005) LPA and S1P increase corneal epithelial and endothelial cell transcellular resistance. *Invest Ophthalmol Vis Sci.* 46, 1927–1933.

42. Tiruppathi, C., Malik, A.B., Del Vecchio, P.J., Keese, C.R., Giaever, I. (1992) Electrical method for detection of endothelial cell shape change in real time: Assessment of endothelial barrier function. *Proc Natl Acad Sci U S A.* 89, 7919–7923.

43. Ciambrone, G.J., Liu, V.F., Lin, D.C., McGuinness, R.P., Leung, G.K., Pitchford, S. (2004) Cellular dielectric spectroscopy: A powerful new approach to label-free cellular analysis. *J Biomol Screen.* 9, 467–480.

44. Leung, G.T.H., McGuiness, R., Verdonk, E., Michelotti, J.M., Liu, V.F. (2005) Cellular dielectric spectroscopy: A label free technology for drug discovery. *JALA.* 10(4), 258–269.

45. Verdonk, E., Johnson, K., McGuinness, R., Leung, G., Chen, Y.W., Tang, H.R., Michelotti, J.M., Liu, V.F. (2006) Cellular dielectric spectroscopy: A label-free comprehensive platform for functional evaluation of endogenous receptors. *Assay Drug Dev Technol.* 4, 609–619.

46. Liebmann, C., Graness, A., Boehmer, A., Kovalenko, M., Adomeit, A., Steinmetzer, T., Nürnberg, B., Wetzker, R., Boehmer, F.D. (1996) Tyrosine phosphorylation of Gsalpha and inhibition of bradykinin-induced activation of the cyclic AMP pathway in A431 cells by epidermal growth factor receptor. *J Biol Chem.* 271, 31098–31105.

47. Peters, M.F., Knappenberger, K.S., Wilkins, D., Sygowski, L.A., Lazor, L.A., Liu, J., Scott, C.W. (2007) Evaluation of cellular dielectric spectroscopy, a whole-cell, label-free technology for drug discovery on Gi-coupled GPCRs. *J Biomol Screen.* 12, 312–319.

48. Yu, N., Atienza, J.M., Bernard, J., Blanc, S., Zhu, J., Wang, X., Xu, X., Abassi, Y.A. (2006) Real-time monitoring of morphological changes in living cells by electronic cell sensor arrays: An approach to study G protein-coupled receptors. *Anal Chem.* 78, 35–43.

49. Fang, Y. (2006) Label-free cell-based assays with optical biosensors in drug discovery. *Assay Drug Dev Technol.* 4, 583–595.
50. Fang, Y., Li, G., Ferrie, A.M. (2007) Non-invasive optical biosensor for assaying endogenous G protein-coupled receptors in adherent cells. *J Pharmacol Toxicol Methods.* 55, 314–322.
51. Lee, P.H., Gao, A., van Staden, C., Ly, J., Salon, J., Xu, A., Fang, Y., Verkleeren, R. (2008) Evaluation of dynamic mass redistribution technology for pharmacological studies of recombinant and endogenously expressed g protein-coupled receptors. *Assay Drug Dev Technol.* 6, 83–94.
52. Fang, Y., Ferrie, A.M. (2008) Label-free optical biosensor for ligand-directed functional selectivity acting on β(2) adrenoceptor in living cells. *FEBS Lett.* 582, 558–564.
53. Dodgson, K., Gedge, L., Murray, D.C., Coldwell, M. (2009) A 100 K well screen for a muscarinic receptor using the Epic label-free system—a reflection on the benefits of the label-free approach to screening seven-transmembrane receptors. *J Recept Signal Transduct Res.* 29(3–4), 163–172.

CHAPTER 12

Screening for Allosteric Modulators of G Protein-Coupled Receptors

CHRISTOPHER LANGMEAD

Heptares Therapeutics Ltd., Welwyn Garden City, UK

12.1. INTRODUCTION

G protein-coupled receptors (GPCRs) mediate responses to a wide range of stimuli, from small molecule neurotransmitters and hormones to peptides and photons of light. Their wide expression profile and roles in major physiological functions have made them prime targets for marketed drugs [1].

Most drugs targeting GPCRs, whether agonists or antagonists, interact with the same binding site as that for the endogenous ligand for the receptor—the "orthosteric" binding site. The location of this site varies across the receptor family: In Class A monoamine receptors, such as muscarinic acetylcholine or dopamine receptors, it is usually located within the transmembrane domain bundle [2]. For Class C receptors, such as the gamma amino butyric acid ($GABA_B$) or metabotropic glutamate (mGlu) receptors, the orthosteric binding site is formed by the large N-terminal domain [3].

Historically, high-throughput screens (HTSs) for GPCRs were run using radioligand binding assays employing a radiolabeled version of the endogenous ligand or, more often, a synthetic competitive antagonist. Such an assay format inherently biased the resultant hits toward compounds that interacted with the same site as the radioligand. However, the advent of functional assays as the screening paradigm of choice in the past decade has seen an increase in the detection of compounds that interact with allosteric sites on GPCRs. Allosteric ligands (from the Greek *allos*, meaning *other*, and *stereos*, meaning *shape*) can bind to sites on GPCRs that are topographically distinct from the

GPCR Molecular Pharmacology and Drug Targeting: Shifting Paradigms and New Directions, Edited by Annette Gilchrist

orthosteric site such that the receptor is able to accommodate two ligands simultaneously. Allosteric binding sites have been identified on many GPCRs, including adenosine [4], muscarinic acetylcholine [5], dopamine [6], chemokine [7], calcium-sensing receptors [8], and mGlu [9] receptors. Upon binding, allosteric ligands are able to alter the conformation of the orthosteric site to modify orthosteric agonist activity as described below.

Allosteric modulators possess a number of advantages as potential drugs. By virtue of being the binding site for the endogenous agonist or transmitter, orthosteric binding sites are likely to be subject to evolutionary pressure to remain conserved across receptor families; however, allosteric binding sites are unlikely to be subject to such pressure and are likely to display divergence within a receptor family. This offers the prospect of selectively targeting individual receptor subtypes where it has not yet been possible to design orthosteric drugs of sufficient selectivity to avoid unwanted side effects, for example, subtype-selective drugs for muscarinic acetylcholine receptors.

As will be seen below, allosteric modulators display an inherent saturability to their effects, which makes them potentially safer as drugs than agonist molecules. Finally, by virtue of their subtly enhancing or reducing the activity of the endogenous ligand at a GPCR, allosteric modulators are capable of maintaining the temporal and spatial resolution of receptor signaling [10]. This is in contrast to direct agonist molecules, which will cause blanket activation of receptors so long as the drug is exposed to the receptor population.

In the late 1990s, the pharmaceutical industry switched from using radioligand binding screening assays to functional assays as a result of the availability of high-throughput generic signaling assays. This switch enabled allosteric ligands to be more routinely identified; any compound that perturbed the action of an agonist (or activated the receptor in its own right) could be detected, irrespective of the location of its binding site on the receptor. In the past 5 years, the pharmaceutical industry has seen the approval for market of its first two allosteric modulators of GPCRs: Cinacalcet, a positive allosteric modulator of the calcium-sensing receptor, was approved in 2004 for hyperparathyroidism [11], and in 2007, maraviroc, a negative allosteric modulator of the chemokine receptor CCR5, was approved as an HIV entry inhibitor [12]. Many companies are now actively pursuing allosteric modulators of GPCRs as novel therapies for a whole range of disease indications.

A key part of the drug discovery process is the identification of compounds, which display activity at the receptor of interest. This typically occurs over a number of stages. In a HTS, compounds are typically assessed at the receptor of interest at a single concentration ("single shot"). Subsequently, at the stage of hit deconvolution or lead optimization, fewer compounds are tested in full concentration–response curve format. Finally, a small number of late-stage compounds may be assessed in more detailed mechanism of action studies. This chapter seeks to critically discuss the methods used to screen for and characterize allosteric modulators of GPCRs—many of the issues (such as choice of assay format) apply to all stages of the screening process, whereas others (such as novel analytical methods) are primarily aimed at later-stage characterization of compounds.

12.2. THE ALLOSTERIC TERNARY COMPLEX MODEL, RADIOLIGAND BINDING, AND AFFINITY

In order to design and critique methods for screening for allosteric modulators, it is necessary to understand the macromolecular basis of the interaction of allosteric ligands with a GPCR. The simplest model of a three-way interaction between a GPCR, an orthosteric ligand, and an allosteric modulator is the allosteric ternary complex model (ATCM) proposed by Stockton et al. [13], which is illustrated in Fig. 12.1. Both an orthosteric ligand (A) and an allosteric ligand (B) can interact with a receptor (R), forming either binary species (AR, BR) or a ternary complex (ARB). K_A and K_B denote the equilibrium dissociation constants of AR and BR, respectively. The symbol α denotes the cooperativity factor and is a quantitative measure of the maximal, reciprocal alteration of affinity of A and B for their respective binding sites when both ligands bind simultaneously to form the ternary complex, ARB.

In short, the value of α dictates whether the allosteric ligand (B) has a positive or negative effect on the binding of the orthosteric ligand (A) and vice versa. Values of $\alpha > 1$ denote positive allosteric modulation of affinity, whereas values of $\alpha < 1$ denote negative allosteric modulation. A value of $\alpha = 1$ represents a special situation whereby the binding of the allosteric modulator does not alter the affinity of the orthosteric probe and vice versa.

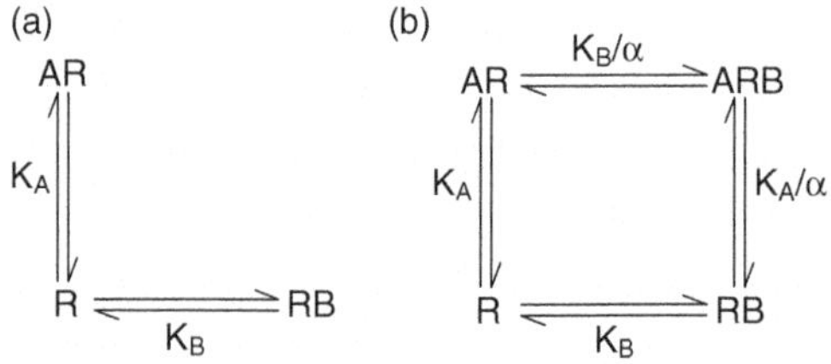

Figure 12.1 Orthosteric (a) and allosteric ternary complex (b) models. Both schemes describe the interaction of two compounds, A and B, with a receptor, R. These ligands interact with R with equilibrium dissociation constants, K_A and K_B, respectively. Where both ligands compete for the orthosteric site, the binding of each ligand is mutually exclusive. However, where the second ligand, B, is an allosteric modulator, both ligands can interact with the receptor to form an allosteric ternary complex (ARB). The magnitude and direction of the allosteric effect on ligand binding affinity is quantified by α. Values of $\alpha > 1$ denote positive cooperativity (positive modulation), whereas values of $\alpha < 1$ (but >0) denote negative cooperativity (negative modulation). Neutral binding cooperativity at equilibrium occurs when $\alpha = 1$.

The effect of allosteric modulators on the binding of a radioligand is shown in Fig. 12.2. A positive allosteric modulator produces a concentration-dependent, saturable increase in the affinity (Fig. 12.2a) of the radioligand, which is more readily visualized on a semilogarithmic plot (Fig. 12.2c). Conversely, a negative allosteric modulator produces a concentration-dependent,

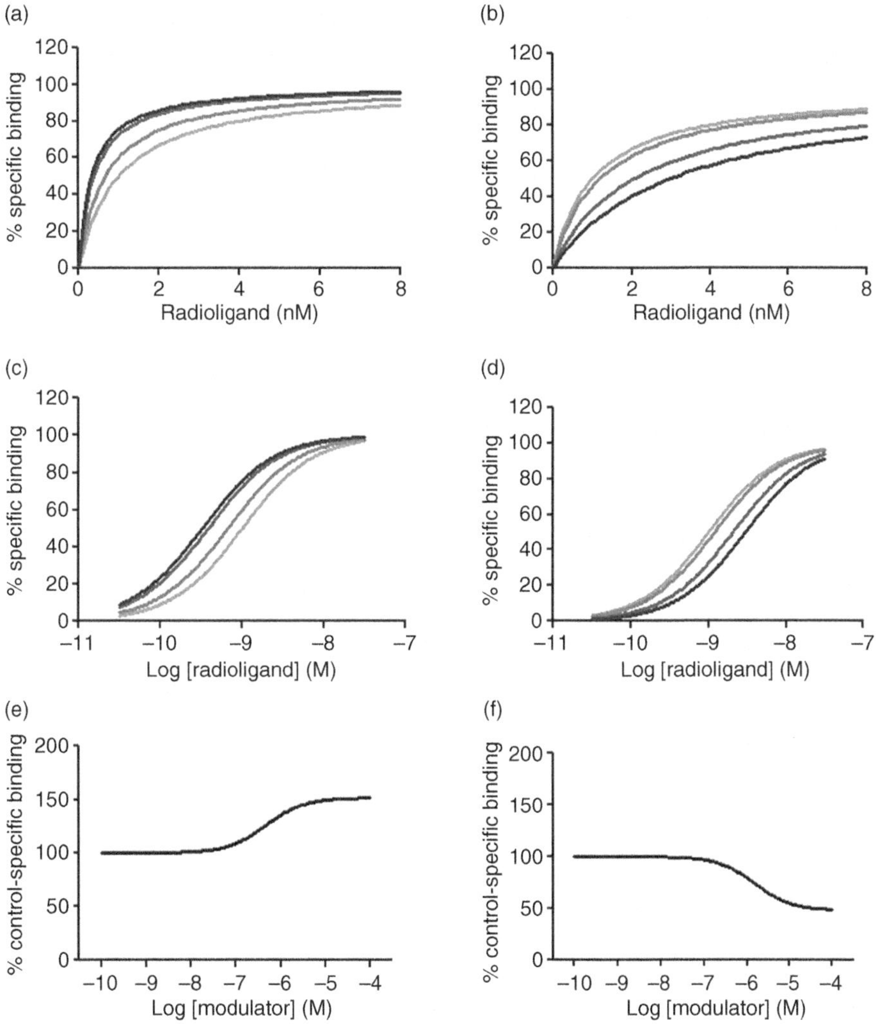

Figure 12.2 (a) Effect of a range of concentrations of positive allosteric modulator ($K_B = 1\,\mu M$; $\alpha = 3$) on the saturation binding of a radioligand ($K_D = 1\,nM$) according to the ATCM; (c) transformed into a semilog plot. (b) and (d) show the same for a negative allosteric modulator ($K_B = 1\,\mu M$; $\alpha = 0.3$). Titration curves for the positive (e) and negative (f) allosteric modulators on the equilibrium binding of a single concentration ($1 \times K_D$) of the same radioligand.

saturable decrease in the affinity of the radioligand, similarly represented in Fig. 12.2b,d. It is possible to analyze such data sets according to the ATCM to generate estimates of the affinity of the allosteric modulator (K_B) and the cooperativity (α) between the modulator and the orthosteric radioligand probe.

However, to run what is in effect a Schild analysis using a radioligand binding assay is unusual and not particularly cost-effective in a drug discovery setting. It is much more common to use a single concentration of a radioligand and run a titration curve of a test compound (often referred to as "competition binding" as, historically, such assays were used to screen analogues of endogenous ligands). The effects of both positive and negative allosteric modulators in such an assay design are shown in Fig. 12.2e,f. Note that a positive modulator yields an enhancement of the level of a bound radioligand, whereas the negative modulator inhibits the binding of the radioligand. These curves can also be analyzed to quantify the affinity and cooperativity of the modulator. It is important to note that the negative modulator is unable to fully inhibit the specific binding of the radioligand. This effect (and the definition of the maximum asymptote for the positive allosteric modulator) is key to determining the degree of cooperativity between the modulator and the radioligand. If the negative modulator was to fully inhibit the specific binding of the radioligand, it would be impossible to distinguish such a compound from a simple competitive antagonist (as the degree of negative cooperativity, $\alpha \rightarrow 0$). If a compound is truly a negative allosteric modulator, this can be resolved by evaluating the test compound against multiple, higher concentrations of radioligand (e.g., $1 \times K_D$, $10 \times K_D$), but typically, this method is less suitable for higher throughput assessment of compounds.

However, there are a number of circumstances whereby an allosteric modulator may be missed using a radioligand binding protocol as described above. First, an allosteric modulator of neutral affinity cooperativity would neither inhibit nor potentiate the binding of a radioligand and would appear as an inactive compound. An example of such a compound is tetra-W84, which does not alter equilibrium [^{3}H]-*N*-methyl scopolamine (NMS) binding at muscarinic M_2 receptors [14]. However, tetra-W84 significantly retards the rate of both association and dissociation of the radioligand, revealing its allosteric mechanism of action (the effect is equal on both association and dissociation; hence, there is no net effect on equilibrium binding). Measuring effects of compounds on orthosteric radioligand dissociation is a very useful way of unmasking an allosteric mechanism at GPCRs as the only way that a compound can alter the rate of dissociation of a preformed receptor–ligand complex is via a topographically distinct site on the receptor [14]. This is partially amenable to a higher throughput screening assay using a two-point kinetic format [14] but is limited to merely confirming an allosteric mechanism rather than providing information on whether a compound is a positive, negative, or neutral modulator.

It is also possible that a compound that appears inactive in an equilibrium radioligand binding assay may possess effects that could be detected in functional assays (e.g., effects on signaling efficacy or direct agonism/inverse agonism—see Section 12.3), and indeed, this was one of the main drivers for the switch from radioligand binding to functional assay formats. For example, a number of allosteric modulators of mGlu receptors do not perturb equilibrium binding of orthosteric radiolabels, for example, CPCCOEt (vs. [^{3}H]-glutamate binding at mGluR1; [15]) and CDPPB (vs. [^{3}H]-quisqualic acid binding at mGluR5; [16]), yet are well-described allosteric modulators in functional assays. Their mechanism of action is discussed further in Section 12.3.

The second possibility concerns the choice of orthosteric radioligand used. For GPCRs where the endogenous ligand is of relatively low affinity or labile, for example, acetylcholine at muscarinic receptors, it is common for the radioligand of choice to be a high-affinity competitive antagonist. However, one of the hallmarks of allosteric interactions is that although the *affinity* of an allosteric modulator for a receptor is an intrinsic property of the molecule, its *cooperativity* is a property unique to the orthosteric/allosteric ligand pair. Invariably, drug discovery efforts aimed toward identifying allosteric modulators are seeking compounds that perturb the binding or function of the endogenous agonist for the receptor in question. In practice, the phenomenon of "probe-dependence" can result in misleading results if the orthosteric radiolabel is not the endogenous ligand—for example, staurosporine is a positive allosteric modulator of [^{3}H]-NMS binding at the M_1 muscarinic receptor but a negative modulator of acetylcholine binding [17]. Thus, wherever possible, the endogenous ligand for the receptor should be used when screening for allosteric modulators.

12.3. BEYOND AFFINITY—FUNCTIONAL ASSAYS, EFFICACY, AND ALLOSTERIC AGONISM

The drawbacks of screening for allosteric modulators using a radioligand binding assay are highlighted in Section 12.2. For these reasons, the most common method to identify such molecules is to run a functional assay to measure a signal transduction event downstream of receptor activation, for example, [^{35}S]-GTPγS binding, cAMP accumulation, inositol phosphate accumulation, intracellular Ca^{2+} flux, or reporter gene activation. Among the earliest known examples of allosteric modulators of GPCRs were compounds such as gallamine, alcuronium, and brucine, which were shown to display saturable shifts in the potencies or affinities of muscarinic receptor ligands [13, 17, 18]. It was assumed that any effects on potency of agonists in a functional assay (akin to the graphs in Fig. 12.2b,c) simply reflected a change in agonist affinity according to the ATCM. Thus, such data were subjected to analysis according to the ATCM to generate estimates of modulator affinity (K_B) and cooperativity with respect to the orthosteric agonist (α).

Exemplifying such a situation is the effect of PD117975 on *R*-(-)-*N*6-(2-phenylisopropyl)-adenosine (R-PIA)-mediated stimulation of extracellular regulated kinase (ERK)1/2 phosphorylation in Chinese hamster ovary (CHO) cells recombinantly expressing the adenosine A_1 receptor [19]. PD117975 is a positive allosteric modulator, producing a saturable, leftward shift in the concentration–response curve to R-PIA; the data set is readily amenable to analysis according to the ATCM. However, there are now many examples of allosteric interactions which cannot be adequately explained using the ATCM. For example, the molecule SIB1893 increases both the potency and the maximal response to L-AP4 in a [^{35}S]-GTPγS binding assay at mGluR4 [20]. Such an effect on maximal agonist response cannot be accommodated by a model that considers only effects of allosteric ligands on orthosteric agonist affinity. Similarly, the negative allosteric modulator of mGluR1, CPCCOEt, reduces the maximal response to glutamate in an inositol phosphate accumulation assay in a concentration-dependent fashion without affecting equilibrium [^{3}H]-glutamate binding to the receptor [15]. Neither of these observations is consistent with CPCCOEt modulating the affinity of glutamate to binding at mGluR1. It is clear from these data sets that these allosteric modulators must be capable of changing the efficiency of the agonist–receptor complex to produce a stimulus. SIB1893 clearly increases the efficacy of L-AP4 to stimulate [^{35}S]-GTPγS binding at mGluR4, either alone or in addition to effects on L-AP4 affinity, while CPCCOEt reduces the efficacy of glutamate at mGluR1.

Furthermore, there is now increasing evidence that allosteric ligands are capable of altering the equilibrium between inactive and active GPCR in the absence of an orthosteric ligand. For example, the negative allosteric modulator at mGluR5, MPEP, reduces the constitutive activity of the receptor in a concentration-dependent manner, that is, MPEP acts as an inverse agonist [21]. Furthermore, several receptors have now been shown to interact with allosteric ligands, which display positive efficacy, for example, ASLW and RSVM at CXCR4 [22], PD81723 at adenosine A1 [23], and AMN082 at mGluR7 [24]. On a practical note, care should be taken to determine whether any apparent "allosteric agonism" is not simply a product of positive modulation of endogenous agonist present in the assay; this phenomenon is common for receptor subtypes for which the endogenous agonist is often found in cell culture media or produced by the cells themselves (e.g., glutamate, adenosine).

However, it is clear that allosteric ligands are capable of modulating affinity and/or efficacy of orthosteric ligands as well as altering the equilibrium between inactive and active receptor in their own right. Such a range of behaviors was predicted by Hall [25], who produced a thermodynamically complete allosteric two-state model that incorporates all of the behaviors described above. However, despite the completeness of the model, it has too many parameters for analysis of any practical data set. More recent efforts have sought to generate semiempirical models to describe the effects of allosteric modulator function. By combining the ATCM with the operational model of agonist action [26], a model has been constructed, which encompasses the modulation of orthosteric agonist efficacy by an allosteric modulator, in addition to any effects on affinity [27]. The operational model of agonism effectively dissects agonist potency into its component parameters, affinity (K_A)

(a)

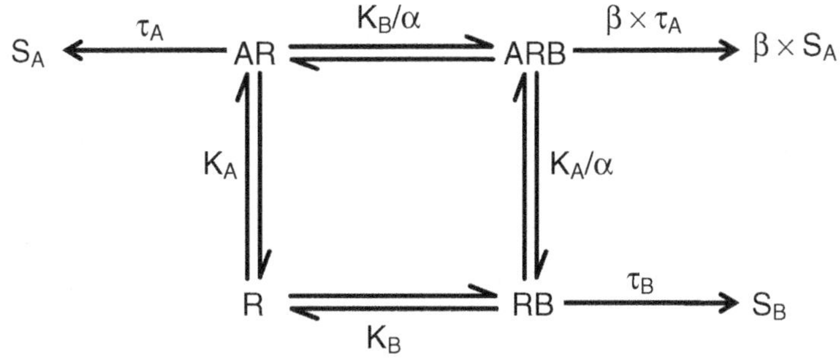

(b)

$$\text{Response} = \text{Basal} + \frac{(\text{Em} - \text{Basal})(\tau_A[A](K_B + \alpha\beta[B]) + \tau_B[B]K_A)^n}{([A]K_B + K_AK_B + K_A[B] + \alpha[A][B])^n + (\tau_A[A](K_B + \alpha\beta[B]) + \tau_B[B]K_A)^n}$$

$$\text{Response} = \text{Basal} + \frac{(\text{Em} - \text{Basal})(([A](K_B + \alpha\beta[B])) + \tau_B \cdot \text{EC}_{50})^n}{(([A](K_B + \alpha\beta[B])) + \tau_B[B] \cdot \text{EC}_{50})^n + (\text{EC}_{50} \cdot (K_B[B])^n)}$$

Figure 12.3 (a) Extension of the ATCM to produce an operational model of allosterism. This model accounts for the interaction of an orthosteric (A) and allosteric (B) ligand with a receptor (R), governed by equilibrium dissociation constants, K_A and K_B, respectively. It also incorporates receptor activation mediated by both orthosteric (τ_A) and allosteric (τ_B) ligands such that the stimulus can be produced by three species, AR, BR, and ARB. Allosteric modulation of affinity is governed by the affinity cooperativity factor, α; modulation of efficacy is governed by the empirical factor, β, which is the ratio of τ_A values in the absence and presence of allosteric ligand. Em, Basal, and n represent the maximal system response, the response in the absence of agonist and the transducer function slope, respectively. (b) A simplified version of the operational model in which it is assumed that the orthosteric agonist, A, is a full agonist (maximal response is equal to Em). The model yields a parameter, $\alpha\beta$, which represents the net cooperative effect of the allosteric modulator on affinity and efficacy. EC_{50} represents the midpoint of the orthosteric agonist concentration–response curve.

and efficacy (τ, tau). The term τ is a cell or tissue-dependent expression of the ability of the agonist–receptor complex to produce a stimulus. In the new model described by Kenakin et al., the "efficacy cooperativity" parameter is neither thermodynamic nor bidirectional, but simply represents the ratio of orthosteric agonist efficacies (τ and $\beta\tau$) in the absence and presence of the modulator, respectively; the ratio (β) is the efficacy cooperativity factor. This model was later extended to include the ability of the allosteric ligand to mediate receptor activation in its own right (Fig. 12.3; [28]). Using this empirical efficacy cooperativity factor, these operational models of allosterism have been effectively used to discriminate the mechanism of action of several allosteric ligands in functional assays [29–31].

It is important to recognize a role for radioligand binding studies in defining the mechanism of action of allosteric modulators. As previously discussed, radioligand binding is limited to detection of effects of modulators on orthosteric ligand *affinity*, but this can be a useful attribute when seeking to dissect out effects on affinity and efficacy. For example, the muscarinic M_4 receptor positive allosteric modulator, LY2033298, causes a 35-fold left shift in the potency of acetylcholine to stimulate Ca^{2+} mobilization in CHO cells stably coexpressing the muscarinic M_4 receptor and a promiscuous G protein [31]. Three-way radioligand binding studies using [^{3}H]-NMS demonstrated that LY2033298 increased the affinity of acetylcholine by approximately 40-fold, suggesting that the positive modulation elicited by this compound was due to changes in agonist affinity, rather than efficacy. Conversely, similar studies have shown that the effects of LY487379 as a positive allosteric modulator of mGluR2 are mediated by changes in glutamate efficacy, rather than affinity [30].

In the context of determining the mechanism of action of an allosteric modulator, it is important to consider the influence of the biological reagent on the outcome. The most common paradigm for screening is to use a recombinant cell line expressing the receptor of interest. Historically, researchers have sought to develop cell lines that express high levels of receptor (often at much greater levels than would be found in a native environment), with an eye toward increasing the size of the signal-to-noise ratio in assays and potentially decreasing the amount of biological/chemical reagent required to run the assay. This approach works very well for radioligand binding assays and functional assays seeking orthosteric antagonist hits. However, as shown below, the choice of assay end point and receptor expression level can have a marked effect on the outcome when screening for allosteric ligands, particularly positive modulators. Figure 12.4a models the effect of varying concentrations of a positive modulator of an agonist response according to the operational model of allosterism (Fig. 12.3a; [28]) under three different assay conditions, represented by τ_A (orthosteric agonist efficacy) values of 1, 10, or 100. These different values could correspond to alternative assay formats, successively more distal to receptor activation (and hence subject to progressively greater signal amplification). Alternatively, they could represent the same assay under increasing levels of receptor expression.

Under conditions of either low receptor expression or in an assay paradigm proximal to receptor activation ($\tau = 1$), the positive modulator causes both a leftward shift and an increase in the maximal agonist response (Fig. 12.4a). As either the level of receptor expression or the signal amplification increases, the profile of both the agonist and positive modulator changes. At the highest τ value (100), the agonist is significantly more potent than at the lower τ values. Furthermore, the maximal agonist response is increased to the maximal level of the assay system ("E_m" in the terms of the operational model). Under these conditions, the positive modulation of efficacy cannot manifest as an increase in the maximal agonist response and so contributes to a further increase in agonist potency. In the absence of prior knowledge, this profile may incorrectly be taken as evidence of affinity-only modulation. In the early stages

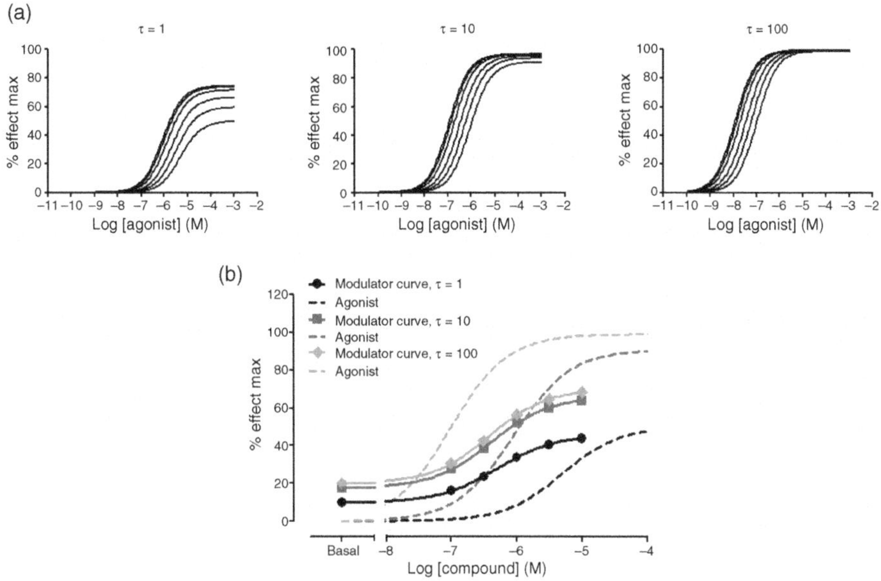

Figure 12.4 (a) The effect of a positive modulator of affinity and efficacy on agonist concentration–response curves under conditions of varying stimulus–response coupling (τ values of 1, 10, and 100). Data are modeled using the operational model of allosterism (Fig. 12.3a) using the following parameters: $pK_A = 5.0$, $pK_B = 6.0$, $\alpha = 3$, $\beta = 3$, $\tau_B = 0$, Basal = 0, Em = 100, n = 1. (b) The resultant positive modulator titration curves that would be generated using an EC_{20} of agonist. The variation in stimulus–response coupling has little or no effect on absolute modulator titration curve potency but becomes apparent when the modulator curves (solid lines) are examined with respect to their respective orthosteric agonist curves (broken lines).

of drug discovery programs, this may not be an issue as the assay is simply being used to compare the activities of different compounds. However, an understanding of the mechanism of action for the allosteric modulator may be important for late-stage compounds that are being taken forward into efficacy models.

It is interesting to note that in isolation, positive modulator titration curves (see below for detailed description) from each of these models show very similar EC_{50} values (Fig. 12.4b). This suggests that in a structure–activity relationship (SAR) screening situation, the stimulus–response coupling of the assay (i.e., the "τ" value for the orthosteric agonist) is not particularly important, as similar modulator titration curves will be produced irrespective of the efficiency of stimulus–response coupling. It is only when the modulator titration curves are examined in relation to their respective orthosteric agonist concentration–response curve (dashed lines in Fig. 12.4b) that it becomes obvious that there are differences.

12.4. ALLOSTERIC MODULATOR TITRATION CURVES

In a modern context of high-throughput screening, running multiple curves to screen a single compound is considered inefficient. Therefore, it has become commonplace for researchers to simplify assay design. Rather than running multiple curves to an agonist in the presence of varying concentrations of a putative modulator (as described above), it is now commonplace to select a single concentration of agonist and examine the effects of a range of putative modulator concentrations on the response. For example, a researcher seeking a negative allosteric modulator may design an assay to look for inhibition of an EC_{80} concentration of agonist. Conversely, an assay designed to look for a positive allosteric modulator may utilize an EC_{20} concentration of agonist to allow for potentiation of the response (Fig. 12.5). Such curves are generally referred to as modulator concentration–response or titration curves, and are analyzed according to simple logistic functions to determine IC_{50} or EC_{50} values (Fig. 12.5). These potency values are then used as the basis for driving the SAR, a key part of a drug discovery program.

Clearly, the reduction in the number of data points in going from an assay of multiple curves to a single curve diminishes the information content that an assay yields. However, simply using the potency value for a compound disregards other information that provides detail pertaining to a modulator's mechanism of action. A previous study has examined the relationship between the properties of positive allosteric modulators according to the simple ATCM and the resultant modulator titration curves [32]. It is clear that for positive modulators, the potency of the titration curve is dictated by both its affinity and the strength of the cooperativity between the agonist and the modulator. This is akin to agonist potency being a product of its affinity and efficacy. For a modulator of fixed affinity, increasing values of α result in higher potencies. Conversely, as the cooperativity value tends to a value of 1 (neutral cooperativity), the resultant pEC_{50} value will tend to the pK_B value for the modulator [32]. By extension, for a negative allosteric modulator, the pIC_{50} of the modulator titration curve will tend to the pK_B value for progressively stronger degrees of negative cooperativity (as $\alpha \rightarrow 0$).

It is possible to extend this relationship using a simplified version of the operational model of allosterism discussed in the Section 12.3. This model, shown in Fig. 12.3b, is based on the full version in Fig. 12.3a. It makes the assumption that the orthosteric agonist being used is a full agonist and that it is not possible for its maximal response to be enhanced or reduced. Therefore, it is limited to the analysis of positive allosteric modulators, where any effects of efficacy will be manifest as a leftward shift in the agonist curve. It estimates a net cooperativity parameter, $\alpha\beta$, which represents the product of the modulator's effects on affinity and efficacy. In addition to the modulator affinity (K_B), it also incorporates the ability of the allosteric ligand to activate the receptor in its own right, governed by the parameter τ_B. If the allosteric modulator exhibits no intrinsic agonist activity ($\tau_B = 0$), then the model is essentially

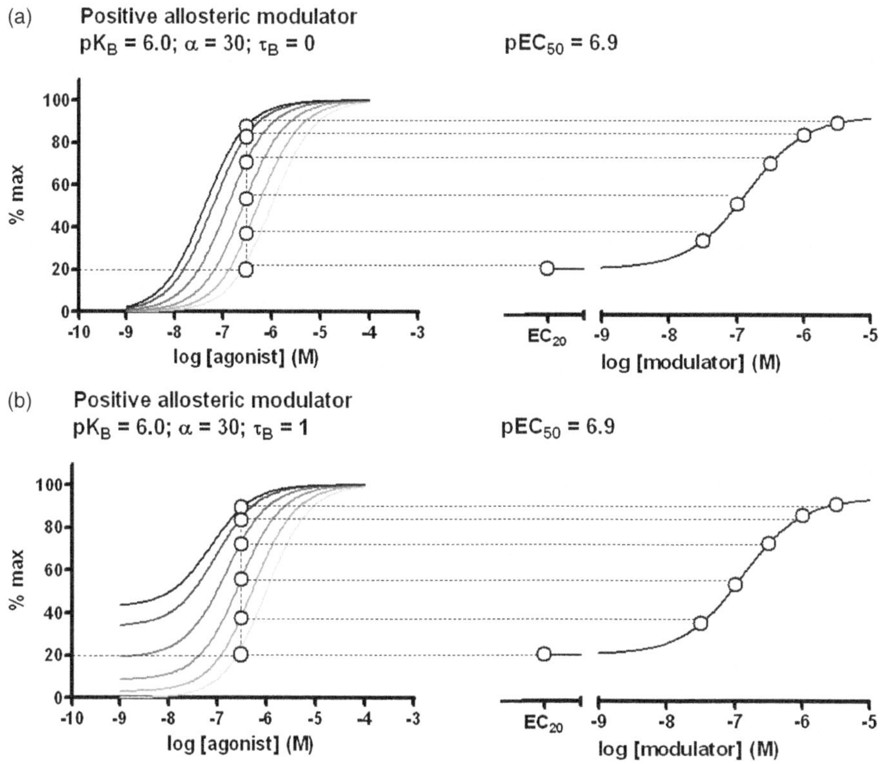

Figure 12.5 (a) Effect of varying concentrations of a positive allosteric modulator on an agonist concentration–response curve, as modeled using the simplified operational model of allosterism in Fig. 12.3b. Also shown is the resultant modulator titration curve that the positive modulator would produce if screened against an EC_{20} of agonist, fitted using a four-parameter logistic equation. For the simulation, the following parameters were used: $pEC_{50} = 6.0$, $pK_B = 6.0$, $\alpha\beta = 30$, $\tau_B = 0$, B = 30 nM – 3 μM. (b) Effect of varying concentrations of a similar positive allosteric modulator that displays intrinsic agonist activity ($\tau_B = 1$) on an agonist concentration–response curve. Also shown is the resultant modulator titration curve that the modulator would produce if screened against an EC_{20} of agonist, fitted using a four-parameter logistic equation.

identical to the ATCM, except that the net cooperativity factor, αβ, replaces the affinity cooperativity factor, α.

Figure 12.5a shows the effect of a positive allosteric modulator with a $pK_B = 6.0$, a net cooperativity, $\alpha\beta = 30$, and no intrinsic agonist activity ($\tau_B = 0$) on an agonist concentration–response curve. Also shown is the resultant modulator titration curve, assuming an EC_{20} of the agonist is used. The fitted pEC_{50} (6.9) is greater than the affinity (6.0) as described above. Figure 12.5b shows the effect of a similar positive allosteric modulator but which also displays a significant degree of intrinsic agonist activity ($\tau_B = 1$) on an agonist concentration–response curve. It is interesting to note that the presence or absence of intrinsic agonist activity in a positive modulator molecule makes no discernible difference to the EC_{50} of the resultant titration curve (Fig. 12.5). This suggests that a modulator titration curve format is not well suited for detecting intrinsic agonist activity in a molecule.

A previous study [32] highlighted a method for analyzing modulator titration curves in conjunction with the orthosteric agonist curve to yield estimates of modulator affinity and cooperativity. However, this method assumes that the modulator behaves according to the ATCM (which, as we have already seen, is often not the case). However, it is relatively simple to extend this analysis to the simplified operational model of allosterism. By recasting the model in Fig. 12.3b such that the modulator concentration ([B]) is the independent variable on the x-axis and that the orthosteric agonist concentration ([A]) is constrained to a fixed value (FixAg), the model can be used to analyze modulator titration curves in conjunction with the orthosteric agonist curve.

Figure 12.6a shows the recast model, and Fig. 12.6b demonstrates the analysis of a positive modulator titration curve in conjunction with the agonist

(a)

$$\text{Response} = \text{Basal} + \frac{(\text{Em} - \text{Basal})(([A](K_B + \alpha\beta[\text{FixAg}])) + \tau_B \cdot EC_{50})^n}{(([A](K_B + \alpha\beta[\text{FixAg}])) + \tau_B[\text{FixAg}] \cdot EC_{50})^n + (EC_{50} \cdot (K_B[\text{FixAg}])^n)}$$

(b)

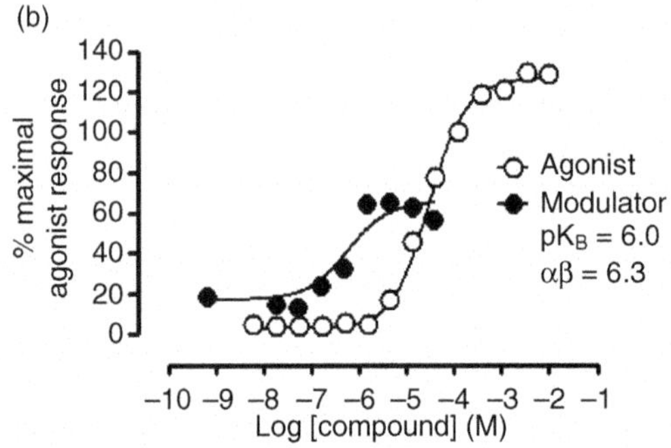

Figure 12.6 (a) The simplified operational model of allosterism, recast such that the modulator concentration, B, is the independent variable and the orthosteric agonist concentration is recast as a constant, FixAg. (b) Analysis of a positive allosteric modulator titration curve in conjunction with the orthosteric agonist concentration–response curve according to the recast equation. Using the global shared analysis feature of GraphPad Prism 5 (GraphPad Software, La Jolla, CA), the values of EC_{50}, Basal, Em, K_B, αβ, and transducer function slope (n) are shared across both data sets, whereas the orthosteric agonist concentration, FixAg, is set at 4 μM. Note that in the analysis, the values of EC_{50}, Basal, Em, and transducer function slope refer to the agonist concentration–response curve, rather than the modulator curve.

concentration–response curve. The allosteric modulator potentiates an EC_{20} of agonist (4 μM). Analysis according to the operational model yields an estimate of affinity ($pK_B = 6.0$) and net cooperativity ($\alpha\beta = 6.3$). This is a relatively simple method for analyzing modulator titration curves, which yields significant information that can be used to characterize allosteric modulators in a relatively high-throughput manner.

However, due to the limited number of data points compared to a full "Schild-type" approach, there are limitations to this analysis. First, it is generally not possible to estimate the degree of intrinsic agonist activity of an allosteric modulator (τ_B) using this approach, as demonstrated with the examples in Fig. 12.5. In the analysis example shown in Fig. 12.6b, the value of τ_B was determined in a separate study and shown to be zero, and, thus, constrained as such. Therefore, this value needs to be independently determined in order for this analysis to work correctly. Second, by the nature of the model, it is assumed that the maximal asymptote of the modulator titration curve cannot exceed the maximal asymptote of the orthosteric agonist curve (as the agonist is assumed to be full). If this does happen, then the model cannot be used. Finally, for highly positive or highly negative cooperativity values, the maximum/minimum asymptote of the modulator titration curve approaches either the maximum asymptote of the agonist curve or zero, respectively, which makes an accurate estimation of both the K_B and $\alpha\beta$ values very difficult.

However, even without applying this methodology, a useful empirical marker of modulator function is the maximum asymptote (for a positive modulator) or the minimum asymptote (for a negative modulator) of the titration curve. If a given positive modulator yields a titration curve with a higher maximal response than for another modulator, then theory suggests that it will have stronger degree of positive cooperativity with the agonist than the second modulator (whether affinity or efficacy mediated). Thus, simply using the fitted EC_{50} (or IC_{50}) value in conjunction with the maximal (or minimal) response should provide sufficient information with which to drive an SAR program.

12.5. THE IMPACT OF FUNCTIONAL ASSAY FORMAT ON ALLOSTERIC MODULATOR SCREENING

In the sections above, functional assays have been spoken about as a homogenous group of assays in which a whole range of allosteric effects can be detected. It is true that all of the functional assays described can detect allosteric modulation of agonist affinity, efficacy, and allosteric agonism. However, the specific assay format chosen can greatly impact the results that are obtained. This section will discuss the issues associated with the use of transient read systems (such as intracellular Ca^{2+} flux) versus accumulation assay formats (exemplified by [^{35}S]-GTPγS binding) to screen for positive allosteric modulators of GPCRs.

Measurement of intracellular Ca^{2+} flux is a very popular screening platform due to the relatively inexpensive cost of reagents and its applicability to most GPCRs via either native coupling (for Gq/11-coupled receptors) or coupling through chimeric or promiscuous G proteins [33]. A screen in the modulator titration curve format for a positive allosteric modulator may be run as follows: In brief, cells would be exposed to varying concentrations of test compound while being monitored for 1–2 min for fluorescence levels (the "agonist" read). This read would detect any agonist activity intrinsic to the compound. After an incubation period (typically between 2 and 30 min), the cells would be exposed to a fixed EC_{20} concentration of orthosteric agonist and monitored for fluorescence again (the "modulator" read). This read will detect any allosteric modulatory activity of the compounds, as evidenced by an increase in the activity of the EC_{20} concentration relative to control. This "dual-read" protocol is common for screening for positive allosteric modulators.

Contrast this method with a $[^{35}S]$-GTPγS assay in which the agonist, modulator, and receptor (in the form of a membrane preparation) are all allowed to pre-equilibrate for up to 1 h prior to the start of the reaction with a radiolabeled, nonhydrolyzable form of GTP ($[^{35}S]$-GTPγS). This incubation is allowed to continue for generally anywhere between 30 min and 2 h before being terminated, and the amount of bound radioligand is determined using scintillation spectroscopy. Both of these formats will give the user information about the modulatory and agonist activity of a test compound. However, whereas the full range of activity of the compound will be apparent in the single output of the $[^{35}S]$-GTPγS assay, in the calcium assay, any agonist activity of the compound will be highlighted only in the first read, and any modulatory activity will be apparent only in the second read.

Already this has made the holistic determination of a compound's activity more difficult as there are two separate data sets that need to be considered. More importantly, it leads to an important question—as the two activities have now been separated into two reads, does the first read impact on the output of the second? Figure 12.7 addresses this question with a data set from a GPCR positive allosteric modulator program screened in a Ca^{2+} flux assay on the FLIPR platform (Molecular Devices, Sunnyvale, CA) using a similar protocol to the one described above. Figure 12.7a–c shows the data for 10 positive allosteric modulators in the first "agonist" read, triaged according to their level of agonist activity. Figure 12.7a highlights two compounds with little or no agonist activity, Fig. 12.7b shows compounds with partial agonist activity, and Fig. 12.7c shows compounds with full agonist activity.

Figure 12.7d then shows the effects of these compounds in the second, "modulator," read. Compound 11 (one of the compounds with no agonist activity in the first read) produces a concentration-dependent leftward shift in the agonist concentration–response curve, with a small increase in the maximal agonist response, consistent with positive allosteric modulation. Figure 12.7e shows the effect of a different modulator on an agonist response. Compound 10, which displayed a full agonist response in the first read, also

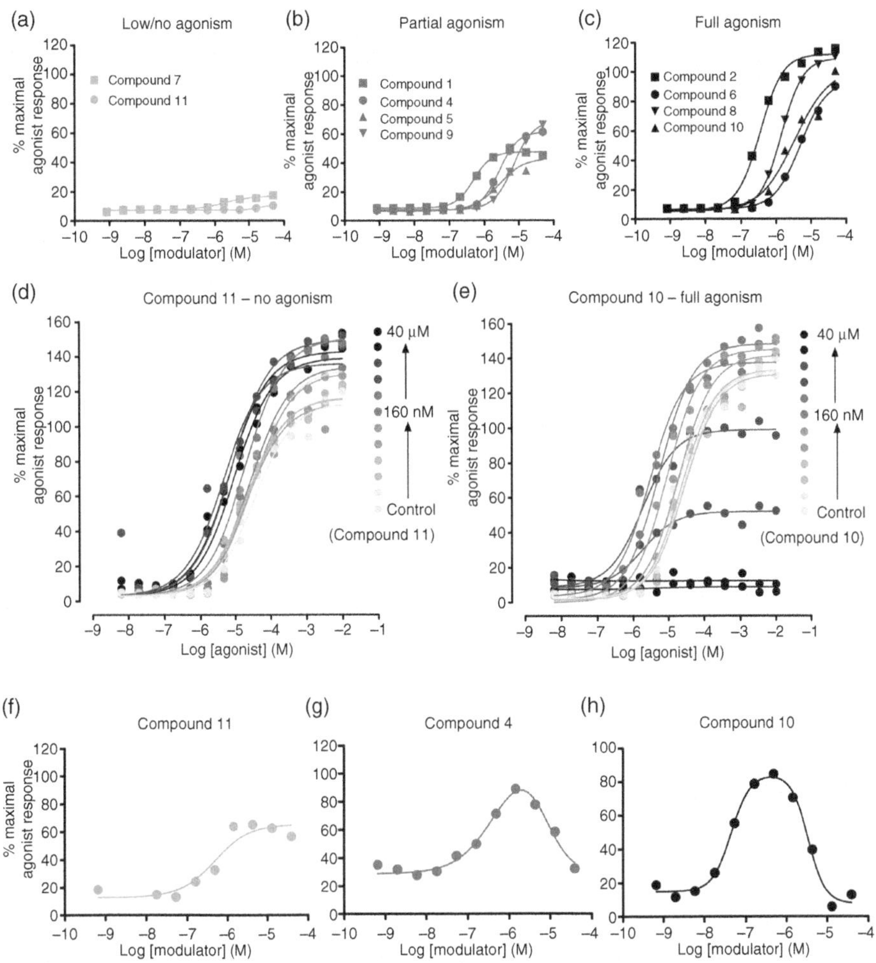

Figure 12.7 Effect of 10 positive allosteric modulators on intracellular calcium mobilization measured on a FLIPR in a CHO cell line stably expressing a GPCR and a promiscuous $G\alpha_{16}$ protein in the absence of orthosteric agonist. Modulator responses are shown, triaged according to their level of agonist response in the first read: (a) low/no response, (b) partial agonist, and (c) full agonist. The effect of a range of concentrations of Compound 11 (d) and Compound 10 (e) on agonist-stimulated calcium mobilization as determined using FLIPR technology. The results are shown from the second, "modulator," read as described in the text. Both compounds produce leftward shifts and increases in the maximal agonist response, consistent with positive allosteric modulation. However, at the highest modulator concentrations (which produced a response in the first, "agonist" read), Compound 10 ablates the orthosteric agonist response. Resultant modulator titration curves are shown for Compound 11 (f), Compound 4 (g), and Compound 10 (h).

causes a leftward shift in the agonist concentration–response curve and a small increase in the maximal agonist response. However, at the highest modulator concentrations, the maximal agonist response is markedly reduced or even abolished. Closer inspection reveals that it is the agonist responses in the presence of the concentrations of modulator which evoked a significant agonist response in the first read that are affected (Fig. 12.7c,e). This provides clear evidence that activity in the first read has negatively impacted the second read. This effect could be due to a number of reasons, including receptor desensitization, but it is most likely due to the depletion of intracellular calcium stores during the "agonist" read; the stores simply do not have sufficient time to recover prior to the second challenge. This is even more clearly shown when the results of the second read are visualized in a modulator titration curve format. Figure 12.7f–h shows the modulator titration curves (based on a fixed agonist concentration of 4 μM) for Compounds 11, 4, and 10, which display no, partial, and full agonism in the first read, respectively.

Compound 11 potentiates the response to 4 μM agonist in a concentration-dependent fashion (Fig. 12.7f). Compounds 4 and 10 also potentiate the agonist response at low modulator concentrations, but at higher modulator concentrations, the response is progressively diminished, yielding bell-shaped concentration–response curves (Fig. 12.7g,h). For these curves, it is impossible to accurately quantify their modulatory effect. If the agonist activity did not adversely affect the second read response, what would the effect have been at modulator concentrations greater than 1 μM? Would the modulator titration curve have reached a plateau or continued to increase? Unfortunately, it is impossible to say, and such uncertainty adversely affects the assessment of these compounds in an SAR program. Furthermore, on a practical level, it leaves a question of how best to analyze the modulator titration curves. Should points be excluded or should the experimenter try to fit a biphasic concentration–response curve? Both are likely to require manual intervention, and neither are attractive options. Unless an experimenter can argue that a transient response system more faithfully reflects the physiological signaling of the receptor under test, then the obvious answer is to avoid transient read assay formats and use accumulation assay formats, such as [^{35}S]-GTPγS binding or cAMP. However, as long as platforms such as Ca^{2+} flux or aequorin remain popular, these questions will remain.

It is also important to highlight the pleiotropic nature of allosteric effects depending on the pathway monitored. The phenomenon of "biased-signalling," "agonist-trafficking," or "functional-selectivity" is well described in the literature [34], and allosteric modulators have been shown to exhibit such effects. For example, the allosteric modulator of chemoattractant receptor-homologous molecule expressed on Th2 cells (CRTH2), 1-(4-ethoxyphenyl)-5-methoxy-2-methylindole-3-carboxylic acid (Compound 1), displays neutral cooperativity with respect to the agonist prostaglandin-D2 (PDG2) in assays of G protein activation (i.e., it does not affect the agonist concentration–response curve in assays of PDG2-stimulated [^{35}S]-GTPγS binding). However,

Compound 1 is a potent *negative* allosteric modulator of PDG2-stimulated recruitment of β-arrestin, a G protein-independent functional end point [35]. Unless the disease-relevant signal transduction pathway for a given receptor is known, then there is no way to know which is the most appropriate end point to measure; rather, the assay decision is generally driven by other factors such as cost, assay robustness, and availability of platforms and reagents.

Differential signaling through divergent effector systems is worth considering when using chimeric or promiscuous G proteins. As discussed above, this is a common practice as it allows non-Gq/11-coupled GPCRs to couple to the PLC-IP_3-Ca^{2+} pathway for detection of intracellular calcium using platforms such as FLIPR or aequorin. However, this introduces a non-native coupling system, which may be subject to differential regulation. For example, both the orthosteric agonist, L-AP4, and the allosteric agonist, AMN082, inhibit forskolin-stimulated cAMP accumulation in CHO cells stably expressing rat mGluR7 and the promiscuous G protein, $G\alpha_{15}$ [36]. In contrast, only L-AP4 stimulates Ca^{2+} mobilization in the same cell line, whereas AMN082 is without effect [36]. If a HTS had been run using the $G\alpha_{15}$ calcium assay alone, then AMN082 would have not been identified. In a similar vein, several studies have recently shown that allosteric agonists may not cause GPCR desensitization in the same manner as orthosteric agonists [22, 37]. This array of novel pharmacologies must be considered when selecting a screening assay—for example, the recruitment of β-arrestin is now often used as screening paradigm for GPCRs [38]. Clearly, it is not going to be possible to cover all bases by running multiple assay formats for screening compounds against a single target, but researchers should be aware of these caveats when developing assays for the routine assessment of their compounds.

12.6. TAKING ADVANTAGE OF STRUCTURAL UNDERSTANDING OF ALLOSTERIC BINDING SITES

The sections above have discussed methods for screening allosteric modulators using wild-type receptors. However, one of the attributes of allosteric binding sites is that they are topographically distinct from orthosteric sites. The distinction between allosteric and orthosteric sites is obvious for some GPCRs, such as for mGluRs and the $GABA_B$ receptor, where the orthosteric site is formed by the large N-terminal domain and allosteric sites have been identified in the transmembrane bundle [3]. For other GPCRs, such as muscarinic acetylcholine receptors, allosteric binding sites at the top of the transmembrane domain and in the extracellular loops (ECLs) [39] are much closer to the orthosteric binding site, which is found slightly deeper in the transmembrane bundle.

These structural differences have enabled the use of truncated, chimeric, or mutated GPCRs to aid in the identification of allosteric ligands. The mGluRs are an excellent example in this regard. Positive [40, 41], negative [36], and neutral [42] allosteric modulators have been identified for receptors within this family and have been shown to interact with multiple binding sites in the transmembrane domain [24, 40, 43]. DFB is a positive allosteric modulator of mGluR5, which shows no intrinsic allosteric agonism. However, in a mGluR5 construct in which the N-terminal glutamate-binding domain was removed, DFB was shown to act as a full agonist [43]. In a similar fashion, the negative allosteric modulator, MPEP, appeared as an inverse agonist at the N-terminally truncated receptor. Such effects of allosteric modulators have also been demonstrated in N-terminal truncates of $GABA_B$ and calcium-sensing receptors [44, 45]. The ability of the transmembrane domain of mGluRs to activate cellular effector systems can be exploited at a screening level; it is easy to imagine that it might be considered attractive to express an N-terminally truncated Family C receptor and perform a screen to detect agonists, rather than run a more complex screen for positive modulators at the wild-type receptor.

A similar example of exploiting the structural basis of allosteric binding sites comes from the muscarinic M_1 receptor. The orthosteric binding site across muscarinic acetylcholine receptors shows a high degree of conservation, and as such, subtype-selective agonists for this family have proven hard to identify [46]. Recently, several agonists have been discovered that display unprecedented levels of selectivity for the muscarinic M_1 subtype [47–49]. AC-42 exemplifies this class of molecule, and studies with chimeric M_1–M_5 muscarinic receptors suggested that the N-terminus/TM1 and ECL3/TM7 regions are key to the selective activity of AC-42 at the M_1 receptor [47]. This region is clearly distinct from the orthosteric binding site, which is formed by residues in TM3 and TM6 [50]. One of the key residues in the orthosteric binding site is Tyr381 within TM6. Mutation of this residue to Ala greatly reduces the affinity and potency of orthosteric agonists, such as acetylcholine and carbachol [47, 51]. However, this mutation leaves the activity of agonists, such as AC-42, relatively unaffected [47] and even potentiates the activity of others [49]. Thus, activity at the Tyr381Ala mutation is a useful marker of atypical agonism, which may prove to be allosteric and/or selective. The examples given above are somewhat specific to the receptors involved. However, the increasing understanding of the structural basis of allosteric regulation will potentially increase the number of surrogate assays that can be performed to identify allosteric ligands.

12.7. SUMMARY AND FUTURE DIRECTIONS

The diverse range of allosteric modulators now being identified is challenging for those involved in drug discovery. This chapter has sought to explore some of the issues faced in screening for allosteric modulators of GPCRs, ranging from the choice of assay format to the methods for data analysis.

The Section 12.1 discussed the benefits and drawbacks of radioligand binding. While it is clear that radioligand binding does not represent a good approach for primary screening for novel allosteric modulators, the importance of the technique should not be underestimated, as it is the only method that can be used to detect allosteric effects on affinity without potentially confounding effects of modulation of efficacy. Given the prevalence of high-throughput functional screening assays, the wider effects of allosteric modulators were considered: Efficacy modulation and allosteric agonism, in addition to affinity modulation, were explored. There are now a range of newer analytical models, incorporating the operational model of agonist action, which has been shown to be highly useful in dissecting the mechanism of action of allosteric modulators exhibiting this wide range of behaviors. The availability of such analytical protocols will greatly aid the understanding of allosteric modulator function.

Also reviewed were the consequences of using different types of functional assays. This is a relatively new consideration for screening allosteric ligands, but as shown above, using transient read assays (such as intracellular calcium flux) as the screening assay compared to an accumulation format (such as cAMP or [^{35}S]-GTPγS binding) can markedly change the output for compounds that display intrinsic allosteric agonism. This may have implications for an SAR program and should be carefully considered when designing a screening program for allosteric modulators.

To increase throughput and avoid running multiple concentration–response curves for a single compound, the use of modulator titration curves as a screening paradigm is now popular. It is apparent that such curves can yield significant information about an allosteric ligand's mechanism of action, and a method of analysis for titration curves is presented and discussed. However, there are limitations to the amount of mechanistic information that can be obtained from a single concentration–response curve. This highlights an interesting conundrum for screening—is it worth reducing the amount of data points per compound to achieve a higher throughput and consequently reduce the level of information obtained? Or, is it better to incur the time and costs, and have more data points per compound and get a greater degree of understanding of a compound's profile? Clearly, there is a finite amount that a single concentration–response will yield; this will not change. However, there have been great improvements in the number of data points that can be screened in a single assay; 1536-well plate screening is commonplace in most HTS facilities. Therefore, it is not difficult to imagine that in the future, it may be possible to perform more quantitative screens, even screens that elucidate full allosteric modulator mechanism of action. This quantitative approach has already been attempted for enzyme targets [52] and could well represent the future of high-throughput compound screening for allosteric modulators of GPCRs.

REFERENCES

1. Insel, P.A., Tang, C.M., Hahntow, I., Michel, M.C. (2007) Impact of GPCRs in clinical medicine: Monogenic diseases, genetic variants and drug targets. *Biochim Biophys Acta*. 1768, 994–1005.

2. Kobilka, B.K. (2007) G protein coupled receptor structure and activation. *Biochim Biophys Acta*. 1768, 794–807.
3. Pin, J.P., Galvez, T., Prezeau, L. (2003) Evolution, structure, and activation mechanism of family 3/C G-protein-coupled receptors. *Pharmacol Ther*. 98, 325–354.
4. Gao, Z.G., Kim, S.K., Ijzerman, A.P., Jacobson, K.A. (2005) Allosteric modulation of the adenosine family of receptors. *Mini Rev Med Chem*. 5, 545–553.
5. Conn, P.J., Jones, C.K., Lindsley, C.W. (2009) Subtype-selective allosteric modulators of muscarinic receptors for the treatment of CNS disorders. *Trends Pharmacol Sci*. 30, 148–155.
6. Schetz, J.A. (2005) Allosteric modulation of dopamine receptors. *Mini Rev Med Chem*. 5, 555–561.
7. Leach, K., Charlton, S.J., Strange, P.G. (2007) Analysis of second messenger pathways stimulated by different chemokines acting at the chemokine receptor CCR5. *Biochem Pharmacol*. 74, 881–890.
8. Hu, J. (2008) Allosteric modulators of the human calcium-sensing receptor: Structures, sites of action, and therapeutic potentials. *Endocr Metab Immune Disord Drug Targets*. 8, 192–197.
9. Kew, J.N. (2004) Positive and negative allosteric modulation of metabotropic glutamate receptors: Emerging therapeutic potential. *Pharmacol Ther*. 104, 233–244.
10. Christopoulos, A. (2002) Allosteric binding sites on cell-surface receptors: Novel targets for drug discovery. *Nat Rev Drug Discov*. 1, 198–210.
11. Shoback, D.M., Bilezikian, J.P., Turner, S.A., McCary, L.C., Guo, M.D., Peacock, M. (2003) The calcimimetic cinacalcet normalizes serum calcium in subjects with primary hyperparathyroidism. *J Clin Endocrinol Metab*. 88, 5644–5649.
12. Fatkenheuer, G., Pozniak, A.L., Johnson, M.A., Plettenberg, A., Staszewski, S., Hoepelman, A.I., Saag, M.S., Goebel, F.D., Rockstroh, J.K., Dezube, B.J., Jenkins, T.M., Medhurst, C., Sullivan, J.F., Ridgway, C., Abel, S., James, I.T., Youle, M., van der Ryst, E. (2005) Efficacy of short-term monotherapy with maraviroc, a new CCR5 antagonist, in patients infected with HIV-1. *Nat Med*. 11, 1170–1172.
13. Stockton, J.M., Birdsall, N.J., Burgen, A.S., Hulme, E.C. (1983) Modification of the binding properties of muscarinic receptors by gallamine. *Mol Pharmacol*. 23, 551–557.
14. Kostenis, E., Mohr, K. (1996) Two-point kinetic experiments to quantify allosteric effects on radioligand dissociation. *Trends Pharmacol Sci*. 17, 280–283.
15. Litschig, S., Gasparini, F., Rueegg, D., Stoehr, N., Flor, P.J., Vranesic, I., Prézeau, L., Pin, J.P., Thomsen, C., Kuhn, R. (1999) CPCCOEt, a noncompetitive metabotropic glutamate receptor 1 antagonist, inhibits receptor signaling without affecting glutamate binding. *Mol Pharmacol*. 55, 453–461.
16. Kinney, G.G., O'Brien, J.A., Lemaire, W., Burno, M., Bickel, D.J., Clements, M.K., Chen, T.B., Wisnoski, D.D., Lindsley, C.W., Tiller, P.R., Smith, S., Jacobson, M.A., Sur, C., Duggan, M.E., Pettibone, D.J., Conn, P.J., Williams Jr., D.L. (2005) A novel selective positive allosteric modulator of metabotropic glutamate receptor subtype 5 has in vivo activity and antipsychotic-like effects in rat behavioral models. *J Pharmacol Exp Ther*. 313, 199–206.
17. Lazareno, S., Popham, A., Birdsall, N.J. (2000) Allosteric interactions of staurosporine and other indolocarbazoles with N-[methyl-(3)H]scopolamine and acetylcho-

line at muscarinic receptor subtypes: Identification of a second allosteric site. *Mol Pharmacol.* 58, 194–207.

18. Jakubik, J., Bacáková, L., el-Fakahany, E.E., Tucek, S. (1995) Subtype selectivity of the positive allosteric action of alcuronium at cloned M1-M5 muscarinic acetylcholine receptors. *J Pharmacol Exp Ther.* 274, 1077–1083.
19. May, L.T., Leach, K., Sexton, P.M., Christopoulos, A. (2007) Allosteric modulation of G protein-coupled receptors. *Annu Rev Pharmacol Toxicol.* 47, 1–51.
20. Mathiesen, J.M., Svendsen, N., Bräuner-Osborne, H., Thomsen, C., Ramirez, M.T. (2003) Positive allosteric modulation of the human metabotropic glutamate receptor 4 (hmGluR4) by SIB-1893 and MPEP. *Br J Pharmacol.* 138, 1026–1030.
21. Porter, R.H., Jaeschke, G., Spooren, W., Ballard, T.M., Büttelmann, B., Kolczewski, S., Peters, J.U., Prinssen, E., Wichmann, J., Vieira, E., Mühlemann, A., Gatti, S., Mutel, V., Malherbe, P. (2005) Fenobam: A clinically validated nonbenzodiazepine anxiolytic is a potent, selective, and noncompetitive mGlu5 receptor antagonist with inverse agonist activity. *J Pharmacol Exp Ther.* 315, 711–721.
22. Sachpatzidis, A., Benton, B.K., Manfredi, J.P., Wang, H., Hamilton, A., Dohlman, H.G., Lolis, E. (2003) Identification of allosteric peptide agonists of CXCR4. *J Biol Chem.* 278, 896–907.
23. Figler, H., Olsson, R.A., Linden, J. (2003) Allosteric enhancers of A1 adenosine receptors increase receptor-G protein coupling and counteract Guanine nucleotide effects on agonist binding. *Mol Pharmacol.* 64, 1557–1564.
24. Mitsukawa, K., Yamamoto, R., Ofner, S., Nozulak, J., Pescott, O., Lukic, S., Stoehr, N., Mombereau, C., Kuhn, R., McAllister, K.H., van der Putten, H., Cryan, J.F., Flor, P.J. (2005) A selective metabotropic glutamate receptor 7 agonist: Activation of receptor signaling via an allosteric site modulates stress parameters in vivo. *Proc Natl Acad Sci U S A.* 102, 18712–18717.
25. Hall, D.A. (2000) Modeling the functional effects of allosteric modulators at pharmacological receptors: An extension of the two-state model of receptor activation. *Mol Pharmacol.* 58, 1412–1423.
26. Black, J.W., Leff, P. (1983) Operational models of pharmacological agonism. *Proc R Soc Lond B Biol Sci.* 220, 141–162.
27. Kenakin, T., Jenkinson, S., Watson, C. (2006) Determining the potency and molecular mechanism of action of insurmountable antagonists. *J Pharmacol Exp Ther.* 319, 710–723.
28. Leach, K., Sexton, P.M., Christopoulos, A. (2007) Allosteric GPCR modulators: Taking advantage of permissive receptor pharmacology. *Trends Pharmacol Sci.* 28, 382–389.
29. Nawaratne, V., Leach, K., Suratman, N., Loiacono, R.E., Felder, C.C., Armbruster, B.N., Roth, B.L., Sexton, P.M., Christopoulos, A. (2008) New insights into the function of M4 muscarinic acetylcholine receptors gained using a novel allosteric modulator and a DREADD (designer receptor exclusively activated by a designer drug). *Mol Pharmacol.* 74, 1119–1131.
30. Langmead, C.J. (2009) *Screening for Positive Allosteric Modulators of Metabotropic Glutamate Receptors.* Lille, France: Society of Biomolecular Sciences Conference.
31. Chan, W.Y., McKinzie, D.L., Bose, S., Mitchell, S.N., Witkin, J.M., Thompson, R.C., Christopoulos, A., Lazareno, S., Birdsall, N.J., Bymaster, F.P., Felder, C.C. (2008)

Allosteric modulation of the muscarinic M4 receptor as an approach to treating schizophrenia. *Proc Natl Acad Sci U S A*. 105, 10978–10983.

32. Langmead, C.J. (2007) Screening for positive allosteric modulators: Assessment of modulator concentration-response curves as a screening paradigm. *J Biomol Screen*. 12, 668–676.
33. Kostenis, E., Waelbroeck, M., Milligan, G. (2005) Techniques: Promiscuous Gα proteins in basic research and drug discovery. *Trends Pharmacol Sci*. 26, 595–602.
34. Urban, J.D., Clarke, W.P., von Zastrow, M., Nichols, D.E., Kobilka, B., Weinstein, H., Javitch, J.A., Roth, B.L., Christopoulos, A., Sexton, P.M., Miller, K.J., Spedding, M., Mailman, R.B. (2007) Functional selectivity and classical concepts of quantitative pharmacology. *J Pharmacol Exp Ther*. 320, 1–13.
35. Mathiesen, J.M., Ulven, T., Martini, L., Gerlach, L.O., Heinemann, A., Kostenis, E. (2005) Identification of indole derivatives exclusively interfering with a G protein-independent signaling pathway of the prostaglandin D2 receptor CRTH2. *Mol Pharmacol*. 68, 393–402.
36. Suzuki, G., Tsukamoto, N., Fushiki, H., Kawagishi, A., Nakamura, M., Kurihara, H., Mitsuya, M., Ohkubo, M., Ohta, H. (2007) In vitro pharmacological characterization of novel isoxazolopyridone derivatives as allosteric metabotropic glutamate receptor 7 antagonists. *J Pharmacol Exp Ther*. 323, 147–156.
37. Davis, A.A. (2005) *The Atypical Agonist AC260584 Activates M1 Muscarinic Acetylcholine Receptor Without Inducing Adaptive Changes*. Washington, DC: Society for Neuroscience Conference.
38. McGuinness, D., Malikzay, A., Visconti, R., Lin, K., Bayne, M., Monsma, F., Lunn, C.A. (2009) Characterizing cannabinoid CB2 receptor ligands using DiscoveRx PathHunter beta-arrestin assay. *J Biomol Screen*. 14, 49–58.
39. Matsui, H., Lazareno, S., Birdsall, N.J. (1995) Probing of the location of the allosteric site on m1 muscarinic receptors by site-directed mutagenesis. *Mol Pharmacol*. 47, 88–98.
40. Chen, Y., Goudet, C., Pin, J.P., Conn, P.J. (2008) N-{4-Chloro-2-[(1,3-dioxo-1,3-dihydro-2H-isoindol-2-yl)methyl]phenyl}-2-hy droxybenzamide (CPPHA) acts through a novel site as a positive allosteric modulator of group 1 metabotropic glutamate receptors. *Mol Pharmacol*. 73, 909–918.
41. Galici, R., Echemendia, N.G., Rodriguez, A.L., Conn, P.J. (2005) A selective allosteric potentiator of metabotropic glutamate (mGlu) 2 receptors has effects similar to an orthosteric mGlu2/3 receptor agonist in mouse models predictive of antipsychotic activity. *J Pharmacol Exp Ther*. 315, 1181–1187.
42. Rodriguez, A.L., Nong, Y., Sekaran, N.K., Alagille, D., Tamagnan, G.D., Conn, P.J. (2005) A close structural analog of 2-methyl-6-(phenylethynyl)-pyridine acts as a neutral allosteric site ligand on metabotropic glutamate receptor subtype 5 and blocks the effects of multiple allosteric modulators. *Mol Pharmacol*. 68, 1793–1802.
43. Goudet, C., Gaven, F., Kniazeff, J., Vol, C., Liu, J., Cohen-Gonsaud, M., Acher, F., Prézeau, L., Pin, J.P. (2004) Heptahelical domain of metabotropic glutamate receptor 5 behaves like rhodopsin-like receptors. *Proc Natl Acad Sci U S A*. 101, 378–383.
44. Binet, V., Brajon, C., Le Corre, L., Acher, F., Pin, J.P., Prézeau, L. (2004) The heptahelical domain of GABA(B2) is activated directly by CGP7930, a positive allosteric modulator of the GABA(B) receptor. *J Biol Chem*. 279, 29085–29091.

45. Ray, K., Tisdale, J., Dodd, R.H., Dauban, P., Ruat, M., Northup, J.K. (2005) Calindol, a positive allosteric modulator of the human Ca(2+) receptor, activates an extracellular ligand-binding domain-deleted rhodopsin-like seven-transmembrane structure in the absence of Ca(2+). *J Biol Chem*. 280, 37013–37020.
46. Langmead, C.J., Watson, J., Reavill, C. (2008) Muscarinic acetylcholine receptors as CNS drug targets. *Pharmacol Ther*. 117, 232–243.
47. Spalding, T.A., Trotter, C., Skjaerbaek, N., Messier, T.L., Currier, E.A., Burstein, E.S., Li, D., Hacksell, U., Brann, M.R. (2002) Discovery of an ectopic activation site on the M(1) muscarinic receptor. *Mol Pharmacol*. 61, 1297–1302.
48. Lebon, G., Langmead, C.J., Tehan, B.G., Hulme, E.C. (2009) Mutagenic mapping suggests a novel binding mode for selective agonists of M1 muscarinic acetylcholine receptors. *Mol Pharmacol*. 75, 331–341.
49. Sur, C., Mallorga, P.J., Wittmann, M., Jacobson, M.A., Pascarella, D., Williams, J.B., Brandish, P.E., Pettibone, D.J., Scolnick, E.M., Conn, P.J. (2003) N-desmethylclozapine, an allosteric agonist at muscarinic 1 receptor, potentiates N-methyl-D-aspartate receptor activity. *Proc Natl Acad Sci U S A*. 100, 13674–13679.
50. Goodwin, J.A., Hulme, E.C., Langmead, C.J., Tehan, B.G. (2007) Roof and floor of the muscarinic binding pocket: Variations in the binding modes of orthosteric ligands. *Mol Pharmacol*. 72, 1484–1496.
51. Ward, S.D., Curtis, C.A., Hulme, E.C. (1999) Alanine-scanning mutagenesis of transmembrane domain 6 of the M(1) muscarinic acetylcholine receptor suggests that Tyr381 plays key roles in receptor function. *Mol Pharmacol*. 56, 1031–1041.
52. Inglese, J., Auld, D.S., Jadhav, A., Johnson, R.L., Simeonov, A., Yasgar, A., Zheng, W., Austin, C.P. (2006) Quantitative high-throughput screening, a titration-based approach that efficiently identifies biological activities in large chemical libraries. *Proc Natl Acad Sci U S A*. 103, 11473–11478.

CHAPTER 13

Ultra-High-Throughput Screening Assays for GPCRs

PRIYA KUNAPULI

In Vitro Sciences, External Discovery and Preclinical Sciences, Merck & Co. West Point, PA

13.1. INTRODUCTION

One of the largest gene families in the human genome is that of the cell surface receptors called G protein-coupled receptors (GPCRs), named after their functional role in activating heterotrimeric G proteins upon agonist stimulation. The human genome encodes approximately 1000 genes of the GPCR superfamily [1], representing ~45% of all molecular drug targets in the human genome, and is thus one of the largest families of druggable targets [2]. The genomics effort in the past decade has led to the identification of a number of GPCRs, some of which remain "orphan," to date, with unknown endogenous ligand and function [3, 4]. Significant research effort in this postgenomic era is aimed at deorphanizing these receptors and understanding their physiological role and potential therapeutic value.

GPCRs constitute a superfamily of cell surface receptors with a common motif of seven membrane-spanning domains. Agonist stimulation initiates a cascade of signals that involve the activation of the heterotrimeric GTP binding proteins (G proteins) [5, 6], resulting in second messenger-dependent modulation of various effector systems [7] and feedback regulation of G protein coupling by receptor desensitization and endocytosis [8, 9].

GPCRs are divided into three classes based on sequence similarity and the nature of the endogenous ligands that activate them: Class 1 GPCRs include rhodopsin-like receptors activated by biological amines, prostanoids, neuropeptides, chemokines, and so on. Class 2 receptors include secretin-like receptors with biological ligands like glucagon, secretin, parathyroid hormone, calcitonin, calcitonin gene-related peptide (CGRP), and so on. Class 3 receptors include

GPCR Molecular Pharmacology and Drug Targeting: Shifting Paradigms and New Directions, Edited by Annette Gilchrist

metabotropic glutamate-like receptors. In addition, many GPCRs remain "orphan" with no identified endogenous ligand. These "orphan" receptors provide additional opportunities for the pharmaceutical industry in the identification of novel targets for drug discovery.

GPCRs share common functional features in addition to a common structural motif. The traditional model of GPCR signal transduction involves the GPCR, a G protein, and an effector system, typically modulated by the Gα subunit or Gβγ subunit of the activated G protein. Ligand stimulation of the GPCR results in stabilization of conformational changes in the receptor, promoting receptor coupling with the heterotrimeric G proteins. Based on the preferential coupling of the activated receptor with the Gα subunit of choice (Gαs, Gαi, Gαq, etc.), GPCR activation results in the modulation of a variety of cellular effectors. For example, GPCRs that couple to Gαs upon activation result in the stimulation of adenylyl cyclase, with a resultant increase of intracellular 3′,5′-cyclic adenosine monophosphate (cAMP); GPCRs that couple to Gαi upon activation result in the inhibition of adenylyl cyclase, with a resultant decrease of intracellular cAMP; and GPCRs that couple to Gαq result in the increase of intracellular Ca^{2+} from internal stores. In addition to this traditional route of signal transduction via the Gα proteins, GPCRs have also been shown to activate signal transduction cascades through the Gβγ subunits, which are released upon G protein activation and subsequent dissociation of the Gα from the Gβγ [10]. Common examples of cellular effectors, which transduce the signal from an activated cell surface GPCR into intracellular second messengers, are adenylyl cyclase, phosphodiesterase, phospholipase A2, guanylyl cyclase, phospholipase C, phosphatidyl-inositol-3kinase, ion channels, and so on, which modulate intracellular second messengers like cAMP, diacylglycerol, inositol triphosphate, cyclic guanosine monophosphate (cGMP), arachidonic acid, intracellular Ca^{2+}, and other ions.

Other common characteristics of GPCRs include their mechanism of regulation. As a mechanism of terminating the signal transduction cascade after activation in order to preserve the transient nature of the signaling event, activated GPCRs are phosphorylated by GPCR kinases (GRKs) (homologous desensitization) [8] and/or by second messenger protein kinases like protein kinase A (PKA) and PKC (heterologous desensitization) within minutes of agonist stimulation [11]. The activated and phosphorylated GPCRs are then a substrate for the binding of the β-arrestin family of proteins, which effectively uncouple the GPCR from the Gα protein, thereby terminating the second messenger signaling process [9]. The phosphorylated and arrestin-bound GPCRs are then targeted for recycling or endocytosis.

Since GPCRs represent one of the largest gene families, it is not surprising that 45% of the currently marketed therapies target GPCRs [2, 12]. With the identification of several orphan GPCRs through the sequencing of the human genome, the pharmaceutical industry continues to explore therapeutic opportunities in this target class to treat diverse pathologies. The first step in the identification of novel therapies for GPCR targets is the identification of ligands that activate or inhibit GPCR function. Identification of GPCR ligands in a cost-efficient and time-effective manner can be accomplished by conducting a high-throughput screen (HTS) for the target GPCR of choice against large collections of small molecule chemical compounds present in corporate compound libraries (typically >1 MM individual chemical compounds).

Given the characteristics shared among this molecular target class, common GPCR features are often leveraged in establishing a suitable assay for the implementation of HTS of GPCR targets. Traditionally, HTS is conducted in a plate format, with each assay plate comprising 96 or more individual wells for measuring the aggregate reaction of the cell population in each well. Over the past decade, the demands on the HTS industry have been on a constant rise. The identification of many more molecular targets as a result of the human genome project, paralleled by the vast increases in the size of corporate compound libraries (by virtue of technological advances in compound synthesis including combinatorial and parallel synthesis, the inclusion of natural products in screening libraries, merging of drug archives, etc.), has resulted in a large number of molecular targets to be screened against an even larger and ever-increasing collection of compounds each year. In addition, the need to conserve difficult-to-obtain biological reagents, and to prevent compound library exhaustion, compounded by the necessity to reduce cost and time, have become prime drivers for the HTS industry to move toward miniaturization. Miniaturized HTS assays, or ultra-high-throughput screening (uHTS), are conducted in lower volume, higher assay plate densities, as shown in Fig. 13.1.

This chapter will describe popular methodologies to measure GPCR function for HTS and uHTS. There are a number of assays currently available to study GPCRs, including biochemical ligand binding and functional cell-based assays (Table 13.1). The latter category, albeit more complex, offers some obvious advantages over traditional ligand binding assays. Functional cell-based assays typically include second messenger and reporter gene assays, which depend directly or indirectly on the cellular signaling cascade initiated upon receptor activation (Fig. 13.2). More recently, high-content cell imaging assays and other cell-based assays monitoring receptor trafficking and regulation are becoming increasingly popular.

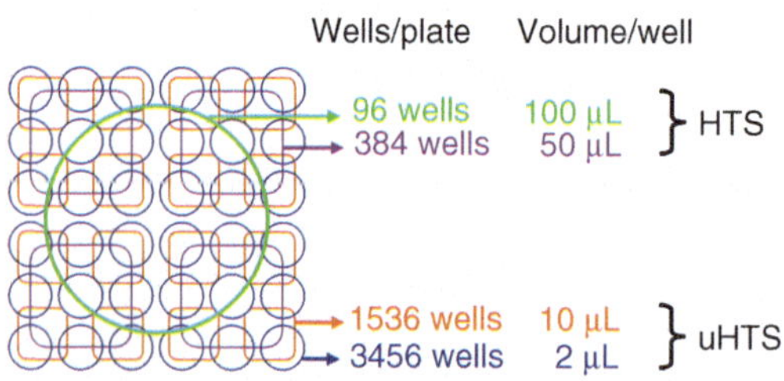

Figure 13.1 Assay plate formats and volumes for HTS and uHTS. Comparison of 96-, 384-, 1536-, and 3456-well plate formats for HTS and uHTS.

TABLE 13.1 GPCR Assays for HTS and uHTS

Biochemical Assays	Cell-Based Assays
Binding assays	Binding assays
Receptor binding filtration assays (96 well)	Whole cell binding (96, 384, 1536 well)
Receptor binding SPA assay (96, 384, 1536 well)	
Functional assays	Functional assays
GTPγS filter binding (96 well)	Second messenger assays
GTPγS SPA binding (96, 384, 1536 well)	cAMP assay (96, 384, 1536, 3456 well)
	Ca^{2+} (96, 384, 1536, 3456 well)
	IP1 (96, 384, 1536 well)
	Reporter gene assays
	Luciferase (96, 384, 1536, 3456 well)
	BLA (96, 384, 1536, 3456 well)
	Translocation assays
	Transfluor (96, 384, 1536 well)
	TANGO (96, 384, 1536, 3456 well)
	PathHunter (96, 384, 1536, 3456 well)
	MOCA (96, 384 well)
	Other
	Surefire (ERK assay)
	BRET

BRET, bioluminescence resonance energy transfer assay; ERK, extracellular signal-regulated kinase assay (PerkinElmer, Waltham, MA); MOCA, multipurpose original cellular assay (Patobios, Toronto, Canada).

13.2. ASSAY TYPES FOR GPCRs IN uHTS

13.2.1. Radioligand Displacement Assays

The radioligand displacement assay was one of the earliest assays used to study GPCRs. For nearly 50 years, the radioligand displacement assays, also known as competition binding assays, were used by the pharmaceutical industry as a preferred screening method to identify candidate drug molecules capable of binding to GPCRs, the largest group of pharmaceutical drug targets involved in a variety of physiological disorders.

These assays, performed with purified membrane preparations from cells [13] or tissues, or on whole cells [14], require prior knowledge of the receptor ligand (agonist or antagonist) and the availability of a radiolabeled ligand. The typical filter binding assay involves the incubation of cell membranes containing GPCRs with the cognate radioligand in the presence of unlabelled compound. The radioligand bound to the GPCRs of interest on the cell membranes is then separated from the free unbound ligand by rapid filtration through glass fiber filters. The amount of radioligand bound to the target receptor is

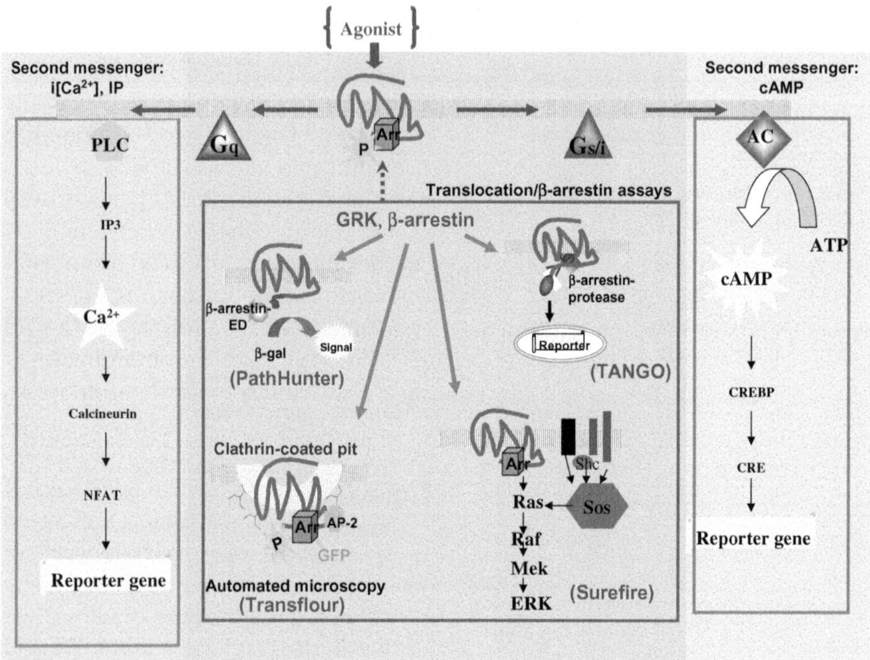

Figure 13.2 Functional cell-based assays for GPCRs for HTS and uHTS. A schematic representation of second messenger and β-arrestin translocation assays utilizing a variety of detection technologies ranging from fluorescent dyes to reporter genes to microscopy for detection of signal.

quantitated by scintillation counting, based on the conversion of the energy from radioactive decay into light photons that can be detected using the photomultiplier tubes in scintillation counters. Filtration binding assays using cell membranes can be used to quantitate GPCR expression (Bmax), as well as the affinity of the radioligand (Kd) for the GPCR. The traditional version of the radioligand binding assay requires many washes of the GPCR-containing cell membranes on the filter to wash out any nonspecifically associated signal, and is thus used primarily in traditional 96- or 384-well plate formats in HTS but is not adaptable for uHTS in 1536 or higher plate densities due to limitations with filtration technologies.

A modified version of the radioligand binding assay, called the "scintillation proximity assay" (SPA), paved the way for an automated approach to HTS for GPCRs. In SPA assays, the scintillant used to measure radioactivity is enclosed within a polystyrene bead coated with wheat germ agglutinin to capture cell membranes on the surface of the beads by nonspecific interaction between the glycosylated proteins on the cell membranes and wheat germ agglutinin. Thus, radioligand-bound GPCRs on cell membranes can be brought into close proximity to the SPA beads containing the scintillant, measured by an appropriate detector. SPA assays are homogeneous (addition only) assays and hence offer a significant advantage over the traditional filtration assays for radioligand binding in HTS laboratories. In addition, with the advent of the charge-coupled device (CCD) camera-based imagers, SPA assays are now amenable for uHTS in 1536-well plate formats. The main disadvantages of the SPA assay are related to the use of radioactive ligands and potential limitations associated with the specific activity of these radioligands. In addition, there is no straightforward method to determining Kd and/or Bmax from the SPA assays.

The advent of fluorescently labeled ligands offers an attractive alternative to radioactive binding assays for GPCRs. For example, the use of cyanine dye-labeled ligands ([Cy] ligands) with laser scanning imaging (LSI) of cells settled at the bottom of a microtiter plate enables the specific detection of the interaction of the fluorescently labeled ligand with the GPCR on the cell surface. In this approach, the need for physical separation of the GPCR-bound ligand from the "free" unbound ligand is accomplished by optical discrimination of solid–liquid phase partitioning of the fluorescent ligand (i.e., optical discrimination between the fluorescently labeled ligand on the cell surface vs. fluorescently labeled ligand in the surrounding buffer in the well), thereby circumventing the wash step commonly used in filter binding assays [15]. This methodology is HTS and uHTS compatible in 1536-well plate formats using detectors like the FMAT (Applied Biosystems, Foster City, CA) and Acumen Explorer (TTP Labtech., Cambridge, MA).

Fluorescence polarization (FP) is another methodology that uses fluorescently labeled ligands to measure binding to GPCRs in a homogeneous assay amenable to HTS and uHTS. FP is based on the principle that small molecules (e.g., fluorescently labeled ligands) rotate or tumble faster in solution that larger molecules (e.g., GPCRs on cell membranes) [16].

13.2.2. Functional Assays

While the past success of the receptor binding assays cannot be understated, advances in assay methodologies that measure GPCR function have begun to circumvent limitations of the ligand binding assay, such as the inability to specifically screen for agonists versus antagonists, the inability to identify allosteric modulators and potentiators, or the requirement of a high-affinity labeled ligand. The use of functional assays for primary HTS in place of traditional receptor binding assays has been a paradigm shift in the pharmaceutical industry. In this scenario, functional information on the activity of newly identified structural classes of compounds is made available early on in the lead identification and lead optimization process, while the more traditional receptor binding assays with lower throughput are used as secondary assays for follow-up, with smaller subsets of compounds to confirm specific interaction with the target GPCR of interest.

With the current trend in the pharmaceutical industry focusing on assays that measure downstream effects of receptor activation, a wide variety of second messenger and reporter gene assays have been developed and made commercially available for HTS. These functional biochemical and/or cell-based assays are exploited as appropriate for HTS and uHTS as described in the following sections.

The categories of GPCR functional assays are as follows:

1. assays to measure G protein activation,
2. indirect downstream measure of GPCR activation by reporter gene assays,
3. second messenger assays indicative of effector activation,
4. receptor regulation/trafficking assays, and
5. other downstream signaling assays.

GTPγS Functional Biochemical Assay Activation of a GPCR by an agonist stabilizes a conformational change in the receptor, resulting in enhanced interaction between the GPCR and the Gα subunit of the heterotrimeric G protein. Activation of the relevant G protein by the GPCR results in guanine nucleotide exchange on the α subunit (exchange of guanosine diphosphate [GDP] for guanosine triphosphate [GTP]) [7], which can be detected by measuring the binding of a radiolabeled nonhydrolyzable GTP analog (guanosine 5′-O-[gamma-thio]triphosphate) (GTPγ^{35}S) to the Gα subunit. The colocalization of the GTPγ^{35}S-bound α subunit with membrane-bound receptors upon GPCR activation can be measured by the classical filter binding assay (separation of bound from unbound) and also by SPA (similar to radioligand binding assays) (Fig. 13.3).

While the GTPγS assay can potentially be applicable to all classes of GPCRs irrespective of the signaling pathway activated by the ligand–receptor interaction, the higher affinity of the Gαi subclass of heterotrimeric G alpha proteins to GPCRs, coupled with higher expression of Gαi proteins in mammalian cells, makes the GTPγ^{35}S assay most suitable for Gαi-coupled GPCRs [17]. Recently, the SPA GTPγS assay has been shown to be miniaturized into 1536-well plate format for uHTS of Gαi-coupled GPCRs [18] (Fig. 13.3). It is important to note that the ability to dispense SPA beads reliably into 1536-well plate format for uHTS is critical to the successful use of this assay type for HTS. Furthermore, while the GTPγS is a functional assay proximal to the GPCR of interest, it requires a priori knowledge of the receptor ligand (i.e., is not suitable for orphan receptors where the ligand is unknown) and is a radioactive assay with limited application suitable primarily to Gαi-coupled GPCRs.

Reporter Gene Assays With the increasing popularity of functional assays to overcome some of the limitations of traditional ligand binding assays, reporter gene assays have made a significant impact on biological assays since the 1990s, especially in HTS laboratories. Reporter genes with readily measurable phenotypes have been used to measure the effects of signal transduction cascades originating from cell surface GPCRs on gene expression [19]. These assays offer high sensitivity, reliability, convenience, and adaptability to large-

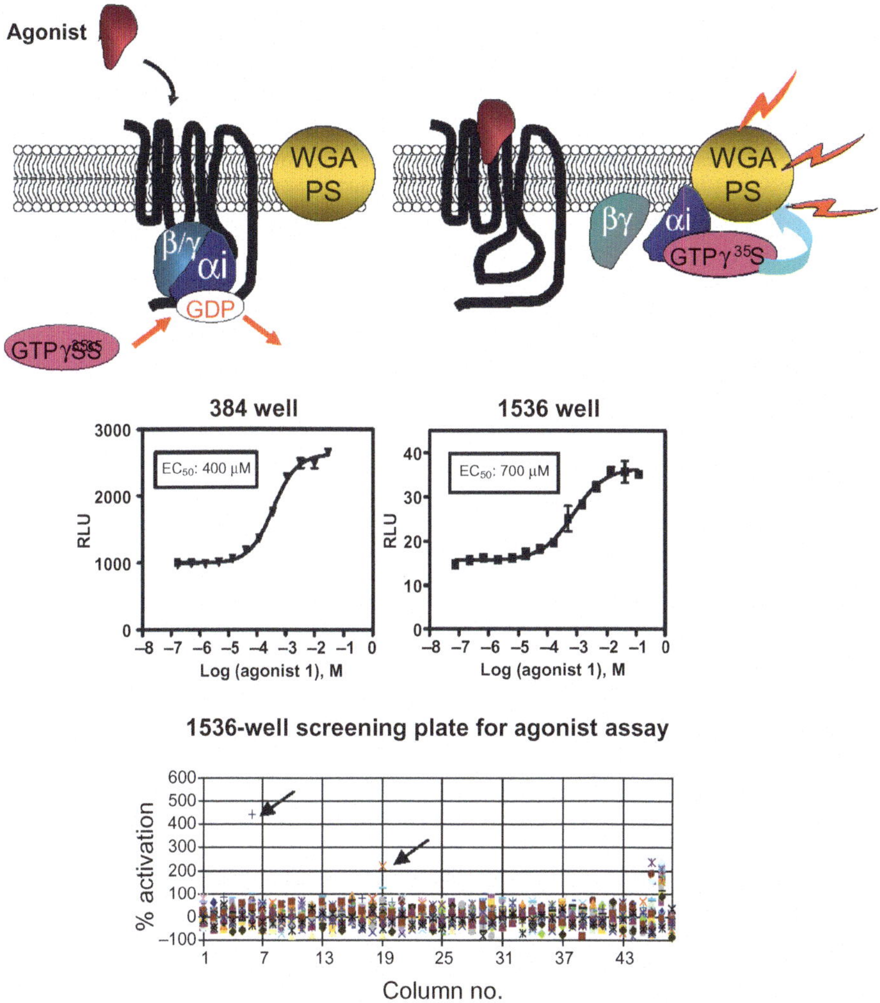

Figure 13.3 Schematic representation of GTPγS SPA assay for uHTS in 1536-well plate format.

scale measurements [20], as required for HTS. Commonly used reporter gene assays include chloramphenicol acetyltransferase (CAT) [21], β-galactosidase [22, 23], secreted alkaline phosphatase [24, 25], luciferase [26, 27], green fluorescent protein (GFP) [28], and β-lactamase (BLA) [29]. Table 13.2 summarizes the key features of some popular reporter gene assays and their suitability for use in HTS and miniaturized uHTS settings.

Among the reporter gene assays, the BLA reporter gene assay is ideally suited for miniaturized uHTS and was the first of its kind to be introduced in the late 1990s when other reporter gene assays were amenable only to lower

TABLE 13.2 Reporter Gene Assays for Measuring GPCR Function in HTS and uHTS

Reporter Gene Technology	Amplification	HTS Plate Density (Wells/Assay Plate)	Ratiometric	FACS Compatible	Stable Cell Lines
BLA	+	96, 384, 1536, 3456	Yes	Yes	Yes
Luciferase	+	96, 384, 1536, 3456	No	No	Yes
CAT	+	96	No	No	No
Secreted alkaline phosphatase	+	96, 384	No	No	Yes
Green fluorescence protein	–	96, 384, 1536	No	Yes	Yes
β-galactosidase	+	96, 384, 1536, 3456	No	Yes	Yes

FACS, fluorescence-activated cell sorting.

throughput assay formats (6–96-well formats) and required multiple washes, transfer steps, cell lysis, as well as longer assay time (>24 h). Development of a cell-permeable fluoregenic BLA substrate, CCF2/AM [29], was a key step in the use of BLA as a reporter gene in mammalian cells [30]. The CCF2 substrate/dye (or CCF4/AM) is cell permeable and consists of coumarin and fluorescein moieties connected by a β-lactam-containing cephalosporin core. Within cells, excitation of CCF2 (or CCF4) at 405 nm leads to fluorescence resonance energy transfer (FRET) from the coumarin moiety to the fluorescein derivative, resulting in green light emission at 530 nm [29]. In the presence of BLA reporter protein in the cell, the substrate is cleaved at the β-lactam ring, spatially separating the coumarin and fluorescein moieties, thereby disrupting the FRET. Hence, excitation of coumarin at 405 nm would result in blue fluorescence emission in the absence of FRET, detected at 460 nm [13]. Thus, basal unstimulated cells appear green by fluorescence microscopy due to FRET, while cells producing BLA appear blue (Fig. 13.4). The BLA activity, reported as a ratiometric readout of 460 nm/530 nm emission, offers a significant advantage for HTS, primarily, to normalize for viable cell numbers per well [31].

The BLA reporter gene assay is well suited for the study of cell surface GPCRs [31]. Receptor signaling through intracellular second messengers can be linked to corresponding modulation of the BLA gene via appropriate promoters. In the case of GPCRs coupled to the Gαq pathway, the BLA gene is engineered under the control of the nuclear factor of activated T cells (NFAT) [32] promoter, which is responsive to changes in intracellular Ca^{2+}. Receptor

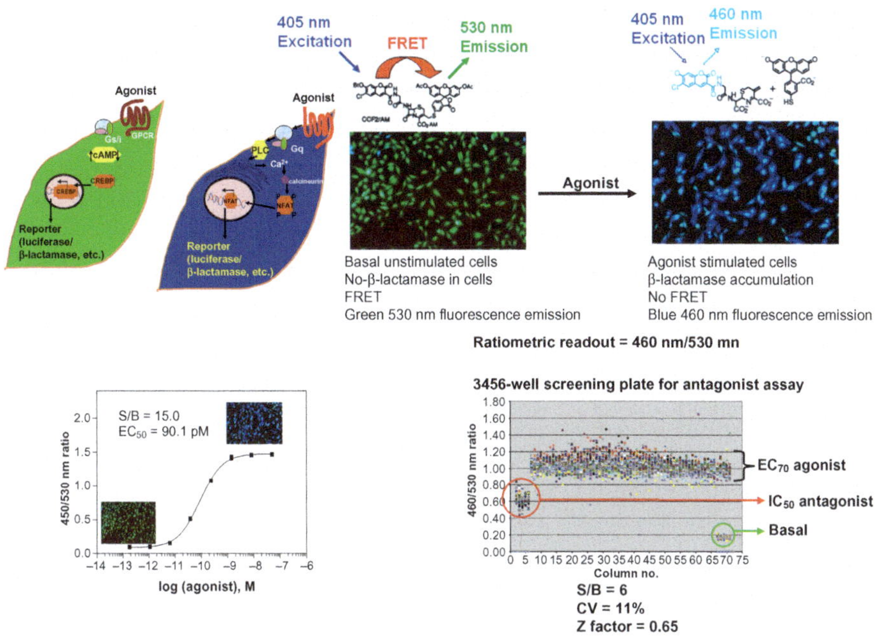

Figure 13.4 Principles and utilization of the BLA reporter gene assay for the detection of GPCR antagonists in 3456-well plate format.

activation and subsequent increase in i[Ca^{2+}] results in the activation of the Ca^{2+}-dependent phosphatase, calcineurin (Fig. 13.4). Activated calcineurin, in turn, dephosphorylates and activates the cytoplasmically located inactive NFAT. Dephosphorylated and activated NFAT translocates into the nucleus and initiates transcription of NFAT-regulated genes (i.e., the BLA gene in engineered cells). For GPCRs coupled to modulation of intracellular cAMP by stimulation or inhibition of adenylyl cyclase via the G proteins Gαs or Gαi, respectively, the BLA gene may be engineered under the control of the cAMP response element, CRE. Thus, Gαs-coupled GPCRs increase intracellular cAMP, resulting in activation of the CRE binding protein (CREB), increased CRE activity, and BLA gene activation [33, 34], while Gαi-coupled GPCRs would result in a decrease in intracellular cAMP, CRE activity, and BLA gene transcription.

Additional details on the use of the BLA reporter gene technology to select and isolate functional GPCR clones and optimize the assay for HTS or uHTS, assay limitations, as well as example assay protocols for uHTS are discussed in Kunapuli [31].

Recent improvements in luciferase reporter gene technology, including improvements in substrates, have now enabled this reporter gene technology to be utilized more broadly for HTS and uHTS. Moreover, the availability of dual Glo luciferase systems using firefly and Renilla luciferase enzymes with appropriate substrate kinetics now provides the opportunity to normalize the luminescence readout in luciferase reporter gene assays to the number of cells per well, similar to the BLA technology. In addition, the luminescence mode of detection is sometimes favored over the fluorescence mode for HTS due to the potential for interference of colored compounds with fluorescent readout.

Although the reporter gene assays were beneficial for providing a functional response to GPCR activation, these assays are relatively long (several hours) with readouts that are quite distal from the actual target GPCR, allowing for potential compound interference at various stages of the signaling pathway when used in HTS [13, 31]. Hence, hit funneling strategies for post-primary screening should be planned appropriately, with the use of multiple upstream assays, such as a second messenger assay, in addition to a receptor binding assay to confirm direct compound interaction with the target GPCR of interest.

Second Messenger Assays

cAMP The concept of transmembrane signaling via GPCRs, which transduces extracellular signals into intracellular messages, was established with the discovery of cAMP by Sutherland et al. [35]. Specifically, binding of an agonist to a GPCR could either increase (via coupling through Gαs) or decrease (via coupling through Gαi) the rate at which cAMP is generated in cells through activation or inhibition of adenylyl cyclase, respectively [7]. There are a variety of assays designed to directly measure levels of cAMP. The original cAMP assays almost 50 years ago consisted of radiolabeling of adenine nucleotides, incubating with hormones, followed by separation on Dowex columns (Sigma-Aldrich, St. Louis, MO). The current cAMP detection methodologies have come a long way in offering a wide selection of assay types, detection, and plate formats suitable for all modes including uHTS. The currently popular cell-based cAMP assays used as a measure of GPCR activation are all based on competition between endogenously produced cellular cAMP and exogenously added labeled cAMP for interaction with anti-cAMP antibodies using a variety of detection technologies [36, 37].

The Flashplate (PerkinElmer) assay is a modification of the original radioimmunoassay using cAMP antibodies and radiolabeled cAMP in a classical enzyme-linked immunosorbant assay (ELISA), converted into a plate mode with scintillant coated wells of the assay plate for implementation in HTS laboratories. While successfully used in HTS, this assay type has limitations in further miniaturization beyond the 384-well plate format, in addition to the disadvantages of using large quantities of radioactivity for HTS.

The HTRF (Cisbio, France) and Lance (PerkinElmer) assays are based on time-resolved FRET (TR-FRET) between a labeled antibody to cAMP and a labeled cAMP molecule. The Lance cAMP technology utilizes energy transfer from europium to an acceptor like Alexa Fluor 647. This principle has been utilized in a competitive binding assay to measure cellular cAMP. The signal in this assay is inversely proportional to the amount of cAMP. TR-FRET assays exhibit relatively low background and high signal/background, and are thereby suitable for HTS. The HTRF assay is similar to the TR-FRET and measures competition between biotinylated labeled cAMP and endogenous cAMP using ratiometric detection of fluorescence at 665 nm and 620 nm.

The enzyme fragment complementation technology uses the principle of complementation between an inactive fragment (enzyme donor, ED) tagged with cAMP and a larger complimentary portion of the β-galactosidase enzyme. Competition between endogenous cAMP and ED-labeled cAMP for binding to an anti-cAMP antibody determines the amount of the reconstituted β-galactosidase active enzyme generated, which is measured by substrate conversion into a fluorescent or luminescent product. This technology has successfully been miniaturized into 1536-well and 3456-well plate formats for uHTS for both Gαs and Gαi-coupled GPCRs [38, 39]. In addition, the availability of the assay technology in luminescence mode offers the opportunity to overcome fluorescence interference often encountered during HTS of large sample collections.

Another technology to measure cellular cAMP is the cAMP-Glo™ assay from Promega (Madison, WI), which is a homogeneous, bioluminescent assay suitable for uHTS in 3456-well plate format. This assay is based on the activation of PKA by cAMP, thereby decreasing the cellular adenosine triphosphate (ATP), and consequent decrease in light production in a coupled luciferase reaction.

Other detection modalities for measuring cAMP generation as a means of monitoring GPCR activation include FP technology, electrochemiluminescence technology (ECL) by Meso Scale Discovery (Gaithersburg, MD), and AlphaScreen technology by PerkinElmer using beads coated with anti-cAMP antibodies. The latter two technologies, albeit being HTS compatible, require a specific detector for the measurements.

Given these wide varieties of cAMP assays currently available to study GPCRs in HTS, the selection of an assay type for use in general laboratory or in HTS mode should be made with care based on assay sensitivity, dynamic range, the use of whole cell or membrane assays, and so on. Some other critical considerations are the use of the assay for Gαs-coupled versus Gαi-coupled receptors (Gαi-coupled receptors are typically more challenging to analyze), detection of agonists or antagonists, and the need for pharmacological agents like forskolin for Gαi-coupled receptors (to increase basal cAMP levels, allowing visualization of the actions of a Gi-coupled GPCR agonist, which will decrease basal cAMP). Among these different scenarios for measuring cellular cAMP downstream of GPCR activation, detection of antagonists of a Gαi-coupled GPCR is by far the most challenging for uHTS, with the need for titrating the optimal amount of both forskolin and agonist stimulation that would allow for a sufficient assay window (signal/basal) while maintaining

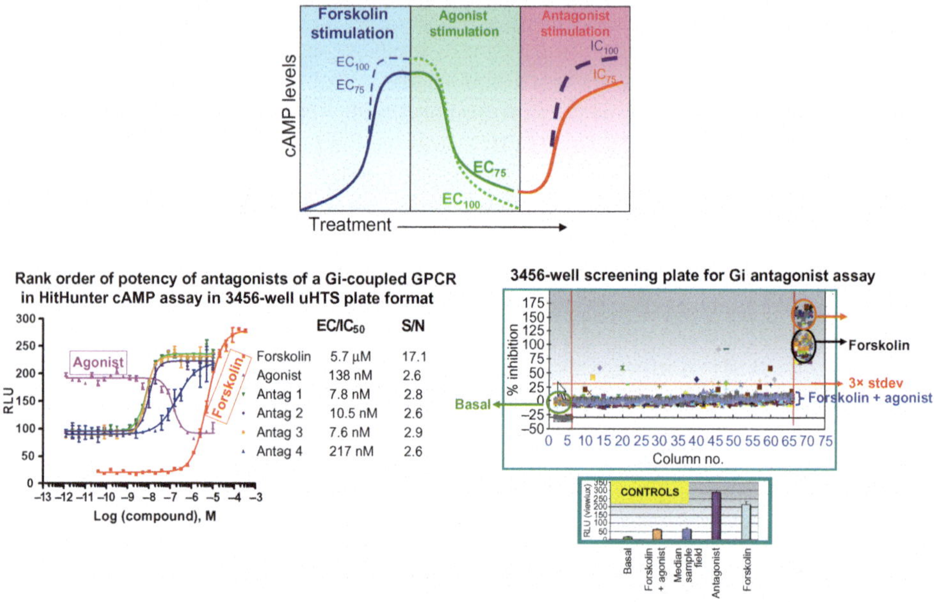

Figure 13.5 cAMP assays for Gαi-coupled GPCRs in 3456-well plate format for uHTS.

suitable assay sensitivity (Fig. 13.5). Some cAMP technologies are being used routinely in this complex setting for uHTS in 3456-well plate format [39].

Calcium Upon receptor activation, GPCRs known to couple to the Gαq class of G proteins result in the activation of phospholipase Cβ, with inositol phosphates (InsP) turnover at the plasma membrane, followed by an increase in intracellular calcium (i[Ca^{2+}]) [40–42]. The transient increase in intracellular calcium from internal stores, like the endoplasmic reticulum, can be detected by fluorescent dyes sensitive to Ca^{2+} ions, bioluminescent photoproteins like aequorin, or Photina (PerkinElmer), as well as reporter genes via the NFAT transcription factor (see previous section on BLA reporter gene).

Bioluminescent photoproteins, such as aequorin, can be used as a reporter for GPCR-mediated Ca^{2+} signaling, as an alternate to direct Ca^{2+} measurements [43]. Although aequorin is an indirect, albeit sensitive, "reporter" of i[Ca^{2+}] and is dependent on transcription and translation (often transient) of the apoaequorin gene, this technology differs from the classical reporter gene technology in that the reporter gene transcription/translation does not depend on cellular activation or the signaling pathway of interest.

Measurements of intracellular Ca^{2+} with fluorescent Ca^{2+} dyes upon GPCR activation has been a popular method for studying Gq-coupled GPCRs for HTS since the identification of FURA-2 [44]. More recently, newer generation

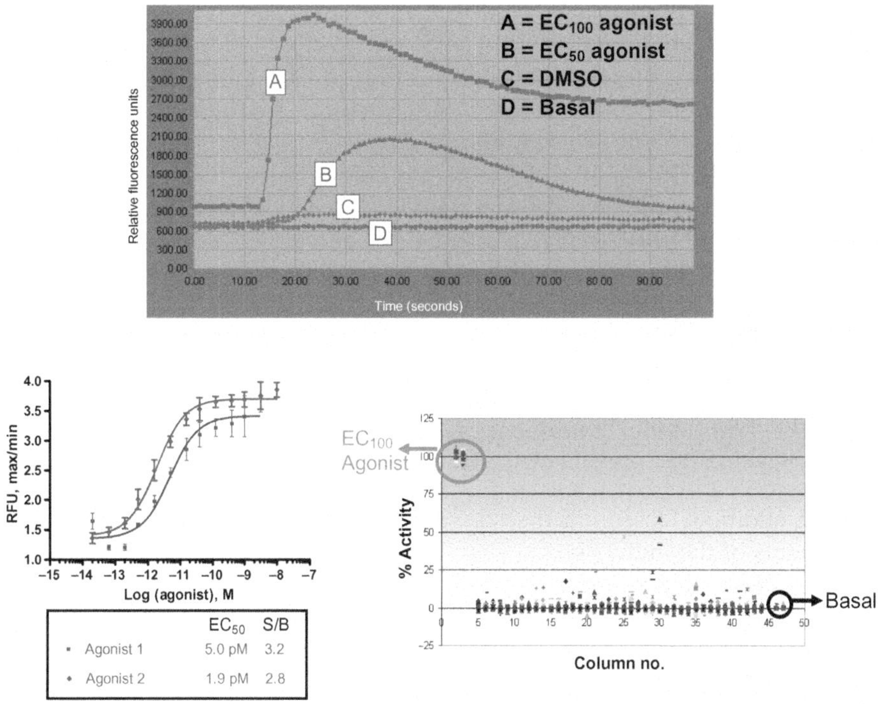

Figure 13.6 Second messenger Ca^{2+} screen in 1536-well plate format for Gαq-coupled GPCR.

calcium indicators like Fluo4 offer greater sensitivity and larger changes in fluorescence intensity upon calcium binding and are widely used in HTS. In the past few years, this methodology has been miniaturized into a 1536-well screening format with the development of CCD-based plate imaging detectors like the fluorescent imaging plate reader (FLIPR, Molecular Devices, Sunnyvale, CA) (Fig. 13.6). In addition, fluorescent dye-based calcium assays have also evolved into homogenous formats with "no-wash" dye protocols (primarily aimed at reducing the background fluorescence in the wells) for further adaptability in uHTS [45].

In addition to being a common methodology for studying Gαq-coupled GPCRs, the measurement of intracellular Ca^{2+} has also been utilized for numerous orphan GPCRs with unknown second messenger signaling through the use of chimeric (Gqi5) and/or promiscuous G proteins like Gα15 and Gα16 signaling [31, 46].

Inositol Phosphates GPCR signaling through Gαq, Gαo, or Gβγ activate phospholipase Cβ, which hydrolyzes phosphatidyl-inositol-4,5-bisphosphate (PIP2) to diacylglycerol and inositol 1,4,5-triphosphate [47]. InsP3 binds to calcium channels on the endoplasmic reticulum, causing a release of the internal Ca^{2+} stores within the cell. Unlike assays to measure intracellular Ca^{2+}, few assays exist to measure InsP3 that are suitable for automated HTS, in part due to its inherently short half-life. The HitHunter InsP3 assay (DiscoveRx, Fremont, CA) is a competitive binding assay wherein cellular InsP3 displaces a fluorescent derivative of InsP3 from a binding protein. This assay is based on the principles of FP. An alternative HTRF assay from CisBio is based on the measurement of IP1, a downstream metabolite of IP3, which accumulates in cells following $G\alpha q$ receptor activation and is stable in the presence of LiCl, thus allowing a suitable format for a functional HTS assay. The HTRF assay for measuring IP1 is relatively new but potentially applicable even for uHTS in 1536-well plate formats.

Receptor Trafficking/Translocation Assays Recently, cell imaging assays monitoring GPCR trafficking have become increasingly popular. GPCR trafficking assays are independent of receptor signaling and are thus ideally suited for orphan receptors. In addition, these assays provide a valuable measure of receptor desensitization, an important feature for the use of GPCR agonists as potential therapeutic agents. The most popular GPCR imaging assays are based on the principles of receptor desensitization and internalization monitored directly or indirectly by GFP [28].

Elucidation of the mechanism of regulation of GPCR function by receptor desensitization in the 1990s laid the foundation for the receptor internalization/trafficking assay. The approach is unique, being independent of the second messenger signaling modulated by the receptor–ligand interaction. GPCR desensitization (waning of the receptor responsiveness with time) is mediated primarily by two protein families: the GRKs and the arrestins [48]. Agonist stimulation of GPCRs promotes the phosphorylation of serine/threonine residues located predominantly in the carboxyl-terminal tail and/or the third intracellular loop of the receptor by the family of GRKs. The activated, phosphorylated GPCRs are a substrate for the arrestin family of proteins, which translocate from the cytoplasm to the receptors at the plasma membrane. Arrestin binding to the activated and GRK-phosphorylated receptors effectively uncouples the receptor–G protein interaction, thereby terminating receptor signaling [49].

In addition to offering a novel mechanism to study GPCR activation, β-arrestin/GPCR interaction appears to be a distinct signaling cascade independent of classic G protein signaling. Recent publications have shown that a "ligand bias" may channel a compound to preferentially block the β-arrestin signaling cascade over the classic G protein-evoked second messenger signaling cascade [50].

The arrestin proteins bound to the activated and phosphorylated receptor subsequently target GPCRs for endocytosis via clathrin-coated pits [51]. The movement of activated GPCRs from the plasma membrane into either pits or vesicles can be monitored as an indication of receptor activation and forms the basis of a functional cell-based assay that employs a fluorescent microscope. Such methodology is also amenable for HTS/uHTS with the advent of automated microscopy and is now being used in high-content screening (HCS) in HTS laboratories. These HCS assays are typically conducted in 384-well plate format, although advances in cellular imaging detectors have enabled miniaturization into 1536-well plate format with readers like the MDC Ultra (Molecular Devices, Sunnyvale, CA), Evotec Opera (PerkinElmer), and so on. An advantage to microscopy-based technologies like Transfluor™ (Molecular Devices) is their ability to analyze potential toxicity associated with screening compounds, as the technology is essentially based on visual inspection of cellular morphology. The use of specific algorithms to monitor cell shape and size, nuclear shape and size, and so on can be used to infer compound toxicity [52] (Fig. 13.7).

Among the different GPCR functional cell-based assays currently available, the receptor trafficking assay offers value as a receptor "proximal"

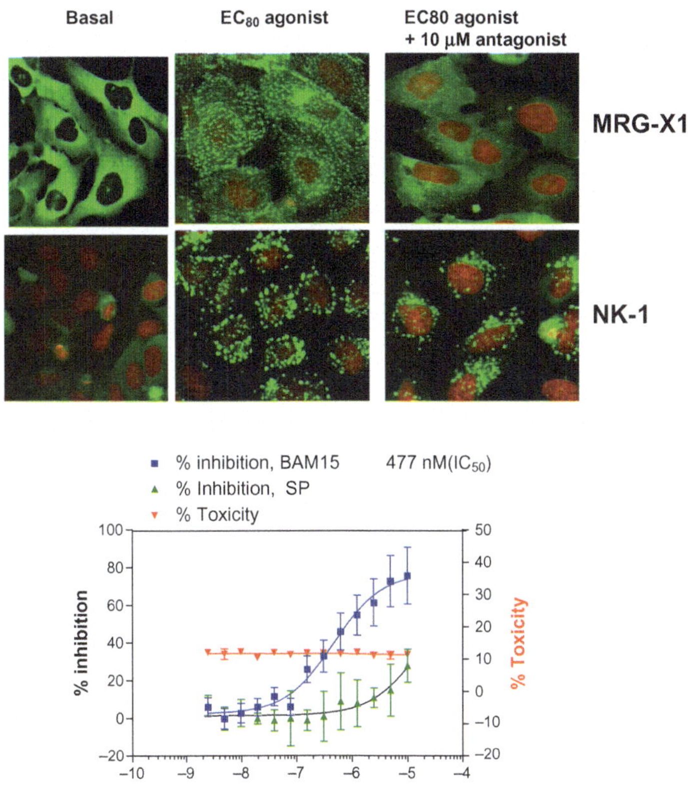

Figure 13.7 High-content cell imaging GPCR trafficking assay measuring receptor specificity and toxicity.

readout (unlike the reporter gene assays) and is particularly advantageous for receptors that lack a suitable radioligand for use in receptor binding assays. In addition, trafficking assays can be used for most GPCRs without prior knowledge of the agonist-induced signaling cascade. Another advantage of this functional assay is the potential to identify different categories of ligands: classical competitive antagonists (also identified by the receptor binding assay) as well as allosteric modulators (i.e., compounds that do not inhibit the binding of a radioligand to the receptor) that presumably inhibit receptor function by binding to sites on the receptor distinct from the agonist binding pocket.

The most direct method to monitor GPCR trafficking in cells is to express the GPCR of interest recombinantly as a chimera with GFP in the carboxyl-terminal tail and monitor the fluorescence localization in the cell upon receptor activation. The disadvantage of this method is the use of a recombinant fusion receptor, albeit at the C-tail [52].

An indirect method to measure GPCR activation using localization of an arrestin-GFP chimera was initially developed by Norak Biosciences (Morrisville, NC) and commercialized by MDC (Molecular Devices) as the Transfluor assay. The primary advantage of this assay over the direct GFP labeling of the receptor is the expression of native unaltered receptors in the recombinant cells. In this assay, the arrestin-GFP fluorescence is localized in the cytoplasm as a diffuse signal when receptors are inactive at the plasma membrane. Upon receptor activation, the arrestin-GFP first translocates to the activated receptors at the plasma membrane and is then subsequently internalized into either small pits near the plasma membrane (as occurs with the MRG-X1 receptor) [52] or larger vesicles (as occurs with the NK-1 receptor) [52]. Thus, the assay is based on the affinity of the arrestin for the activated and phosphorylated receptor [53] (Fig. 13.7). An advantage of the Transfluor assay is that it does not require manipulation of the receptor sequence. Several GPCR screening campaigns have been conducted with this indirect GFP tracking mode [52].

β-arrestin translocation to the activated receptor can also be measured by the reporter gene assay technology, TANGO (Fig. 13.2). The TANGO assay system from Invitrogen (Madison, WI) is a cell-based assay technology for monitoring protein–protein interaction and is applicable to study GPCRs based on the interaction of β-arrestin with the activated receptor. In this assay, the GPCR of interest is expressed as a recombinant fusion protein expressing the tetracyclin-controlled transactivator (tTA) in the C-terminal tail of the receptor with a protease consensus sequence at the intersection of the C-terminal tail of the GPCR and the tTA site. The β-arrestin molecule is expressed as a fusion with a protease. Upon GPCR activation, β-arrestin binding to the receptor brings the protease in close proximity to the protease cleavage site, thereby cleaving and releasing the tTA transcription factor, which translocates to the nucleus and transcribes the downstream reporter gene (luciferase or BLA). The TANGO reporter gene assay is suitable for HTS and uHTS.

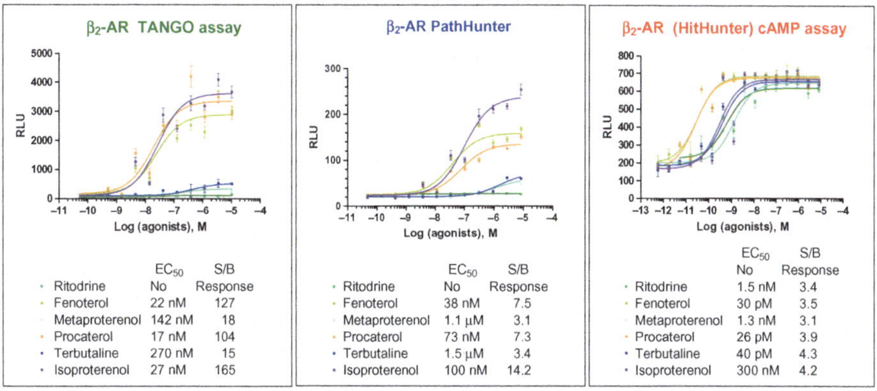

Figure 13.8 Rank order of potency of β_2-adrenergic receptor ligands in TANGO, PathHunter, and HitHunter cAMP assays in 3456-well plate format.

The PathHunter β-arrestin assay from DiscoveRx is based on the principles of enzyme fragment complementation (similar to the cAMP assay). The target GPCR is expressed as a recombinant fusion protein with a fragment of the β-galactosidase enzyme and is coexpressed with a β-arrestin fusion protein containing the complementary portion of the β-galactosidase enzyme. GPCR activation and consequent translocation of β-arrestin to the cell membrane results in complementation of the two parts of the β-galactosidase enzyme to regenerate active enzyme capable of cleaving a chemiluminescent substrate that can be detected as light emission. The amount of light produced in this system is directly correlated to the amount of the receptor that is activated.

Both the TANGO and the PathHunter assays offer limited off-target activity during HTS, since they both utilize recombinant tagged receptors; moreover, both approaches are applicable for miniaturized screening in 3456-well plate formats (Fig. 13.8). It is important to bear in mind that comparing different GPCR assay technologies may reveal differences in compound sensitivities (IC_{50} or EC_{50} values). This may be due, in part, to the inherent differences in assay sensitivities or the differences in assay conditions (preincubation of compounds vs. coincubation), buffer and serum concentrations, incubation time and temperature, presence or absence of washes, and so on (Table 13.3).

13.3. SUMMARY

There are currently a myriad of GPCR assays to consider when embarking on a HTS campaign—biochemical or cell based, functional, or binding, and so on. While all HTS assays would benefit from the use of appropriate downstream assays to eliminate off-target activities, functional assays effectively cast a larger net during HTS, identifying compounds with specific and nonspecific effects. Compounds identified by binding assays, however, do not always

TABLE 13.3 Comparison of GPCR Antagonist Potencies among Different HTS Assays

Sample No.	IC_{50} (nM)			
	Whole Cell Binding	FLIPR	BLA	Transfluor
8	320	103	50	730
9	ND	157	64	2800
10	182	187	80	550
11	ND	209	90	6782
12	1220	904	124	2200
13	607	751	140	3600
14	1430	220	163	2200
15	630	334	165	4800
16	ND	1500	200	5200
17	ND	492	209	10,000
18	1290	460	244	5200
19	2400	3300	252	8800
20	1850	1300	290	6300
21	1180	2700	375	29,000
22	2510	9800	418	13,500
23	1850	632	556	5600

Assay Plate Format Sample	Receptor Binding 384 Well IC_{50} (nM)	HitHunter cAMP 1536 Well IC_{50} (nM)	PathHunter 3456 Well IC_{50} (nM)
1	11	2.5	23
2	3	0.85	13
3	18	5.3	9
4	18	2.3	200
5	0.49	6.8	176
6	9.3	165	24
7	6.1	31	24
8	739	111	31
9	865	259	177
10	28	33.5	170
11	255	781	171
12	83	7.5	599

translate into having functional efficacy or altering cellular physiology in a predictable manner. The assay of choice for HTS or uHTS should be selected with care depending on the signaling cascade and availability of activating ligands, with preference given to more proximal readouts (second messenger vs. reporter gene). Recent advances in GPCR biology also point to additional signaling cascades originating from activated GPCRs in addition to the classical second messenger-driven signaling [50]. The physiological effects driven by each of these signaling pathways need to be recognized and taken into account in the selection of suitable assays for HTS and uHTS.

ACKNOWLEDGMENTS

I would like to thank Dr. Eric Johnson for his critical review of HTS technologies for GPCRs and to the staff of Automated Biotechnology who contributed to the work mentioned in this review.

REFERENCES

1. Venter, J.C., Adams, M.D., Myers, E.W., Li, P.W., Mural, R.J., Sutton, G.G., Smith, H.O., Yandell, M., Evans, C.A., Holt, R.A., Gocayne, J.D., Amanatides, P., Ballew, R.M., Huson, D.H., Wortman, J.R., Zhang, Q., Kodira, C.D., Zheng, X.H., Chen, L., Skupski, M., Subramanian, G., Thomas, P.D., Zhang, J., Gabor Miklos, G.L., Nelson, C., Broder, S., Clark, A.G., Nadeau, J., McKusick, V.A., Zinder, N., Levine, A.J., Roberts, R.J., Simon, M., Slayman, C., Hunkapiller, M., Bolanos, R., Delcher, A., Dew, I., Fasulo, D., Flanigan, M., Florea, L., Halpern, A., Hannenhalli, S., Kravitz, S., Levy, S., Mobarry, C., Reinert, K., Remington, K., Abu-Threideh, J., Beasley, E., Biddick, K., Bonazzi, V., Brandon, R., Cargill, M., Chandramouliswaran, I., Charlab, R., Chaturvedi, K., Deng, Z., Di Francesco, V., Dunn, P., Eilbeck, K., Evangelista, C., Gabrielian, A.E., Gan, W., Ge, W., Gong, F., Gu, Z., Guan, P., Heiman, T.J., Higgins, M.E., Ji, R.R., Ke, Z., Ketchum, K.A., Lai, Z., Lei, Y., Li, Z., Li, J., Liang, Y., Lin, X., Lu, F., Merkulov, G.V., Milshina, N., Moore, H.M., Naik, A.K., Narayan, V.A., Neelam, B., Nusskern, D., Rusch, D.B., Salzberg, S., Shao, W., Shue, B., Sun, J., Wang, Z., Wang, A., Wang, X., Wang, J., Wei, M., Wides, R., Xiao, C., Yan, C., Yao, A., Ye, J., Zhan, M., Zhang, W., Zhang, H., Zhao, Q., Zheng, L., Zhong, F., Zhong, W., Zhu, S., Zhao, S., Gilbert, D., Baumhueter, S., Spier, G., Carter, C., Cravchik, A., Woodage, T., Ali, F., An, H., Awe, A., Baldwin, D., Baden, H., Barnstead, M., Barrow, I., Beeson, K., Busam, D., Carver, A., Center, A., Cheng, M.L., Curry, L., Danaher, S., Davenport, L., Desilets, R., Dietz, S., Dodson, K., Doup, L., Ferriera, S., Garg, N., Gluecksmann, A., Hart, B., Haynes, J., Haynes, C., Heiner, C., Hladun, S., Hostin, D., Houck, J., Howland, T., Ibegwam, C., Johnson, J., Kalush, F., Kline, L., Koduru, S., Love, A., Mann, F., May, D., McCawley, S., McIntosh, T., McMullen, I., Moy, M., Moy, L., Murphy, B., Nelson, K., Pfannkoch, C., Pratts, E., Puri, V., Qureshi, H., Reardon, M., Rodriguez, R., Rogers, Y.H., Romblad, D., Ruhfel, B., Scott, R., Sitter, C., Smallwood, M., Stewart, E., Strong, R., Suh, E., Thomas, R., Tint, N.N., Tse, S., Vech, C., Wang, G., Wetter, J., Williams, S., Williams, M., Windsor, S., Winn-Deen, E., Wolfe, K., Zaveri, J., Zaveri, K., Abril, J.F., Guigó, R., Campbell, M.J., Sjolander, K.V., Karlak, B., Kejariwal, A., Mi, H., Lazareva, B., Hatton, T., Narechania, A., Diemer, K., Muruganujan, A., Guo, N., Sato, S., Bafna, V., Istrail, S., Lippert, R., Schwartz, R., Walenz, B., Yooseph, S., Allen, D., Basu, A., Baxendale, J., Blick, L., Caminha, M., Carnes-Stine, J., Caulk, P., Chiang, Y.H., Coyne, M., Dahlke, C., Mays, A., Dombroski, M., Donnelly, M., Ely, D., Esparham, S., Fosler, C., Gire, H., Glanowski, S., Glasser, K., Glodek, A., Gorokhov, M., Graham, K., Gropman, B., Harris, M., Heil, J., Henderson, S., Hoover, J., Jennings, D., Jordan, C., Jordan, J.,

Kasha, J., Kagan, L., Kraft, C., Levitsky, A., Lewis, M., Liu, X., Lopez, J., Ma, D., Majoros, W., McDaniel, J., Murphy, S., Newman, M., Nguyen, T., Nguyen, N., Nodell, M., Pan, S., Peck, J., Peterson, M., Rowe, W., Sanders, R., Scott, J., Simpson, M., Smith, T., Sprague, A., Stockwell, T., Turner, R., Venter, E., Wang, M., Wen, M., Wu, D., Wu, M., Xia, A., Zandieh, A., Zhu, X. (2001) The sequence of the human genome. *Science*. 291, 1304–1351.

2. Drews, J. (2000) Drug discovery: A historical perspective. *Science*. 287, 1960–1964.
3. Civelli, O., Nothacker, H.P., Saito, Y., Wang, Z., Lin, S.H., Reinscheid, R.K. (2001) Novel neurotransmitters as natural ligands of orphan G protein-coupled receptors. *Trends Neurosci*. 24, 230–277.
4. Marchese, A., George, S.R., Kolalowski, L.F., Lynch, K.R., O'Dowd, B.F. (1999) Novel GPCRs and their endogenous ligands: Expanding the boundaries of physiology and pharmacology. *Trends Pharmacol Sci*. 20, 370–375.
5. Neer, E.J. (1995) Heterotrimeric G proteins: Organizers of transmembrane signals. *Cell*. 80, 249–257.
6. Gautam, N., Downes, G.B., Yan, K., Kisselev, O. (1998) The G protein βγ complex. *Cell Signal*. 10, 447–455.
7. Gilman, A.G. (1987) G proteins: Transducers of receptor generated signals. *Annu Rev Biochem*. 56, 615–649.
8. Lefkowitz, R.J. (1993) G protein-coupled receptor kinases. *Cell*. 74, 409–412.
9. Krupnick, J.G., Benovic, J.L. (1998) The role of receptor kinases and arrestins in G protein-coupled receptor regulation. *Annu Rev Pharmacol Toxicol*. 38, 289–319.
10. Clapham, D.E., Neer, E.J. (1997) G protein βγ subunits. *Annu Rev Pharmacol Toxicol*. 37, 167–203.
11. Hausdorff, W.P., Caron, M.G., Lefkowitz, R.J. (1990) Turning off the signal: Desensitization of the beta-adrenergic receptor function. *FASEB J*. 4, 2881–2889.
12. Nambi, P., Aiyar, N. (2003) G protein-coupled receptors in drug discovery. *Assay Drug Dev Technol*. 1, 305–310.
13. Kunapuli, P., Ransom, R., Murphy, K.L., Pettibone, D., Kerby, J., Grimwood, S., Zuck, P., Hodder, P., Lacson, R., Hoffman, I., Inglese, J., Strulovici, B. (2003) Development of an intact cell reporter gene beta-lactamase assay for G protein-coupled receptors for high throughput screening. *Anal Biochem*. 314, 16–29.
14. Kunapuli, P., Lee, S., Zheng, W., Alberts, M., Kornienko, O., Mull, R., Hwang, J.I., Simon, M.I., Strulovici, B. (2003) Identification of small molecule antagonists of the human MRG-X1 receptor. *Anal Biochem*. 351, 50–61.
15. Zuck, P., Lao, Z., Skwish, S., Glickman, F., Yang, K., Burbaum, J., Inglese, J. (1999) Ligand-receptor binding measured by laser-scanning imaging. *Proc Natl Acad Sci U S A*. 96, 11122–11127.
16. Allen, M., Reeves, J., Mellor, G. (2000) High throughput fluorescence polarization: A homogeneous alternative to radioligand binding for cell surface receptors. *J Biomol Screen*. 5, 63–70.
17. Milligan, G. (2003) Principles: Extending the utility of [^{35}S]GTP gamma S binding assays. *Trends Pharmacol Sci*. 24, 87–90.
18. Johnson, E.N., Shi, X., Cassaday, J., Ferrer, M., Strulovici, B., Kunapuli, P. (2008) A 1536-well [^{35}S]GTPγS scintillation proximity binding assay for ultra-high throughput screening of an orphan Gαi-coupled GPCR. *Assay Drug Dev Technol*. 6, 327–337.

19. Hill, S.J., Baker, J.G., Rees, S. (2001) Aequorin as a reporter gene in yeast. *Curr Opin Pharmacol.* 1, 526–532.

20. Naylor, L.H. (1999) Reporter gene technology: The future looks bright. *Biochem Pharmacol.* 58, 749–757.

21. Gorman, C.M., Moffatt, L.F., Howard, B.H. (1982) Recombinant genomes which express chloramphenicol acetyl transferase in mammalian cells. *Mol Cell Biol.* 2, 1044–1051.

22. Jain, V.K., Magrath, I.T. (1991) A chemiluminescent assay for quantitation of β-galactosidase in femtogram range: Application to quantitation of β-galactosidase in lacZ-transfected cells. *Anal Biochem.* 199, 119–124.

23. Chen, W., Shields, T.S., Stork, P.J.S., Cone, R.D. (1995) A calorimetric assay for measuring activation of Gs- and Gq-coupled signaling pathways. *Anal Biochem.* 226, 349–354.

24. Berger, J., Hauber, J., Hauber, R., Geiger, R., Cullen, B.R. (1988) Secreted placental alkaline phosphatase: A powerful new quantitative indicator of gene expression in eukaryotic cells. *Gene.* 66, 1–10.

25. Durocher, Y., Perret, S., Thibaudeau, E., Gaumond, M.H., Kamen, A., Stucco, R., Abramovitz, M. (2000) A reporter gene assay for high throughput screening of G-protein-coupled receptors stably or transiently expressed in HEK293 EBNA cells grown in suspension culture. *Anal Biochem.* 284, 316–326.

26. Williams, T.M., Burlein, J.E., Ogden, S., Kricka, L.J., Kant, J.A. (1989) Advantages of firefly luciferase as a reporter gene: Application to the interleukin-2 gene promoter. *Anal Biochem.* 176, 28–32.

27. Fitzgerald, L.R., Mannan, I.J., Dytko, G.M., Wu, H.L., Nambi, P. (1999) Measurement of responses from Gi-, Gs-, or Gq-coupled receptors by a multiple response element/cAMP response element-directed reporter assay. *Anal Biochem.* 275, 54–61.

28. Kain, S.R. (1999) Green fluorescent protein (GFP): Applications in cell-based assays for drug discovery. *Drug Discov Today.* 4, 304–312.

29. Zlokarnik, G., Negulescu, P.A., Knapp, T.E., Mere, L., Burres, N., Feng, L., Whitney, M., Roemer, K., Tsein, R.Y. (1998) Quantitation of transcription and clonal selection of single living cells with β-lactamase as reporter. *Science.* 279, 84–88.

30. Zlokarnik, G. (2000) Fusions to β-lactamase as a reporter for gene expression in live mammalian cells. *Methods Enzymol.* 326, 221–241.

31. Kunapuli, P. (2006) Aurora assays. In *Handbook of Assay Development in Drug Discovery: Drug Discovery Series 5*, ed. L.K. Minor, pp. 319–342. Boca Raton, FL: CRC Press, Taylor and Francis Group.

32. Rao, A., Luo, C., Hogan, P.G. (1997) Transcription factors of the NFAT family: Regulation and function. *Annu Rev Immunol.* 15, 70747.

33. Fink, J.S., Verhave, M., Kasper, S., Tsukada, T., Mandel, G., Goodman, R.H. (1998) The CGTCA sequence motif is essential for biological activity of the vasoactive intestinal peptide gene cAMP-regulated enhancer. *Proc Natl Acad Sci U S A.* 85, 6662–6666.

34. Gonzales, G.A., Montminy, M.R. (1989) Cyclic AMP stimulates somatostatin gene transcription by phosphorylation of CREB at serine 133. *Cell.* 59, 675–680.

35. Sutherland, E.W., Robinson, G.A., Butcher, R.W. (1968) Some aspects of the biological role of adenosine 3″ 5″-monophosphate (cAMP). *Circulation.* 37, 279–306.

36. Williams, C. (2004) cAMP detection methods in HTS: Selecting the best from the rest. *Nat Rev Drug Discov*. 3, 125–135.
37. Kasila, P., Harney, H. (2006) The evolution of cAMP assays. In *Handbook of Assay Development in Drug Discovery: Drug Discovery Series 5*, ed. L.K. Minor, pp. 243–268. Boca Raton, FL: CRC Press, Taylor and Francis Group.
38. Weber, M., Ferrer, M., Zheng, W., Inglese, J., Strulovici, B., Kunapuli, P. (2004) A 1536-well cAMP assay for Gs- and Gi-coupled receptors using enzyme fragmentation complementation. *Assay Drug Dev Technol*. 2, 39–49.
39. Weber, M., Muthusubramaniam, L., Murray, J., Hudak, E., Kornienko, O., Johnson, E.N., Strulovici, B., Kunapuli, P. (2007) Ultra high throughput screening for antagonists of a Gi-coupled receptor in a 2.2 ul 3456-well plate format cAMP assay. *Assay Drug Dev Technol*. 5, 117–125.
40. Austin, C.E., Faussner, A., Robinson, H.E., Chakravarty, S., Kyle, D.J., Bathon, J.M., Proud, D. (1997) Stable expression of the human kinin B1 receptor in Chinese hamster ovary cells. Characterization of ligand binding and effector pathways. *J Biol Chem*. 272, 11420–11425.
41. Mathis, S.A., Criscimagna, N.L., Leeb-Lundberg, L.M.F. (1996) B1 and B2 kinin receptors mediate distinct patterns of intracellular Ca^{2+} signaling in single cultured vascular smooth muscle cells. *Mol Pharmacol*. 50, 128–139.
42. Marsh, K.A., Hill, S.J. (1994) Des-Arg9-bradykinin-induced increases in intracellular calcium ion concentration in single bovine tracheal smooth muscle cells. *Br J Pharmacol*. 112, 934–938.
43. Tanahashi, H., Ito, T., Inouye, S., Tsuji, F.I., Sakaki, Y. (1990) Photoprotein aequorin: Use as a reporter enzyme in studying gene expression in mammalian cells. *Gene*. 95, 249–255.
44. Brady, A.E., Limbird, L.E. (2002) G protein-coupled receptor interacting proteins: Emerging roles in localization and signal transduction. *Cell Signal*. 14, 297–309.
45. Hodder, P., Mull, R., Cassaday, J., Berry, K., Strulovici, B. (2004) Miniaturization of intracellular calcium functional assays to 1536 well plate format using a fluorometric imaging plate reader. *J Biomol Screen*. 9, 417–426.
46. Liu, J., Conklin, B.R., Blin, N., Yun, J., Wess, J. (1995) Identification of a receptor/G-protein contact site critical for signaling specificity and G-protein activation. *Proc Natl Acad Sci U S A*. 92, 11642–11646.
47. Berridge, M.J., Dawson, R.M.C., Downer, C.P., Heslop, J.P., Irvine, R.F. (1983) Changes in the levels of inositol phosphates after agonist-dependent hydrolysis of membrane phosphoinositide. *Biochem J*. 212, 473–482.
48. Ferguson, S.S. (2001) Evolving concepts in G protein-coupled receptor endocytosis: The role in receptor desensitization and signaling. *Pharmacol Rev*. 53, 1–24.
49. Gurevich, V.V., Dion, S.B., Onorato, J.J., Ptasienski, J., Kim, C.M., Sterne-Marr, R., Hosey, M.M., Benovic, J.L. (1995) Arrestin interactions with G protein-coupled receptors. Direct binding studies of wild type and mutant arrestins with rhodopsin, β2-adrenergic and M2 muscarinic cholinergic receptors. *J Biol Chem*. 270, 720–731.
50. Violin, J.D., Lefkowitz, R.J. (2007) β-arrestin-biased ligands at seven-transmembrane receptors. *Trends Pharmacol Sci*. 28, 416–422.
51. Krupnick, J.G., Goodman Jr., O.B., Keen, J.H., Benovic, J.L. (1997) Arrestin/clathrin interaction. Localization of the clathrin binding domain of non visual arrestins to the carboxy terminus. *J Biol Chem*. 272, 15011–15016.

52. Lee, S., Howell, B., Kunapuli, P., Inglese, J. (2006) Cell imaging assays for G protein-coupled receptor internalization: Application to high throughput screening. *Methods Enzymol.* 414, 79–98.
53. Oakley, R.H., Hudson, C.C., Cruickshank, R.D., Meyers, D.M., Payne Jr. R.E., Rhem, S.M., Loomis, C.R. (2002) The cellular distribution of fluorescently labeled arrestins provides a robust, sensitive and universal assay for screening G protein-coupled receptors. *Assay Drug Dev Technol.* 1, 21–30.

CHAPTER 14

New Techniques to Express and Crystallize G Protein-Coupled Receptors

JAMES C. ERREY and FIONA H. MARSHALL
Heptares Therapeutics Ltd., Hertfordshire, UK

14.1. INTRODUCTION

G protein-coupled receptors (GPCRs) are integral membrane-spanning proteins with an extracellular N-terminal domain, 7-α-helical membrane-spanning domains connected by intracellular and extracellular loops, and, finally, an intracellular C-terminal domain. GPCRs are important signal transduction proteins, and study of their function is usually carried out in native tissue or in recombinant cellular systems. A true understanding of GPCR function at a molecular level requires high-resolution three-dimensional (3D) structures of the different functional conformations across the superfamily of receptors. High-resolution crystal structures are also required to enable the range of structure-based drug design methodologies, which are routinely applied to soluble proteins to be used on GPCRs.

Despite a huge effort over many years, progress in obtaining GPCR structures, as well as other membrane proteins, has been painstakingly slow. To date, atomic structures have been obtained for 178 transmembrane proteins using X-ray crystallography (http://blanco.biomol.uci.edu/Membrane_Proteins_xtal.html). The majority of these structures are obtained from bacteria and archaebacteria, which do not have GPCRs, only the ancestral protein bacteriorhodopsin. In total, atomic structures have been solved for only four different GPCRs (Table 14.1), three of which have been published in 2008. There are several ways to obtain 3D structures of proteins: X-ray crystallography,

GPCR Molecular Pharmacology and Drug Targeting: Shifting Paradigms and New Directions, Edited by Annette Gilchrist

TABLE 14.1 List of GPCR Crystal Structures

Rhodopsin: bovine ROS	2.8 Å	1F88	[1]
Rhodopsin: bovine ROS	2.6 Å	1L9H	[32]
Rhodopsin: bovine ROS	2.65 Å	1GZM	[33]
Rhodopsin: bovine ROS	2.2 Å	1U19	[35]
Rhodopsin: recombinant bovine rhodopsin mutant, N2C/D282C	3.4 Å	2J4Y	[36]
Rhodopsin, photoactivated: bovine ROS	3.8–4.15 Å	2I37	[43]
Opsin: bovine ROS	2.9 Å	3CAP	[202]
Rhodopsin: squid	2.5 Å	2Z73	[38]
Rhodopsin: squid	3.7 Å	2ZIY	[37]
Turkey β_1AR (StaR engineered for stability) Complex with cyanopindolol	2.7 Å	2VT4	[4]
Human β_2AR Fab5 complex Complex with carazolol	3.4/3.7 Å	2R4R 2R4S	[3]
Engineered human β_2AR T4L replaces third ICL. Complex with carazolol	2.4 Å	2RH1	[2]
Engineered human β_2AR T4L replaces third ICL. E122W stability mutation Complex with timolol	2.8 Å	3D4S	[9]
Human A_{2a} adenosine receptor, in complex with a high-affinity subtype-selective antagonist ZM241385	2.6 Å	3EML	[5]

TABLE 14.2 Hurdles Preventing Crystallization of GPCRs and How They Are Being Overcome

Crystallization Step	Problem with GPCRs	Solution	Reference
Protein expression	Low expression in native tissues	High-level recombinant expression systems	
Protein purification	Instability Low yield of functional protein	Engineering stability	[9, 135, 136, 203]
Detergent solubilization	Instability in detergent	Engineering stability Lipid bicelles Lipidic cubic phase	[9, 135, 136, 203–205]
Crystallization	Structural heterogeneity	*E. coli* expression Mutagenesis of sites for PTM	
	Conformational heterogeneity	Ligand binding Conformational stabilization	[135, 136]
	Low polar surface area	Antibody complex T4L fusion	[2, 3, 164]
	Small crystals	Synchrotron microfocus beam	[186, 206]

TABLE 14.3 Comparative Heterologous Expression Levels of the Human β_2AR

Expression Host	Strain	N-Terminus	C-Terminus	Expression Level (pmol/mg)	Minimal Level of Expression Required (pmol/mg) [133, 137]	Scalability	Reference
E. coli	XL1-blue	MalE	His6	3	13	High	[138]
Yeast	*P. pastoris*	α-factor signal His6		6	5	High	[139]
Yeast	*S. cerevisiae*	Ste2		115	5	High	[129]
Insect	Sf9			73	26	Medium	[95]
Insect	Sf9 stable			3–4	26	Medium	[140]
Mammalian	Stable HEK239		1D4 epitope	220	26	Low	[141]
Cell Free	*E. coli* S30 extract	TrxA	His6	150 μg/mL[a]	N/A	Low	[115]

[a] Expression levels in μg/mL of CF expression reaction mixture.

electron cryomicroscopy (cryo-EM), and nuclear magnetic resonance (NMR) spectroscopy. X-ray crystallography using synchrotron radiation sources is the most powerful method for obtaining high-resolution structures of GPCRs, and this chapter will focus exclusively on this area. However, many of the methodological improvements described will also benefit EM and NMR studies.

Until very recently, the only GPCR that had been crystallized was the photopigment from retinal rod cells—rhodopsin. This receptor is unique in terms of its natural abundance within rod membranes and its stability in detergent solutions. The structure, which was first obtained in 2000 [1], has proved extremely valuable as a template for homology modeling of other Family A receptors despite the fact that the homology is less than 30%. The use of rhodopsin as a basis for modeling Family B and Family C, however, is questionable, due to the lack of sequence homology across these families.

The last 2 years have seen several major breakthroughs in obtaining high-resolution structures of GPCRs, which have resulted in the publication of two structures of the β_2-adrenergic receptor (β_2AR) [2, 3], the structure of the closely related β_1-adrenergic receptor (β_1AR) [4] and the adenosine A_{2a} receptor [5]. The new approaches, which are now being applied in the field of GPCR crystallography (Table 14.2), provide hope that the next decade will see a wealth of new GPCR structures that will transform our understanding of GPCR function and guide novel drug discovery.

In this chapter, we discuss the technical challenges that have hindered the production of stable functional receptor protein for structural studies. We will review the history and impact of the structures of rhodopsin, and we will describe the new developments that have resulted in the recent β-adrenergic (βAR) and adenosine receptor structures along with the applicability of these approaches to other GPCRs. We will also consider the new insights obtained from the new structures in terms of understanding ligand binding and receptor function, and we conclude by discussing likely future development in this rapidly growing area of GPCR biology.

14.2. KEY PROBLEMS LIMITING PRODUCTION OF 3D GPCR STRUCTURES

A prerequisite in obtaining protein crystals is the availability of large quantities of correctly folded pure protein in a single form. Rhodopsin represents >90% of the protein in its native tissue, the retinal rod outer segments (ROSs), and is thus easy to obtain in large quantities. In contrast, the vast majority of GPCRs are expressed at only low levels in native tissue and are more difficult to purify to homogeneity against a background of other membrane and cellular proteins. There is a wide range of recombinant overexpression systems that have been used for GPCRs. These have been very useful for studying GPCR function by conventional means, such as radioligand binding and measurements of signal transduction where the receptor is retained within the cell membrane environment. However, much higher expression levels are required for crystallographic studies, but this is only of value if it results in the purification of correctly folded protein.

The natural environment for GPCRs is the lipid bilayer of the plasma membrane. Detergents are used to solubilize receptors and remove them from the lipid bilayer. During the purification of GPCRs, it is often necessary to include the presence of lipids to ensure that the protein remains in its functional conformation [6]. The selection of detergents, in particular, those that are most compatible with crystallization, is fraught with difficulty and often results in a loss of function. Selection of the right detergent and advances in the use of suitable detergent/lipid combinations have been one of the critical steps in obtaining crystals of GPCRs. The choice of detergent and the lipid requirement varies considerably from one receptor to another and must be determined experimentally for each receptor protein [7, 8]. Many membrane proteins have specific lipid binding sites, which may be required to preserve correct function. In addition, some receptors also have cholesterol binding sites, which may also be of functional consequence [9].

A crucial requirement following the expression and purification of any GPCR is the demonstration that the protein has retained its native properties and correct folding. Although biophysical techniques such as circular dichroism have been used to assess the secondary structure of soluble proteins, algorithms developed from these structures are poor when applied to the spectra of membrane proteins [10]. A better measure of the function of purified GPCRs is to measure directly the ability of the protein to bind radiolabeled or fluorescently labeled ligands, which preferably bind within the transmembrane domains. Ideally, this assay should be sufficiently quantitative enough to enable a calculation of the percentage of active protein within the purified preparation. For example, if radioligand binding is used, then the Bmax in pmol/mg of protein can be compared to the absolute amount of protein obtained, as determined by amino acid analysis.

For many years, the availability of purified GPCRs in large quantities was considered the major hurdle in obtaining X-ray structures. Considerable progress has been made in this area as outlined in this chapter. Even with sources of pure protein to hand, the growth of well-ordered 3D crystals required for crystallography is another major hurdle. During crystallization, the aim of the crystallographer is to slowly form ordered (crystalline) arrays of protein from solution. This occurs in two stages: nucleation and growth. Nucleation is commonly initiated by increasing the concentration of a precipitant in a drop of protein solution by vapor diffusion. During this stage, changes in the protein–lipid detergent complex can result in aggregation or phase separation, which reduces crystal formation. Variable stoichiometries of lipids bound to the proteins result in nonuniform building blocks, which prevent highly ordered 3D crystals. A careful choice of detergents and conditions must be made. Here, a fundamental problem arises in the crystallization of GPCRs. As discussed above, membrane protein structure is dependent on the surrounding lipid.

Choices of detergents and lipids that are suitable for retaining function during solubilization are often not compatible with growing diffraction quality crystals. Detergents must be the correct size to enable protein/protein interactions without causing protein denaturation. The extramembrane regions of GPCRs, which are disordered and flexible, are usually removed or truncated for the purposes of crystallization, but proteins with small extramembrane domains are difficult to crystallize due to the low polar surface area available for crystal lattice contacts. Methods to overcome these issues are discussed in detail below, but essentially, they involve either engineering the receptor protein to increase their stability in detergent and/or using more native lipid environments for crystallization.

In addition to the homogeneity of protein/lipid complexes being a prerequisite for the formation of ordered crystals, it is also essential that the proteins themselves be uniform both in structure and conformation. Posttranslational modifications (PTMs), which vary depending on the expression system used, include glycosylation, phosphorylation, and palmitoylation. These can be relatively easily eliminated by truncation of extramembrane regions or by specific site-directed mutagenesis of the receptor.

GPCRs naturally exist in multiple conformations ranging from the fully inactive ground state (R) to the fully activated state (R*), which couples to G proteins and elicits signal transduction. For any given receptor, an equilibrium exists between these two states that, in cells, determine the basal level of activity. Once solubilized, a receptor may continue to exist in multiple conformations and transition between states, which can contribute to instability and resultant unfolding. Different conformations will have different levels of stability. The flexibility of GPCRs, which is fundamental to their signaling function, represents one of the greatest impediments to maintaining stability of solubilized receptors and the formation of crystals.

Once crystals are obtained, a further problem emerges. GPCR crystals, in particular, those grown in lipidic cubic phase, are extremely small compared with crystals usually obtained from soluble proteins. The small size and radiation sensitivity limits collection of data on conventional synchrotron beam lines, which are in the order of 50 μm.

14.3. HISTORY OF GPCR STRUCTURES

14.3.1. Early Studies on Rhodopsin

Rhodopsin is a unique GPCR with a highly specialized function for the detection of light. Rhodopsin has been an invaluable template upon which the structure and function of all other GPCRs have been modeled. Rhodopsin differs from other GPCRs in that the ligand, 11-*cis*-retinal, is covalently linked to the receptor protein (opsin) at Lys296 in transmembrane helix 7 (TM7) via a protonated Schiff base, where it acts as a full inverse agonist to hold the receptor in the inactive conformation [11]. Upon absorption of a photon of light, 11-*cis*-retinal is isomerized to all-*trans*-retinal; this process converts the ligand from an inverse agonist to a full agonist. The conversion of 11-*cis* to all-*trans*-retinal results in a conformational change in the opsin structure that alters its state from inactive to active, allowing it to couple to the G protein tranducin (G_t). This activated form of the receptor, which is analogous to the R* state in other GPCRs, is known as metarhodopsin II (MII). Between rhodopsin and MII, there exist a number of short-lived but distinct photointermediates, which can be studied due to changes in their λ_{max} of light absorption. The first intermediate is bathorhodopsin, which thermally relaxes to the blue-shifted intermediate (BSI), followed by lumirhodopsin, and then metarhodopsin I (MI). In MI, the all-*trans*-retinal remains bound in a protonated Schiff base linkage. During the transition of MI to MII, the all-*trans*-retinylidene Schiff base becomes deprotonated. MII is actually a heterogeneous form of several photoactivated conformations. These include MIIa and MIIb, which exist in a pH-dependent equilibrium regulated by proton uptake at Glu134 in the (E/D)RY motif. It is only MIIb that is capable of activating G_t. Eventually, the Schiff base is hydrolyzed, and all-*trans*-retinal is reduced by retinol dehydrogenase to all-*trans*-retinol [12–14], leaving the free opsin (Fig. 14.1). Free opsin has some basal activity in the absence of ligand, and, since this would essentially result in the sensation of light when there was none, there is a rapid transformation of opsin back to rhodopsin through binding a new molecule of the inverse agonist 11-*cis*-retinal. This enables rod cells to maintain a zero level of activity and a low activation threshold.

Some of the properties of GPCRs that make them difficult to crystallize do not apply to rhodopsin, and thus, until very recently, rhodopsin has been the only structural template for the entire GPCR family. The most important feature of rhodopsin, which has enabled its crystallization, is its inherent stability. Recombinant rhodopsin can be heated to greater than 50°C in detergent before losing activity; in contrast, most other GPCRs lose activity below 30°C in detergent (Fig. 14.6). Covalent binding of an inverse agonist is one reason for the stability of the protein. A second differentiating feature of rhodopsin is its abundance in native tissue. Rhodopsin represents >90% of protein within the ROS membrane. Here, the protein is packed into dense arrays interspersed with phospholipids and cholesterol. A single mouse retina contains ~650 pmol of rhodopsin [13]. Bovine retinas have provided an abundant source of material for purification of protein. As a result of the stability of the protein in complex with 11-*cis*-retinal, purification can be achieved simply by a selective solubilization of bovine outer rod segment membranes with, for example, alkylthioglucosides and divalent cations, resulting in a highly purified and concentrated preparation [15].

The first transmembrane protein structure to be determined from two-dimensional (2D) crystals was that of bacteriorhodopsin [16–19]. This consisted of seven transmembrane helices lying approximately perpendicular to

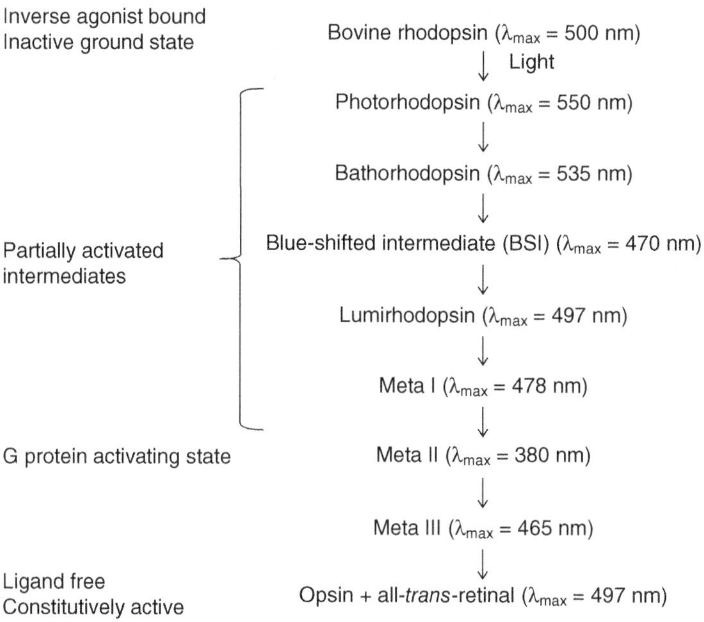

Figure 14.1 Light cycle of rhodopsin. Rhodopsin is held in an inactive ground state through the binding of the inverse agonist 11-*cis*-retinal. Absorption of light leads to photoisomerization, resulting in multiple intermediate forms, which ultimately result in the active form of the receptor capable of activating G proteins known as meta II. The isomerized ligand all-*trans*-retinylidene is released from opsin as all-*trans*-retinal leaving the apo receptor, which shows some degree of constitutive activity and therefore represents a partially active form of the receptor.

the membrane surface (Fig. 14.2). Although bacteriorhodopsin was used to model GPCRs, before the structure of rhodopsin was determined, the proteins have no sequence homology and the arrangement of the helices is somewhat different.

The first suggestion that rhodopsin consisted of a bundle of 7-α-helical transmembrane domains came from circular dichroism studies of rhodopsin octyl glucoside (OG) [20, 21]. This was supported by the publication of the full primary sequence of the protein determined by chemical sequencing. Hydropathy plots of the 348 amino acid sequence indicated six clear and one less clear hydropathic transmembrane domains [22, 23].

Although, at this time, great strides were being made in both the rhodopsin field and the study of other GPCRs, most notably the β2AR in Lefkowitz's group, surprisingly, the fields were not brought together until the cloning of β2AR in 1986, which clearly demonstrated the structural homology between adrenergic and later other GPCRs with rhodopsin [24, 25].

Figure 14.2 Balsa wood model of the structure of bacteriorhodopsin [18].

The first structural data on any GPCR was obtained by the team of Schertler and Henderson in 1993 [26] using cryo-EM of 2D crystals of bovine rhodopsin. A projection map at 9 Å resolution was constructed in which 4-transmembrane helices could be clearly observed perpendicular to the membrane with a continuous area of density suggesting a further 3-transmembrane helices. Further data collection and the analysis of tilted images of bovine rhodopsin resulted in a 3D projection map, which was resolved to 9 Å in a planar direction and 47 Å vertical resolution. From this structure, four out of the seven helices were well resolved, while the others formed an "arc shaped" feature due to their more tilted orientation within the plane of the membrane [27]. The arrangement of helices in this paper differed significantly from that of bacteriorhodopsin. Subsequent to the 2D structures of bovine rhodopsin, two additional projection maps of frog rhodopsin were obtained at the higher resolution of 7 Å and 6 Å [28].

Although 3D crystals from X-ray diffraction provide a higher resolution than can be obtained from 2D and 3D projection maps, the advantage of cryo-EM is the higher proportion of lipid present in the structure and the ability to gain information about the orientation of the protein relative to the bilayer. In 2003, Krebs et al. obtained a much higher resolution (5.5 Å) structure of bovine rhodopsin from 2D crystals [29]. In this structure, the crystals had a symmetry in which neighboring molecules in the structure were upside down but overlapped in the center of the membrane plane. This enabled the position of the center of the membrane to be defined together with the orientation of the molecules relative to the membrane plane.

14.3.2. Higher Resolution Structures of Bovine Rhodopsin Using X-Ray Crystallography

The most important breakthrough in understanding GPCR structure came in 2000 with the publication of the first high-resolution (2.8 Å) 3D crystal structure utilizing X-ray diffraction of the ground state of bovine rhodopsin [1]. This high-quality and detailed structure revolutionized our understanding of GPCRs and provided a template of sufficient quality to model other GPCRs. Although, at that time, considerable information had been generated from 2D crystals, obtaining conditions for growing 3D crystals suitable for X-ray diffraction proved considerably more difficult. After numerous trials varying the solubilization conditions, pH, temperature, as well as the concentrations of various additives, a critical step forward came with the use of a mixed micelle system containing nonyl β-D-glucoside (NG) and heptane-1,2,3-triol (HPTO). This combination enabled highly purified rhodopsin to be obtained by a single-step extraction from rod membranes, provided a stabilizing environment for rhodopsin, and also altered the phase separation boundary in the detergent/precipitant system, which facilitated the formation of crystals. Crystals had to be grown at a low temperature for several months in the dark. A practical problem when working with rhodopsin crystals is that exposure to visible light, even at low temperature, bleached the crystals and caused them to decompose within a few minutes, thus making analysis of the crystallization plates under a light microscope difficult.

In this first 3D structure, all 198 residues, which constitute the seven transmembrane helices (TM), were clearly defined. The dimensions of rhodopsin were found to be ~75 Å perpendicular to the membrane, 45 Å wide, and ~35 Å thick. The transmembrane helices do not all lie parallel, but rather, helices 1, 4, 6, and 7 are bent at proline residues, most notably in TM4, which is bent close to the extracellular end, and in TM6, which is bent close to the center at Pro267. An important region of Class A GPCRs found within the transmembrane domains is the highly conserved tripeptide sequence Glu134, Arg135, Tyr136 known as the (D/E)R(Y/W) motif on the end of TM3. This area is involved in several hydrogen bonds with surrounding residues, in particular, Glu134 forms a salt bridge with Arg135, which is also connected to Glu247 and Thr251 in TM6. This interaction was later named the "ionic lock" and was proposed to play an important role in stabilizing the ground state of rhodopsin. Later evidence suggested that disruption or breakage of the ionic lock was a key conformational change, which occurs during receptor activation [30].

An unusual feature of the new structure, which had not been predicted, was the presence of an additional helical region at the start of the C-terminal tail but clearly distinct from TM7. This short helix (H8) lies perpendicular to TM7 and thus parallel with the membrane. Another interesting region in the extra-membrane regions of rhodopsin is the second extracellular loop, part of which folds into the center of the protein. Arg177 to Glu181 (β3) forms a β-sheet, which interacts in an antiparallel fashion with another β-sheet formed from Ser186 to Asp190 (β4). β4 is next to, and forms part of, the 11-*cis*-retinal binding pocket. Adjacent to this, Cys187 forms a disulfide bond with Cys110, located at the extracellular end of TM3. This disulfide bond is conserved across the majority of Family A GPCRs. The extracellular domains, made up of the N-terminal 33 amino acids and the three extracellular loops, form a close association with each other, which is now known as the plug or cap.

Another region that is highly conserved in GPCRs is the NPxxY (NPVIY in rhodopsin) sequence close to the end of TM7. The side chains of two polar residues in the region, Asn302 and Tyr306, project into the protein, and there is likely an interaction between Tyr306 and Asn73 in TM2, which is also highly conserved. In addition, Asn302 may interact with a water molecule near Asp83, thereby providing a network of contacts among TM2, TM3, and TM7.

The 2000 structure of rhodopsin also provided a detailed picture of the ligand 11-*cis*-retinal binding site with the Schiff base linkage to Lys296 in TM7 and the retinylidene group located toward the extracellular side. The counter-ion for the Schiff base is Glu113, which is highly conserved in vertebrate visual pigments. The β-ionone ring at one end of the retinal structure is surrounded by side changes from TM5 and TM6 (Met207, His211, Ph212, Tyr268, and Ala269). From this ring, the retinylidene group runs parallel to TM3, which provides various side chains forming the binding pocket including Glu113. Additional residues from TMs 1, 2, and 7 (Tyr43, Met44, Leu47, Thr94, and Phe293) surround the Schiff base. As mentioned above, the second extracellular loop also provides a contribution to the binding site, in particular, Glu181 and Tyr191 form part of the retinylidene binding site (Fig. 14.3).

Although this structure represented the inactive conformation of the receptor, a number of deductions could be made about likely changes occurring upon receptor activation. The model of rhodopsin obtained from the structure demonstrated that isomerization of the 11-*cis*-retinal to an all-*trans* configuration would result in a movement of the β-ionone ring toward TM2 and a likely displacement of the Schiff base/C9/C13 regions. Movement of TM3 within the binding pocket would translate into a reorientation of the ERY motif and a possible breaking of the ionic lock. It was also suggested that activation could disrupt the network of intramolecular interactions, leading to rearrangements of the helical bundle and movements in both TM3 and TM6. This structure was further refined (1HZX) [31], giving better resolution to some of the amino acids missing in the first structure, and was extended again to a 2.6-Å structure (1L9H), where additional water molecules, including some close to the ligand binding site, were seen [32].

In 2004, a new crystal form of rhodopsin was obtained to 2.65 Å [33]. Overall, this structure not only agreed well with previous structures but also provided significant new information. The most striking differences between

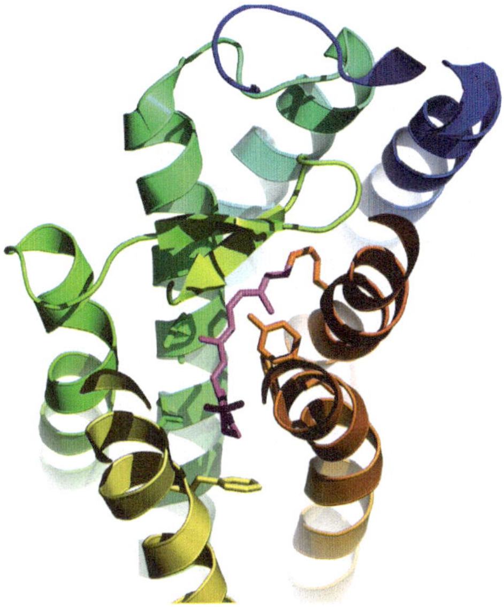

Figure 14.3 The ligand binding site of rhodopsin containing the chromophore 11-*cis*-reinylidene (pink) and the residues of the receptor providing points of interaction, in particular, Lys296, which couples to the chromophore via the protonated Schiff base. Transmembrane helices are colored as follows: TM1, blue; TM2, turquoise; TM3, dark green; TM4, light green; TM5, yellow; TM6, orange; TM7, red.

the structures were the C3 loop and the cytoplasmic ends of TM5 and TM6. In the new structure, they were extended by one turn at the cytoplasmic end, thereby raising the C3 loop above the surface of the membrane. This was an important finding since the C3 loop is known to be involved in G protein coupling. Another interesting feature of this structure was the presence of an ordered molecule of the detergent LDAO (*N*,*N*-dimethyldodecylamine-*N*-oxide) bound to the kink in TM6. The LDAO may be contributing to stabilizing the inactive protein conformation. The addition of LDAO during solubilization was a critical step in obtaining better ordered crystals.

The new structure also contained additional ordered water molecules, which allowed an extended hydrogen bonding network to be identified. In particular, Asp83 and Trp265 in the retinal binding pocket were linked by ordered H-bond networked to Met257 and Asn302, which is in the NPxxY motif at the end of TM7.

While obtaining this structure of native rhodopsin, the same group [34] also examined the effect of a variety of steps to modify the rhodopsin protein with a view to facilitating crystallization. These included protease treatment to remove the C-terminus and portions of the third intracellular loop (ICL) and reductive methylation of exposed lysines. Phosphorylation conditions were also controlled so as to produce homogeneously diphosphorylated species rather than a mixture of any number of up to seven phosphates, which is normally obtained from ROS membranes.

In the same year, Okada et al. [35] achieved a dramatic improvement in the resolution of rhodopsin to 2.2 Å using a new method of purification and crystallization with the single detergent heptylthioglucoside. This structure gave a complete definition of the protein backbone, including the extended helical structure of TM6. The structure of the 11-*cis*-retinal ligand within the binding site was more precisely defined, demonstrating a significant negative pretwist of the C11–C12 bond. This new observation was important as the C11–C12 bond is involved in the activation of the receptor by light via photoisomerization of the 11–12 bond from *cis* to *trans*.

The first structure of a recombinantly expressed GPCR was published in 2007 [36]. Although rhodopsin is a relatively stable receptor, it still did not express well in recombinant systems, such as in bacteria. Furthermore, expression in other systems resulted in heterogeneously glycosylated proteins. Cultured cells do not produce 11-*cis*-retinal so this must be added exogenously prior to solubilization. Standfuss et al. expressed a thermostably engineered mutant form of rhodopsin in COS cells containing an additional disulfide bond between the N-terminus and the third extracellular loop (Asn2Cys/Asp282Cys). Mutation of the asparagine residues (Asn2 and Asn15) (Asn15Asp) also removed both N-linked glycosylation sites, thereby avoiding potential carbohydrate heterogeneity. The crystals obtained were very small (5 µm × 5 µm × 90 µm) and required data collection with a microdiffractometer. This allowed diffraction to ~3.5 Å at a much higher success rate than would normally be achieved with native rhodopsin crystals.

14.3.3. Squid Rhodopsin

In 2008, two structures of rhodopsin from the Japanese flying squid *Todarodes pacificus* were published. This was of wider significance than might first be supposed since invertebrate rhodopsin, unlike the previously solved bovine rhodopsin, signal through a Gq-type G protein rather than transducin. Protein was purified from rhabdomeric membranes isolated from squid retina and treated with proteases to remove the unique C-terminal proline-rich extension of squid rhodopsin. Shimamura et al. [37] crystallized the protein in dodecyl maltoside and obtained a 3.7 Å structure, while Murakami and Kouyama [38] used OG and obtained a structure to 2.5 Å. The overall structure and arrangement of the seven helices was similar to that previously found for bovine rhodopsin; however, a major difference was that helices TM5 and TM7 protrude an additional 25 Å further into the cytoplasm. In addition, as well as having H8 in the C-terminal tail, as is found in bovine rhodopsin, the squid structure contained an additional helix, H9, which is after H8 in the amino acid sequence and lies close to the extended helix TM6. Between helices H8 and H9 is a short helical loop, which dips into the hydrophobic membrane region. These additional cytoplasmic structures are likely to be involved in binding of Gq, and it is possible that other Gq-linked receptors may interact with their G protein via a similar mechanism perhaps with the G protein adjacent to the helices [39].

Another significant difference found between the squid and bovine rhodopsin structures is the retinal binding pocket. The side chains interacting with retinal are altered compared to those in bovine rhodopsin, and the retinal polyene chain lies in a more linear fashion. Asn87 and Tyr111 replace Glu113 and Gly89 as the possible hydrogen bonding partner of the Schiff base in the dark state, while the putative counterion Glu180, which is highly conserved in other visual pigments, is too far away to have a direct interaction.

An important feature of the invertebrate eye is that it detects the polarization of light. In the Murakami and Kouyama structure, rhodopsin forms two intermembrane dimers arranged in a tetrameric structure in which four retinylidene chromophores are oriented parallel with one another. This might provide a mechanism for the detection of the polarization plane. The final difference is that in squid rhodopsin, covalently bound retinal is not released from the lysine side chain after photoisomerization but may be isomerized back to 11-*cis* configuration within the protein.

14.3.4. Activated Opsin and Binding to G Proteins

The crystal structures described so far revealed detailed molecular structure of a GPCR in the inactive ground state but did not provide any information about the mechanism of activation or subsequent rearrangements of the helical domains upon activation. The first data on activated states of rhodopsin came from 2D crystals of the photointermediate MI, which were obtained by cryo-EM [40]. Although these structures were all of rather low resolution (2.7–5.5 Å), they still enabled comparison with the dark (ground) state of rhodopsin. In these structures, no large (rigid body) movements or rotations of helices had occurred. The 2D map revealed movements of the side chains close to the retinal binding pocket. For example, a clear density bulge was present in the middle of TM6 facing the β-ionone ring in the MI structure, which was not present in the ground state. This is close to the location of Trp265, suggesting movement of this residue in the MI state. 3D crystal structures of bathorhodopsin and lumirhodopsin were obtained by trapping these photolyzed states at low temperature [41, 42]. In the higher resolution 3D structure of lumirhodopsin, the main difference was in the middle of TM3, suggestive of an outward movement of the polypeptide backbone. However, this change was not propagated to the intracellular side associated with G protein activation. In the lumirhodopsin structure, retinal is almost completely in the all-*trans* conformation. The β-ionone ring is displaced in the direction toward TM3 and TM4. Displacement of the β-ionone ring causes a number of perturbations of surrounding residues, which results in the breakage or weakening of electrostatic restraints between helices. These include hydrogen bonding sites, which are highly conserved across other GPCRs including Asn55–Asp83, and hydrogen bonding to Asn302 in the NPxxY motif.

Rhodopsin is inherently unstable in its photoactivated deprotonated state so Salom et al. [43] attempted to get round this problem by selecting crystals in the ground state that remained stable upon exposure to light at room temperature. Most crystals lose diffraction when illuminated; however, the group was able to identify crystallization conditions that generated two crystal forms that could withstand photoactivation. One was a rhombohedral structure that diffracted to 3.7 Å but lost resolution upon light, whereas the other was a trigonal form that diffracted to 4.1–4.2 Å before and after photoactivation. Interestingly, these crystals could still activate G protein when dissolved in detergent, although only to a limited extent. These were the first crystals to contain all-*trans*-rhodopsin and represented an MII-like intermediate stage in the activation of the receptor. Unfortunately, the low resolution of the structures meant that locations of the main-chain carbon atoms could not be precisely determined and locations of the side chains could not be resolved. Nevertheless, some conclusions could be drawn; both cytoplasmic loops became more disordered upon photoactivation, indicating movement in this area; in addition, as seen in the previous structure, there were no large displacements of the individual helices.

The lack of movement of individual helices in these structures did not fit with those predicted from other indirect methods, such as electron pair spin resonance (EPR) studies on spin-labeled cysteine mutants [44, 45]. It is probable that the crystal lattice limits the magnitude of changes observed in these structures or that the additional energetic barriers have not been overcome to enable the formation of the functionally active conformation despite the presence of all-*trans*-retinal and the deprotonated Schiff base.

11-*cis*-retinal is a strong inverse agonist, which holds rhodopsin in its inactive state. Subsequent to receptor activation, all-*trans*-retinal is released from its binding site, and a new light-sensitive rhodopsin is generated through the binding of a new 11-*cis*-retinal molecule. The transiently formed apoprotein, in the absence of any ligand, is opsin. In the absence of 11-*cis*-retinal, opsin can bind and activate G protein, and is therefore considered a constitutively active receptor [46, 47]. Like other ligand-free GPCRs, opsin is very unstable and difficult to purify. In 2008, Park et al. [48] were able to purify native opsin from bovine rod disk membranes. These formed colorless crystals in β-D-octylglucopyranoside, which changed to red upon addition of 11-*cis*-retinal, a process that is also observed with ROS membranes. A 2.9 Å structure was determined, which represented the first activated structure of a GPCR. In this structure, the protein is arranged as a dimer with a TM1/8 interface. There was very little difference in the arrangement of TM1–TM4 in opsin compared to the ground state of rhodopsin. Larger changes were found for TM5–TM7, in particular, at the ends of the helices, resulting in a rearrangement of both the second and third cytoplasmic loops, C2 and C3, respectively. There was also a significant movement of TM5 at the cytoplasmic end toward TM6. TM6 itself was moved outward from the center of the helix bundle, and the cytoplasmic end of TM6 was shifted by as much as 6–7 Å outward, pivoting at Trp265.

In this partially activated form of opsin, the ionic lock between Asp135 and Glu247 was broken, and two new interactions stabilized the ends of TM5 and TM6. There was also a deviation of TM7, which includes the NPxxY motif, causing Tyr306 to rotate into the helix. A surprising finding in this structure was the appearance of two openings close to the retinal binding pocket: one between the extracellular ends of TM5 and TM6, and one between TM1 and TM7. Intriguingly, these two openings may be the entrance and exit route for retinal channeling into and out of the binding site. Prior to this structure, it was hypothesized that the extracellular cap might be displaced to allow entry to the ligand binding site. This entry route to the binding site between TM5 and TM6 may be a general route for other Family A GPCRs that bind hydrophobic ligands. Although this is a partially active form of the receptor, the relationship between this structure and the fully activated R* state is not clear. This awaits the crystal structure of MII in complex with a G protein.

To date, the most convincing data on the transitions, which occur upon activation, come not from X-ray crystallography, but from a new version of EPR known as double electron–electron resonance (DEER). Using this method, 16 pairs of nitroxide spin labels were introduced, one pair at a time, by site-directed mutagenesis and chemical modification, into the cytoplasmic ends of the helices of rhodopsin [49, 50]. Interspin distances were then measured in the inactive and light-activated state. This study was facilitated by the development of an optimized protocol, which included detergent (dodecyl maltopyranoside, DDM), low pH, and high concentrations of glycerol together with rapid cooling in liquid nitrogen to trap the MII conformation. The key findings of these experiments were the outward displacement of TM6 by 5 Å. Smaller movements were also observed in TM1, TM7, and the C-terminal domain. The negative data in this study were also important, namely, that no changes were observed in TM2, TM3, TM4, and TM5, which form part of a core helical bundle that does not change upon photoactivation.

Subsequent to the publication of the opsin structure, the same group [51] was able to obtain a structure of activated opsin in complex with an 11-amino acid peptide derived from the extreme C-terminus of the transducin $G\alpha_t$ subunit (Gα-CT) solved to 3.2 Å. In the structure, the Gα-CT is seen binding in an α-helical conformation to a crevice formed by the movement of TM5 and TM6. An important feature of the structure is the role of Arg135 from the E(D)RY motif. As described above, in the activated opsin structure, the ionic lock is broken, and Arg135 no longer interacts with Gly134 but instead interacts with Tyr233 in TM5. This causes the arginine side chain to swing into the center of the binding crevice for Gα-CT, where it interacts with the carbonyl of Cys347 on the G protein C-terminus. As in the opsin structure, Tyr306 (TM7) extends into the helical bundle below Arg135, where it may act to stabilize the position of TM6.

A particularly interesting feature of the Scheerer et al. structure is the long-range changes in the receptor through to the ligand binding site as a result of binding of the G protein peptide. A network of stabilizing interactions is formed among Lys296 on TM7 in the retinal binding site, Ser186 and Glu181 in the E2 loop, and Tyr268 in TM6. Thus, it appears that both ligand and/or G protein can stabilize the activated state of the receptor. This structure has provided the first evidence for a model of signal transduction whereby the C-terminal α5 helix of $G\alpha_t$ acts as a "transmission rod," such that binding of the C-terminal tail to the receptor alters the juxtaposition of the α5 helix and the remainder of the G protein, resulting in receptor-catalyzed GDP release. It remains to be determined how the active receptor first recognizes and binds the GDP-bound G protein.

14.3.5. Rhodopsin as a Model for Other GPCRs

The years of research that have culminated in the various structures of rhodopsin described here have not only benefited research into the mechanism and disorders of vision. In the absence of other GPCR structures, they have also provided an invaluable model and template for the entire GPCR family of receptors. Rhodopsin continues to be the model receptor, leading the way in our understanding of GPCR function at the molecule level. However, despite sharing many common features, seven transmembrane domains, conserved sequence motifs, and the ability to couple to G proteins, rhodopsin is less than 25% homologous to most other Family A GPCRs and has no homology with the other GPCR families, such as the secretin (Family B) and metabotropic (Family C) receptors. Nevertheless, pioneering work in the development of methods for obtaining crystal structures of rhodopsin, including methods for purification, selection of detergents, and conditions for crystallization, as well as the construction of stabilized mutants, has paved the way for the breakthroughs in obtaining the structures of other GPCRs, which are now described. The availability of other GPCR structures will enable us to determine just how good a model rhodopsin has been.

14.4. THE SEARCH FOR OTHER GPCR STRUCTURES

14.4.1. Expression of Recombinant Receptors

In the absence of tissues expressing high quantities of homogenous native receptors, much of the work leading up to the crystallization of other GPCRs has focused on their expression. The expression of membrane proteins is notoriously difficult and has been one of the major bottlenecks in the structural biology of GPCRs, which typically require milligram quantities of active protein. A number of expression systems have been developed for the expression of GPCRs, each with their own particular advantages and disadvantages. To date, the majority of the work has focused on four main expression systems: bacterial (*Escherichia coli*), yeasts (*Saccharomyces cerevisiae, Schizosaccharomyces pombe, Pichia pastoris*), insect cells (*Spodoptera frugiperda* Sf9, Sf21, *Trichoplusia ni* Hi5, and *Drosophila Schneider* S2), and mammalian cells (Chinese hamster ovary [CHO], HEK, COS-1, etc.). However, other systems have been investigated; these include cell-free (CF) expression systems and the more unusual use of whole organisms (*Drosophila melanogaster*, *Xenopus laevis*, and silkworm).

Bacterial Expression While a number of systems are available for heterologous protein production, *E. coli* remains one of the most attractive because of its ease of use with respect to both the cloning and the scale up of protein expression. Even as *E. coli* has served as a great tool in the study of soluble cytosolic proteins, the ability to translate this to more complex membrane proteins has been limited for a number of reasons.

E. coli is unable to perform the majority of PTMs (glycosylation, phosphorylation, and fatty acid acylation), some of which have been shown to be critical for functional expression of GPCRs [52]. However, this is not a universal observation, and for some GPCRs, PTMs have been shown to have little or no effect on ligand binding [53] or protein expression [54]. Furthermore, the reductive environment of the bacterial periplasm can affect the correct folding of receptors, where formation of disulfide bridges is required for an active conformation.

The lipidic composition of bacterial membranes is very different from that of eukaryotic cells, and this can sometimes affect receptor stability and binding properties [55, 56]. The composition of bacterial inner membranes are richer in phosphatidylethanolamine (PE) and phosphatidylglycerol (PG) lipids [57] rather than the phosphatidylcholine (PC) found in eukaryotic cells. In addition, cholesterol is entirely absent from bacteria, while it forms 40% of the plasma membrane in eukaryotic cells. This may represent a limiting factor in recovering functional GPCRs when expressed in a bacteria-based heterologous system.

For the correct insertion and subsequent folding of transmembrane proteins to occur upon translation, they have to interact with the cells' endogenous translocation machinery in order to be inserted into the cell membrane. When GPCRs are expressed in *E. coli*, the addition of an N-terminal signal sequence (MalE, OmpA) can help to successfully target protein to the inner membrane. However, high levels of expression can overwhelm the bacterial translocation pathway and subsequently effect cell viability and functional protein expression levels. The folding of heterologous membrane proteins may also be affected by the lack of appropriate chaperones, although the degree of influence that they play is unclear, with conflicting reports [58–60].

Other properties that may hinder protein expression levels observed in *E. coli* are the low genomic GC content when compared to mammalian genes and the existence of rare codons. While the expression of the human leukotriene B4 receptor BLT_1 was significantly improved using a synthetic BLT_1

cDNA—with codons optimized for *E. coli*—this resulted in the expression of an insoluble protein aggregate [61]. In the case of soluble expression of the rat neurotensin receptor (NTR), the efficiency of transcription and translation (codon usage) did not seem critical, and it was receptor insertion into the cytoplasmic membrane that seemed to be rate limiting [62].

There are two possible strategies that may be used for the expression of GPCRs in a bacterial expression system: (a) expression of functional, membrane-inserted receptors, and (b) expression of incorrectly folded, aggregated protein to which a refolding strategy is applied to obtain a functionally active receptor.

Initial successes in using bacterial expression to achieve the soluble expression of functionally active GPCR involved the use of fusion proteins. The first successful use of this approach was with the expression of the β_2AR as an N-terminal fusion with β-galactosidase, resulting in expression levels of 0.4 pmol/mg [63]. Subsequently, a number of attempts to improve upon this have been performed by looking at the effect of the promoter and fusion partners. Some of the pioneering work in this field by Grisshammer and coworkers led to the development of systems where the NTR receptor can be expressed at levels of 15 pmol/mg [62, 64] and the human adenosine A_{2a} receptor at levels approaching 34 pmol/mg [65]. This was achieved by optimizing the expression construct to include a maltose binding protein (MBP) fusion partner with an N-terminal signal peptide and the replacement of the tac promoter with a weaker lac promoter [64]. The exact role of the fusion partner is unclear; however, the MBP could help drive the correct insertion of the fused GPCR into the membrane through its translocation to the periplasm [56]. Further modification to improve upon the MBP fusion system has involved the generation of a triple-protein fusion construct (MBP–GPCR–thioredoxin [Trx]) that appears to further stabilize the receptor, and improve expression and purification [66–68]. Other factors influencing the expression levels are the *E. coli* strains used and the growth temperature [62, 67, 69, 70].

While, in general, the bacterial expression of GPCRs results in relatively low levels of soluble protein, relatively high levels of insoluble protein expression have been observed. The overexpression of heterologous proteins can lead to the formation of inclusion bodies—high-density bodies of almost pure but misfolded protein, which are resistant to proteolysis and easy to isolate. Expression of GPCRs in inclusion bodies reduces the risk of toxicity to the cell during expression and has the potential to produce high levels of protein [70]. While a number of advances have been made in the development of refolding technologies for GPCRs, such as the use of amphipols [71, 72], our ability to capitalize on the high levels of protein expressed has been limited. Although a few examples of successful refolding from inclusion bodies exist, including opioid receptors [73], bacteriorhodopsin [74], $5HT_4$ [75], and BLT_1 [61], the ability to successfully refold GPCRs has been limited. Moreover, the generation of inclusion bodies is poorly understood; despite the assumption that the expression of large hydrophobic proteins should readily form inclusion bodies, this is not always the case. A number of techniques have been developed to facilitate GPCR expression in inclusion bodies, including the use of strong promoters and high-copy-number plasmids [69, 70] (pGEX and pET types), codon optimization [61], fusions to ketosteriod isomerase [75] (which is commonly used to direct the formation of inclusion bodies), and glutathione-*S*-transferase [76] (known to drastically affect protein expression levels), yet there is no general strategy for the systematic production of GPCRs from inclusion bodies. Expression studies have been able to improve the levels of expression of GPCRs in inclusion bodies; however, this is only the first step in the purification pathway, and further steps involving protein solubilization, purification, and renaturation are required.

The efficiency of refolding depends on the competition between protein refolding and aggregation. The BLT_1 and $5HT_4$ receptors, for example, were solubilized with harsh chaotropic agents and detergents (urea and sodium dodecyl sulfate [SDS]). The receptors were then refolded by solvent exchange using a solid Ni-NTA matrix, resulting in protein expression levels of ~0.5 mg/L [61, 75]. One of the most crucial factors in obtaining active, refolded protein was the composition of detergent/lipid micelles in which the GPCR was reconstituted. $5HT_4$, for example, was reconstituted in mixed dimyristoylphosphatidylcholine (DMPC)/3-[(3-cholamidopropyl)dimethylammonio]-1-propanesulfonate (CHAPS) micelles in the presence of cholesterol [75]. As our understanding of the structural properties of membrane proteins and their interactions with their lipid environment improves, better refolding techniques will surely follow.

While *E. coli* has been successfully used to express a wide range of GPCRs from the three major classes [70], there are limitations. In general, smaller proteins tend to lend themselves to higher, more soluble levels of expression, especially in a bacterial system. This observation has been supported by a study that compared the expression of 100 GPCRs in *E. coli*. Data from this study showed that proteins of ≤54 kD can be more easily expressed than larger ones [70]. Other prokaryotic expression systems, which have been used for the expression of GPCRs, include *Haloferax volcanii* [77]; however, this system suffers from proteolysis. *Halobacterium salinarum* [78] has been used as it produces high levels of the GPCR ortholog bacteriorhodopsin [79]. Although the initial studies with *H. salinarum* resulted in relatively low levels of expression [80], improvements to the levels of protein expression resulted in crystals of a fusion protein of the β_2AR, which diffracted at a low resolution [78]. Although, *Lactococcus lactis* has been extensively used for the expression of a number of membrane proteins [81], no GPCRs have been reported to be successfully expressed using this system. Although bacterial expression has been used to produce a number of GPCRs with varying degrees of success, new bacterial expression systems better able to cope with membrane proteins have been keenly sought. The photosynthetic bacterium *Rhodobacter sphaeroides* has been used for the production of human GPCRs. The advantage of utilizing *R. sphaeroides* is that the cell has a much larger membrane surface area per cell compared to other expression hosts. A system whereby overexpression of recombinant receptors placed under the control of the moderately strong and highly regulated superoperonic photosynthetic promoter pufQ has been developed [82]. Using this host, the human angiotensin AT_{1a} receptor has been expressed at levels of up to 12 pmol/mg. Although a relatively low level of expression, this system offers a practical alternative to functionally express GPCRs, using a system that can be scaled up [82].

Optimizing conditions for the overexpression and purification of membrane proteins is a laborious and time-consuming process. With the large number of variables affecting expression levels of GPCRs, methods need to be developed to systematically investigate each variable. This process can be accelerated using membrane protein–GFP fusions [83], which allow direct monitoring and visualization of membrane proteins at any stage during overexpression, solubilization, and purification [84]. With the short time required for plasmid construction and expression, an *E. coli* expression platform provides the ideal host for such a screen. Using GFP–GPCR fusion expression, optimization of both the human central cannabinoid receptor (CB_1) [58] and the bradykinin receptor (BK_2) [59] has been rapidly achieved.

Yeast Expression Yeast systems contain a number of desirable attributes required for the high level expression of GPCRs needed for structural studies. Like *E. coli*, yeasts grow quickly, and are easy and inexpensive to grow. They can be cultured to high cell densities, and scale up can be achieved using fermentation technology [85]. Furthermore, yeast expression systems allow for isotopic labeling and nonnatural amino acid incorporation [86]—both important factors when considering an expression system for structural studies.

While yeast systems possess a number of the same benefits as do bacterial systems, they also have several advantages. Yeast has compartmentalized organelles, allowing for more natural protein expression and folding, with subsequent insertion into the plasma membrane. Furthermore, unlike *E. coli*, they can perform the majority of PTMs, although glycosylation is substantially different to that observed in mammalian cells [87], and there are examples of GPCRs that are not glycosylated in yeast [88]. This phenomenon may constitute a problem for GPCRs that require correct glycosylation to obtain correctly folded protein and may therefore represent a limitation for yeast-based expression of GPCRs. However, the introduction of yeast strains engineered to provide a mammalian glycosylation profile may represent a solution to this problem [87]. A number of different strains of yeast have been used for the overexpression of GPCRs, including *S. cerevisiae* [86], *P. pastoris* [89], and *S. pombe* [88]. The process has been further simplified with the availability of a number of commercial expression systems, such as the *P. pastoris* system of Invitrogen (Carlsbad, CA).

There are a number of drawbacks associated with yeast that can prevent the efficient expression and purification of GPCRs. Like *E. coli*, the yeast lipidic environment, which is crucial for the efficient functional expression of GPCRs [90], is different from that of mammalian cells. Yeasts have much lower cholesterol and a higher ergosterol content [91], which can dramatically affect the ligand binding activity of mammalian GPCRs. Furthermore, loss of receptor ligand binding activity is often seen during the solubilization process when GPCRs are removed from the native membrane environment and reconstituted in a detergent micelle [92]. In some cases, the loss in activity can be recovered by cosolubilization with mammalian lipids, which have been shown to specifically bind to GPCRs [9] and stabilize the active confirmation of the receptor [93, 94].

Other problems that exist relate to the inherent mechanical strength of the yeast cell wall, making cell lysis challenging and often requiring harsher cell lysis processes to be used. Proteolysis is another problem of yeast expression systems that, while not limited to yeasts, can present a problem for the efficient expression and purification of GPCRs in yeast. However, protease-deficient strains can minimize such problems [88]. In contrast to bacterial expression, where the upper limit of expression is ≤54 kD, no such limits have been observed with respect to size and expressibility of GPCRs in yeast. Class C receptors (95–102 kD) have been successfully expressed in *P. pastoris* [70]. Furthermore, of the 100 GPCRs studied, 94% were expressed in yeast compared with only 48% that could be expressed in *E. coli* [70].

Insect Expression After *E. coli* and yeast, insect cell expression systems have provided the greatest number of structures, with more than 1200 structures in the RCSB protein database (PDB). To date, insect cell expression has provided the gold standard expression system for the structural determination of GPCRs. The recent structural determination of the human β_2AR and A_{2a} adenosine receptors [2, 3, 5] was achieved using *S. frugiperda* Sf9 cells, while the structural determination of the turkey β_1AR [4] was achieved using *T. ni* Hi5 cells.

The relative success of insect cell expression in the structural biology of GPCRs relates to their ability to generate multimilligram quantities of high-quality protein. A wide range of insect cells are commercially available, including *S. frugiperda* Sf9, Sf21, Hi5, and *D. Schneider* S2 cells, allowing screening of different cells to improve expression levels. As Hi5 cells have a high capacity for the expression of membrane proteins, they may have the potential to be useful for the expression of GPCRs [73]. Insect cells have been used to express a variety of different GPCRs [95] with very respectable yields; for example, the turkey β_1AR has been expressed at levels of 7 mg/L in Hi5 cells, while expression in Sf9 cells yielded 1.25 mg/L [96]. The eukaryotic, baculovirus-based expression system has been used for over 20 years and relies on the viral transfection of insect cells to achieve protein production. Subsequently, a number of technological advances have resulted in the establishment of commercially available systems, which are more efficient and user friendly than the original system.

As with yeast expression, insect cell expression provides a platform in which most PTMs required for functional protein can be achieved. In some cases, insect cell expression results in heterogeneous protein glycosylation; for example, when the D2 dopamine receptor was expressed in Sf9 cells, large amounts of unglycosylated receptor were produced [97]. Expression of the human formyl peptide receptor in insect cells also resulted in the production of immature and incompletely glycosylated forms of the receptor [98]. Therefore, there has been considerable interest in the modification of insect cells, such as the Mimic™ (Invitrogen) Sf9 insect cells, which have a more homogeneous human-like glycosylation machinery [99]. While an insect cell expression system provides a more native intracellular environment compared with *E. coli* and yeast, differences do exist in their lipidic membrane environments. Insect cells are typically grown at 27°C, and the types of lipid required to maintain membrane fluidity at this temperature are different to that of a native mammalian cell membrane. Insect cell membranes are very low in cholesterol and have no phosphatidyl serine in their plasma membranes. In addition, they have a comparatively high phosphatidyl inositol content, and following infection, an enrichment in phosphatidylcholine is observed [100]. In GPCRs, an altered lipid environment can lead to modifications in ligand binding, as was observed for the oxytocin receptor in which heterogeneous expression of low-affinity and high-affinity receptors was noted. The addition of cholesterol to the growth media resulted in a more homogeneous receptor population with a shift toward high-affinity ligand binding [101].

Insect cells are essentially free of endogenous GPCRs [102], which provide a low background environment for ligand binding assays. However, Sf9 cells contain endogenous G proteins that can couple to heterologously expressed GPCRs, allowing both agonists and antagonists to be assayed. While insect cell expression appears to contain a number of the desirable attributes required in an expression system for use in structural studies of GPCRs, a number of drawbacks exist. These drawbacks prevent insect cells from being the universal system required for large-scale structural studies of GPCRs. Even with recent improvements to the baculovirus expression system, it can take up to 1 month to generate large quantities of recombinant high-titer baculovirus. Furthermore, baculovirus-induced expression is normally a transient process requiring the constant generation of large quantities of recombinant baculovirus.

One of the major benefits of the insect cell expression system is the ability to scale up protein production using relatively simple culture techniques in a biosafety level 1 laboratory environment [103]. Insect cells are semiadherent, allowing growth under attached conditions (rollers, microcarriers) or in suspension. However, culture media are more complex and more expensive than the media used in the culture of bacteria and yeast.

Mammalian Expression Mammalian cells have been used to express a wide range of GPCRs' different classes [70]. Mammalian cells have all the cellular machinery required to correctly translate, fold, modify, and insert the protein into the cell membrane. Obtaining complex incomplete or heterogeneous glycosylation can, however, be a problem when receptors are overexpressed at high levels [104].

Heterologous protein production can be achieved through transient or stable expression, and a wide variety of cell lines are available (e.g., CHO, HEK, COS-1, etc.). Transient expression of GPCRs has been widely used for a number of years, although its use as a tool in structural biology has been limited due to cost and scalability. Transiently transfected cells usually express protein through the cytomegalovirus (CMV) promoter after 48–72 h post-transfection, with a steady decrease thereafter. This system provides a quick and relatively easy method of screening protein expression and allows for rapid site-directed mutagenesis. While lipid-based transfection methods have proved impractical for large-scale protein expression, other methods, such as the use of recombinant virus (e.g., adenovirus) [105] or Semliki Forest virus (SFV) [106], have shown promise. To date, much of the work carried out on viral transfections has centered around the use of SFV, which has a very high rate of success in the expression of GPCRs [70]. A limitation of the SFV system is the requirement of virus propagation, isolation, and determination of virus titer. In particular, virus titer is an important factor in relation to expression levels.

Stably transected mammalian cell systems offer a solution to some of the problems suffered by using transiently transfected cells. Stable cell lines provide a constant source of recombinant protein, although their generation can be particularly time consuming. Like the transient transfection cell system, the gene of interest is placed under the control of a strong promoter, such as the CMV promoter. The use of inducible promoters may be particularly advantageous with respect to membrane proteins as constitutive expression can lead to cell toxicity [107]. In order to obtain a stable cell line, the expression construct is stably integrated into the host's cell genome; this typically requires the use of a selective marker (e.g., antibiotic resistance) as integration events are rare. Advances in generating stable cell lines have made this process simpler, for example, the lentivirus [108] and the Flp-In T-REx™ expression system (Invitrogen), which allows the generation of stable mammalian cell lines exhibiting tetracycline-inducible expression by placing a tetracycline-inducible promoter into the genome via Flp recombinase-mediated DNA recombination at the FLP recombinase target (FRT) site [109].

The use of GFP coupled with fluorescence-activated cell sorting (FACS) has allowed the rapid selection of high protein expressers within mixed cell populations. The use of internal ribosome entry site (IRES)-GFP technology has improved upon regular GFP fusion selection by placing an IRES downstream of the promoter and the coding sequence for the "gene of interest," followed by the coding sequence for GFP. This allows for a single bicistronic messenger RNA encoding both genes to be produced. The two separate proteins are then translated from the same message, and their expression levels are thereby coupled. The use of IRES-GFP provides a monitor for the levels of target protein expression. This has been successfully used to improve the expression levels of rat $5HT_{2c}$, whereby levels of 140–160 pmol/mg of membrane protein have been achieved [110].

A number of issues have prevented the widespread use of mammalian cell culture as a tool for GPCR structural biology, including cost and scalability. Complex media, requiring the addition of expensive antibiotics, needed to maintain the expression of selective markers have been a major stumbling block. Problems also exist in scaling up cell production. While mammalian cells, such as HEK293, are able to grow in suspension—making them amenable to fermentation—overall yields can be poor as cells struggle to survive in suspension cultures [73]. Anchorage-dependent cells can be grown in cell factories or on microcarrier beads, although issues relating to cost and practicality can make obtaining large quantities of cells difficult. To date, few examples exist of GPCRs that have been expressed and purified from mammalian cell cultures to quantities required for structural studies. One exceptional example is opsin, which has been expressed and purified using a tetracycline-inducible stable HEK293S cell line, resulting in levels of protein expression up to 10 mg/L [107].

CF Expression Since the 1950s, *in vitro* transcription translation (IVT) has been successfully used to express soluble cytosolic proteins [111]. However, until recently, its application in expressing more complex mammalian membrane proteins has been limited [112]. Some of the advantages of CF protein expression include that it does not depend on cellular integrity and does not require complex culture conditions. Although a number of CF systems exist, the three main sources are rabbit reticulocyte extract [113], wheat germ extract [114], and *E. coli* extract [115]. CF expression systems are typically used in batch expression, and the basic system contains all of the high molecular weight machinery required for transcription and translation. Modifications to the system, whereby the cellular machinery is compartmentalized within a semipermeable membrane to allow continuous feeding of low molecular weight precursors into the system [116], have increased both protein production yields (levels up to ~10 mg/mL) [117] and rates [118]. This also allows the system to be used for isotopic labeling studies [119]. One of the factors that has limited the use of CF expression of membrane proteins has been compatibility issues relating to the detergent required to maintain the structural integrity of the membrane protein and the effects that the detergent has on transcription and translation. As a result, protein production requires extensive optimization efforts prior to scale-up. Recent advances in technology and increased availability of high-quality commercial CF systems have led to a renewed interest in using IVT as a viable system for the expression of GPCRs for structural biology. A number of GPCRs, including the olfactory receptor 17–4 (at yields of up to 0.3 mg/mL) [114] and the human β_2AR (1 mg/mL) [115], have been successfully expressed using CF expression. However, a number of questions still surround the use of CF expression as a robust, cost-effective method to express GPCRs. A clear advantage to CF expression of GPCRs is its demonstrated compatibility with detergents. This has been a stumbling block in the development of new technologies such as nanodiscs [120] (high-density lipoprotein partials), which provides a more natural membrane environment for the incorporation GPCRs [121]. The high cost, and, in turn, scalability, of CF expression is a major problem; commercially available CF expression systems can be extremely expensive [122] such that large-scale expression is not cost-effective when compared with other, more established, expression systems.

Other Expression Systems Despite the majority of the work to date having focused on the main expression systems previously outlined, a number of other strategies for GPCR expression have been investigated. One such strategy has been the use of whole organisms to express GPCRs. A wide range of organisms have been used to express GPCRs, including *Xenopus oocytes*, which were used to express the pituitary thyrotropin-releasing hormone receptor [123]. The requirement of mRNA microinjection has likely precluded this approach from large-scale production. The *D. melanogaster* metabotropic glutamate receptor has been expressed in *Drosophila* photoreceptor cells at levels of 170 μg/g of fly heads (equivalent to approximately 3000 flies) [124], a level that is at least threefold higher than those achieved with conventional baculovirus systems [56]. However, the task of scaling up is significantly easier for insect cells than for whole flies. More recently, transgenic silkworms have been used to express the human μ-opioid receptor to levels of 150–250 ng per silkworm, comparable to the levels of protein obtained using Sf9 cells [125]. Transgenic animals have also been used as a method to express GPCRs. Adenovirus-mediated expression of the chemokine receptor CXCR1 in transgenic mice yielded approximately 1 mg of homogenous rabbit CXCR1 from 20 mice livers—an amount which would normally require 5 L of cell culture media [126]. While whole organism expression has been successfully employed to express GPCRs, the ability to scale up production remains a problem.

14.4.2. Factors Influencing GPCR Overexpression

Although the choice of expression system is the key factor in obtaining sufficient quantities of high-quality GPCR, a number of other factors have to be considered and modulated to optimize GPCR expression levels. For structural studies, the development of constructs that are compatible with the method of structural determination is crucial. The presence of large loops and extended N- and C-termini can hamper crystallization, increase proteolysis, and reduce the overall protein yield [96]. Expression levels can also be improved using mutagenesis. For example, the expression of turkey β_1AR was significantly improved following a Cys116Leu mutation [127]. Directed evolution is another alternative given that a 12-fold increase in expression of the NTR receptor was achieved over wild-type receptor [128], and as previously mentioned, the use of fusion partners has also been used to increase expression levels [64]. While modifying the construct can lead to increases in expression levels and reduced heterogeneity, care has to be taken to not dramatically affect the pharmacology of the receptor in a detrimental way.

Finally, manipulation of culture conditions also offers a route to improve expression levels. Optimization of the culture conditions through the addition of additives, such as ligands [65, 129], amino acids [130], and DMSO [130, 131], and by altering temperature [130] can increase expression by nearly 10-fold [130].

14.4.3. Summary

GPCRs can be produced using a wide variety of expression systems from simple bacteria expression to whole organisms. Table 14.3 gives an overview of the different expression systems that have been used to express the human β_2AR. Although a universal system for the expression of GPCRs has been sought, we are not much closer to achieving this goal then we were a decade ago. Significant steps have, however, recently been taken to improve the expression systems currently available.

In general, GPCRs are not expressed at particularly high levels (0–10 mg/L) when compared with soluble cytosolic proteins (>10 mg). For most structural studies, large quantities of homogeneous protein are required. In view of this, bacterial systems that have been the expression system of choice for much of structural biology over the past 30 years would seem to be a good choice; however, for GPCRs, this does not seem to be the case [70] due to poor levels of fully functional protein expressed. Empirical studies [70, 95] suggest that more complex expression systems such as yeast, insect, and mammalian cells are most likely to succeed in the soluble expression of GPCRs at quantities high enough for structural studies. When choosing an appropriate expression system, three factors have to be considered: cost, scalability, and the yield of homogeneous protein. Typically, the amount of biomass generated per liter of media is less in mammalian and insects cells than in bacteria and yeast [132]. Hence, a cost compromise has to be sought between growth media and scalability. The question of yield relates both to the quantity and quality of protein produced. The gold standard for assessing the quality of protein is by measuring the ligand binding properties of the protein. Thus, the Bmax values, which are relative measurements of specific activity expressed in moles of ligand bound to the protein divided by the amount of total protein, offer the best means to measure the quality of the protein. Obtaining sufficient quantities of high-quality, homogeneous protein is especially important for use in X-ray crystallography. Heterogeneous populations of protein can arise from a number of sources, such as PTMs and misfolded protein. Expression systems that both minimize heterogeneity and maintain a high level of expression are ideal candidates for structural studies. To date, the only expression system that has been able to successfully express GPCRs (non-rhodopsin) to sufficient levels in a state that has allowed their structural determination through X-ray crystallography has been insect cell expression [4, 5, 9]. Although the baculovirus/insect cell system does not offer the perfect system for the expression of GPCRs, the problems that it suffers with can be sufficiently dealt with to allow structural determination. For example, PTM heterogeneity can be addressed through combining chemical or enzymatic treatment with protein mutagenesis [4, 5, 9].

Currently, a systematic approach to expression studies is required in order to obtain a chemically and structurally homogeneous protein in sufficient quantities for structural studies. Grisshammer and Tate [133] proposed a minimal production of 1 mg of recombinant protein per 5 L of culture medium. While this level of protein expression is the ideal target, this has to be viewed as a minimum, and levels approaching that obtained for soluble proteins have to be sought (>5 mg/L). For future structural genomics studies, a more rational approach will be required whereby a better understanding of the physiological interactions of the GPCR with the host cell would allow of heterogeneous protein expression. With the development of new technologies such as GFP-aided screening [84], nanodiscs [134], and thermostabilization [135, 136], we are one step closer to generic approaches for expressing GPCRs for structural studies.

14.5. PROTEIN PURIFICATION AND SOLUBILIZATION

The first step in any protein purification is to develop a robust protocol to extract the protein of interest from the host cell/culture media. GPCR purification typically requires an initial cell lysis step, which in itself is governed by the type of expression system used. Due to the mechanical strength of their cell walls, both bacterial and yeast cells require a greater degree of force to achieve efficient cell lysis in comparison with insect and mammalian cells. In order to afford efficient cell lysis in bacteria and yeast, cells typically require the use of a French press or a bead beater. In contrast, mammalian and insect cells require much less mechanical force and can be lysed using fluidizers and freeze/thaw methods. When considering the method for lysis, care has to be paid to prevent denaturation of the GPCR through sheer force or thermal denaturation. Furthermore, the inclusion of protease inhibitors (e.g., pepstatin, leupeptin) to prevent proteolytic degradation of GPCRs upon lysis is of particular importance [142].

The large-scale expression of GPCRs typically results in the generation of a large amount of biomass of which only a small percentage contains the protein of interest. The generation of membrane preparations offers an effective method to crudely purify the GPCR away from the majority of the cytosolic protein. The most effective method to generate high-quality membrane preparations is through the use of ultracentrifugation and stringent washing of the membrane preparation with high salt and acid/alkaline washes. The generation of high-quality membrane preparation was exemplified in the recent structural determination of the human A_{2a} structure, where the membrane preparation was washed up to nine times with a high osmotic buffer containing 1.0 M NaCl; this subsequently facilitated a one-step immobilized metal affinity chromatography (IMAC) purification [5].

As with the purification of all membrane proteins, GPCRs need to be extracted from the lipidic environment that they were expressed in and subsequently solubilized into a surrogate detergent. The key to this step is establishing the critical solubilization concentration (CSC), which is the minimal concentration of detergent required to disrupt the cell membrane into micellar dispersion. In addition, it is essential to use a detergent that does not inactivate the ligand binding properties of the receptor. It is assumed that if ligands can still bind to a detergent solubilized protein, then it is likely that the 3D structure is in a physiologically relevant state [143].

One of the factors that has aided the study of membrane proteins is the large and diverse range of detergents that are available. Detergents can be classified into three main classes: nonionic, zwitterionic, and ionic. Nonionic are characterized by their uncharged, hydrophilic head group and include mild detergents such as digitonin, DDM, and Triton X-100 (Sigma-Aldrich, St. Louis, MO). Zwitterionic detergents contain both a positive and a negative charge in their hydrophilic head group and include detergents such as Fos-choline 12 (Anatrace, Maumee, OH), CHAPS, and LDAO. Anionic detergents are characterized by their charged hydrophilic head groups and include detergents such as SDS, sodium cholate, and decyltrimethylammonium chloride (HDTCI). When considering the ideal detergent for use in structural studies, a number of points have to be considered. The most important requirement is that the detergent has to be able to solubilize the receptor without affecting the structural integrity and activity of the protein. Ideally, the detergent should be easily removed from the protein suspension and has to be compatible with any subsequent chromatography steps (e.g., low UV absorbance). In addition, to make protein production reproducible, the detergent has to be commercially available on a large scale at a reasonable cost. Most importantly, from a structural point of view, detergents need to be well-defined homogenous compounds. This final point has prevented the use of polymeric detergents such as Triton X-100 and the Tween series (Sigma-Aldrich) [143].

As detergents and membrane proteins interact in an unpredictable manner, it is impossible to anticipate which is the most appropriate detergent to use in the purification process; this needs to be determined experimentally. However, within the three main classes of detergents previously described, they can be subdivided into detergents that are mild enough to maintain activity and structural integrity, and those that are harsh with a higher probability of protein denaturation [144]. Structurally, detergents can be split into two sections, the head group and the tail group, both of which contribute to the nature of the detergent. In general, as the size of the head group is reduced or the charge becomes modulated from neutral to zwitterionic and then to ionic, the harshness of the detergent increases. Furthermore, in general, as the length of the tail group decreases, so does the mildness of the detergent [145]. While a detergent may be considered structurally mild, the relationship between the detergent and membrane protein is a complex one, and even those detergents that are considered mild can result in receptor denaturation, inferring a specific "chemical effect" of the detergent that can be impossible to predict. An example of such an interaction can be observed in the stabilized turkey β_1AR, which is stable in decyl maltopyranoside (DM) but is less stable in the milder polyoxyethylene [136].

When solubilizing a receptor, detergent concentrations of 0.5–2% w/v are typically required, using detergent/protein ratios of 1:1 to 3:1 [146]. The concentration of the detergent is an important factor as detergent-solubilized GPCRs exist as protein detergent complexes (PDC). At low concentrations, detergents exist as monomers, and as the concentration increases above their critical micellar concentration (CMC), they form structures called micelles. The size and shape of micelles depend on the type, size, and stereochemistry of the detergent, as well as the aqueous environment that they occupy (ionic strength, pH, etc.) [147]. For the purification of all membrane proteins, it is vital to maintain the concentration above the CMC; otherwise, the micelle surrounding the membrane protein can disassemble, leading to protein aggregation and inactivation. While maintaining a level of detergent above the CMC is an important factor in protein stability, the addition of other agents can also dramatically enhance their stability. As previously mentioned, lipids can play an important role in maintaining the stability and activity of GPCRs [9, 101]. During the detergent solubilization process, lipids are coextracted along with the protein. Lipids surrounding detergent-solubilized membrane proteins are often important in maintaining structural stability, so their loss can be an important factor in preventing crystal formation [148, 149]. Therefore, maintaining a balance between the detergent used and the amount of lipids copurifying with the receptor can be critical. As loss of lipid during the purification process can have a detrimental impact on protein stability, some studies have remedied the problem by addition of exogenous lipid. The exact role that the lipid plays in the purification process is often unknown, and the addition of lipid can be an empirical one. However, it has been suggested that lipids can bind to specific sites on GPCRs [9], and the addition of cholesterol can provide a large hydrophobic surface for lipid retention [132]. In purifying the human A_{2a} receptor, it was noted that the addition of cholesterol hemisuccinate (CHS) was crucial in maintaining activity of the protein during the IMAC purification [65].

The choice of detergent is a key consideration for maintaining the structural integrity of detergent-solubilized GPCRs, but stability can also be improved by the addition of salts (e.g., ammonium sulfate, NaCl, and $MgCl_2$), reducing agents (e.g., tris[2-carboxyethyl]phosphine [TCEP] and 2-mercaptoethanol), osmolytes (e.g., glycerol, trehalose, and histidine), and ligands (e.g., theophylline). For example, optimization of the purification of bovine rhodopsin from ROS membranes was found to require the addition of alkylthioglucosides to an appropriate detergent/lipid balance in combination with a divalent cation (zinc) [15]. Thus, careful optimization of the extraction and purification conditions is essential to ensure that the purified receptor is folded in a native conformation.

As GPCRs are purified as a PDC, regular protein chromatography techniques can be complicated due to the presence of lipids and detergents that can affect the performance of the chromatographic matrices. Fortunately, the most widely used technique for purifying recombinant proteins, IMAC, performs well in most neutral detergents [65, 66, 96]. Both nickel [96] and cobalt [5] matrices have been used for batch purification of GPCRs with good success. However, it is noteworthy that the position of the His-tag (N- or C- terminal) can significantly effect the efficiency of GPCR binding to the resin [65], and the tag may have to be removed prior to protein crystallization [150]. A number of other affinity chromatography techniques have also been investigated for the purification of GPCRs. The use of other affinity tags, such as strep [64] and biotin tags [151], have been used with varying degrees of success. Antibody affinity resins directed toward FLAG [92], TAP [152], and 1D4 [153] tags have been explored. Lectin-based chromatography [154] (e.g., wheat germ agglutinin and heparin) can be used to purify GPCRs [34] and is particularly useful where problems exist with respect to glycan heterogeneity. More classical techniques such as ion exchange (e.g., Q-sepharose, Sigma-Aldrich) [139] and size exclusion chromatography (SEC) (e.g., Superdex S200, Sigma-Aldrich) [96] can also be used, although the efficiency of protein separation by SEC is often poor in detergents with large micelles. One of the most useful techniques in GPCR purification is the use of ligand affinity chromatography. Ligand affinity columns have been used for a number of years and have allowed GPCRs to be purified from native sources [155], a task that would ordinarily be particularly challenging due to their low natural abundance. Resins with covalently bound ligands, such as xanthine amine congener (XAC) [156], 3-(2′-aminobenzhydry1oxy)-tropane (ABT) [155], and alprenolol [157], have all been used to purify GPCRs. The major advantage of ligand affinity chromatography is that only functional receptors will bind to the resin. Therefore, the process is particularly useful where a mixed population of active and inactive GPCRs are present. Although affinity chromatography has a number of benefits, it can suffer from low binding and poor elution. In addition, the columns must usually be made in the laboratory since affinity matrices are not commercially available.

The aim of any protein purification methodology to produce receptors for structural studies is to develop a robust protocol that allows the selective purification of homogeneous, monodisperse protein with a respectable yield. In order to maintain strict quality control, measures need to be put in place to assess the quality of the purified receptor. The use of techniques such as SEC provides a good indication of the monodisperse nature of the purified protein. However, the Bmax value is arguably the most important indicator of protein quality, and over the course of the purification process, a significant increase in this value should be observed. For example, assuming a 1:1 stoichiometry of protein to ligand, a GPCR with a molecular weight of 50 kD should have a Bmax value of 20 nmol/mg. This is often extremely challenging to obtain, but theoretical maximum activity has been obtained for some GPCRs, including the neurotensin receptor [158].

In the future, further developments in purification techniques will have to be made to help accommodate the need to purify significant quantities of more complex and challenging GPCRs for structural studies. The development of such technologies will be driven in part by new detergents, which are necessary for the purification and crystallization of some of the more unstable GPCRs. Moreover, the use of surface plasmon resonance (SPR) technologies to screen different conditions and ligands provides a great opportunity to improve and speedup the whole protein purification process [159].

14.5.1. Choice of Detergents for Structural Studies

Integral membrane proteins like GPCRs are adapted to function within lipid bilayers, making them difficult to manipulate, and are generally considered to be the most difficult type of protein to crystallize. For successful crystallization of a membrane protein to occur, a number of issues have to be addressed to generate and maintain the crystal lattice. To date, most of the membrane protein structures that have been solved by X-ray crystallography have involved the use of nonionic and zwitterionic detergents. Of the structures that have been solved, the vast majority have involved the use of alkyl maltosides, glucosides, dimethyl *N*-oxides (e.g., LDAO), and polyoxyethylene glycols (e.g., $C_{12}E_8$) [145]. Furthermore, the majority of the detergents tend to have an alkyl chain length of between C7 and C12, which are thought to mimic the approximate thickness of the hydrophobic portion of the cell membrane [160]. A number of considerations have to be made when assessing which detergent to use in the crystallization process. The crystal lattice of a membrane protein has to accommodate a number of very different components, including protein, lipid, detergent, and aqueous buffer. With each component having very different physical properties, conditions must be found, which allow a complex network of interactions to exist. A thorough characterization of the PDC with respect to the number of bound detergent and lipid molecules can be very helpful. Controlling the concentration of the detergent is also key, especially as some filter-based protein concentration methods can also concentrate the detergent, leading to receptor denaturation and phase separation during crystallizations. The amount of detergent required for successful crystallization can be a very fine balance, with crystallization often occurring around the phase separation boundaries [160].

When considering the structure of the PDC, it can be viewed as protein surrounded by a flexible and dynamic detergent belt. A key factor in obtaining strong lattice interactions is the size of the detergent micelle in relation to the size of the protein. GPCRs often have only small hydrophilic domains outside the membrane, and these can be occluded by detergents with large micelles (e.g., DDM). Ideally, crystallization should be performed using detergents with smaller micelles (e.g., OG, NG), but these detergents are often more denaturing than DDM, resulting in denaturation of the receptor upon solubilization.

Therefore, a balance has to be struck between shielding the hydrophobic domains of the GPCR from the aqueous environment and, at the same time, fostering interactions between the hydrophilic domains of the protein, thus driving lattice formation. Typically, this means that a protein is purified into a mild detergent, such as DM, and is then exchanged into a detergent with a smaller micelle size, such as octyl-thioglucopyranoside [4]. However, detergents themselves can also play a more direct role in lattice formation, with detergent molecules becoming more ordered at specific locations on the hydrophobic surface of the membrane protein and driving lattice formation [161].

Trying to optimize the crystallization process can be a time-consuming process, as this requires screening various detergents and additives, such as small amphiphiles, which can be used to reduce the number of detergent molecules in PDC [162]. In some cases, crystallization can only occur through a combination of detergents, where one detergent makes specific interactions within the protein which may stabilize it, while a second detergent allows crystal contacts to form [144]. Bovine rhodopsin is an example of a protein that has been successfully crystallized using a mixture of detergents. Rhodopsin was initially purified with LDAO and was then exchanged into C_8E_4 immediately prior to crystallization [34].

A great deal of momentum has been achieved in the crystallization of GPCRs, with the recent development of new strategies for their crystallization, including antibodies and bicelles [3], fusion proteins and lipidic cubic phase [2, 5], and thermostabilization [4]. The structure of the human β_2AR [3] provided the first major breakthrough in the structural biology of GPCRs since the first structure of rhodopsin published nearly a decade prior [1]. The structure of β_2AR was determined after crystallization of the receptor, with crystal contacts being mediated by an antibody Fab fragment bound to IC3. The use of bicelles was first applied to membrane protein crystallization to determine the structure of bacteriorhodopsin [163]. Bicelles are large lipid detergent disks that can be used like pure detergents for solubilizing membrane proteins. Bicelles can be formed from a mixture of either DMPC and CHAPS or from DMPC/DHPC. Their major advantage is that the high lipid content significantly stabilizes GPCRs, and their low viscosity means that standard crystallization robots can be used. The disadvantage of bicelles is that receptor crystallization may be prevented due to the large size.

14.5.2. Crystallization Chaperones

To obtain well-ordered crystals of membrane proteins, it is often necessary to include a protein that will act as a crystallization chaperone. Such proteins include antibodies [164] or DARPins [165], or smaller more crystallizable proteins may be inserted into the target protein of interest in the form of a fusion protein. These chaperones facilitate crystallization by using a number of different mechanisms. First, the bound chaperone increases the overall hydrophilicity of the complex. This can be important since many membrane proteins contain relatively small hydrophilic domains. Crystal contacts are usually formed between polar surfaces, which protrude from the detergent micelle. Second, the chaperone can assist in crystallization by further stabilizing the protein in a particular conformation. Finally, co-crystallization with a protein of known structure allows the protein structure to be more easily determined. This is because molecular replacement can be used, as opposed to the more difficult and time-consuming strategies for phase determination using heavy atom derivatives or proteins labeled with selenomethionine.

Monoclonal antibodies' fragments (Fv or Fab) have been used as a tool in the structural determination of a number of membrane proteins, including cytochrome c [166] and the voltage-gated K^+ channel KvAP [167]. To be useful, antibodies must selectively recognize only the correctly folded membrane protein, ideally, in a relatively nonflexible region so that the antibody protein complex is rigid. Antibodies can be raised by traditional immunization methods; however, this is problematic for a number of reasons. Native antibodies are not suitable for crystallization since they consist of a number of domains linked by flexible regions. Instead, monovalent antibodies should be obtained by proteolytic cleavage. Such Fab fragments have been successfully used for crystallization [168]. Immunization requires availability of a suitable antigen. It is often difficult to obtain purified GPCRs of sufficient quality and quantity to use as an immunogen, especially if the protein is required for crystallization. Instead, peptide fragments corresponding to particular regions of the GPCR, such as the N- or C-terminus, are often used as the immunogen. However, the use of peptide fragments usually results in flexible antibody–protein complexes that are of doubtful utility in structure determination. A more successful strategy is to immunize with the complete receptor in a stabilized form, such as in a proteoliposome. In the case of the β_2AR [169], the ICL3 linking the cytoplasmic ends of transmembrane segments (TM) 5 and 6 was shown to be the most flexible region, as defined by protease susceptibility. A monoclonal antibody was generated by immunizing mice with the β_2AR reconstituted into proteoliposomes. Antibodies were selected that preferentially bound to the native protein rather than to protein denatured by SDS. The antibodies bound to ICL3 and stabilized a particular conformation, as defined by the binding of a conformationally selective fluorescent dye. For crystallization, a Fab fragment of this antibody was generated by protease cleavage.

Recombinant antibody technologies provide a useful approach to rapidly generating protein binders [170, 171]. Techniques such as phage display or the selected lymphocyte antibody method (SLAM) [172] use less protein and allow conditions to be adjusted to increase selection of antibodies suitable for crystallization. Such an approach was used to generate high-affinity Fabs, which bound to the conformational epitopes of the citrate carrier CitS, a transporter with 11 transmembrane domains [171].

A number of alternative scaffolds, such as affibodies [173] and repeat proteins, represent recent alternatives to antibodies. Of particular note are DARPins [165, 174], which contain multiple domains involved in protein–protein interactions. The ankyrin repeats of 33-amino acid residues are assembled into highly complex variable domains and can bind to a variety of different target molecules. Pluckthun and coworkers have developed a library based upon ankyrin repeat proteins, which represents an extensive repertoire of structures that can be screened for binding to any protein of interest. DARPins have been used to crystallize proteins from several different families, including the membrane protein component of the multidrug exporter AcrB [165].

Potentially, the most rapid and generic approach to increasing the polar surface area of membrane proteins involves the use of fusion proteins. The main limitation to using fusion proteins is in building the link between the fusion and target protein. If the link between the fusion and target protein is too long and flexible, this will have a negative impact on crystallization. In contrast, if the link is too short or constrained, the structure of the target protein may be distorted. Such an approach was used in attempts to crystallize the 12 transmembrane domain protein, lac permease. To crystallize lac permease, a number of fusion proteins, including cytochrome b562 and T4 lysozyme (T4L) [175], were inserted into the loops between the transmembrane helices. The insertion of cytochrome b562 into lac permease resulted in the formation of 2D crystals [176].

T4L has also been utilized as a fusion protein in the crystallization of GPCRs. Specifically, T4L replaced most of the third ICL of the β_2AR and the adenosine A_{2a} receptor [5, 142]. Through careful optimization of the constructs, residues 2–161 of T4L were inserted between residues 231 and 262 of the β_2AR to mimic the predicted distance of 15.9 Å between TM5 and TM6, as determined from the structure of rhodopsin [35]. Following on from this initial success, the same approach was applied to the human adenosine A_{2a} receptor, in which residues 209–221 from the ICL3 loop were replaced with T4L. In both cases, the introduction of T4L into the protein resulted in high-resolution structures of 2.4 Å and 2.6 Å for the β_2AR [2] and A_{2a} [5] receptors, respectively (Fig. 14.4). As insertion of T4L into the protein does not substantially increase protein stability, the fusion proteins must be crystallized in the stabilizing environment of the lipidic cubic phase. The main problem with T4L insertion is that it appears to affect the conformation of the receptor. In addition, the fusion prevents coupling to G proteins and therefore limits full pharmacological characterization of the engineered protein.

Overall, the use of crystallization chaperones, such as antibodies and fusion proteins, has made a major contribution to the field of GPCR structural studies, especially when combined with other methods such as *in cubo* crystallization.

14.6. *IN CUBO* CRYSTALLIZATION

In contrast to the original human β_2AR [2] and turkey β_1AR [4], which were crystallized using hanging drop vapor diffusion, both the β_2AR and A_{2a} T4L

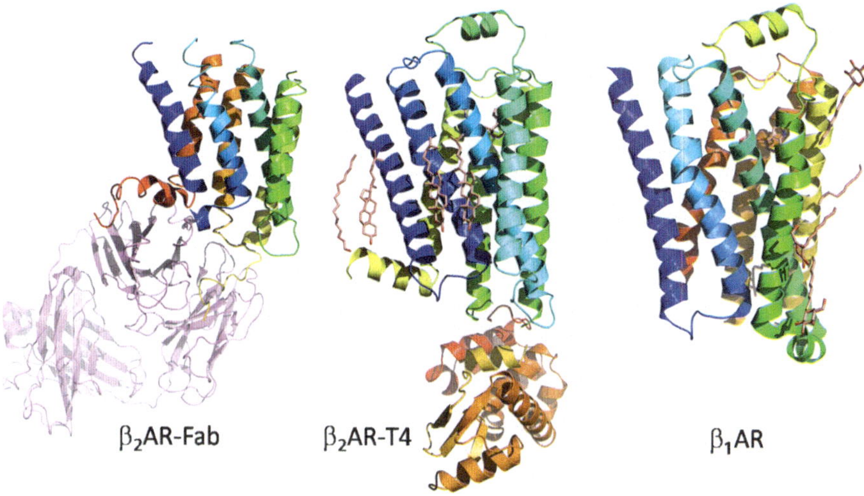

Figure 14.4 Comparison of the structures of the βAR showing the positions of the antibody and T4L insertion used to facilitate crystallization.

fusions were crystallized using the cubic phase method to grow crystals. To date, 54 structures have been determined from crystals grown using the cubic phase method, which represents approximately 10% of the total membrane protein structures deposited in the PDB, although the majority of these are homologues and mutants of bacteriorhodopsin [177]. Moreover, the use of the cubic phase method has resulted in some of the highest resolution GPCR structures to date (e.g., halobacterial bacteriorhodopsin was solved to 1.43 Å) [178]. Cubic phase has also been shown to be compatible with detergent-free membrane protein crystallization in which bacteriorhodopsin crystals have been directly grown from native cell membranes without exposure to any detergent [179]. While cubic phase is a relatively recent development in protein crystallography, it offers a real opportunity to further develop the field of GPCR structural determination. The basic premise behind *in cubo* crystallization is that membrane proteins crystallize in a more native-like lipid bilayer, as opposed to a detergent micelle. This relies on the protein being incorporated into an appropriate lipidic matrix [180], without altering its native structure, allowing diffusion in three dimensions to allow for nucleation and crystal growth (Fig. 14.5) [177, 181].

The lipidic matrix of choice is a bicontinuous cubic phase of monoacylglycerols (MAGs) and water. MAGs are an important intermediate in fat metabolism that adopts a remarkable variety of liquid crystal phases when dispersed in water; this makes them an ideal matrix choice. In both the recent β_2AR and A_{2a} T4L fusion structures, the MAG used was monoolein, which was substantially supplemented with cholesterol. While cholesterol provides an active component required to stabilize proteins, the lipid also provides a large number

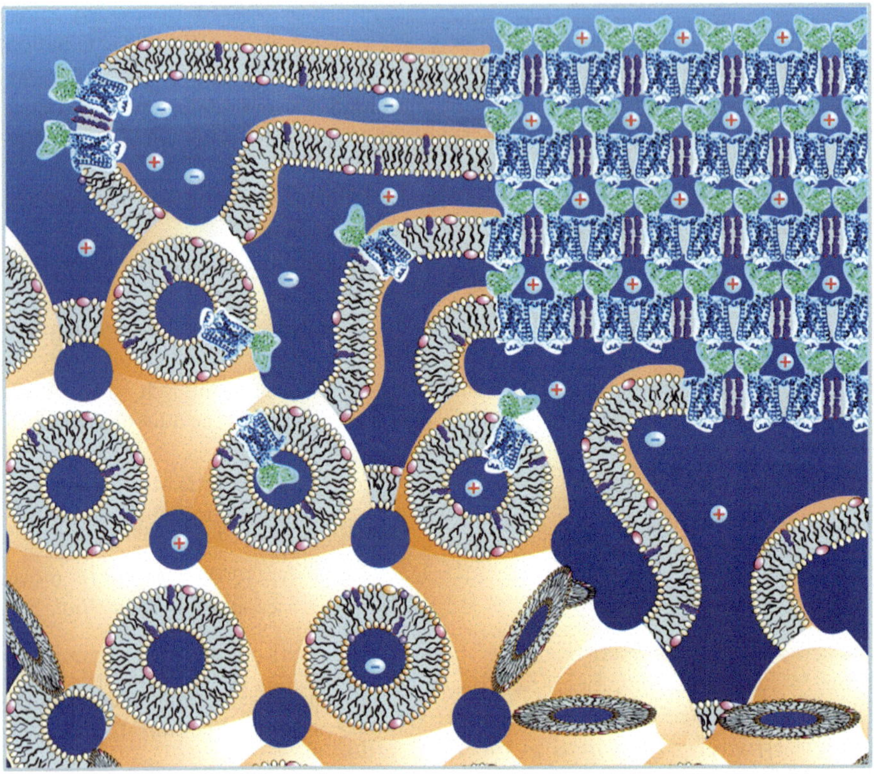

Figure 14.5 Schematic representation of the events proposed to take place during the crystallization of a GPCR from the lipidic cubic mesophase. Crystallization is initiated with the protein being reconstituted into the highly curved bilayers (bottom left-hand quadrant of the figure). The addition of "precipitants" then shift the equilibrium away from stability in the cubic membrane, leading to phase separation wherein protein molecules diffuse from the continuous bilayered reservoir of the cubic phase by way of the lamellar portal (left upper quadrant of figure) to lock into the lattice of the advancing crystal face (right upper quadrant of figure). Salt (positive and negative signs) facilitates crystallization by charge screening. Co-crystallization of the protein with native lipid (cholesterol) is shown in this illustration. As much as possible, the dimensions of the lipid, detergent, native membrane lipid, protein (β_2AR–T4L; PDB code 2RH1), bilayer, and aqueous channels have been drawn to scale. The lipid bilayer is approximately 40-Å thick. The figure was kindly provided by Professor Martin Caffrey, University of Limerick [207].

of the crystal-packing interactions between symmetry-related protein molecules, as observed for the β_2AR. The interfaces between the β_2AR molecules contained six cholesterol and two covalently bound palmitic acid molecules, with the latter further illustrating the requirement for a homogeneous protein. This further highlights the role in which PTMs can play in facilitating crystallization [2]. Although monoolein has typically been the most favored MAG used in cubic phase, the use of more rationally designed MAG matrices, all of which allow the formation of the requisite cubic mesophase at room temperature required for crystallization, has been sought [177].

The unique structure of the cubic phase means that it differs in a number of ways from that of the more classic solution phase (hanging drop vapor diffusion crystallization). Upon reconstitution into the bilayer of the cubic phase, integral membrane proteins such as GPCRs become confined to the 3D network of the curved bilayers, through which they diffuse freely (Fig. 14.5). Concurrently, small molecules, such as precipitants and buffers, diffuse through the adjacent network of aqueous channels [182, 183]. Diffusion conditions may be further complicated around the crystal, where the lipid phase forms a stack of lamellar layers [183]. In contrast, during solution–phase crystallization, both the protein molecules and the small molecule precipitants freely diffuse through the same 3D space.

It is thought that *in cubo* crystallization is mediated through the development of a series of overlapping concentration gradients within the porous mesophase. Upon reconstitution, the protein diffuses within the plane of the cubic phase bilayer, and, as precipitant is added to the mesophase, this triggers phase separation. Under conditions favoring crystallization, one of the separated phases is enriched in protein, facilitating protein–protein and protein–lipid contacts, which nucleate and develop into crystals [177]. The *in cubo* model also includes a lamellar conduit between the bulk cubic phase acting as a protein reservoir and the face of the crystal [184].

For the future development of GPCR crystallization, the use of new and emerging technologies will be required in order to maintain this momentum. The development of new techniques to set up crystallization trays with minimal amounts of protein, such as microfluidic protein crystallization systems [185], will dramatically increase the range of conditions that can be screened with relatively small quantities of protein. Furthermore, the application of *in cubo* crystallization will likely allow further structural determination of GPCRs. Finally, the development of microdiffraction technology [186], which has been a crucial component in all of the recent GPCR structures [2, 4, 5], will hopefully become more widely available in the future.

14.7. ENGINEERING RECEPTOR STABILITY

Poor stability of GPCRs in detergent has been a key limitation in obtaining well-diffracting crystals of GPCRs and, indeed, other membrane proteins. In order to prevent the protein from irreversibly forming a stable nonnative conformation outside the membrane bilayer, which may lead to complete denaturation and aggregation, the protein must be thermodynamically stable in a functionally relevant conformation. In the case of GPCRs, such conformations include the spectrum from the inactive ground state R through to the fully active R* or R* bound to its G protein (R*G). It is now clear that receptors can occupy more than one active R* state, and this results in agonist-directed trafficking of the receptor to specific signaling pathways [187]. To obtain crystals of GPCRs, this conformational flexibility must be reduced such that the receptor is trapped in a single conformational state. The simplest way to accomplish this is by the addition of compounds that bind only to particular receptor conformations [188]. This alone, however, has not been sufficient to enable crystallization. An alternative approach is to identify stabilizing mutations, which can strengthen the interactions in the correctly folded form of the protein, thus slowing inactivation [189].

The use of thermostabilizing mutations in membrane proteins was first demonstrated for two bacterial proteins, diacylglycerol kinase (DGK) and bacteriorhodopsin [190, 191]. Thermostabilizing mutations in membrane proteins are surprisingly common and can be identified by random or systematic mutagenesis. It appears that around 5–10% of random mutants will increase the stability of a membrane protein. In comparison, very few mutants can be found that stabilize soluble proteins. This suggests that membrane proteins have not evolved to be stable, perhaps because this would reduce the flexibility required for protein function or negatively impact protein folding, membrane insertion, or turnover [189].

The melting point of a protein can be used as a measure of thermostability. Melting point curves can be constructed by heating the protein and measuring the extent of unfolding with techniques such as circular dichroism or by differential scanning calorimetry. Alternatively, there are a number of fluorescent dyes that can detect protein unfolding by, for example, binding to exposed cysteine residues [192]. There are two main problems in using such techniques for constructing thermostability curves of GPCRs. First, these methods require pure protein samples, and, in many cases, GPCRs are too unstable to purify. Second, none of these methods takes into account the requirement for the protein to retain its native pharmacology. An important breakthrough in this area was the combination of measuring thermostability in conjunction with radioligand binding on detergent-solubilized receptors to determine at what temperature a GPCR would lose its ability to bind ligand [136]. This method also had the advantage of identifying mutations that directly affected ligand binding or pharmacology. Such mutations would be undesirable in a crystal structure of a receptor, in particular, if it was to be used for structure-based analysis of ligand binding and drug design (Fig. 14.6).

Construction of such a thermostability curve for rhodopsin using absorption spectroscopy as a measure of retinal binding gave an apparent T_m of 37°C in n-octyltetraoxythylene [36]. The thermostabilized mutant of rhodopsin has an apparent T_m of 47°C. Since these receptors were able to be crystallized, this T_m represents a goal for stabilizing receptors prior to crystallization. Analysis of a wide range of GPCRs using such methods demonstrates a very wide spectrum of stabilities. One of the most stable GPCRs is the human β_2AR, and this may be one reason why this receptor has proved such a useful prototype for the study of GPCR function. β_2AR was the first receptor to be cloned [24] (following purification from native tissue) and was the first GPCR other than

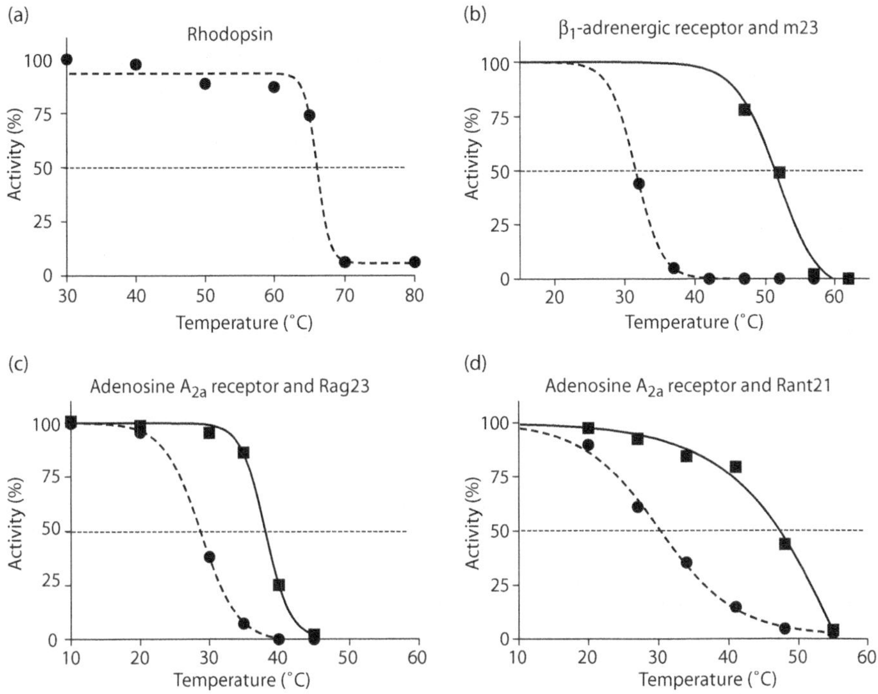

Figure 14.6 Thermostability of purified GPCRs and their StaRs. (a) Rhodopsin thermostability; (b) comparison of the stability of the turkey β_1AR (circles, dashed) and the stabilized turkey β_1AR(squares, solid) (m23) [136]; (c) comparison of the stability of the adenosine A_{2a} receptor (circles, dashed) and the stabilized agonist A_{2a} receptor Rag23 (squares, solid); (d) comparison of the stability of the adenosine A_{2a} receptor (circles, dashed) and the stabilized antagonist A_{2a} receptor Rant21 (squares, solid) [135].

rhodopsin to be crystallized [3]. Many other receptors are extremely unstable in detergent and must be kept at <5°C to retain activity. Thus far, it is not apparent from the sequence of the receptor what causes this wide range in thermostability, and so it remains impossible to predict the stability of any individual receptor. Indeed, very closely related receptors may have widely divergent stabilities. For example, the apparent T_m of the turkey β_1AR receptor is about 20°C higher than that of human β_1AR (Dr. C. Tate, pers. comm.).

A complete alanine-scanning mutagenesis of a GPCR will identify approximately 20 mutations that are thermostabilizing. However, each of these alone will not provide sufficient stabilization necessary to enable crystallization. Stabilizing mutations must be combined to give the maximum effect, both in terms of thermostabilization and conformational stabilization. In addition, although alanine substitutions may be tried initially, other amino acid replacements may increase stability even further. The possible number of combinations to be selected from 20 stabilizing mutations is therefore very high. So far, the optimal combinations are reached through trial and error, although in the future, a suitable random mutagenesis method may be developed to identify stable mutant receptors. Alternatively, as our knowledge and understanding of the rules that govern stabilization of receptors increase, we may be able to rationally select individual residues or networks of residues within the receptor sequence that will lead to stabilization.

The first non-rhodopsin GPCR to be engineered to include thermostabilizing mutants for the purposes of crystallography was the turkey β_1AR [136]. The most stable mutant identified (β_1AR-m23) contained six-point mutations and resulted in an apparent T_m 21°C higher than the native protein. In the presence of the antagonist alprenolol, this receptor was as stable as bovine rhodopsin, giving researchers some degree of confidence that crystallization would be achievable. Thermostability measurements were coupled with an antagonist [^{3}H]-dihydroalprenolol binding assay. Following an alanine scan, 18 residues were identified, which increased stability while maintaining antagonist binding. Each of these was mutated to alternative amino acids, of which five improved stability further. During the recombination process, a number of mutations proved to be additive in their stabilizing properties. The final set of mutations that were selected was Arg68Ser, Met90Val, Tyr227Ala, Ala282Leu, Phe327Ala, and Phe338Met [136]. GPCRs that have been engineered in this way have been called "StaRs" or "*sta*bilised *r*eceptors," and are now being utilized for crystallography and structure-based drug discovery (http://www.heptares.com).

Pharmacological characterization of the β_1AR StaR known as βAR-m23 indicated that the receptor had been stabilized in the antagonist R state, since the receptor showed no difference in antagonist binding affinity compared to the native receptor but had a significantly reduced affinity for agonists. Expression of the β_1AR StaR in a stable cell line demonstrated that the receptor, unlike its wild-type counterpart, showed no basal activity, presumably due to its inability to form R* in the absence of ligand [4]. Although in the absence of ligand, the β_1AR StaR receptor preferred the R state, upon addition of sufficient concentrations of agonist, the receptor could couple to G proteins and increase cAMP to the same level as wild-type receptor. This ability to trap the receptor in a specific conformation during the stabilization process is also a considerable advantage in the production of diffraction quality crystals. The identification of thermostabilizing mutants allowed expression and purification of large quantities of β_1AR StaR using either *E. coli* or baculovirus expression. Once purified, the receptor remained stable and functional in a wide range of detergents including short-chain detergents, such as OG, LDAO, and nonyl-glucoside (NG), which are useful in crystallizing membrane proteins. Stabilization of the β_1AR receptor in this way enabled the generation of crystals in which the receptor was bound to the high-affinity ligand cyanopindolol in octyl-thioglucoside, which gave excellent diffraction to 2.7 Å, resulting in the first structure of the β_1AR [4]. This important structure is described in more detail below.

The approach of thermostabilizing mutations has also been used to stabilize the adenosine A_{2a} receptor [135]. In this case, the receptor was stabilized in both agonist and antagonist conformations by carrying out the mutagenesis selection process using either the antagonist [^{3}H]-ZM241385 or the agonist [^{3}H]-5′-*N*-ethylcarboxamidoadenosine (NECA). Interestingly, a different set of stabilizing mutations was identified in the agonist assay compared to the antagonist assay. Recombination of the most stabilizing mutants resulted in the generation of the antagonist StaR Rant21, containing four mutations, which was stabilized by 17°C compared to the wild-type receptor, and the agonist StaR Rag23, containing five mutations, which was stabilized by 9°C. In radioligand binding studies, the adenosine A_{2a}-stabilized receptor Rant21 bound antagonists with similar affinity to the wild-type receptor, but agonist affinity was dramatically reduced, consistent with the receptor being stabilized in an antagonist conformation. Radioligand binding studies of Rag23 showed that the agonist-stabilized adenosine A_{2a} receptor had a slightly increased affinity for agonists and a weaker affinity for antagonists when compared to the wild-type receptor.

The mechanism by which mutations are able to confer stabilization on GPCRs is not clear. In the case of β_1AR StaR, it was clear from the comparison of the β_1AR and β_2AR structures that there were no obvious changes resulting from the mutagenesis. Thermostabilizing mutations can be found in all areas of the receptor, including the transmembrane domains, N-terminal and C-terminal domains, as well as in the intracellular and extracellular loops. Those in the transmembrane domains may help to stabilize intrahelical interactions or, alternatively, for those mutations that appear to be outward facing, may alter the protein/lipid interface. Mutations within the transmembrane domains may also be involved in altering the equilibrium between the active and inactive forms, and as a result, confer conformational stabilization. Several stabilizing mutations are located at the ends of the helices, which may be a site of initiation for unfolding. A number of studies have suggested that GPCRs behave as two-domain proteins consisting of TM1–5 and TM6–7 since truncated receptors containing TM1–5 can generate functional receptors when coexpressed with their corresponding C-terminal portion [193–195]. The loop between these domains (IC3) represents a flexible region, which may be the site of initiating receptor unfolding. Stabilization or masking of this area by mutagenesis, antibody binding, or formation of a fusion protein may all act in a similar manner, to reduce unfolding in this region of the receptor.

14.8. STRUCTURES OF THE β_2AR

The many developments described above finally culminated in the publication of two crystal structures of the β_2AR in 2007, followed by the structure of the β_1AR in 2008. It is fitting and not surprising that the β_2AR proved to be the first nonopsin GPCR to be crystallized. β_2AR has been the prototypical GPCR since its cloning, based on protein sequencing of the receptor purified from native tissue in 1986 [24]. Many of the methods originally developed for purification from native tissue, such as the alprenolol affinity column [157], were also used in the isolation of pure protein from recombinant systems required in the crystallization. A thorough understanding of the receptor, which came from years of single-minded study by researchers such as Robert Lefkowitz [196, 197] and Brian Kobilka [198, 199], paved the way for strategies to modify the receptor sequence to facilitate crystallization. A further reason behind β_2AR being the first receptor crystallized (and indeed cloned) is that it is inherently more stable upon purification than most other GPCRs. It therefore remains to be seen whether the strategies for solving the β_2AR structure will be transferable to other less stable or poorly expressed GPCRs.

The first structure of the β_2AR in complex with the partial inverse agonist carazolol was obtained using a monoclonal antibody, which binds to the third ICL as described above [3]. This structure had a relatively low resolution (3.4/3.7 Å) at the cytoplasmic ends and in the transmembrane domains, while the extracellular regions could not be resolved. Nevertheless, it was clear that the overall structure was, as expected, very similar to rhodopsin. This was closely followed by a much higher resolution (2.4 Å) structure of an engineered β_2AR in which the unstructured sequence in IC3 was replaced with the sequence from the enzyme T4L [2] to form a fusion protein (β_2AR–T4L), again crystallized in the presence of carazolol.

The high-resolution structure of β_2AR demonstrated its remarkable similarity to rhodopsin with respect to the arrangement of the helical bundles. As might be predicted, the largest differences were observed in the helices contributing to the ligand binding site. For example, TM3 and TM5 contain the catecholamine binding residues and are the most conformationally shifted. In β_2AR, TM3 pivots out of the ligand binding site and has a 4 Å displacement relative to rhodopsin at the cytoplasmic end, while TM5 is closer to the binding pocket. Another major difference is in TM1, which, in β_2AR, is shifted by 7 Å at the extracellular end. This results in a wider opening at the ligand binding site. The entrance to the ligand binding site is also altered in β_2AR when compared to rhodopsin in that there is no "cap" formed between the N-terminal domain and extracellular loops. Instead, the second extracellular loop forms a short α-helical structure, which sits separate from the transmembrane domains and is stabilized by both an intra-loop disulfide bond and a disulfide bond between the loop and the top of TM3. This helical structure of β_2AR is positioned so as to allow access to the ligand binding pocket from the extracellular environment.

The carazolol binding site in β_2AR is located in a region corresponding to the retinal binding site in rhodopsin. Interpretation of the conformation of the receptor associated with binding of carazolol in the T4L fusion protein is complicated by a number of factors. The T4L fusion appears to tilt the equilibrium of at least part of the receptor into an agonist-like state [142].

This is seen as an increase in the affinity of agonists and partial agonists, such as isoproterenol and salbutamol, compared to the wild-type receptor. In contrast, the affinity of antagonists is unchanged. This may be due to the T4L protein essentially forcing the bottom of TM5 and TM6 into the more open position normally observed in the active state. Such a change would be expected to cause constitutive activation of the receptor; however, this cannot be determined in the T4 fusion proteins since introduction of the lysozyme prevents G protein coupling. Instead, the authors used a fluorescent probe attached to Cys265 as a reporter for ligand-induced conformational changes. Data from this reporter system supported the fact that the introduction of T4L resulted in a constitutively active mutant (CAM)-like receptor that could be further activated by agonists. Under normal conditions, β_2AR shows some basal activation, and, in these systems, carazolol behaves as a partial inverse agonist reducing basal activity but not resulting in a fully "off" receptor. This is analogous to 11-*cis*-retinal-induced conformation found in the dark state of rhodopsin. However, in the β_2AR–T4L structure, the ionic lock found in the dark state of rhodopsin is not found, as TM3 and TM6 are further apart, and the salt bridge between these helices is unable to form. Since β_2AR was the first receptor to be crystallized after rhodopsin, it was not immediately clear whether the broken ionic lock was due to the partially active state of the receptor, an artifact of T4L fusion, or a difference between rhodopsin and other receptors.

As seen in the rhodopsin structure, the β_2AR also contains an array of water molecules, which provide a network of hydrogen bonds, from the base of the ligand binding pocket through to the cytoplasmic face of the receptor. This network is postulated to play a role in propagating conformational changes, from the ligand binding site to regions involved in G protein coupling.

The T4L approach was also successfully used to produce a co-crystal structure of β_2AR with the beta blocker timolol [9]. This receptor was thermostabilized by the addition of a single point mutation (Glu122Trp). A particularly interesting feature of this structure was the presence of two cholesterol molecules bound to the receptor. Although cholesterol molecules were also found in the first structure, the biological relevance of these was questionable as they could have been an artifact of crystal packing. In the timolol structure, the receptors were packed as monomers in an antiparallel orientation but still retained two distinct cholesterol binding sites suggestive of a physiologically relevant interaction. It is known that cholesterol hemisuccinate (CHS) has a stabilizing effect on detergent-solubilized β_2AR. In the timolol β_2AR structure, two molecules of cholesterol were found bound to clefts formed by TM1, TM2, TM3, and TM4. The interactions on TM4 may represent a highly conserved cholesterol binding consensus motif since Trp158 ($^{4.50}$using the Ballesteros–Weinberg numbering system) [200] is one of the most highly conserved residues in Family A and is involved in an interaction with the sterol ring of one of the cholesterol residues. Other residues that are conserved and involved in cholesterol binding, which may contribute to a consensus motif, are Arg151 ($^{4.39-4.43}$R or K), Ile154 ($^{4.46}$I, V, or L), and Tyr70 ($^{2.41}$F or Y).

The recognition that thermostability is one of the most important factors governing purification and crystallization of GPCRs culminated in the first crystal structure of the β_1AR [4]. This receptor contained the six-point mutations identified by Serrano-Vega [136] and described above, which allowed crystallization of the receptor bound to the antagonist cyanopindolol in octylthioglucoside. Modifications of this receptor resulted in a protein that had an antagonist-stabilized conformation as defined by a reduction in agonist binding and a loss of basal activity when expressed in cells. Therefore, the receptor represents a more truly inactive conformation than that of β_2AR–T4L. Interestingly, in this structure, as was found in β_2AR–T4L, the ionic lock is broken, suggesting that although inactive, the state of this receptor is not equivalent to that of dark-state rhodopsin.

Overall, the structure of the stabilized turkey β_1AR was very close to that of β_2AR, with which it shares 67% of identity in amino acid sequence in the transmembrane domains. This similarity was encouraging from a technical point of view as it demonstrated that the two markedly differing approaches resulted in similar structures with no evidence for changes in backbone conformation due to the mutations or T4 fusion. The structures of the extracellular loops were also similar between the two receptors. In the β_1AR structure, a sodium ion binding site can be observed at the end of the second extracellular loop, which may help to stabilize its helical conformation.

The ligand binding sites in β_1AR and β_2AR are highly conserved, and the 15 residues involved directly in the binding of cyanopindolol to β_1AR are identical to those involved in the binding of the related ligand carazolol to β_2AR (Fig. 14.7). Since neither of these ligands is particularly selective for β_1AR over β_2AR, it is not immediately obvious from these structures how some ligands can be highly selective for one receptor over the other. Such differences may be due to residues around, but not in, the ligand binding site, such as Val172$^{4.56}$ and Phe325$^{7.35}$ in β_1AR, which correspond to Thr164$^{4.56}$ and Tyr308$^{7.35}$ in β_2AR. Another significant difference is seen in the amino acids around the entrance to the binding site, which have a different charge distribution and may contribute to ligand selectivity. As well as the residues in the transmembrane domain, there are also contributions to the ligand binding site from the second extracellular loop, which folds down into the entrance of the ligand binding pocket. This region does show differences between the two receptors. It is hoped that as more receptors are crystallized, we will see additional structures with the receptors in complex with more selective ligands and that such structures will provide a more detailed molecular understanding of the reasons for selectivity.

One of the most notable new features seen in the β_1AR was a well-defined helix in the second ICL, which interacts directly via Tyr149 with Asp138$^{3.49}$ of the DRY motif at the end of TM3. These regions are thought to play a key

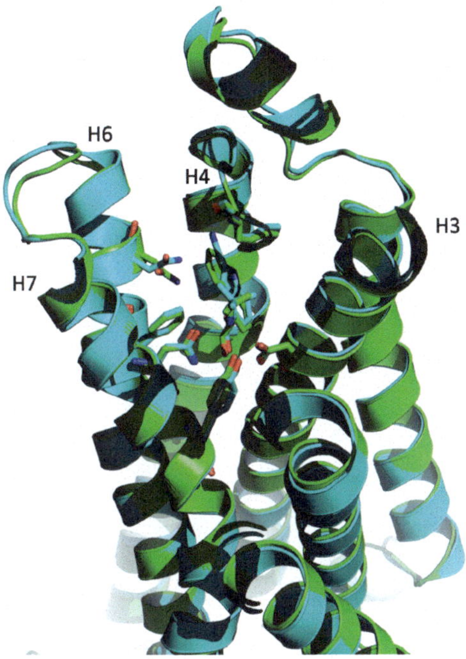

Figure 14.7 Comparison of the ligand binding sites of β_1AR and β_2AR showing the close homology between the two receptors and the key side chains involved in ligand binding. The β_1AR is shown in blue and the β_2AR is in green.

role in G protein coupling. The absence of this structure in β_1AR–T4L is likely to be due to local perturbations caused by the presence of the T4L fusion.

14.9. THE ADENOSINE A_{2a} RECEPTOR

The T4L fusion strategy in combination with the lipidic cubic phase crystallization method that resulted in the structure of the β_2AR has also been applied successfully to the adenosine A_{2a} receptor. This suggests that the T4L technology is transferable across different subfamilies of Class A GPCRs. In this structure, obtained to 2.6 Å resolution, the receptor is crystallized in complex with the high-affinity antagonist ZM241385. Incorporation of the T4L moiety shifts the conformation to an agonist-like state as defined by an increase in the affinity of agonist and no change in antagonist affinity.

The adenosine A_{2a}–T4L structure had a number of features not seen in previous structures. Although the overall arrangement of the helices was

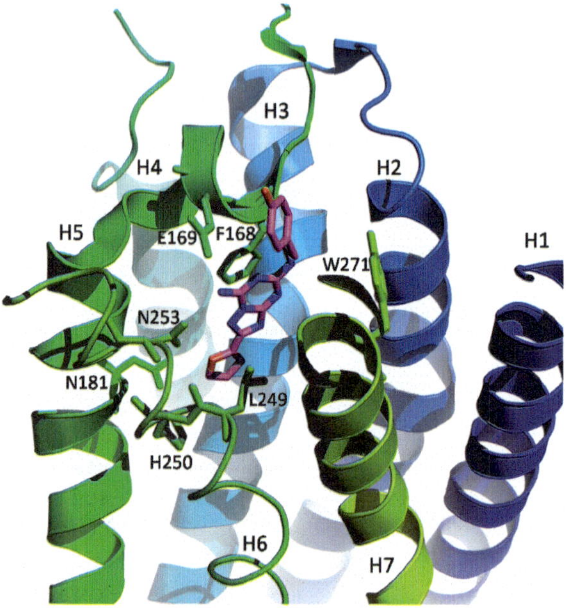

Figure 14.8 The binding site of the antagonist ZM241385 (pink) in the adenosine A_{2a} receptor showing its vertical position within the A_{2a} binding pocket.

similar to the other structures, a change in the relative positions of the helices results in a movement of the antagonist binding site. TM6 and TM7 are shifted closer to the binding pocket by 7 Å and 3 Å, respectively, and there is a lateral shift in TM3 toward TM5 by 3 Å. In contrast to carazolol and cyanopindolol, the antagonists for the β-receptors, ZM241385 lies in a plane perpendicular to the membrane (Fig. 14.8). The bicyclic core of this compound interacts with the aromatic ring of $Phe168^{5.29}$ in the second extracellular loop, while $Trp246^{6.48}$ is associated with stabilizing the furan ring common in many adenosine A_{2a} antagonists. The binding pocket also includes four ordered water molecules.

The extracellular regions of the A_{2a} receptor differ markedly from the β-receptors. The second extracellular loop does not form an α-helix but rather a random coil constrained by three disulfide bonds. A further disulfide bond links the third extracellular loop with the top of TM6, producing a kinked structure. This array of disulfide bonds helps to keep an open entrance to the ligand binding pocket, presumably allowing access of the natural ligand from the extracellular space.

As with the βAR structures, the ionic lock is also absent in the A_{2a} structure. The absence of the lock in three structures starts to bring into question the relevance of the ionic lock as a mechanism for stabilizing the inactive state of GPCRs, other than rhodopsin. Instead, the adenosine A_{2a} receptor and β_1AR contain a network of hydrogen bond interactions between Asp101$^{3.49}$ and TM2, and the second ICL. The role of Arg102$^{3.50}$ in this situation may be to stabilize the deprotonated Asp101$^{3.49}$.

14.10. CONCLUSIONS AND FUTURE DEVELOPMENTS

Technological developments in the expression and purification of GPCRs, together with methods for engineering more stable or crystallizable proteins, have resulted in dramatic advances in the number of GPCR structures over the last 2 years. The scene is now set for a plethora of new structures that will drive our understanding of GPCR diversity and function. The advent of structures will allow modern techniques in structure-based drug discovery to significantly enhance the design of improved drugs. Structures will be useful to assist in the design of more selective compounds, which has been a particular problem for many CNS drugs targeted at the highly conserved biogenic amine receptors. In addition, the optimization of drugs at peptide receptors has proved difficult as such molecules tend to have high molecular weights and be associated with poor drug-like properties such as bioavailability or pharmacokinetics. The availability of structures for peptide receptors may allow compounds with a higher ligand efficiency [201] and improved drug-like properties to be developed.

Further structures are now needed across the different classes of Family A, including costructures with ligands of different chemical classes. Hopefully, the recent developments for this family will also enable structures of Family B and C GPCRs. Most importantly, we await structures of receptors in active conformations, preferably with an agonist, bound in complex with a G protein.

ACKNOWLEDGMENTS

The authors would like to thank Dr. Nathan Robertson, Professor Martin Caffrey, and Dr. Chris Tate for kindly providing some of the figures used in the chapter. The authors are also grateful to Dr. Chris Tate and Dr. Corin Wing for their critical reading of this manuscript.

REFERENCES

1. Palczewski, K., Kumasaka, T., Hori, T., Behnke, C.A., Motoshima, H., Fox, B.A., Le Trong, I., Teller, D.C., Okada, T., Stenkamp, R.E., Yamamoto, M., Miyano, M. (2000) Crystal structure of rhodopsin: A G protein-coupled receptor. *Science.* 289, 739–745.

2. Cherezov, V., Rosenbaum, D.M., Hanson, M.A., Rasmussen, S.G., Thian, F.S., Kobilka, T.S., Choi, H.J., Kuhn, P., Weis, W.I., Kobilka, B.K., Stevens, R.C. (2007) High-resolution crystal structure of an engineered human β_2-adrenergic G protein-coupled receptor. *Science.* 318, 1258–1265.
3. Rasmussen, S.G., Choi, H.J., Rosenbaum, D.M., Kobilka, T.S., Thian, F.S., Edwards, P.C., Burghammer, M., Ratnala, V.R., Sanishvili, R., Fischetti, R.F., Schertler, G.F., Weis, W.I., Kobilka, B.K. (2007) Crystal structure of the human β_2 adrenergic G-protein-coupled receptor. *Nature.* 450, 383–387.
4. Warne, T., Serrano-Vega, M.J., Baker, J.G., Moukhametzianov, R., Edwards, P.C., Henderson, R., Leslie, A.G., Tate, C.G., Schertler, G.F. (2008) Structure of a β_1-adrenergic G-protein-coupled receptor. *Nature.* 454, 486–491.
5. Jaakola, V.P., Griffith, M.T., Hanson, M.A., Cherezov, V., Chien, E.Y., Lane, J.R., Ijzerman, A.P., Stevens, R.C. (2008) The 2.6 angstrom crystal structure of a human A_{2A} adenosine receptor bound to an antagonist. *Science.* 322, 1211–1217.
6. Seddon, A.M., Curnow, P., Booth, P.J. (2004) Membrane proteins, lipids and detergents: Not just a soap opera. *Biochim Biophys Acta.* 1666, 105–117.
7. Hunte, C. (2005) Specific protein-lipid interactions in membrane proteins. *Biochem Soc Trans.* 33, 938–942.
8. Lacapere, J.J., Pebay-Peyroula, E., Neumann, J.M., Etchebest, C. (2007) Determining membrane protein structures: Still a challenge! *Trends Biochem Sci.* 32, 259–270.
9. Hanson, M.A., Cherezov, V., Griffith, M.T., Roth, C.B., Jaakola, V.P., Chien, E.Y., Velasquez, J., Kuhn, P., Stevens, R.C. (2008) A specific cholesterol binding site is established by the 2.8 A structure of the human β_2-adrenergic receptor. *Structure.* 16, 897–905.
10. Wallace, B.A., Lees, J.G., Orry, A.J., Lobley, A., Janes, R.W. (2003) Analyses of circular dichroism spectra of membrane proteins. *Protein Sci.* 12, 875–884.
11. Okada, T., Ernst, O.P., Palczewski, K., Hofmann, K.P. (2001) Activation of rhodopsin: New insights from structural and biochemical studies. *Trends Biochem Sci.* 26, 318–324.
12. Ballesteros, J., Palczewski, K. (2001) G protein-coupled receptor drug discovery: Implications from the crystal structure of rhodopsin. *Curr Opin Drug Discov Devel.* 4, 561–574.
13. Palczewski, K. (2006) G protein-coupled receptor rhodopsin. *Annu Rev Biochem.* 75, 743–767.
14. Ridge, K.D., Palczewski, K. (2007) Visual rhodopsin sees the light: Structure and mechanism of G protein signaling. *J Biol Chem.* 282, 9297–9301.
15. Okada, T., Takeda, K., Kouyama, T. (1998) Highly selective separation of rhodopsin from bovine rod outer segment membranes using combination of bivalent cation and alkyl(thio)glucoside. *Photochem Photobiol.* 67, 495–499.
16. Grigorieff, N., Ceska, T.A., Downing, K.H., Baldwin, J.M., Henderson, R. (1996) Electron-crystallographic refinement of the structure of bacteriorhodopsin. *J Mol Biol.* 259, 393–421.
17. Henderson, R., Schertler, G.F. (1990) The structure of bacteriorhodopsin and its relevance to the visual opsins and other seven-helix G-protein coupled receptors. *Philos Trans R Soc Lond B Biol Sci.* 326, 379–389.

18. Henderson, R., Unwin, P.N. (1975) Three-dimensional model of purple membrane obtained by electron microscopy. *Nature.* 257, 28–32.
19. Leifer, D., Henderson, R. (1983) Three-dimensional structure of orthorhombic purple membrane at 6.5 A resolution. *J Mol Biol.* 163, 451–466.
20. Albert, A.D., Litman, B.J. (1978) Independent structural domains in the membrane protein bovine rhodopsin. *Biochemistry.* 17, 3893–3900.
21. Hargrave, P.A., Fong, S.L. (1977) The amino- and carboxyl-terminal sequence of bovine rhodopsin. *J Supramol Struct.* 6, 559–570.
22. Hargrave, P.A., McDowell, J.H., Curtis, D.R., Wang, J.K., Juszczak, E., Fong, S.L., Rao, J.K., Argos, P. (1983) The structure of bovine rhodopsin. *Biophys Struct Mech.* 9, 235–244.
23. Ovchinnikov Yu, A. (1982) Rhodopsin and bacteriorhodopsin: Structure-function relationships. *FEBS Lett.* 148, 179–191.
24. Dixon, R.A., Kobilka, B.K., Strader, D.J., Benovic, J.L., Dohlman, H.G., Frielle, T., Bolanowski, M.A., Bennett, C.D., Rands, E., Diehl, R.E., Mumford, R.A., Slater, E.E., Sigal, I.S., Caron, M.G., Lefkowitz, R.J., Strader, C.D. (1986) Cloning of the gene and cDNA for mammalian beta-adrenergic receptor and homology with rhodopsin. *Nature.* 321, 75–79.
25. Strader, C.D., Sigal, I.S., Dixon, R.A. (1989) Structural basis of beta-adrenergic receptor function. *FASEB J.* 3, 1825–1832.
26. Schertler, G.F., Villa, C., Henderson, R. (1993) Projection structure of rhodopsin. *Nature.* 362, 770–772.
27. Unger, V.M., Schertler, G.F. (1995) Low resolution structure of bovine rhodopsin determined by electron cryo-microscopy. *Biophys J.* 68, 1776–1786.
28. Schertler, G.F., Hargrave, P.A. (1995) Projection structure of frog rhodopsin in two crystal forms. *Proc Natl Acad Sci U S A.* 92, 11578–11582.
29. Krebs, A., Edwards, P.C., Villa, C., Li, J., Schertler, G.F. (2003) The three-dimensional structure of bovine rhodopsin determined by electron cryomicroscopy. *J Biol Chem.* 278, 50217–50225.
30. Ballesteros, J.A., Jensen, A.D., Liapakis, G., Rasmussen, S.G., Shi, L., Gether, U., Javitch, J.A. (2001) Activation of the β_2-adrenergic receptor involves disruption of an ionic lock between the cytoplasmic ends of transmembrane segments 3 and 6. *J Biol Chem.* 276, 29171–29177.
31. Teller, D.C., Okada, T., Behnke, C.A., Palczewski, K., Stenkamp, R.E. (2001) Advances in determination of a high-resolution three-dimensional structure of rhodopsin, a model of G-protein-coupled receptors (GPCRs). *Biochemistry.* 40, 7761–7772.
32. Okada, T., Fujiyoshi, Y., Silow, M., Navarro, J., Landau, E.M., Shichida, Y. (2002) Functional role of internal water molecules in rhodopsin revealed by X-ray crystallography. *Proc Natl Acad Sci U S A.* 99, 5982–5987.
33. Li, J., Edwards, P.C., Burghammer, M., Villa, C., Schertler, G.F. (2004) Structure of bovine rhodopsin in a trigonal crystal form. *J Mol Biol.* 343, 1409–1438.
34. Edwards, P.C., Li, J., Burghammer, M., McDowell, J.H., Villa, C., Hargrave, P.A., Schertler, G.F. (2004) Crystals of native and modified bovine rhodopsins and their heavy atom derivatives. *J Mol Biol.* 343, 1439–1450.

35. Okada, T., Sugihara, M., Bondar, A.N., Elstner, M., Entel, P., Buss, V. (2004) The retinal conformation and its environment in rhodopsin in light of a new 2.2 A crystal structure. *J Mol Biol.* 342, 571–583.

36. Standfuss, J., Xie, G., Edwards, P.C., Burghammer, M., Oprian, D.D., Schertler, G.F. (2007) Crystal structure of a thermally stable rhodopsin mutant. *J Mol Biol.* 372, 1179–1188.

37. Shimamura, T., Hiraki, K., Takahashi, N., Hori, T., Ago, H., Masuda, K., Takio, K., Ishiguro, M., Miyano, M. (2008) Crystal structure of squid rhodopsin with intracellularly extended cytoplasmic region. *J Biol Chem.* 283, 17753–17756.

38. Murakami, M., Kouyama, T. (2008) Crystal structure of squid rhodopsin. *Nature.* 453, 363–367.

39. Schertler, G.F. (2008) Signal transduction: The rhodopsin story continued. *Nature.* 453, 292–293.

40. Ruprecht, J.J., Mielke, T., Vogel, R., Villa, C., Schertler, G.F. (2004) Electron crystallography reveals the structure of metarhodopsin I. *EMBO J.* 23, 3609–3620.

41. Nakamichi, H., Okada, T. (2006) Crystallographic analysis of primary visual photochemistry. *Angew Chem Int Ed Engl.* 45, 4270–4273.

42. Nakamichi, H., Okada, T. (2006) Local peptide movement in the photoreaction intermediate of rhodopsin. *Proc Natl Acad Sci U S A.* 103, 12729–12734.

43. Salom, D., Lodowski, D.T., Stenkamp, R.E., Le Trong, I., Golczak, M., Jastrzebska, B., Harris, T., Ballesteros, J.A., Palczewski, K. (2006) Crystal structure of a photoactivated deprotonated intermediate of rhodopsin. *Proc Natl Acad Sci U S A.* 103, 16123–16128.

44. Hubbell, W.L., Altenbach, C., Hubbell, C.M., Khorana, H.G. (2003) Rhodopsin structure, dynamics, and activation: A perspective from crystallography, site-directed spin labeling, sulfhydryl reactivity, and disulfide cross-linking. *Adv Protein Chem.* 63, 243–290.

45. Farrens, D.L., Altenbach, C., Yang, K., Hubbell, W.L., Khorana, H.G. (1996) Requirement of rigid-body motion of transmembrane helices for light activation of rhodopsin. *Science.* 274, 768–770.

46. Lamb, T.D., Pugh Jr., E.N. (2004) Dark adaptation and the retinoid cycle of vision. *Prog Retin Eye Res.* 23, 307–380.

47. Vogel, R., Siebert, F. (2001) Conformations of the active and inactive states of opsin. *J Biol Chem.* 276, 38487–38493.

48. Park, P.S., Lodowski, D.T., Palczewski, K. (2008) Activation of G protein-coupled receptors: Beyond two-state models and tertiary conformational changes. *Annu Rev Pharmacol Toxicol.* 48, 107–141.

49. Altenbach, C., Kusnetzow, A.K., Ernst, O.P., Hofmann, K.P., Hubbell, W.L. (2008) High-resolution distance mapping in rhodopsin reveals the pattern of helix movement due to activation. *Proc Natl Acad Sci U S A.* 105, 7439–7444.

50. Huber, T., Sakmar, T.P. (2008) Rhodopsin's active state is frozen like a DEER in the headlights. *Proc Natl Acad Sci U S A.* 105, 7343–7344.

51. Scheerer, P., Park, J.H., Hildebrand, P.W., Kim, Y.J., Krauss, N., Choe, H.W., Hofmann, K.P., Ernst, O.P. (2008) Crystal structure of opsin in its G-protein-interacting conformation. *Nature.* 455, 497–502.

52. Kaushal, S., Ridge, K.D., Khorana, H.G. (1994) Structure and function in rhodopsin: The role of asparagine-linked glycosylation. *Proc Natl Acad Sci U S A.* 91, 4024–4028.
53. Fukushima, Y., Oka, Y., Saitoh, T., Katagiri, H., Asano, T., Matsuhashi, N., Takata, K., van Breda, E., Yazaki, Y., Sugano, K. (1995) Structural and functional analysis of the canine histamine H2 receptor by site-directed mutagenesis: N-glycosylation is not vital for its action. *Biochem J.* 310, 553–558.
54. van Koppen, C.J., Nathanson, N.M. (1990) Site-directed mutagenesis of the m2 muscarinic acetylcholine receptor. Analysis of the role of N-glycosylation in receptor expression and function. *J Biol Chem.* 265, 20887–20892.
55. Lagane, B., Gaibelet, G., Meilhoc, E., Masson, J.M., Cezanne, L., Lopez, A. (2000) Role of sterols in modulating the human mu-opioid receptor function in *Saccharomyces cerevisiae*. *J Biol Chem.* 275, 33197–33200.
56. Mancia, F., Hendrickson, W.A. (2007) Expression of recombinant G-protein coupled receptors for structural biology. *Mol Biosyst.* 3, 723–734.
57. Mozharov, A.D., Shchipakin, V.N., Fishov, I.L., Evtodienko Yu, V. (1985) Changes in the composition of membrane phospholipids during the cell cycle of *Escherichia coli*. *FEBS Lett.* 186, 103–106.
58. Skretas, G., Georgiou, G. (2008) Genetic analysis of G protein-coupled receptor expression in *Escherichia coli*: Inhibitory role of DnaJ on the membrane integration of the human central cannabinoid receptor. *Biotechnol Bioeng.* 102, 357–367.
59. Link, A.J., Skretas, G., Strauch, E.M., Chari, N.S., Georgiou, G. (2008) Efficient production of membrane-integrated and detergent-soluble G protein-coupled receptors in *Escherichia coli*. *Protein Sci.* 17, 1857–1863.
60. Butz, J.A., Niebauer, R.T., Robinson, A.S. (2003) Co-expression of molecular chaperones does not improve the heterologous expression of mammalian G-protein coupled receptor expression in yeast. *Biotechnol Bioeng.* 84, 292–304.
61. Baneres, J.L., Martin, A., Hullot, P., Girard, J.P., Rossi, J.C., Parello, J. (2003) Structure-based analysis of GPCR function: Conformational adaptation of both agonist and receptor upon leukotriene B4 binding to recombinant BLT1. *J Mol Biol.* 329, 801–814.
62. Grisshammer, R., Duckworth, R., Henderson, R. (1993) Expression of a rat neurotensin receptor in *Escherichia coli*. *Biochem J.* 295, 571–576.
63. Marullo, S., Delavier-Klutchko, C., Eshdat, Y., Strosberg, A.D., Emorine, L. (1988) Human b_2-adrenergic receptors expressed in *Escherichia coli* membranes retain their pharmacological properties. *Proc Natl Acad Sci U S A.* 85, 7551–7555.
64. Tucker, J., Grisshammer, R. (1996) Purification of a rat neurotensin receptor expressed in *Escherichia coli*. *Biochem J.* 317, 891–899.
65. Weiss, H.M., Grisshammer, R. (2002) Purification and characterization of the human adenosine A_{2a} receptor functionally expressed in *Escherichia coli*. *Eur J Biochem.* 269, 82–92.
66. Grisshammer, R., Tucker, J. (1997) Quantitative evaluation of neurotensin receptor purification by immobilized metal affinity chromatography. *Protein Expr Purif.* 11, 53–60.
67. Harding, P.J., Attrill, H., Ross, S., Koeppe, J.R., Kapanidis, A.N., Watts, A. (2007) Neurotensin receptor type 1: *Escherichia coli* expression, purification, characterization and biophysical study. *Biochem Soc Trans.* 35, 760–763.

68. Furukawa, H., Haga, T. (2000) Expression of functional M_2 muscarinic acetylcholine receptor in *Escherichia coli*. *J Biochem.* 127, 151–161.
69. Bane, S.E., Velasquez, J.E., Robinson, A.S. (2007) Expression and purification of milligram levels of inactive G-protein coupled receptors in *E. coli*. *Protein Expr Purif.* 52, 348–355.
70. Lundstrom, K., Wagner, R., Reinhart, C., Desmyter, A., Cherouati, N., Magnin, T., Zeder-Lutz, G., Courtot, M., Prual, C., Andre, N., Hassaine, G., Michel, H., Cambillau, C., Pattus, F. (2006) Structural genomics on membrane proteins: Comparison of more than 100 GPCRs in 3 expression systems. *J Struct Funct Genomics.* 7, 77–91.
71. Dahmane, T., Damian, M., Mary, S., Popot, J.-L., Banères, J.-L. (2009) Amphipol-assisted in vitro folding of G protein-coupled receptors. *Biochem.* 48(27), 6516–6521.
72. Pocanschi, C.L., Dahmane, T., Gohon, Y., Rappaport, F., Apell, H.J., Kleinschmidt, J.H., Popot, J.L. (2006) Amphipathic polymers: Tools to fold integral membrane proteins to their active form. *Biochemistry.* 45, 13954–13961.
73. Stanasila, L., Pattus, F., Massotte, D. (1998) Heterologous expression of G-protein-coupled receptors: Human opioid receptors under scrutiny. *Biochimie.* 80, 563–571.
74. Huang, K.S., Bayley, H., Liao, M.J., London, E., Khorana, H.G. (1981) Refolding of an integral membrane protein. Denaturation, renaturation, and reconstitution of intact bacteriorhodopsin and two proteolytic fragments. *J Biol Chem.* 256, 3802–3809.
75. Baneres, J.L., Mesnier, D., Martin, A., Joubert, L., Dumuis, A., Bockaert, J. (2005) Molecular characterization of a purified 5-HT4 receptor: A structural basis for drug efficacy. *J Biol Chem.* 280, 20253–20260.
76. Kiefer, H., Krieger, J., Olszewski, J.D., Von Heijne, G., Prestwich, G.D., Breer, H. (1996) Expression of an olfactory receptor in *Escherichia coli*: Purification, reconstitution, and ligand binding. *Biochemistry.* 35, 16077–16084.
77. Patenge, N., Soppa, J. (1999) Extensive proteolysis inhibits high-level production of eukaryal G protein-coupled receptors in the archaeon *Haloferax volcanii*. *FEMS Microbiol Lett.* 171, 27–35.
78. Jaakola, V.P., Rehn, M., Moeller, M., Alexiev, U., Goldman, A., Turner, G.J. (2005) G-protein-coupled receptor domain overexpression in *Halobacterium salinarum*: Long-range transmembrane interactions in heptahelical membrane proteins. *Proteins.* 60, 412–423.
79. Dunn, R., McCoy, J., Simsek, M., Majumdar, A., Chang, S.H., Rajbhandary, U.L., Khorana, H.G. (1981) The bacteriorhodopsin gene. *Proc Natl Acad Sci U S A.* 78, 6744–6748.
80. Bartus, C.L., Jaakola, V.P., Reusch, R., Valentine, H.H., Heikinheimo, P., Levay, A., Potter, L.T., Heimo, H., Goldman, A., Turner, G.J. (2003) Downstream coding region determinants of bacterio-opsin, muscarinic acetylcholine receptor and adrenergic receptor expression in *Halobacterium salinarum*. *Biochim Biophys Acta.* 1610, 109–123.
81. Niu, Y., Kong, J., Xu, Y. (2008) A novel GFP-fused eukaryotic membrane protein expression system in *Lactococcus lactis* and its application to overexpression of an elongase. *Curr Microbiol.* 57, 423–428.

82. Roy, A., Shukla, A.K., Haase, W., Michel, H. (2008) Employing *Rhodobacter sphaeroides* to functionally express and purify human G protein-coupled receptors. *Biol Chem.* 389, 69–78.

83. Drew, D.E., von Heijne, G., Nordlund, P., de Gier, J.W. (2001) Green fluorescent protein as an indicator to monitor membrane protein overexpression in *Escherichia coli*. *FEBS Lett.* 507, 220–224.

84. Drew, D., Lerch, M., Kunji, E., Slotboom, D.J., de Gier, J.W. (2006) Optimization of membrane protein overexpression and purification using GFP fusions. *Nat Methods.* 3, 303–313.

85. Singh, S., Gras, A., Fiez-Vandal, C., Ruprecht, J., Rana, R., Martinez, M., Strange, P.G., Wagner, R., Byrne, B. (2008) Large-scale functional expression of WT and truncated human adenosine A_{2A} receptor in *Pichia pastoris* bioreactor cultures. *Microb Cell Fact.* 7, 28.

86. Huang, L.Y., Umanah, G., Hauser, M., Son, C., Arshava, B., Naider, F., Becker, J.M. (2008) Unnatural amino acid replacement in a yeast G protein-coupled receptor in its native environment. *Biochemistry.* 47, 5638–5648.

87. Hamilton, S.R., Gerngross, T.U. (2007) Glycosylation engineering in yeast: The advent of fully humanized yeast. *Curr Opin Biotechnol.* 18, 387–392.

88. Sander, P., Grunewald, S., Reilander, H., Michel, H. (1994) Expression of the human D2S dopamine receptor in the yeasts *Saccharomyces cerevisiae* and *Schizosaccharomyces pombe*: A comparative study. *FEBS Lett.* 344, 41–46.

89. Feng, W., Cai, J., Pierce Jr., W.M., Song, Z.H. (2002) Expression of CB2 cannabinoid receptor in *Pichia pastoris*. *Protein Expr Purif.* 26, 496–505.

90. Escriba, P.V., Wedegaertner, P.B., Goni, F.M., Vogler, O. (2007) Lipid-protein interactions in GPCR-associated signaling. *Biochim Biophys Acta.* 1768, 836–852.

91. Schneiter, R., Toulmay, A. (2007) The role of lipids in the biogenesis of integral membrane proteins. *Appl Microbiol Biotechnol.* 73, 1224–1232.

92. Kim, T.K., Zhang, R., Feng, W., Cai, J., Pierce, W., Song, Z.H. (2005) Expression and characterization of human CB1 cannabinoid receptor in methylotrophic yeast *Pichia pastoris*. *Protein Expr Purif.* 40, 60–70.

93. Gimpl, G., Fahrenholz, F. (2002) Cholesterol as stabilizer of the oxytocin receptor. *Biochim Biophys Acta.* 1564, 384–392.

94. Burger, K., Gimpl, G., Fahrenholz, F. (2000) Regulation of receptor function by cholesterol. *Cell Mol Life Sci.* 57, 1577–1592.

95. Akermoun, M., Koglin, M., Zvalova-Iooss, D., Folschweiller, N., Dowell, S.J., Gearing, K.L. (2005) Characterization of 16 human G protein-coupled receptors expressed in baculovirus-infected insect cells. *Protein Expr Purif.* 44, 65–74.

96. Warne, T., Chirnside, J., Schertler, G.F. (2003) Expression and purification of truncated, non-glycosylated turkey beta-adrenergic receptors for crystallization. *Biochim Biophys Acta.* 1610, 133–140.

97. Boundy, V.A., Lu, L., Molinoff, P.B. (1996) Differential coupling of rat D2 dopamine receptor isoforms expressed in *Spodoptera frugiperda* insect cells. *J Pharmacol Exp Ther.* 276, 784–794.

98. Wenzel-Seifert, K., Seifert, R. (2003) Critical role of N-terminal N-glycosylation for proper folding of the human formyl peptide receptor. *Biochem Biophys Res Commun.* 301, 693–698.

99. Harrison, R.L., Jarvis, D.L. (2006) Protein N-glycosylation in the baculovirus-insect cell expression system and engineering of insect cells to produce "mammalianized" recombinant glycoproteins. *Adv Virus Res.* 68, 159–191.

100. Marheineke, K., Grunewald, S., Christie, W., Reilander, H. (1998) Lipid composition of *Spodoptera frugiperda* (Sf9) and *Trichoplusia ni* (Tn) insect cells used for baculovirus infection. *FEBS Lett.* 441, 49–52.

101. Gimpl, G., Klein, U., Reilander, H., Fahrenholz, F. (1995) Expression of the human oxytocin receptor in baculovirus-infected insect cells: High-affinity binding is induced by a cholesterol-cyclodextrin complex. *Biochemistry.* 34, 13794–13801.

102. Butkerait, P., Zheng, Y., Hallak, H., Graham, T.E., Miller, H.A., Burris, K.D., Molinoff, P.B., Manning, D.R. (1995) Expression of the human 5-hydroxytryptamine1A receptor in Sf9 cells. Reconstitution of a coupled phenotype by co-expression of mammalian G protein subunits. *J Biol Chem.* 270, 18691–18699.

103. Ames, R.S., Fornwald, J.A., Nuthulaganti, P., Trill, J.J., Foley, J.J., Buckley, P.T., Kost, T.A., Wu, Z.N., Romanos, M.A. (2004) BacMam recombinant baculoviruses in G protein-coupled receptor drug discovery. *Recept Channels.* 10, 99–107.

104. Fortin, J.P., Dziadulewicz, E.K., Gera, L., Marceau, F. (2006) A nonpeptide antagonist reveals a highly glycosylated state of the rabbit kinin B1 receptor. *Mol Pharmacol.* 69, 1146–1157.

105. Archer-Lahlou, E., Escrieut, C., Clerc, P., Martinez, J., Moroder, L., Logsdon, C., Kopin, A., Seva, C., Dufresne, M., Pradayrol, L., Maigret, B., Fourmy, D. (2005) Molecular mechanism underlying partial and full agonism mediated by the human cholecystokinin-1 receptor. *J Biol Chem.* 280, 10664–10674.

106. Lundstrom, K., Mills, A., Allet, E., Ceszkowski, K., Agudo, G., Chollet, A., Liljestrom, P. (1995) High-level expression of G protein-coupled receptors with the aid of the Semliki Forest virus expression system. *J Recept Signal Transduct Res.* 15, 23–32.

107. Reeves, P.J., Kim, J.M., Khorana, H.G. (2002) Structure and function in rhodopsin: A tetracycline-inducible system in stable mammalian cell lines for high-level expression of opsin mutants. *Proc Natl Acad Sci U S A.* 99, 13413–13418.

108. Jin, Q., Marsh, J., Cornetta, K., Alkhatib, G. (2008) Resistance to human immunodeficiency virus type 1 (HIV-1) generated by lentivirus vector-mediated delivery of the CCR5{Δ}32 gene despite detectable expression of the HIV-1 co-receptors. *J Gen Virol.* 89, 2611–2621.

109. O'Gorman, S., Fox, D.T., Wahl, G.M. (1991) Recombinase-mediated gene activation and site-specific integration in mammalian cells. *Science.* 251, 1351–1355.

110. Mancia, F., Patel, S.D., Rajala, M.W., Scherer, P.E., Nemes, A., Schieren, I., Hendrickson, W.A., Shapiro, L. (2004) Optimization of protein production in mammalian cells with a coexpressed fluorescent marker. *Structure.* 12, 1355–1360.

111. Winnick, T. (1950) Studies on the mechanism of protein synthesis in embryonic and tumor tissues. I. Evidence relating to the incorporation of labeled amino acids into protein structure in homogenates. *Arch Biochem.* 27, 65–74.

112. Klammt, C., Schwarz, D., Lohr, F., Schneider, B., Dotsch, V., Bernhard, F. (2006) Cell-free expression as an emerging technique for the large scale production of integral membrane protein. *FEBS J.* 273, 4141–4153.

113. Robelek, R., Lemker, E.S., Wiltschi, B., Kirste, V., Naumann, R., Oesterhelt, D., Sinner, E.K. (2007) Incorporation of in vitro synthesized GPCR into a tethered artificial lipid membrane system. *Angew Chem Int Ed Engl.* 46, 605–608.

114. Kaiser, L., Graveland-Bikker, J., Steuerwald, D., Vanberghem, M., Herlihy, K., Zhang, S. (2008) Efficient cell-free production of olfactory receptors: Detergent optimization, structure, and ligand binding analyses. *Proc Natl Acad Sci U S A.* 105, 15726–15731.

115. Ishihara, G., Goto, M., Saeki, M., Ito, K., Hori, T., Kigawa, T., Shirouzu, M., Yokoyama, S. (2005) Expression of G protein coupled receptors in a cell-free translational system using detergents and thioredoxin-fusion vectors. *Protein Expr Purif.* 41, 27–37.

116. Schwarz, D., Junge, F., Durst, F., Frolich, N., Schneider, B., Reckel, S., Sobhanifar, S., Dotsch, V., Bernhard, F. (2007) Preparative scale expression of membrane proteins in *Escherichia coli*-based continuous exchange cell-free systems. *Nat Protoc.* 2, 2945–2957.

117. Endo, Y., Sawasaki, T. (2006) Cell-free expression systems for eukaryotic protein production. *Curr Opin Biotechnol.* 17, 373–380.

118. Kim, D.M., Choi, C.Y. (1996) A semicontinuous prokaryotic coupled transcription/translation system using a dialysis membrane. *Biotechnol Prog.* 12, 645–649.

119. Kigawa, T., Yabuki, T., Yoshida, Y., Tsutsui, M., Ito, Y., Shibata, T., Yokoyama, S. (1999) Cell-free production and stable-isotope labeling of milligram quantities of proteins. *FEBS Lett.* 442, 15–19.

120. Bayburt, T.H., Carlson, J.W., Sligar, S.G. (1998) Reconstitution and imaging of a membrane protein in a nanometer-size phospholipid bilayer. *J Struct Biol.* 123, 37–44.

121. Leitz, A.J., Bayburt, T.H., Barnakov, A.N., Springer, B.A., Sligar, S.G. (2006) Functional reconstitution of β_2-adrenergic receptors utilizing self-assembling Nanodisc technology. *Biotechniques.* 40, 601–602, 604, 606, passim.

122. McCusker, E.C., Bane, S.E., O'Malley, M.A., Robinson, A.S. (2007) Heterologous GPCR expression: A bottleneck to obtaining crystal structures. *Biotechnol Prog.* 23, 540–547.

123. Straub, R.E., Frech, G.C., Joho, R.H., Gershengorn, M.C. (1990) Expression cloning of a cDNA encoding the mouse pituitary thyrotropin-releasing hormone receptor. *Proc Natl Acad Sci U S A.* 87, 9514–9518.

124. Eroglu, C., Cronet, P., Panneels, V., Beaufils, P., Sinning, I. (2002) Functional reconstitution of purified metabotropic glutamate receptor expressed in the fly eye. *EMBO Rep.* 3, 491–496.

125. Tateno, M., Toyooka, M., Shikano, Y., Takeda, S., Kuwabara, N., Sezutsu, H., Tamura, T. (2009) Production and characterization of the recombinant human {micro}-opioid receptor from transgenic silkworms. *J Biochem.* 145, 37–42.

126. Sarmiento, J., Kypreos, K.E., Prado, G.N., Suetomi, K., Stanzel, C., Maxwell, C., Shumate, D., Tandang-Silvas, M.R., Rajarathnam, K., Navarro, J. (2008) Adenovirus mediated expression "in vivo" of the chemokine receptor CXCR1. *J Struct Funct Genomics.* 10, 17–23.

127. Parker, E.M., Kameyama, K., Higashijima, T., Ross, E.M. (1991) Reconstitutively active G protein-coupled receptors purified from baculovirus-infected insect cells. *J Biol Chem.* 266, 519–527.

128. Sarkar, C.A., Dodevski, I., Kenig, M., Dudli, S., Mohr, A., Hermans, E., Pluckthun, A. (2008) Directed evolution of a G protein-coupled receptor for expression, stability, and binding selectivity. *Proc Natl Acad Sci U S A.* 105, 14808–14813.

129. King, K., Dohlman, H.G., Thorner, J., Caron, M.G., Lefkowitz, R.J. (1990) Control of yeast mating signal transduction by a mammalian β_2-adrenergic receptor and Gs alpha subunit. *Science.* 250, 121–123.

130. Andre, N., Cherouati, N., Prual, C., Steffan, T., Zeder-Lutz, G., Magnin, T., Pattus, F., Michel, H., Wagner, R., Reinhart, C. (2006) Enhancing functional production of G protein-coupled receptors in *Pichia pastoris* to levels required for structural studies via a single expression screen. *Protein Sci.* 15, 1115–1126.

131. Shukla, A.K., Reinhart, C., Michel, H. (2006) Dimethylsulphoxide as a tool to increase functional expression of heterologously produced GPCRs in mammalian cells. *FEBS Lett.* 580, 4261–4265.

132. Chiu, M.L., Tsang, C., Grihalde, N., MacWilliams, M.P. (2008) Over-expression, solubilization, and purification of G protein-coupled receptors for structural biology. *Comb Chem High Throughput Screen.* 11, 439–462.

133. Grisshammer, R., Tate, C.G. (1995) Overexpression of integral membrane proteins for structural studies. *Q Rev Biophys.* 28, 315–422.

134. Nath, A., Atkins, W.M., Sligar, S.G. (2007) Applications of phospholipid bilayer nanodiscs in the study of membranes and membrane proteins. *Biochemistry.* 46, 2059–2069.

135. Magnani, F., Shibata, Y., Serrano-Vega, M.J., Tate, C.G. (2008) Co-evolving stability and conformational homogeneity of the human adenosine A_{2a} receptor. *Proc Natl Acad Sci U S A.* 105, 10744–10749.

136. Serrano-Vega, M.J., Magnani, F., Shibata, Y., Tate, C.G. (2008) Conformational thermostabilization of the β_1-adrenergic receptor in a detergent-resistant form. *Proc Natl Acad Sci U S A.* 105, 877–882.

137. Sarramegn, V., Muller, I., Milon, A., Talmont, F. (2006) Recombinant G protein-coupled receptors from expression to renaturation: A challenge towards structure. *Cell Mol Life Sci.* 63, 1149–1164.

138. Hampe, W., Voss, R.H., Haase, W., Boege, F., Michel, H., Reilander, H. (2000) Engineering of a proteolytically stable human β_2-adrenergic receptor/maltose-binding protein fusion and production of the chimeric protein in *Escherichia coli* and baculovirus-infected insect cells. *J Biotechnol.* 77, 219–234.

139. Noguchi, S., Satow, Y. (2006) Purification of human β_2-adrenergic receptor expressed in methylotrophic yeast *Pichia pastoris*. *J Biochem.* 140, 799–804.

140. Kleymann, G., Boege, F., Hahn, M., Hampe, W., Vasudevan, S., Reilander, H. (1993) Human β_2-adrenergic receptor produced in stably transformed insect cells is functionally coupled via endogenous GTP-binding protein to adenylyl cyclase. *Eur J Biochem.* 213, 797–804.

141. Chelikani, P., Reeves, P.J., Rajbhandary, U.L., Khorana, H.G. (2006) The synthesis and high-level expression of a β_2-adrenergic receptor gene in a tetracycline-inducible stable mammalian cell line. *Protein Sci.* 15, 1433–1440.

142. Rosenbaum, D.M., Cherezov, V., Hanson, M.A., Rasmussen, S.G., Thian, F.S., Kobilka, T.S., Choi, H.J., Yao, X.J., Weis, W.I., Stevens, R.C., Kobilka, B.K. (2007) GPCR engineering yields high-resolution structural insights into β_2-adrenergic receptor function. *Science.* 318, 1266–1273.

143. Helenius, A., Simons, K. (1975) Solubilization of membranes by detergents. *Biochim Biophys Acta.* 415, 29–79.
144. Tate, C.G. (2010) Practical considerations of membrane protein instability for purification and crystallisation. *Methods Mol Biol.* 601, 187–203.
145. Prive, G.G. (2007) Detergents for the stabilization and crystallization of membrane proteins. *Methods.* 41, 388–397.
146. Banerjee, P., Joo, J.B., Buse, J.T., Dawson, G. (1995) Differential solubilization of lipids along with membrane proteins by different classes of detergents. *Chem Phys Lipids.* 77, 65–78.
147. Garavito, R.M., Ferguson-Miller, S. (2001) Detergents as tools in membrane biochemistry. *J Biol Chem.* 276, 32403–32406.
148. Lemieux, M.J., Reithmeier, R.A., Wang, D.N. (2002) Importance of detergent and phospholipid in the crystallization of the human erythrocyte anion-exchanger membrane domain. *J Struct Biol.* 137, 322–332.
149. Lemieux, M.J., Song, J., Kim, M.J., Huang, Y., Villa, A., Auer, M., Li, X.D., Wang, D.N. (2003) Three-dimensional crystallization of the *Escherichia coli* glycerol-3-phosphate transporter: A member of the major facilitator superfamily. *Protein Sci.* 12, 2748–2756.
150. Massotte, D., Pereira, C.A., Pouliquen, Y., Pattus, F. (1999) Parameters influencing human mu opioid receptor over-expression in baculovirus-infected insect cells. *J Biotechnol.* 69, 39–45.
151. Mize, G.J., Harris, J.E., Takayama, T.K., Kulman, J.D. (2008) Regulated expression of active biotinylated G-protein coupled receptors in mammalian cells. *Protein Expr Purif.* 57, 280–289.
152. Daulat, A.M., Maurice, P., Froment, C., Guillaume, J.L., Broussard, C., Monsarrat, B., Delagrange, P., Jockers, R. (2007) Purification and identification of G protein-coupled receptor protein complexes under native conditions. *Mol Cell Proteomics.* 6, 835–844.
153. Salom, D., Le Trong, I., Pohl, E., Ballesteros, J.A., Stenkamp, R.E., Palczewski, K., Lodowski, D.T. (2006) Improvements in G protein-coupled receptor purification yield light stable rhodopsin crystals. *J Struct Biol.* 156, 497–504.
154. Liu, Q., Reubi, J.C., Wang, Y., Knoll, B.J., Schonbrunn, A. (2003) In vivo phosphorylation of the somatostatin 2A receptor in human tumors. *J Clin Endocrinol Metab.* 88, 6073–6079.
155. Andre, C., De Backer, J.P., Guillet, J.C., Vanderheyden, P., Vauquelin, G., Strosberg, A.D. (1983) Purification of muscarinic acetylcholine receptors by affinity chromatography. *EMBO J.* 2, 499–504.
156. Nakata, H. (1989) Affinity chromatography of A1 adenosine receptors of rat brain membranes. *Mol Pharmacol.* 35, 780–786.
157. Caron, M.G., Srinivasan, Y., Pitha, J., Kociolek, K., Lefkowitz, R.J. (1979) Affinity chromatography of the beta-adrenergic receptor. *J Biol Chem.* 254, 2923–2927.
158. White, J.F., Grisshammer, R. (2007) Automated large-scale purification of a recombinant G-protein-coupled neurotensin receptor. *Curr Protoc Protein Sci.* Chapter 6, Unit 6.8.
159. Rich, R.L., Miles, A.R., Gale, B.K., Myszka, D.G. (2009) Detergent screening of a G-protein-coupled receptor using serial and array biosensor technologies. *Anal Biochem.* 386, 98–104.

160. Wiener, M.C. (2004) A pedestrian guide to membrane protein crystallization. *Methods.* 34, 364–372.

161. Bond, P.J., Faraldo-Gomez, J.D., Deol, S.S., Sansom, M.S. (2006) Membrane protein dynamics and detergent interactions within a crystal: A simulation study of OmpA. *Proc Natl Acad Sci U S A.* 103, 9518–9523.

162. Gast, P., Hemelrijk, P., Hoff, A.J. (1994) Determination of the number of detergent molecules associated with the reaction center protein isolated from the photosynthetic bacterium *Rhodopseudomonas viridis*. Effects of the amphiphilic molecule 1,2,3-heptanetriol. *FEBS Lett.* 337, 39–42.

163. Faham, S., Bowie, J.U. (2002) Bicelle crystallization: A new method for crystallizing membrane proteins yields a monomeric bacteriorhodopsin structure. *J Mol Biol.* 316, 1–6.

164. Hunte, C., Michel, H. (2002) Crystallisation of membrane proteins mediated by antibody fragments. *Curr Opin Struct Biol.* 12, 503–508.

165. Sennhauser, G., Grutter, M.G. (2008) Chaperone-assisted crystallography with DARPins. *Structure.* 16, 1443–1453.

166. Ostermeier, C., Iwata, S., Ludwig, B., Michel, H. (1995) Fv fragment-mediated crystallization of the membrane protein bacterial cytochrome c oxidase. *Nat Struct Biol.* 2, 842–846.

167. Jiang, Y., Lee, A., Chen, J., Ruta, V., Cadene, M., Chait, B.T., MacKinnon, R. (2003) X-ray structure of a voltage-dependent K+ channel. *Nature.* 423, 33–41.

168. Kwong, P.D., Wyatt, R., Robinson, J., Sweet, R.W., Sodroski, J., Hendrickson, W.A. (1998) Structure of an HIV gp120 envelope glycoprotein in complex with the CD4 receptor and a neutralizing human antibody. *Nature.* 393, 648–659.

169. Day, P.W., Rasmussen, S.G., Parnot, C., Fung, J.J., Masood, A., Kobilka, T.S., Yao, X.J., Choi, H.J., Weis, W.I., Rohrer, D.K., Kobilka, B.K. (2007) A monoclonal antibody for G protein-coupled receptor crystallography. *Nat Methods.* 4, 927–929.

170. Huang, L., Sato, A.K., Sachdeva, M., Fleming, T., Townsend, S., Dransfield, D.T. (2005) Discovery of human antibodies against the C5aR target using phage display technology. *J Mol Recognit.* 18, 327–333.

171. Rothlisberger, D., Pos, K.M., Pluckthun, A. (2004) An antibody library for stabilizing and crystallizing membrane proteins—Selecting binders to the citrate carrier CitS. *FEBS Lett.* 564, 340–348.

172. Lightwood, D.J., Carrington, B., Henry, A.J., McKnight, A.J., Crook, K., Cromie, K., Lawson, A.D. (2006) Antibody generation through B cell panning on antigen followed by in situ culture and direct RT-PCR on cells harvested en masse from antigen-positive wells. *J Immunol Methods.* 316, 133–143.

173. Hogbom, M., Eklund, M., Nygren, P.A., Nordlund, P. (2003) Structural basis for recognition by an in vitro evolved affibody. *Proc Natl Acad Sci U S A.* 100, 3191–3196.

174. Binz, H.K., Amstutz, P., Kohl, A., Stumpp, M.T., Briand, C., Forrer, P., Grutter, M.G., Pluckthun, A. (2004) High-affinity binders selected from designed ankyrin repeat protein libraries. *Nat Biotechnol.* 22, 575–582.

175. Engel, C.K., Chen, L., Prive, G.G. (2002) Insertion of carrier proteins into hydrophilic loops of the *Escherichia coli* lactose permease. *Biochim Biophys Acta.* 1564, 38–46.

176. Zhuang, J., Prive, G.G., Werner, G.E., Ringler, P., Kaback, H.R., Engel, A. (1999) Two-dimensional crystallization of *Escherichia coli* lactose permease. *J Struct Biol.* 125, 63–75.

177. Caffrey, M. (2008) Crystallizing membrane proteins for structure determination: Use of lipidic mesophases. *Annu Rev Biophys.* 38, 29–51.

178. Schobert, B., Cupp-Vickery, J., Hornak, V., Smith, S., Lanyi, J. (2002) Crystallographic structure of the K intermediate of bacteriorhodopsin: Conservation of free energy after photoisomerization of the retinal. *J Mol Biol.* 321, 715–726.

179. Nollert, P., Royant, A., Pebay-Peyroula, E., Landau, E.M. (1999) Detergent-free membrane protein crystallization. *FEBS Lett.* 457, 205–208.

180. Nollert, P., Navarro, J., Landau, E.M. (2002) Crystallization of membrane proteins in cubo. *Methods Enzymol.* 343, 183–199.

181. Caffrey, M. (2000) A lipid's eye view of membrane protein crystallization in mesophases. *Curr Opin Struct Biol.* 10, 486–497.

182. Landau, E.M., Rummel, G., CowanJacob, S.W., Rosenbusch, J.P. (1997) Crystallization of a polar protein and small molecules from the aqueous compartment of lipidic cubic phases. *J Phys Chem B.* 101, 1935–1937.

183. Qutub, Y., Reviakine, I., Maxwell, C., Navarro, J., Landau, E.M., Vekilov, P.G. (2004) Crystallization of transmembrane proteins in cubo: Mechanisms of crystal growth and defect formation. *J Mol Biol.* 343, 1243–1254.

184. Cherezov, V., Clogston, J., Papiz, M.Z., Caffrey, M. (2006) Room to move: Crystallizing membrane proteins in swollen lipidic mesophases. *J Mol Biol.* 357, 1605–1618.

185. Li, L., Mustafi, D., Fu, Q., Tereshko, V., Chen, D.L., Tice, J.D., Ismagilov, R.F. (2006) Nanoliter microfluidic hybrid method for simultaneous screening and optimization validated with crystallization of membrane proteins. *Proc Natl Acad Sci U S A.* 103, 19243–19248.

186. Riekel, C., Burghammer, M., Schertler, G. (2005) Protein crystallography microdiffraction. *Curr Opin Struct Biol.* 15, 556–562.

187. Kenakin, T. (2007) Functional selectivity through protean and biased agonism: Who steers the ship? *Mol Pharmacol.* 72, 1393–1401.

188. Zhou, Y., Wen, J., Bowie, J.U. (1997) A passive transmembrane helix. *Nat Struct Biol.* 4, 986–990.

189. Bowie, J.U. (2001) Stabilizing membrane proteins. *Curr Opin Struct Biol.* 11, 397–402.

190. Lau, F.W., Nauli, S., Zhou, Y., Bowie, J.U. (1999) Changing single side-chains can greatly enhance the resistance of a membrane protein to irreversible inactivation. *J Mol Biol.* 290, 559–564.

191. Zhou, Y., Bowie, J.U. (2000) Building a thermostable membrane protein. *J Biol Chem.* 275, 6975–6979.

192. Alexandrov, A.I., Mileni, M., Chien, E.Y., Hanson, M.A., Stevens, R.C. (2008) Microscale fluorescent thermal stability assay for membrane proteins. *Structure.* 16, 351–359.

193. Kobilka, B.K., Kobilka, T.S., Daniel, K., Regan, J.W., Caron, M.G., Lefkowitz, R.J. (1988) Chimeric α_2-,β_2-adrenergic receptors: Delineation of domains involved in effector coupling and ligand binding specificity. *Science.* 240, 1310–1316.

194. Maggio, R., Vogel, Z., Wess, J. (1993) Coexpression studies with mutant muscarinic/adrenergic receptors provide evidence for intermolecular "cross-talk" between G-protein-linked receptors. *Proc Natl Acad Sci U S A.* 90, 3103–3107.
195. Maggio, R., Vogel, Z., Wess, J. (1993) Reconstitution of functional muscarinic receptors by co-expression of amino- and carboxyl-terminal receptor fragments. *FEBS Lett.* 319, 195–200.
196. Lefkowitz, R.J., Sun, J.P., Shukla, A.K. (2008) A crystal clear view of the β_2-adrenergic receptor. *Nat Biotechnol.* 26, 189–191.
197. Shukla, A.K., Sun, J.P., Lefkowitz, R.J. (2008) Crystallizing thinking about the β_2-adrenergic receptor. *Mol Pharmacol.* 73, 1333–1338.
198. Kobilka, B. (1992) Adrenergic receptors as models for G protein-coupled receptors. *Annu Rev Neurosci.* 15, 87–114.
199. Kobilka, B.K., Deupi, X. (2007) Conformational complexity of G-protein-coupled receptors. *Trends Pharmacol Sci.* 28, 397–406.
200. Ballesteros, J.A., Weinstein, H., Stuart, C.S. (1995) Integrated methods for the construction of three-dimensional models and computational probing of structure-function relations in G protein-coupled receptors. In *Methods in Neurosciences*, Vol. 25, ed. S.C. Sealfon, P.M. Conn, pp. 366–428. San Diego, CA: Academic Press.
201. Hopkins, A.L., Groom, C.R., Alex, A. (2004) Ligand efficiency: A useful metric for lead selection. *Drug Discov Today.* 9, 430–431.
202. Park, J.H., Scheerer, P., Hofmann, K.P., Choe, H.W., Ernst, O.P. (2008) Crystal structure of the ligand-free G-protein-coupled receptor opsin. *Nature.* 454, 183–187.
203. Roth, C.B., Hanson, M.A., Stevens, R.C. (2008) Stabilization of the human β_2-adrenergic receptor TM4-TM3-TM5 helix interface by mutagenesis of Glu122(3.41), a critical residue in GPCR structure. *J Mol Biol.* 376, 1305–1319.
204. Caffrey, M. (2003) Membrane protein crystallization. *J Struct Biol.* 142, 108–132.
205. Landau, E.M., Rosenbusch, J.P. (1996) Lipidic cubic phases: A novel concept for the crystallization of membrane proteins. *Proc Natl Acad Sci U S A.* 93, 14532–14535.
206. Cusack, S., Belrhali, H., Bram, A., Burghammer, M., Perrakis, A., Riekel, C. (1998) Small is beautiful: Protein micro-crystallography. *Nat Struct Biol.* 5 (Suppl.), 634–637.
207. Caffrey, M. (2008) On the mechanism of membrane protein crystallization in lipidic mesophases. *Crystal Growth Des.* 8, 4244–4254.

CHAPTER 15

Structure and Modeling of GPCRs: Implications for Drug Discovery

KIMBERLY A. REYNOLDS,[1] VSEVOLOD KATRITCH,[2] and RUBEN ABAGYAN[3]

[1]Department of Pharmacology, University of Texas Southwestern Medical Center, Dallas, TX

[2]Skaggs School of Pharmacy and Pharmaceutical Sciences, La Jolla, CA

[3]Bioinformatics and Systems Biology Graduate Program, University of California, La Jolla, CA

15.1. INTRODUCTION

Despite a common structural scaffold of seven transmembrane helices, the G protein-coupled receptor (GPCR) superfamily enables recognition of a broad range of extracellular stimuli encompassing odorants, peptides, lipids, biogenic amines, hormones, nucleotides, and even light (Fig. 15.1). Phylogenetic analysis of the human GPCR superfamily has identified 860 distinct receptors, of which approximately 300 are nonolfactory [1, 2]. Given this remarkable diversity in sequence and function, dissecting the relationship between primary amino acid sequence, receptor structure, ligand structure, and ligand activity is non-trivial. In many instances, a single ligand binds multiple receptor types, possibly with differing activities at each. Similarly, a single receptor may recognize a wide array of structurally diverse ligands. Rational GPCR drug discovery, which requires the identification of novel high-affinity GPCR ligands with a predicted specificity and activity, is therefore a particularly complex problem.

As an additional complication, the experimental characterization of GPCR structures is extremely challenging, and the conformational changes associated with ligand binding remain largely unknown. Until recently, the only GPCR to be crystallized was bovine rhodopsin (bRho). The rhodopsin structures supply essential information for constructing three-dimensional

GPCR Molecular Pharmacology and Drug Targeting: Shifting Paradigms and New Directions, Edited by Annette Gilchrist

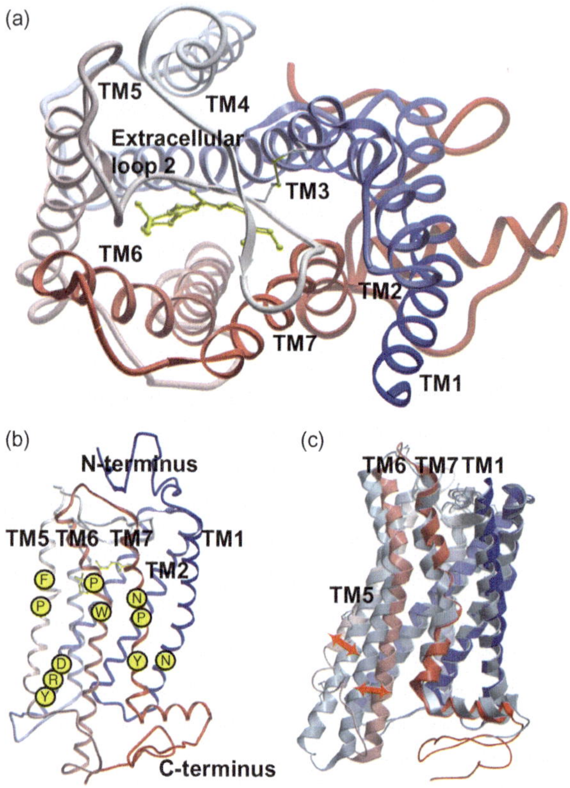

Figure 15.1 General GPCR structure. Panel (a) depicts the extracellular view of bRho (PDB: 1U19). The receptor backbone is shown in ribbon and is color coded from the N- to C-terminus (blue to red). The full inverse agonist retinal is shown in yellow sticks. Panel (b) provides a side view of bRho; several Class A conserved microdomains are indicated with yellow circles. These amino acids are thought to play a role in receptor activation. Panel (c) compares the bRho (N- to C-termini color-coded ribbon) and unliganded opsin structures (pale blue ribbon). The unliganded opsin structure corresponds to an activated conformation.

(3D) GPCR models but are poorly representative of ligand binding pocket structure in some respects. Unlike many GPCRs, rhodopsin does not bind a freely diffusible ligand but rather is covalently linked to the inverse agonist 11-*cis*-retinal, which undergoes photo conversion to the agonist all-*trans*-retinal. Moreover, the retinal binding pocket displays very low sequence identity to the ligand binding pocket of other receptors, including those within the Class A GPCR family.

Due to these limitations, ligand-based methods, such as structure–activity relationship (SAR) models, 2D and 3D pharmacophores, and chemogenomics analysis, have prevailed in GPCR drug discovery [3, 4]. These approaches do not rely on a 3D receptor structure but instead select new small molecules based on similarity to existing agonists or antagonists. However, ligand-based methods are limited in the novelty of the chemical leads identified and are restricted to receptors where a number of known ligands have already been characterized. Therefore, over the past decade, significant efforts have been made at developing computational techniques for predicting GPCR structure and receptor/ligand interactions. Practical application of a structure-based approach to ligand screening and optimization has nonetheless been limited by the low accuracy of these models, which were initially built by homology to bacteriorhodopsin [5] and, later, bRho [6, 7], or constructed de novo [8–11].

The recent publication of high-resolution structures for the β_2-adrenergic (β_2AR) [12–14], β_1-adrenergic (β_1AR) [15], and adenosine A_{2a} ($AA_{2a}R$) [16] receptors in complex with diffusible small molecule ligands ushers in a new era for GPCR modeling. These structures furnish additional templates for homology model construction and a set of test cases for computational protocols. Nonetheless, experimental characterization of additional high-resolution GPCR structures remains a challenging task because of the conformational flexibility intrinsic to these proteins. This will likely limit the number of solved GPCR crystal structures, as well as the achievable resolution for some receptors. Importantly, even when the crystallographic coordinates provide insight into the interactions underlying ligand binding and receptor activation, these observations are restricted to the specific receptor type and co-crystallized ligand. This limited set of structures must be extrapolated to describe the conformation of hundreds of essential GPCRs in complex with tens of thousands of small molecule modulators. The protein databank (PDB)-deposited coordinates themselves do not specify the location of hydrogen atoms and may display inaccuracies in conformation arising from the uncertainty inherent in electron density fitting. Such errors include incorrect orientations of asparagine or glutamine, suboptimal hydrogen bonding networks, or misplaced side chains. In all cases, these errors lead to diminished interpretation of the existing ligand/receptor interactions, as well as poor ligand docking and screening performance. Additional complications arise from the existence of multiple ligand-selective receptor conformational states corresponding to differences in ligand potency and functionality. Without the aid of further computational analysis, a single GPCR structure is unable to address the binding pocket conformational changes necessary for recognition of structurally distinct agonists and antagonists [17].

In the absence of comprehensive structural data, computational approaches can leverage existing structures to more accurately interpret the experimental electron density and receptor coordinates, provide insight into other ligand-bound receptor conformations, and model other receptor types. Such methods maximize the benefit of the crystallographic data to the drug discovery community and will gain importance as they are increasingly validated and a growing number of structures are determined. While approximate, low-resolution models of the full-length receptor are suitable for certain tasks, such as selecting amino acid positions for experimental mutagenesis or fluorescent label placement, atomic level accuracy in a select group of binding pocket side chains is required for ligand pose prediction, lead optimization, and virtual ligand screening (VLS). In each case, the availability of experimental data and

TABLE 15.1 Application of 3D Protein Modeling and Ligand Docking Approaches to GPCR Drug Discovery

	Structural Data Available	Current Targets	What Can Be Done	3D Modeling Methods	Drug Design Applications
	Crystal structure with a lead compound	β_1AR, AA_{2a}R	Refine ligand contacts and H-bonds, predict SAR	Combined energy-based/electron density refinement	Comprehensive 3D model for lead optimization
1–2 Å	Crystal structure with unrelated ligand + SAR data	• β_1AR, β_2AR, AA_{2a}R • GRM1-3-7 (N-terminal domain)	Predict ligand induced fit, binding mode, SAR, including agonists	Flexible receptor docking, MRC, SCARE	3D VLS for new chemical scaffolds, optimization of new leads
Modeling resolution	Structure of a close homolog + SAR data	• Most adrenergic, 5HT, dopamine • Nucleotide like • Most Class C (N-term domain)	Predict ligand binding mode, SAR, and selectivity profile	Homology modeling and flexible docking	3D lead discovery and optimization, with support of ligand-based methods
	Structure of a distant homolog + SAR data	• Most Class A • TM pockets in Classes B and C	Predict ligand binding and selectivity sites, extend SAR	Ligand-guided 3D modeling of flexible receptor	Combined receptor-based and ligand-based drug discovery
3–5 Å	Overall 7TM geometry, little or no ligand binding data	All GPCRs, including orphans	Predict low-resolution model, some ligand–receptor contacts	Global optimization with experimental restraints, de novo methods	Predict low-affinity nonselective ligands, 3D support for ligand-based methods

computational protocol employed will determine the quality and nature of information that may be reliably obtained by receptor modeling (Table 15.1). In the present chapter, we address how computational modeling can augment experimental structure and discuss the strengths and limitations of a structure-based approach in the context of varying resolutions of experimental data. The discussion is generally arranged to proceed from "high-resolution" modeling, wherein extensive experimental data are available and the corresponding models are atomic level in accuracy, to "low-resolution" modeling, where little to no experimental data exist and the resulting predictions are therefore more approximate.

15.2. HIGH-RESOLUTION GPCR MODELING

15.2.1. From Electron Density to a Full Atom Model Suitable for Drug Discovery: Refinement of Existing Crystal Structures

Dramatic progress in GPCR expression, purification, and crystallography techniques has recently resulted in high-resolution structures of β_2AR, β_1AR, and AA_{2a}R in complex with several partial inverse agonists [13–16, 18]. These receptors are presently the only "druggable" GPCRs to have been crystallized and provide a good testing ground for structure-based drug discovery methods. Nonetheless, further computational analysis is required to convert and extend a set of PDB-deposited coordinates to a fully protonated all-atom model suitable for rational drug design. Such 3D model preparation goes beyond standard crystallographic refinement and validation procedures [19, 20], and is usually based upon conformational energy optimization, which may be restrained by (or combined with) electron density map fitting. As functional understanding of ligand binding depends critically on hydrogen bond satisfaction, generation of a fully protonated model focuses on the correct assignment and optimization of polar and ionic interactions. This analysis takes into account: (1) flipping of glutamine and asparagine side-chain orientations; (2) protonation and tautomerization of histidine; (3) appropriate protonation of asparatate, glutamate, and cysteine as defined by the local environment; and (4) placement of rotatable hydrogens in other polar side chains. When recently evaluated, the error in asparagine/glutamine rotamer assignment within the PDB was on the order of 20%, indicating that errors in hydrogen bonding network structure are fairly widespread [21]. Additionally, in conformationally flexible regions displaying high atomic b-factors, the electron density is often not sufficient for correct assignment of side-chain rotamers or ligand conformational state. It should be noted that resolutions as high as 2.2 Å and 2.4 Å, as observed for bRho (PDB: 1U19) and β_2AR (PDB: 2RH1), respectively, may be rather exceptional for GPCRs. High-resolution crystals were obtained for these two receptors due in part to the extreme stability of the dark-adapted bRho state and the picomolar affinity of β_2AR with carazolol. Furthermore, local variations in electron density map quality may result in poorly resolved regions of protein structure. For example, in the antibody-bound complex of β_2AR, the entire region from the ligand binding pocket to the extracellular domains is unresolved [13]. It is anticipated that many GPCR complexes relevant to drug discovery will be characterized at lower resolutions and display more poorly defined ligand binding interactions. Ligand-guided refinement of the receptor ligand binding pocket, wherein the binding pocket side chains and ligand conformation are flexibly optimized under an empirical energy potential, can be employed to optimize the ligand/receptor interactions and provide a more thorough description of ligand recognition.

Previously, we conducted ligand-guided refinement of a full atom β_2AR model derived from the carazolol-bound β_2AR structure (PDB: 2RH1) to further optimize the binding pocket conformation and analyze deviations of the computational predictions from the crystal structure [22]. In this procedure, the binding pocket side chains and carazolol ligand were flexibly optimized using biased probability Monte Carlo sampling as implemented in the Internal Coordinates Mechanics (ICM) software package (Molsoft, La Jolla, CA) [23]. This gave rise to a new model including five additional hydrogen bonds in the ligand binding pocket that are missing or suboptimal in the PDB-deposited coordinates. Importantly, the energy-optimized model calculates a binding pocket conformation and (–)-carazolol geometry nearly identical to the crystal structure, with the exception of the rotameric states of Ser203$^{5.42}$ and Ser204$^{5.43}$ (Fig. 15.2a). The new rotamer of Ser203$^{5.42}$ allows formation of

Figure 15.2 (a) The β_2AR binding pocket in complex with (–)-carazolol. The β_2AR backbone is shown in gray cartoon and (–)-carazolol in yellow sticks. Receptor side chains from the PDB-deposited coordinates are depicted as gray sticks, while side-chain conformations from a ligand-guided model of the β_2AR pocket are shown in orange. The electron density map is contoured at 1.2σ and displayed in blue mesh. (b) The lowest energy conformation of (–)-isoproterenol (yellow sticks) in an agonist-bound β_2AR model. This conformation was generated by LGM with backbone flexibility in the TM5 proline-induced kink (residues 205–210) and a portion of ECL2 loop (191–196). The flexible backbone regions are indicated in green ribbon, and the optimized position of the extracellular portion of TM5 is shown as red ribbon. The original position of TM5 and other static helices are indicated in gray ribbon, and flexible side chains in proximity of the binding site have carbon atoms colored green. Hydrogen bonds are shown with dotted lines. (c) Predicted ligand binding affinities (pK_d^{Pred}) for β_2AR models with flexible and rigid TM5 backbones. Docking results are shown for a diverse set of ligands ranging from inverse agonists (e.g., carazolol) and partial agonists (e.g., dopamine, MAPE) to full agonists (e.g., isoproterenol and epinephrine). Arrows highlight the most pronounced increases in predicted ligand binding affinity from the rigid backbone model to the flexible TM5 model for the full agonists. This increase is much smaller for partial agonists and negligible for antagonists/inverse agonists. The accuracy of affinity predictions is estimated as $R^2 = 0.75$, RMSE = 1.3 pK_d for the rigid backbone model, and $R^2 = 0.89$, RMSE = 0.7 pK_d for the flexible TM5 model.

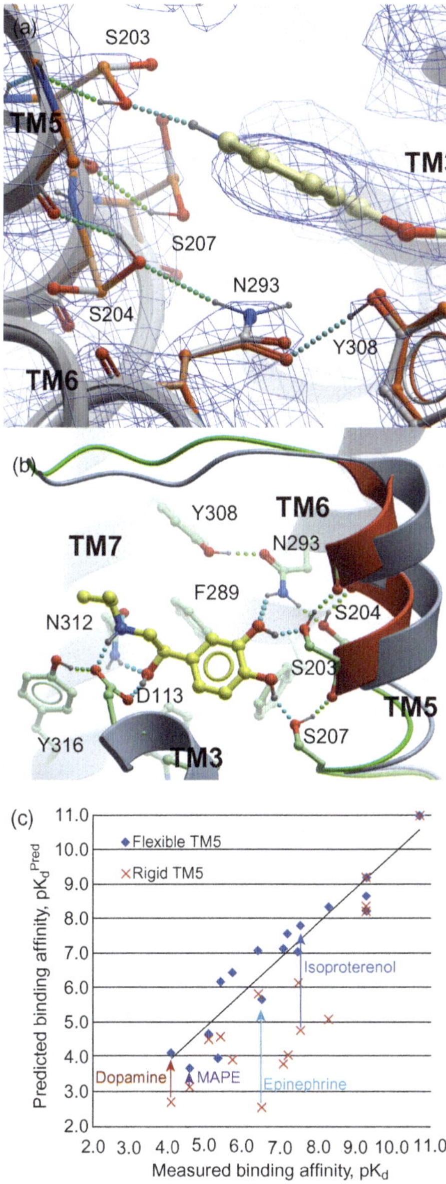

a hydrogen bond to the Tyr199$^{5.38}$ backbone carbonyl and significantly improves its hydrogen bonding distance to the carbazole moiety of carazolol from 3.3 to 2.7 Å. Additionally, the predicted rotamer of Ser204$^{5.43}$ has an improved geometry of hydrogen bonding with the Ala200$^{5.39}$ backbone and displays a new hydrogen bond between the Ser204$^{5.43}$ hydroxyl and Asn293$^{6.55}$ side-chain nitrogen, which is likely to play a role in the basal activity of β_2AR [22]. These optimized conformations of Ser203$^{5.42}$ and Ser204$^{5.43}$ are consistent with the electron densities observed in the crystal structure yet provide an enhanced view of stabilizing interactions within the ligand binding pocket.

In addition to side-chain placement and protonation state, analysis of GPCR structural data requires special attention due to the conformational flexibility inherent to these receptors. The crystallization conditions, crystal contacts, specific mutations, and fusions employed for receptor stabilization affect the final conformation and should be considered when constructing and evaluating a model. For example, biophysical experiments have shown that the magnitude of activation-related helix mobility is decreased in membrane-associated GPCRs relative to detergent-solubilized receptors [24, 25]. In contrast, the limited conformational changes observed between a photoactivated deprotonated rhodopsin intermediate and the dark-adapted ground state rhodopsin may possibly be explained by physiologically relevant dimerization of the receptor in the crystal structures [26, 27]. Later in this chapter, we discuss conformational differences among the presently available GPCR crystal structures and the impact upon the ligand binding pocket architecture.

15.2.2. Ligand-Guided Modeling of Binding Pocket Conformation

For a structural chemist designing a small molecule modulator of a protein, there are two ideal environments. In the first, all receptor conformations and receptor/ligand complexes would have been crystallized, and in the second, reliable prediction of ligand/receptor structure would be possible. Unfortunately, the reality is far from both situations. Only a handful of GPCRs have been crystallized, and this limited amount of structural data must be extrapolated to describe ligand/receptor interactions for thousands of potential complexes. Ligand docking is highly sensitive to receptor conformation, requiring great accuracy in the predicted structural models. A benchmark study for a large test set of enzyme/inhibitor complexes found that conformational deviations greater than 1.5 Å in the protein-active site precluded correct inhibitor docking [28, 29]. As existing GPCR structures for distinct receptor types exhibit an approximately 2.0 Å root mean standard deviation (RMSD) within the helical TM domains (see Section 15.3.2), starting GPCR homology models will often be too distorted for meaningful ligand docking. Additionally, alternate binding modes for the same protein with other ligand types can be difficult to infer from an existing structure. For example, agonist/β_2AR interactions cannot be properly described by the existing antagonist-bound structures. A protocol for refining and adapting the receptor binding pocket conformation for the recognition of diverse ligand types is therefore needed.

The Preconditions Ligand-guided modeling (LGM) of the receptor pocket conformation requires the following components:

1. One or several approximate starting protein models (M_0). These models may be built by homology, collected from crystallographic or NMR structures, derived from an alternative functional state of the same protein (e.g., an agonist-bound model derived from an antagonist-bound conformation), or all of the above.
2. A set of validated ligands for a particular functional state of the protein of interest. Let us call this set L, and divide it into a subset of high-affinity "seed" ligands (L_{seed}) and the rest of the set (L_{test}). At least one potent and validated ligand is required in L_{seed}.
3. A larger set of closely related nonbinders (designated N) to the same functional state of the same protein. This set may contain ligands for other functional states of the same protein or ligands for related proteins. It is acceptable if a small fraction of molecules in the negative set (say, 5–10%) have not been fully annotated and are in fact uncharacterized binders.
4. One or several restraints specifying receptor atoms or residues that must interact with ligand atoms. Specific interatomic interactions between the best reference binders (L_{seed}) and protein pocket atoms are the most desirable, such as the well-characterized hydrogen bonding interaction between Asp 3.32 and the positively charged ligand amine for biogenic amine receptors. Such restraints permit more reliable and accurate positioning of the ligand but are not strictly mandatory.

The Procedure A general outline of the LGM approach is shown in Fig. 15.3. It consists of three steps: model generation (G) by conformational optimization with seed ligand(s) (L_{seed}), an optional model compression step (C), and model selection (S) based upon binder (L_{test}) versus nonbinder (N) discrimination. Steps G and S provide two different ways to incorporate ligand guidance in the procedure. It is important, however, to point out that either step may be omitted. Multiple models (M_0) may be generated without seed ligand, or taken from other sources (experimental or computational) and fed directly into step S [30]. Conversely, if only a single ligand is available (e.g., for a new protein with only one characterized ligand or substrate), step S can be skipped, though L versus N discrimination should preferably be assessed using the single known ligand.

Model Generation (G) During step G, receptor conformation is sampled in the presence of a known high-affinity ligand (L_{seed}). The addition of ligand prevents collapse of the binding pocket and allows the receptor side chains to form stabilizing interactions with the ligand functional groups. Alternatively, the receptor may be sampled in the presence of a "blob" of repulsive density placed within the ligand binding pocket [31]. This technique also prevents pocket collapse and permits reshaping of the pocket. The receptor-flexible docking of ligands was first described in Reference 32 and was based upon

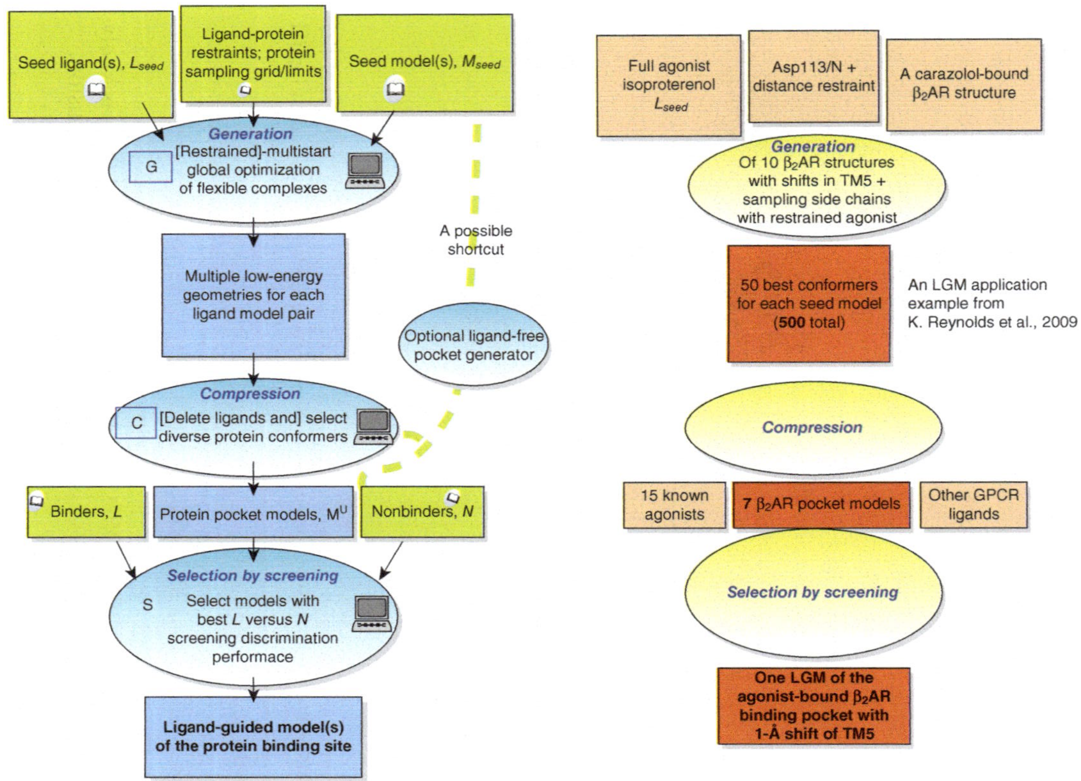

Figure 15.3 The LGM protocol. A general schematic is shown on the left side of the figure, while the right side describes the specific protocol followed in generating agonist-bound β_2AR models for VLS. Experimental data that may be gathered from the literature or outside sources are indicated with small book icons, and computational steps are marked with computers.

previous work with flexible ligand docking in internal coordinates [33, 34]. Cavasotto and Abagyan later applied this procedure to flexibly model kinase/ligand complexes and demonstrated that VLS with a training set can be used as a selection criterion [35].

When generating binding pocket conformations for Class A GPCRs, the following parameters should be considered:

1. Treatment of extracellular loops 2 and 3. These loops may be deleted, conformationally sampled, or modeled after ligand placement (for further discussion, see Section 15.3.3).
2. Conformational sampling of the TM helices: selection of movable helices, choice of possible rigid body moves (helix rotation, translation, tilt), and choice of flexible backbone regions (proline kinks). In the simplest case, the helix conformation is fixed.
3. Selection of side chains for sampling. While the region of sampled side chains should be limited, convergence occurs quickly if internal coordinate mechanics is used for simulation.
4. Choice of distance or positional restraints included during the flexible receptor/flexible ligand docking. If experimental data do not support any specific interatomic restraints, simple nonspecific volume restraints can enforce ligand docking within a known binding pocket.

Compressing the Set of Models (C) The compression step is a straightforward one. Two criteria are applied to reduce the number of conformers: (1) geometrical proximity (typically an RMSD cutoff for pocket heavy atoms ranging between 0.2 and 1.5 Å), and (2) the energy of the conformer, receptor/ligand complex, or a binding energy estimate. The first metric ensures structural diversity, while the second eliminates physically unrealistic or strained conformers. Depending upon the cutoff parameters, compression may reduce the number of conformers dramatically.

Selection by VLS (S) The idea to use VLS enrichment as a selection criterion came from the refinement of weights for the docking score energy terms [36]. In that approach, VLS enrichment was calculated for 25 sets of true ligands (L) and nonbinders (N), and the weights providing the best discrimination were selected. While performing the calculations, we realized that different models of the same protein may differ dramatically in their ability to distinguish between binders and nonbinders. The capacity of a given model to distinguish true ligands from nonbinders is a very sensitive criterion, particularly when the true and decoy molecules are similar in size and chemical nature.

During the selection step (S), all molecules from sets L and N are docked and scored with each nonredundant model form. Following docking, standard measures of VLS performance are calculated and used to select a model.

Commonly used measures include the median rank of the true positives, the hit rate, or the "area-under-the-curve" (AUC). In this case, the "curve" is a plot of the "number of true positives" versus the number of top-scoring molecules.

The outcome of the LGM procedure can be any of three possibilities:

1. Multiple conformers are highly selective. There is no requirement that a single conformer is selected. This outcome is particularly interesting if different conformers are selective with respect to different ligands in L.
2. A single highly selective conformer is identified.
3. No conformers impart acceptable selectivity.

The first outcome leads to a multiple receptor conformation (MRC) set for future VLS, the second possibility to a traditional single conformer VLS, and the last case requires reconsidering the model generation steps.

Applications The first practical application of LGM was reported in 2007 [37]. In this study, LGM was used to generate antagonist-bound models of the androgen receptor (AR) from an agonist-bound crystal structure. Starting from an agonist-bound conformation, and two known antagonists (L_{seed}), thousands of hypothetical antagonist-bound pocket conformers were generated. Each conformer was then tested for the ability to discriminate known AR antagonists from nonbinders in a panel of 88 nuclear receptor ligands. The two AR conformations providing the best enrichment were then employed for VLS of a marketed drugs database. Interestingly, three antipsychotic drugs were identified as "hits" and, subsequently, were experimentally confirmed to exhibit antiandrogenic activity. These nonsteroidal molecules were rationally repurposed to improve AR antagonism and reduce affinity for the dopaminergic and sterotonergic receptors.

LGM has now been applied to several GPCRs. A bRho-based homology model was constructed for melanin-concentrating hormone receptor (MCHR), a target for obesity therapeutics [38]. Sampling of the receptor side chains and ligand conformation for four distinct antagonist/receptor complexes yielded 800 ligand binding pocket conformers. Following clustering, three final models were selected for optimal VLS enrichment on a small test set of ligands. The best models were used to screen a larger compound library; experimental characterization of 129 predicted "hits" identified six novel compounds with low micromolar affinity. In this case, even a relatively distorted initial homology model was sufficient to yield productive conformers. Additionally, no experimental data were available to aid ligand placement, demonstrating that unrestrained LGM can enable the discovery of novel ligand chemotypes, even for receptors with limited experimental data. A similar ICM-based protocol was applied with comparable success to the human muscarinic M2 receptor [39]. Initial homology models were constructed with bRho, and the receptor side chains were sampled with several known antagonists to generate a collection of conformers. More recent work has tested β_2AR-based M2 receptor homology models as starting points [39]. Most recently, LGM was assessed in a set of blind predictions for the $AA_{2a}R$ structure [40]. An initial homology model of $AA_{2a}R$ was constructed using β_2AR as a template. Roughly 400 additional $AA_{2a}R$ conformers were generated from this model using a combination of heavy atom elastic network normal modes analysis, and flexible sampling of the receptor side chains and docked ligand. These conformers were evaluated for ligand/nonbinder discrimination, and the best models were iteratively refined through the LGM process. Remarkably, LGM yielded an $AA_{2a}R$ model that correctly recapitulates more than 40% of the ligand/receptor contacts [41]. The LGM model predicted the largest fraction of correct ligand–receptor contacts of those models evaluated in the blind assessment [40]. Further, this model attains an enrichment factor of 29 for the top 1% of compounds in comparison to an enrichment factor of 20 for the $AA_{2a}R$ crystallographic coordinates for VLS on a large test set of 14,000 GPCR ligands.

15.2.3. Coupling LGM and TM Domain Motions to Capture Binding Site Conformational Changes Necessary for Agonist Recognition

The therapeutic action of many drugs and drug candidates is based upon GPCR activation. Agonist binding at the receptor extracellular domain initiates (or modulates) a complex signal transduction process, wherein dynamic changes in the TM helices and receptor loops are coupled to intracellular binding and activation of G proteins [42,43]. Deciphering the details of agonist/receptor interaction is essential to the discovery of novel agonist chemotypes, but the GPCR activation mechanism remains poorly understood at the atomic level. Though a crystal structure for activated bovine opsin was recently published [44], it is unknown whether the full complement of conformational changes necessary for GPCR activation is represented. All other available GPCR structures are complexed with an antagonist or inverse agonist, and are consistent with an inactive (antagonist bound) conformation. Importantly, useful models for the structure-based design of agonists do not necessitate modeling of the full-length activated receptor but only require an accurate representation of conformational changes in the ligand binding pocket. As discussed above, LGM was successful in generating antagonist-bound pocket models for the MCH and M2 GPCRs. However, due to the nature of activation-associated conformational changes, sampling of the ligand binding pocket side chains is inadequate for fully capturing GPCR/agonist interactions. While side-chain minimization can improve the prediction of agonist/receptor contacts, it was recently found that repositioning key residues of β_2AR ligand binding pocket is alone insufficient to provide agonist-selective enrichment in VLS; selectivity was obtained only when the model was combined with a molecular interaction fingerprint scoring function [45]. More sophisticated methods that allow flexibility within the TM backbone or include sampling of helix orientation are thus needed. In the current section, we discuss the modeling of conformational changes restricted to the β_2AR ligand binding pocket and the sufficiency of these for allowing accurate agonist recognition. Computational approaches for investigating global conformational changes and downstream effects resulting from receptor activation are discussed later in the chapter.

Recently, we constructed an agonist-bound β_2AR model from the β_2AR/carazolol crystal structure by conducting LGM with the full agonist isoproterenol [22]. Flexibility was permitted in the TM5 proline kink, extracellular loop 2 (ECL2), and binding pocket side chains during optimization. Following refinement, an approximately 2-Å tilt of the TM5 extracellular end toward the ligand binding pocket was obtained (Fig. 15.2b). This movement enabled optimal engagement of the ethanolamine "tail" and catechol "head" of isoproterenol with corresponding hydrogen bonding partners in TM3 and TM5, consistent with prior mutagenesis data [46–49]. In comparison, a rigid helix model placed TM3 and TM5 too far apart to allow simultaneous interaction with the agonist "head" and "tail" functional groups. Relative binding energies for a diverse set of compounds ranging from full agonists to inverse agonists were accurately calculated with the TM5 flexible model, while a TM5 rigid model underestimated affinities for several ligands (Fig. 15.2c). As discussed below (see "VLS with Agonist-Selective Models"), incorporating shifts in the TM5 position provides a means to generate agonist-selective models of the receptor binding pocket in VLS. Interestingly, smaller shifts in the TM5 position of β_2AR were predicted for optimal binding of partial agonists like dopamine or salbutamol, while inverse agonists and antagonists (such as carazolol or propranolol) blocked an inward move of TM5. The degree of tilt in the extracellular portion of TM5 then appears to serve as a ligand-dependent "rheostat," regulating the propensity of the receptor for activation. Mobility in TM5 may facilitate differentiation among the spectrum of full and partial agonists, antagonists, and inverse agonists. This model provides a satisfying description of ligand/receptor interactions and suggests a hypothesis for the coupling of ligand structure and receptor activation. However, polar interactions between full agonists, and serine or threonine side chains in positions 5.42, 5.43, and 5.46 of TM5 have been demonstrated for only a select group of GPCRs in the adrenergic, dopamine, and serotonin families. For other GPCR types, the preference for polar side chains in these pocket-exposed TM5 positions is not observed. As such, ligand-induced conformational changes within the binding pocket may vary between receptors and therefore require alternate procedures for modeling activation-associated conformational change.

15.2.4. VLS with High-Resolution Models: Antagonist/Agonist Selectivity

In contrast to the dynamic conformational changes involved during *in vivo* agonist recognition and receptor activation, a single static receptor conformation represents the ligand binding pocket during virtual screening. As a consequence of this rigid receptor representation, induced fit effects are overlooked, and ligands often dock incorrectly to alternate crystallized or modeled conformations of the same receptor. This phenomenon is referred to as the cross-docking problem. Previous estimates based upon a general protein/ligand benchmark set concluded that at most, 50% of ligands cross-dock correctly to a single receptor conformation [50]. This problem may be heightened for GPCRs, which exhibit multiple ligand-dependent conformational states. Even provided with a highly accurate crystallographic conformation, the extent to which a single structure will facilitate recognition of other diverse ligand types is unclear. The ligand types selected and the overall VLS yield then depend critically on the nature of the receptor conformation employed for screening.

VLS with Antagonist-Selective Models The crystal structures of bRho with retinal, β_2AR with carazolol, β_2AR with timolol, β_1AR with cyanopindolol, and AA_{2a}R with ZM241385 correspond to an inactive receptor conformation adapted for antagonist or inverse agonist binding. Homology models constructed from these templates may then prove more effective at retrieving antagonist than agonist compounds [51]. We have also recently shown that the β_2AR/carazolol structure provides high enrichment factors for antagonists, while failing to retrieve agonists [52]. To evaluate docking performance, we constructed a protonated all-atom model from the PDB-deposited coordinates and conducted VLS trials with 1000 (1K) and 14,000 (14K) ligand test sets composed of only small molecule GPCR ligands. For the 1K test set, the PDB-based β_2AR model provided an enrichment factor of 50.9 for antagonist compounds and a hit rate of 80% for the top-scoring 1% of the database (10 compounds). Similarly, an enrichment factor of 36.5 and hit rate of 38.6% were obtained for the top-scoring 1% (140 compounds) of the 14K test set (Table 15.2). High yields of known antagonist and inverse agonist compounds in the hit lists indicated that the PDB-based model is able to retrieve a number of structurally distinct small molecules –80% of the known antagonist and inverse agonist compounds contained in the 1K test set were retrieved in the top-scoring 10%. This demonstrates that a single GPCR conformation can provide excellent antagonist enrichment for a diverse set of compounds, even in the absence of empirical restraints. Recently, Topiol et al. also reported qualitatively successful antagonist VLS results for the β_2AR crystal structure using a proprietary compound database, though values for enrichment factors or yields were not provided [53]. Our quantitative results are consistent with their findings and are extended to a stringent test set that is restricted to only known GPCR ligands.

VLS with Agonist-Selective Models Previously, agonist selectivity in VLS has been demonstrated for a limited number of bRho-based GPCR homology models. In work by Bissantz and colleagues, manual rotations of TM6 were introduced to approximate receptor activation, as indicated by experimental

TABLE 15.2 Summary of VLS Results for Several β_2AR Models

	bRho-Based β_2AR Model, 1K Test Set	PDB-Based β_2AR Model, 1K Test Set	PDB-Based β_2AR Model, 14K Test Set	PDB-Based β_2AR Model, No ECL2, 1K Test Set	β_2AR Agonist Model, TM5 1-Å Shift, 1K Test Set	β_2AR Agonist Model, TM5 1-Å Shift, 14K Test Set	β_2AR Agonist Model, No ECL2, 1K Test Set	Ideal Scores, 1K Test Set	Ideal Scores, 14K Test Set
				Antagonists					
EF/hit rate (1%)	6.4/10	**50.9/80**	**36.5/38.6**	**50.9/80**	6.4/10	1.4/1.4	6.4/10	63.6/100	94.6/100
EF/hit rate (5%)	2.7/4.2	**14.6/22.9**	**10.7/11.3**	**13.3/20.8**	1.3/2.1	1.5/1.6	1.3/2.1	19.9/31.2	20.0/21.1
EF/hit rate (10%)	3.3/5.3	**8.0/12.6**	**6.4/6.8**	**6.7/10.5**	2.0/3.2	1.4/1.4	1.3/2.1	10.0/15.8	10.0/10.6
Yield (10%)	33.3	**80**	**64.2**	**66.7**	20.0	13.5	13.3	100	100
				Agonists					
EF/hit rate (1%)	0/0	0/0	1.6/2.1	0/0	**50.9/80**	**38.4/52.9**	**31.8/50**	63.6/100	72.6/100
EF/hit rate (5%)	5.3/8.3	0/0	3.5/4.9	0/0	**18.6/29.2**	**12.7/17.4**	**14.6/23**	19.9/31.2	20.0/27.6
EF/hit rate (10%)	4.0/6.3	2.7/4.2	3.3/4.5	2.0/3.2	**9.4/14.7**	**7.4/10.2**	**7.4/11.6**	10.0/15.8	10.0/13.8
Yield (10%)	40	26.7	32.6	20	**93.3**	**74.1**	**73.3**	100	100

Two test sets were evaluated: the 1K test set is composed of 954 compounds (15 antagonists/15 agonists), while the 14K test set is composed of 14,006 compounds (148 antagonists/193 agonists).

EF, enrichment factor.

evidence from electron paramagnetic resonance (EPR) and fluorescence spectroscopy studies. This was performed to generate bRho-based homology models of the agonist-bound dopamine D3, δ-opioid, and β_2AR receptors [51]. These models were successful in retrieving full agonists from a decoy set when combined with a scoring scheme that produced consensus hit lists from the results of multiple docking and scoring functions. However, the combination of docking and scoring functions used to generate the hit list was optimized separately for each receptor to yield the best hit rates, and a single, general scoring scheme independent of receptor type was not identified. Additionally, a systematic exploration of the optimum TM6 rotation was not conducted.

In our previous work (described above) with TM5 flexible models of β_2AR, we found that TM5 shifts toward the ligand binding pocket facilitated agonist docking and binding affinity calculations for a select group of compounds. To investigate whether manipulation of TM5 could generate agonist-bound models suitable for predicting binding geometries and relative binding energies for a much larger variety of ligands, we recently tested a series of TM5 shifted models in a VLS application [52]. In this protocol for agonist-bound model generation, TM5 shifts were systematically introduced in 0.5-Å increments, the ligand binding pocket side chains were optimized with the full agonist isoproterenol, and the best receptor conformation was selected by VLS enrichment (Fig. 15.3). This strategy is simply an extension of LGM, where the generation of alternate receptor conformations is modified to incorporate rigid body shifts of the TM domains. While our present strategy focuses on shifts of TM5, future work could incorporate additional shifts, tilts, or rotations of TM5, TM6, or TM7.

A set of seven hypothetical agonist-bound models was evaluated by VLS, and a 1-Å shift of TM5 toward the ligand binding pocket provided the best enrichment factors and hit rates for known agonists. Examination of the binding pocket conformation for this model finds that the TM5 shift allows formation of two hydrogen bonds between the para-hydroxyl of the catecholamine ring, and Ser203 and Ser207 while maintaining the hydrogen bond between the ligand amine and Asp113, similar to the 2-Å tilted model obtained by flexible refinement of TM5 as described above (see Section 15.2.3). Screening with this single pocket conformation achieved agonist enrichment factors of 50.9, 18.6, and 9.4 for cutoffs at the top-scoring 1%, 5%, and 10% of database molecules, respectively (Table 15.2). Inspection of the hit list found that the top-scoring 1% of the database contained a diverse set of ligand chemotypes, including both catecholamine (norepinephrine, dobutamine, dopexamine) and non-catecholamine (salmeterol, pirbuterol, sibenadet, formoterol, and fenoterol) ligands, as well as full (norepinephrine) and partial (salbutamol) agonists. The retrieval of a diverse set of agonist compounds with a single receptor conformation validates this approach for constructing other agonist-selective receptor models for use in VLS.

15.3. CONSTRUCTING AND EVALUATING HOMOLOGY MODELS OF OTHER RECEPTOR TYPES

Thus far, we have focused on the manipulation of crystal structures to describe receptor interactions with other antagonists and agonists, and conduct screening for novel modulators. However, in the vast majority of cases, a 3D model is desired for a GPCR with no available structure. For these receptors, computational modeling can indicate the amino acids lining the binding pocket cavity, suggest ligand/receptor interactions, and potentially allow screening for novel ligand chemotypes. GPCR conformational predictions must be carefully evaluated and interpreted with caution, as two previous eras of comparative modeling—the bacteriorhodopsin period from 1990 to 2000 and the 2000–2008 rhodopsin period—have demonstrated that the best available methods of structure prediction could not accurately capture the correct helical geometry and binding pocket shape for other GPCR types, let alone the conformation of ECL2 or other structurally variable regions [5]. Here we discuss relevant considerations in constructing and evaluating GPCR models for other receptor types and highlight several recent successes in the field.

15.3.1. A Note on De Novo Methods

In the absence of experimentally determined crystallographic coordinates, a number of de novo approaches to modeling GPCR structure have arisen over the past decade [8–11, 54]. Unlike comparative or homology modeling, these methods do not employ a known crystal structure as a template. Instead, de novo methods directly address the protein folding problem and attempt to construct a receptor model from the primary amino acid sequence using a combination of empirical and physical energy terms. These procedures generally begin with a coarse-grained description of the helical bundle and use empirical measures, such as hydrophobic moments or residue contact frequencies, to obtain a general orientation of the relative helix positions. In some cases, low-resolution experimental data arising from the 7.5-Å frog rhodopsin electron density maps, site-directed spin labeling combined with EPR (SDSL-EPR), or fluorescence resonance energy transfer experiments (FRET) may be included as restraints [8, 10]. This is often followed by a higher resolution refinement step that may incorporate rotamer optimization of the receptor side chains, refinement of loop orientations, minimization of the transmembrane helix kinks, and molecular dynamics simulation in the presence or absence of a lipid bilayer. In several cases, de novo models have demonstrated qualitative agreement with experimental mutagenesis results [11, 55, 56], enabled the prediction of new ligand types [57], and allowed calculation of ligand binding affinities with correlation to experimental values [58].

Importantly, the accuracy of de novo methods should be considered in the global context of general protein folding methodologies [59]. In some cases, atomic resolution may be achieved for small, globular, soluble proteins, but in general, de novo model quality has remained poor relative to models built by template-based methods [60]. As few high-resolution seven transmembrane structures exist for de novo method validation, many procedures were evaluated by their accuracy in reconstructing bRho. The Membstruk method of Goddard et al. is reported to reproduce the TM helix backbone heavy atoms of bRho within 2.8 Å, while PREDICT yields a RMSD of 2.9 Å and Bundler predicts a RMSD of 3.2 Å [8, 9, 61]. The most successful of these methodologies is TASSER, which uses a combination of template fragments obtained from the PDB with threading and refinement to obtain a final model [11]. TASSER reproduces bRho with an RMSD of 2.1 Å, even when templates with less than 30% sequence identity (SID) are removed from the PDB. However, examination of the TASSER threading results finds that approximately 72% of other modeled GPCRs use either the bRho, sensory rhodopsin, halorhodopsin, or bacteriorhodopsin structures as a template, suggesting that the best models are obtained from a procedure more akin to homology modeling and refinement, rather that true de novo modeling [11]. Additionally, as in comparative modeling, the success of TASSER still depends strongly on the quality of the template [62]. While the results reported by the above de novo methods are indeed impressive, we note that the crystallographic β_2AR and bRho structures exhibit a backbone heavy atom RMSD of 2.1 Å within the TM domains. Thus, even if no improvement was obtained from the starting template, a homology model of β_2AR derived from bRho would be within the error reported by the de novo methods. The sequence identity between β_2AR and bRho is only 22.8%; for templates with higher sequence identity to the modeled structure, the accuracy of comparative modeling methods is expected to be even greater (Table 15.3). Thus, while de novo methods can provide a useful description of receptor structure, and are critical in refining our understanding of the physical rules governing membrane protein assembly, in the present work, we focus on comparative modeling of other receptor types.

15.3.2. Criteria for Homology Model Template Selection

Since the publication of the first crystallographic structure of bRho in 2000 by Palczewski and colleagues [63], several additional bRho structures, a bovine opsin structure, and one squid rhodopsin structure have been obtained in alternate space groups and activation states [26, 44, 64–67]. Importantly, several new structures of clinically relevant GPCRs were recently published: human β_2AR, turkey β_1AR, and human AA_{2a}R [13–16, 18]. Currently, this provides a total of 14 GPCR structures, 9 of rhodopsin/opsin in various activation states, 3 of β_2AR, 1 of β_1AR, and 1 of A_{2a} [68]. The first step in homology model construction is then selection of an appropriate template structure and alignment of the target and template receptor sequences. When evaluating the available structures as homology modeling templates, the single most important consideration is sequence identity to the target protein. In Table 15.3, sequence identities to β_2AR and bRho are compared for several GPCRs of

TABLE 15.3 Sequence Identities among Several Class A GPCRs

Receptor Name (SwissProt ID)	Receptor Family	Comparison to bRho		Comparison to β_2AR		Comparison to $AA_{2a}R$		ECL2 Length
		SID TM Domains (%)	SID Binding Pocket (%)	SID TM Domains (%)	SID Binding Pocket (%)	SID TM Domains (%)	SID Binding Pocket (%)	
bRho (P02699)	Opsins	100.0	100	22.8	22.0	23.3	14.3	26
β_2-adrenergic (P07550)	Amines	22.8	22.0	100	100	33.5	9.5	25
$AA_{2a}R$ (P29274)	Adenosine	*23.3*	*14.3*	33.5	9.5	100	100	33
α_2-adrenergic (P08913)	Amines	*23.3*	*14.3*	***38.8***	***57.1***	36.4	9.1	22
5HT1A (P08908)	Amines	*22.3*	*14.3*	***42.2***	***57.1***	40.7	40.9	18
Dopamine D1 (P21728)	Amines	*20.9*	*9.5*	***44.2***	***66.7***	46.4	31.8	30
Muscarinic M2 (P08172)	Amines	*27.2*	*14.3*	***31.6***	***23.8***	29.3	31.8	22
Histamine H1 (P35367)	Amines	*19.9*	*19.1*	***37.4***	***38.1***	35.0	40.9	23
Vasopressin 1a (P37288)	Vasopeptides	20.4	23.8	26.2	33.3	29.3	18.2	25
Neurokinin 1 (P25103)	Peptides	27.2	9.5	30.1	28.6	30.0	22.7	25
μ-opioid (P35372)	Opiates	26.2	14.3	30.1	19.1	30.0	18.2	24
Type 2 angiotensin II (P50052)	Chemoattractants	25.7	19.1	26.2	19.1	28.6	22.7	27
Chemokine receptor 5 (P51681)	Chemokine	23.3	14.3	27.2	9.5	29.3	36.4	25

Receptor family annotations are taken from Reference 3.

interest. For the biogenic amine receptors, including the histamine, serotonin, muscarinic, and dopamine receptors, β_2AR or the highly similar β_1AR are clearly superior templates relative to bRho. The sequence identity between β_2AR and the human D1 dopamine receptor is 44.2% in the TM helical domains, and 66.7% when considering only the ligand binding pocket. In comparison, the sequence identity of bRho to D1 within the TM domains and ligand pocket is 20.9% and 9.5%, respectively. For other target receptors, the optimal template choice is less clear. For example, both bRho and β_2AR display approximately 20–25% SID to the human type 2 angiotensin II receptor. Despite relatively low sequence identities, several key sequence/structure motifs conserved among the Class A GPCRs can be used to reliably align the TM domains. When creating a sequence alignment for homology model construction, care should be taken that these motifs align and that gaps are not introduced within the TM domains (Fig. 15.1b). A good overview of these conserved GPCR microdomains and a general homology modeling procedure is provided in Patny et al. [69].

Alongside sequence identity, differences in receptor activation state, helix position, and loop conformation, all affect binding pocket architecture and should be considered during template selection. The structures of bRho covalently bound to the full inverse agonist retinal, β_2AR in complex with the partial inverse agonists timolol and carazolol, β_1AR in complex with the antagonist cyanopindolol, and AA_{2a}R in complex with the antagonist ZM241385 are all consistent with the inactive conformation. As described previously, VLS trials conducted with these structures, or homology models based upon them, may then be biased toward antagonist compounds [51, 52]. In contrast, the 2.9-Å structure of unliganded bovine opsin and the 3.2-Å structure of opsin in complex with a C-terminal peptide of the Gα protein subunit are thought to represent an activated conformation. In both structures, TM5, TM6, and TM7 undergo large rearrangements. Most prominently, the intracellular end of TM6 pivots outward from the helical bundle by approximately 6 Å, allowing interaction with the C-terminal Gα peptide. The helical movements serve to widen the ligand binding pocket, and the overall pocket volume (as calculated using the ICM pocket finder utility [70]) is expanded to 446 Å^3 from 425 Å^3 (Fig. 15.4). Models created using this structure as a template are then anticipated to accommodate slightly larger ligands, and perhaps provide a more accurate description of the agonist-bound receptor state.

In addition to activation-associated conformational differences, significant changes in binding pocket dimension are observed between receptor types (Fig. 15.4). For example, the interhelical distance between TM3 and TM5 is shortened by approximately 2 Å in β_2AR relative to bRho, whereas the distance between TM3 and TM7 is lengthened by approximately 1 Å. This has important implications for ligand binding in β_2AR, as the partial inverse agonists timolol and carazolol are coordinated at one end by a hydrogen bonding network between residues in TM3 and TM7, and at the other end by residues in TM5. Even small changes in helical geometry can considerably

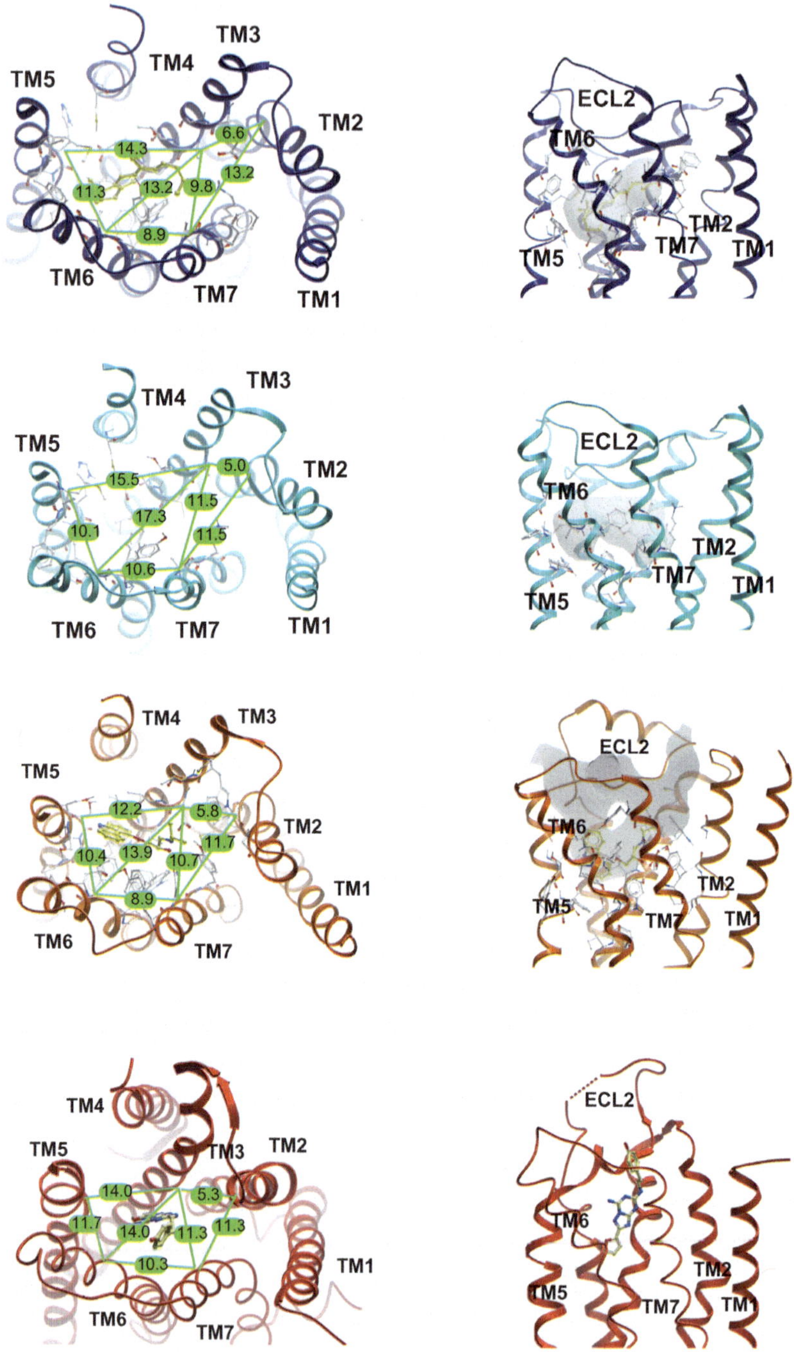
TM3
TM4
TM5
TM2
6.6
14.3
13.2
11.3
13.2
9.8
8.9
TM6
TM7
TM1
ECL2
TM6
TM2
TM5
TM7
TM1
TM4
TM3
TM5
5.0
TM2
15.5
11.5
17.3
10.1
11.5
10.6
TM6
TM7
TM1
ECL2
TM6
TM2
TM7
TM1
TM5
TM4
TM3
TM5
12.2
5.8
TM2
11.7
10.4
13.9
10.7
8.9
TM1
TM6
TM7
ECL2
TM6
TM2
TM5
TM7
TM1
TM4
TM5
TM3
TM2
14.0
5.3
11.7
14.0
11.3
11.3
10.3
TM1
TM6
TM7
ECL2
TM6
TM2
TM5
TM7
TM1

Figure 15.4 Comparison of binding pocket architecture for four GPCR structures. In the topmost panel, bRho (PDB: 1U19) is shown in navy ribbons; in the top middle panel, opsin (PDB: 3CAP) is shown in cyan ribbons; in the bottom middle panel, β_2AR (PDB: 2RH1) is shown in orange ribbons; and in the bottom panel, AA_{2a}R (PDB: 3EML) is shown in red ribbons. Binding pocket side chains and co-crystallized ligand are shown in gray and yellow sticks, respectively. The left column provides an extracellular view of the receptor; the N-terminus and ECL2 are omitted for clarity. The right column shows a side view, and the calculated surface of the binding pocket is shown as a gray surface. Green lines indicate interatomic distances; the distances are measured from the α-carbon atoms of: 117/113/84(TM3) to 211/207/185(TM5), 211/207/185(TM5) to 269/290/250(TM6), 269/290/250(TM6) to 292/312/274(TM7), 292/312/274(TM7) to 90/86/59(TM2), 90/86/59(TM2) to 117/113/84(TM3), 292/312/274(TM7) to 117/113/84 (TM3), and 269/290/250(TM6) to 117/113/84(TM3), where the first number indicates the residue ID in rhodopsin/opsin, the second in β_2AR, and the third in AA_{2a}R.

affect distance- and angle-dependent hydrogen bonding interactions, resulting in incorrect ligand docking. Further, the conformation of ECL2 is dramatically different between the bRho/opsin structures and those of β_1AR/β_2AR and AA_{2a}R. In both bRho and the unliganded opsin, ECL2 forms a β-hairpin structure that caps the ligand binding pocket and partially protrudes into the TM helical domain. In β_1AR and β_2AR, ECL2 is in an α-helical conformation tethered away from the binding pocket by an additional disulfide bond between Cys184 and Cys190. The open conformation of ECL2 observed in β_1AR/β_2AR expands the calculated volume of the ligand binding pocket to 1081 Å^3 relative to 425 Å^3 in bRho (Fig. 15.4). While most β_1AR/β_2AR ligands are accommodated within a well-defined lower pocket of the binding cavity (further within the TM bundle), longer ligands may occupy the top portion and form more extensive contacts with ECL2. Of the present GPCR structures, AA_{2a}R is the only one to describe direct interactions between ECL2 and ligand. The antagonist ZM241385 forms a pi-stacking interaction with loop residue Phe168 and a hydrogen bonding interaction with Glu169. Interestingly, the conformation of ECL2 in A_{2a} lacks the secondary structure elements seen in bRho and β_1AR/β_2AR, but is constrained by a series of three disulfide bonds. This loop, like that of β_1AR/β_2AR, forms an open structure that increases the solvent accessibility of the ligand binding pocket. In selecting a homology modeling template, known mutagenesis and SAR data can serve to define elements of the expected binding pocket geometry and guide the choice of a template more likely to capture previously experimentally defined interactions.

The solution of multiple GPCR structures enables comparison of them in docking, VLS, and homology modeling experiments. To investigate the extent to which bRho-based homology models capture important features of β_2AR structure, Constanzi docked (−)-carazolol to either a bRho-based β_2AR homology model or the β_2AR crystal structure and compared the resulting ligand

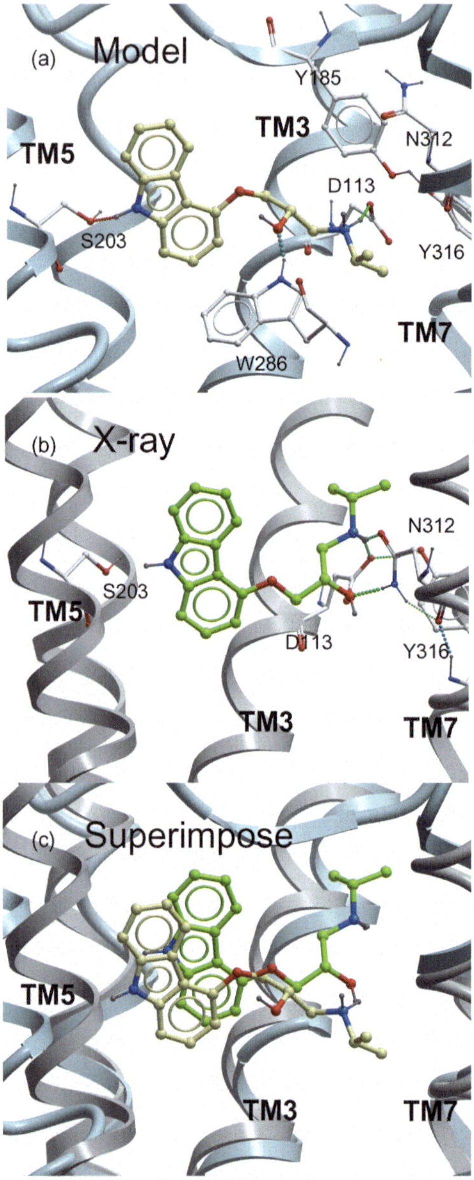

Figure 15.5 Comparison of β_2AR ligand binding pockets from a rhodopsin-based homology model and the β_2AR crystal structure. Panel (a) depicts the homology model, panel (b) depicts the crystallized conformation, and panel (c) depicts the superimposed structures. For all three panels, the receptor is shown in gray ribbon and (–)-carazolol in yellow or green sticks. TM6 is omitted for clarity.

conformations [71]. Importantly, it was found that the conformation of ECL2 in bRho was prohibitive for correct carazolol docking. If ECL2 was modeled using bRho as a template, a 3.7-Å RMSD was obtained between heavy atoms of the docked and crystallographic carazolol conformations. When the ECL2 conformation was instead predicted by a de novo protocol that removed the loop from the ligand binding pocket, the ligand RMSD was improved to 2.9 Å. Similarly, we have constructed a bRho-based β_2AR homology model using standard sequence alignment and homology modeling procedures within ICM (Fig. 15.5) [52]. The backbone conformation of bRho was kept fixed, including all loop regions, and ligand-directed refinement of the binding pocket side chains with fully flexible (–)-carazolol was performed to optimize the ligand binding pocket conformation. Following refinement, the total binding pocket RMSD between the crystal structure and homology model is 5.3 Å, when all ligand and receptor side-chain heavy atoms within 4 Å of the crystallographic conformation of carazolol are included. The RMSD for carazolol alone is 3.7 Å. In the homology model, carazolol is shifted away from the extracellular part of the GPCR and placed deeper into the TM pocket, as a result of a clash between Tyr185 in ECL2 and the crystallographic conformation of carazolol. Additionally, due to the increased distance between TM3 and TM5 in bRho, carazolol is shifted toward TM5 to allow simultaneous hydrogen bond formation with Asp113 and Ser203. Accordingly, when the VLS performance of the bRho-based β_2AR homology model and β_2AR crystal structure were compared, the homology model was found to attain far lower enrichment factors and yields (Table 15.2). These results highlight the inadequacies of the bRho backbone as a homology modeling template, both in the region of ECL2 and in the overall positioning of the TM helices. Other GPCR homology models, whether based on bRho or β_2AR, may suffer from similar discrepancies in correct conformational prediction, particularly for the highly flexible and sequence variable ECL2. Removal of this loop is one means to reduce conformational uncertainty in this region of the receptor structure; this approach and others are discussed later in the next section.

15.3.3. Structure and Modeling of Loop Regions

The seven transmembrane GPCR helices are connected by three intracellular (ICL) and three ECL loops. The shorter loops ICL1, ICL2, and ECL1 are well resolved in the crystal structures and are amenable to homology modeling for most Class A GPCRs. However, modeling of the other loop regions (ICL3, ECL3, and ECL2) is very challenging due to their conformational flexibility, large length variation, and low sequence conservation among GPCRs. For example ICL3, which is closely involved in G protein recognition and signal transduction, has a length ranging from only 4–5 residues (e.g., in CXCR4) to more than 45 residues in the β-adrenergic receptor. The structure of ICL3 in rhodopsin (~12 residues) has been resolved, albeit with low electron density and high B-factor [64]. In the β_2AR and AA_{2a}R crystal structures, the ICL3 conformation remains unknown as the loop is replaced by T4 lysozyme [14, 16]. However, the structure of the β_1AR ICL3 loop has been reported by Schertler's group (Proceedings of the GPCR2009 Congress). The structures of ICL3 for bRho and β_1AR provide a first glance at the potential structural diversity in this region; however, the biological relevance of these conformations will be better assessed in the context of co-crystallized G proteins.

The conformations of ECL3 and ECL2 are well resolved in the available crystal structures of bRho, β_2AR, β_1AR, and AA_{2a}R. Nonetheless, these structures reveal a great diversity between GPCRs in this region, particularly in ECL2 as discussed above (Fig. 15.4). Unlike the cytoplasmic loops, the ECLs are anticipated to play a major role in orthosteric small molecule binding, and ECL2 has been demonstrated to directly impact ligand specificity/selectivity for several GPCRs [72, 73]. As the results of a recent blind GPCR modeling assessment show [40], prediction of the ECL2–TM3 disulfide bond and the Phe168 side-chain conformation in AA_{2a}R are critical for accurate ligand placement in the model. The models that correctly predicted the Phe168 conformation not only correctly predicted the highest number of ligand–receptor contacts [40], but also achieved improved performance in a large-scale VLS benchmark (V. Katritch, M. Rueda, P. Lam, M. Yeager, and R. Abagyan, unpublished data). Unlike the relatively conserved downstream part of ECL2, large variations in length, sequence, and secondary structure elements of ECL2 upstream of the conserved cysteine residue preclude template-based modeling of this region, leaving ab initio modeling as the only available option for most GPCRs.

In the context of small soluble proteins, ab initio loop modeling methods can be quite accurate (RMSD < ~1.0 Å) for short loop lengths of four to six residues [74]. This indicates that several of the smaller GPCR loops may be well approximated using loop prediction methods. However, for loops of longer length, the accuracy rapidly diminishes. In a recent loop prediction benchmark, over 50% of medium length loops (seven to nine residues) had RMSDs of worse than 2 Å even when the lowest RMSD conformation was selected for comparison [74]. For membrane proteins, further inaccuracies arise, as interactions among the receptor loops, receptor termini, transmembrane helical bundle, and phospholipid membrane may affect loop conformation and are frequently not considered during the modeling procedure. These errors may be greater for GPCR homology models, as the end-to-end distances of the helices are uncertain, and may even be inappropriate for the target loop length. Recently, Mehler et al. developed a de novo loop prediction algorithm for membrane proteins that explicitly considers a loop in the context of the full protein and other predicted loop conformations. This procedure also attempts to minimize error by predicting an ensemble of compatible loop conformations rather than a single low energy structure [75]. For shorter loops, such as ICL1 and ECL1, this method was found to be quite effective, though the performance on longer loops has yet to be extensively verified [76].

Recently, "modeling by omission" has been highlighted as an important strategy for error reduction in computational models [30]. In this technique, areas of conformational uncertainty are eliminated to facilitate correct ligand docking and conformational prediction. This tactic follows from a similar strategy implemented in the recently validated Scan with Alanines and REfine (SCARE) algorithm, wherein pairs of side chains are systematically deleted to allow for induced fit docking [50]. Previous work with bRho found that retinal could be correctly docked following elimination of all receptor loops, as well as the N- and C-termini [77]. In a separate study, VLS results were compared for three GPCR homology models wherein ECL2 was either modeled de novo or deleted. In two of the three cases, the ECL2-deleted models outperformed the de novo loop conformations, leading the authors to suggest that ECL2 be eliminated in homology models except in circumstances where extensive experimental restraints are available [78]. We have also assessed the impact of this loop on β_2AR docking by conducting VLS trials with ECL2-deleted versions of the β_2AR crystal structure and agonist-bound TM5 shifted model described earlier in the chapter (Table 15.2). The absence of ECL2 did not affect the docked conformation of either (–)-carazolol with the crystal structure or (–)-isoproterenol with the agonist-bound model. Carazolol docked to the ECL2 truncated version of the receptor with an RMSD of 0.23 Å to the crystallized orientation, while the RMSD of isoproterenol docked to the loop-deleted and intact agonist-bound models was 0.16 Å. Additionally, omission of ECL2 had a relatively small impact on VLS performance (Table 15.2). Antagonist enrichment factors for the top-scoring 1% and 5% of the 1K test set is comparable in the presence and absence of ECL2, though the total yield of antagonists in the top-scoring 10% was decreased, from 80% to 66.7%. Enrichment factors were also slightly decreased for the agonist-bound model in VLS. Nonetheless, VLS in the absence of ECL2 provides high enrichment factors and hit rates, with enrichment factors of 50.9 (antagonists) and 31.8 (agonists) for the top-scoring 1% of the 1K test set. This indicates that omission of ECL2 from receptor homology models is a suitable alternative when: (1) mutagenesis data indicate minimal interactions between ligand and ECL2; and (2) experimental data are unable to provide accurate conformational information.

15.3.4. GPCR Model Validation and Evaluation

Validation of GPCR models by direct comparison to crystallographic data is hindered by a limited number of structures. Nonetheless, the currently available set of GPCR coordinates provides a basic benchmark for testing modeling protocols, and indirect experimental evidence of ligand/receptor interactions can be used to evaluate predictions for uncrystallized receptors. For receptors with several characterized ligands, VLS is an additional metric for assessing binding pocket conformation. Here, we address procedures for validating predicted GPCR conformations in light of the available experimental data.

New modeling protocols should first be tested with several benchmark cases based upon crystallographic data before application to other ligand/receptor complexes. For example, an accurate docking algorithm should reproduce ("self-dock") the conformations of retinal within the bRho ligand binding pocket, carazolol and timolol with the corresponding β_2AR structures, cyanopindolol with β_1AR, and ZM241385 in complex with AA_{2a}R. ICM has been demonstrated to dock retinal to the bRho pocket within an RMSD of 0.2 Å to the crystallographic coordinates [77]. Similarly, (−)-carazolol is docked to β_2AR with an RMSD of 0.3 Å, and calculation of the relative binding affinities of (−)-carazolol and (+)-carazolol with ICM provides a rationalization for β_2AR stereoselectivity [22]. A stringent second test would then be to cross-dock several known ligands to a single structure to check for consistency in ligand binding pose and ligand/receptor interactions. Additionally, if optimization of the receptor side chains is to be employed, it should be verified that the refinement procedure converges on a minimum energy conformation resembling that of the starting crystal structure. Once this has been achieved, the docking and refinement procedure may be confidently applied to other receptor types.

In the absence of a high-resolution structure for comparison, indirect measures can be used to assess model accuracy. One metric of correct binding pocket conformation and docked ligand pose is the presence of intermolecular ligand/receptor hydrogen bonds. Unsatisfied ligand donor and acceptor atoms result in decreased binding affinity and may indicate an incorrectly docked conformation. Additionally, predicted binding pocket conformation should be globally consistent with experimental mutagenesis results. In the ideal case, experimental binding affinities are available for several ligand analogs with a panel of receptor mutants. Such data provide evidence for specific contacts between a particular receptor side chain and ligand functional group, and can even be used to impose distance restraints during LGM. However, mutagenesis data are often subject to alternate interpretations. A receptor mutation that decreases the apparent agonist binding affinity may act directly by removing an important agonist/receptor interaction or indirectly by affecting the receptor activation mechanism. Further, mutagenesis data alone are often insufficient for evaluating ligand pose. Limited experimental results may indicate an approximate binding location without specifying the ligand orientation. The conformation of long and conformationally flexible ligands is particularly difficult to validate solely via mutagenesis experiments. Fentanyl, a powerful analgesic that selectively targets the μ-opioid receptor, contains seven rotatable bonds. Despite numerous modeling studies and the availability of extensive mutagenesis and SAR data, the binding mode and molecular determinants of fentanyl recognition remain poorly defined [79]. In such situations, the comparison of binding modes for multiple docked ligands of varying chemotypes and complementarity of the binding pocket conformation with pharmacophore models can improve confidence in a predicted conformation. For example, when modeling the complex between arylpiperazine ligands and the $5HT_{1A}$ receptor, the known ligand/receptor interaction points are symmetric, preventing unambiguous assignment of correct ligand binding orientation [80]. By docking several conformationally constrained arylpiperazine derivatives and comparing binding conformation with several other diverse ligand chemotypes, Nowak et al. proposed a consistent binding mode for a $5HT_{1A}$ homology model.

Finally, an accurate GPCR model should provide high enrichment factors and hit rates in VLS trials. This demonstrates that the binding pocket conformation allows recognition for a set of diverse antagonists (or agonists) and energetic partition of these ligands from a set of decoys. For many GPCRs, little mutagenesis or SAR data exist to specify probable ligand/receptor interactions, and VLS may be the primary criteria used for model selection and validation (see Section 15.2.2). Several online databases provide compilations of known GPCR ligands that can be used in constructing a VLS test set, including the KiDB, GPCR ligand database (GLIDA), and DrugBank [81–83]. VLS enrichment for known ligands from a random test set has been demonstrated for several GPCR models, including the dopamine D3, cannabinoid CB_2, alpha1a adrenergic, and neurokinin I receptors [84–87]. When available, ligand/receptor interactions or pharmacophore information may be combined with VLS as a prescreening filter, ligand–receptor restraints, or a component of the docked ligand scoring function. This integrated approach facilitates correct ligand docking and enhances VLS enrichment factors, but requires prior experimental knowledge.

15.3.5. Ligand Subtype Selectivity in GPCR Models

Promiscuous ligand binding among GPCR subtypes is a frequent obstacle to the design of safe and effective drugs [88]. This issue is particularly critical for clinically relevant GPCR families (including the adrenergic, adenosine, and dopamine receptors), where several functionally distinct receptor subtypes share a conserved endogenous ligand. For example, the β_1AR, β_2AR, and β_3AR receptors are all activated by the endogenous ligand norepinephrine. β-blocker medications that target the β_1AR receptor mediate a desirable antihypertensive effect and are frequently used in the treatment of high blood pressure. However, cross-reactivity of these compounds with the β_2AR receptor results in airway constriction and asthma-like symptoms. Thus, small molecule antagonists with improved β_1AR versus β_2AR specificity are desired. Optimization of specificity is challenging due to high levels of sequence conservation within the core ligand binding pocket residues. This limits the efficacy of traditional ligand-based pharmacophore and comparative SAR modeling. While minor variations to an endogenous ligand scaffold have been reported to yield subtype specificity without expressly targeting nonconserved regions of the ligand binding pocket [89], these effects are often subtle and achieved at the expense of reduced ligand affinity.

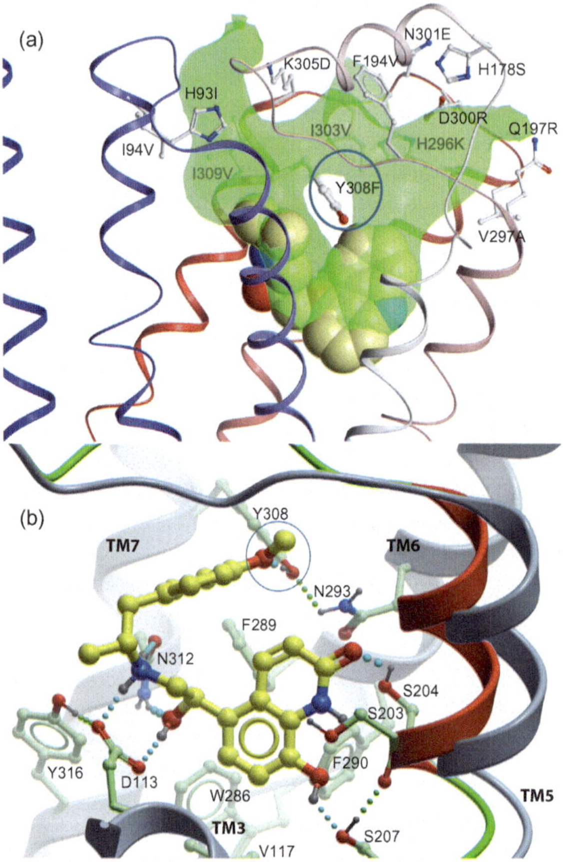

Figure 15.6 (a) Residue conservation in the ligand binding pocket of the β-adrenergic GPCRs. The β_2AR ligand binding cavity as calculated by ICM PocketFinder is shown as a transparent green surface. Side chains lining the cavity are shown as sticks, with labels indicating variations between β_2AR and β_1AR in corresponding positions. The core binding pocket is defined by bound carazolol (shown in CPK), and Y308F (red label) is the only side-chain variation in a 5-Å radius. (b) Predicted binding conformation for the β_2AR-selective agonist TA-2005 (depicted in yellow sticks). This conformation was generated by LGM of the β_2AR binding pocket with flexible side chains and TM5 backbone. The flexible regions are colored green. The hydrogen bond between the *p*-methoxyphenyl group of TA-2005 and Tyr308$^{7.35}$ (circled) is implicated in β_2AR selectivity of TA-2005.

While the core ligand binding pocket residues are typically conserved among members of the same GPCR type, non-subtype-conserved side chains at the periphery of the binding site can be exploited for specificity design. This approach necessitates a detailed understanding of the binding pocket structure and ligand/receptor interactions. For example, comparison of the β_2- and β_1AR receptor structures finds that Tyr308$^{7.35}$ of β_2AR (Phe in the β_1AR) is the only nonconserved side chain in a 5-Å radius of the core binding site (Fig. 15.6a). Interestingly, the β_2AR Y308F mutation has been implicated as a specificity determinant for several synthetic β_2AR ligands, including the high-affinity full agonist TA-2005 [90, 91]. To analyze the basis for TA-2005 specificity at the molecular level, this ligand was docked to a TM5 flexible model of β_2AR (see Section 15.2.3) [22]. The lowest energy conformation of the complex includes a hydrogen bond between the oxygen of the *p*-methoxyphenyl group of TA-2005 and Tyr308$^{7.35}$, in addition to other ligand/receptor interactions common for β_2AR full agonists (Fig. 15.6b). The model predicted approximately 4-fold and 15-fold decreases in affinity for TA-2005 in complex with Y308F and Y308A, in agreement with the values reported [91]. As detailed in Fig. 15.6a, several other nonconserved amino acids are positioned at the extracellular entrance to the β_2AR binding pocket, and these may be targeted for rational design of ligand specificity in the future.

The tendency for common binding core is also prominent for the adenosine receptor family, where a set of shared chemical scaffolds gives rise to highly selective ligands for all four different subtypes (A_1, A_{2a}, A_{2b}, and A_3) [92, 93]. Analysis of the $AA_{2a}R$ crystal structure and homology models for the A_3 and A_1 subtypes reveals conservation of a core receptor binding pocket that interacts with the ligand scaffold and significant side-chain variations on the pocket periphery that confer subtype selectivity (V. Katritch, unpublished data). In particular, replacement of the $AA_{2a}R$ Glu169 side chain, which forms a hydrogen bond with the exocyclic amine in most ligand scaffolds, with Val (as in AA_3R) opens up the ligand binding pocket and permits bulky substitutions at one of the amine hydrogens; such substituted ligands exhibit greater than ~1000-fold selectivity toward AA_3R.

15.4. MODELING GPCR FUNCTIONAL FEATURES—ANALYSIS OF ACTIVATION AND SIGNALING

While accurate modeling of the ligand binding pocket is essential for docking and screening, a global (even if low resolution) understanding of the GPCR activation and signaling mechanism would provide insight into the physiological action of GPCR drugs and identify potential allosteric sites as targets for drug discovery. The downstream effects of ligand recognition are also important for deciphering the structural basis of functional selectivity, wherein distinct ligands are able to induce different patterns of receptor activity [94]. For example, both dopamine and norepinephrine are reasonably good agonists of β_2AR-mediated G protein activation, while only norepinephrine is efficient at inducing β_2AR internalization [13]. Design of functionally selective ligands may result in drug candidates with greater efficacy and fewer side effects, but requires identification of the structural determinants of receptor activation, internalization, and G protein coupling [95]. While modeling of GPCR activation and multimeric GPCR/G protein complexes remains in its infancy, such work provides a platform for the development of further hypotheses.

15.4.1. Activation-Related Conformational Changes in GPCRs

In lieu of a high-resolution crystal structure for a fully activated GPCR, activation-related changes in rhodopsin and other receptors have been extensively probed using indirect biochemical and biophysical methods [94, 96, 97]. These data provide a set of empirical restraints for modeling receptor activation. However, comprehensive modeling of the GPCR activation mechanism has yet to overcome multiple hurdles. GPCR activation is a dynamic, multistage, and multicomponent process, involving not only large-scale rearrangements of loops and local flexibility of TM helices, but also the binding of G proteins, as well as structured water, ions, and lipid moieties. Further, while experimentally derived constraints are frequently employed to facilitate accurate model creation, these data are often inconsistent due to variations of GPCR behavior in different assays, cell types, and culture conditions.

The TM domains exhibit a high degree of structural similarity and include several universally conserved features, such as the D(E)RY motif, suggesting the possibility of a common activation mechanism conserved among GPCR types [43, 96]. Further experimental and theoretical evaluation in different receptors is required to establish those conformational changes that are receptor specific and those which are part of a universal activation mechanism. Indeed, the outward dislocation of the TM6 cytoplasmic end has been indicated by numerous indirect biochemical and spectroscopic studies, as reviewed in Park et al. [43]. The recently solved crystal structures of opsin also confirm a significant 6–7-Å outward displacement of the TM6 cytoplasmic end [44, 98]. This conformational change in TM6 forms a binding site for a G-α C-terminal peptide, which forms additional contacts with the D(E)RY motif [98]. The general requirement for G protein interaction suggests that this motion of the TM6 cytoplasmic portion is a common feature of GPCR activation. However, other conformational rearrangements such as the "rotamer toggle switch" may be less universal. The "toggle switch" mechanism was initially suggested based on an observed correlation between rotamer changes in the conserved amino acid W6.48 and enhancement of the adjacent proline kink in TM6 [96, 99–101]. While the TM6 proline kink angle is dramatically increased in both ligand-free opsin structures, no rotamer switch in W6.48 or other aromatic residues of the binding site was observed [44, 98]. One must therefore be judicious in incorporating particular experimental restraints or structural features during model development.

Despite the complexity of structural rearrangements in GPCRs and the paucity of their structural data availability, several modeling approaches have attempted to reproduce and/or predict activation-related conformational change in different GPCR families. The majority of these studies explicitly use experimental data as structural constraints to guide model optimization or select "best" models from a large set generated by molecular dynamic (MD) or molecular mechanic (MC) calculations. In work by the Fanelli group, comparative MD simulations of several wild-type, inactivated, and constitutively active receptor mutants were performed to reproduce functional features of the receptor active state [102]. This study focused on the D(E)RY ionic lock of the α1b-adrenergic receptor and found that a weakening of interactions with the D(E)RY arginine and an increase in solvent accessibility at the cytoplasmic surface correlated with receptor activation. However, additional biochemical and computational evidence suggests that the D(E)RY motif should be considered in the context of supramolecular complexes with Gα and/or other GPCR homodimers/heterodimers, rather than in calculations considering only a single GPCR monomer [103–105]. New horizons for analysis of activation-related conformational changes in GPCRs have been opened by the recently solved high-resolution structure of ligand-free opsin and the Ops*–Gα-CT peptide complex [44, 98]. The Ops*–Gα-CT complex may also prove useful for deciphering the structural basis of Gα selectivity and building more reliable models of GPCR-G-αβγ activation complexes.

15.4.2. Macromolecular Complexes of GPCRs

GPCR signaling is achieved via interaction with several classes of molecules, including G proteins, arrestin, and receptor activity-modifying proteins (RAMPs) [106]. In addition, many GPCRs may form homodimers and/or heterodimers, though the functional significance of GPCR oligomerization remains unclear [107–109]. Formation of GPCR macromolecular complexes has been studied in situ by atomic force microscopy (AFM) [110, 111], as well as by indirect approaches, such as disulfide cross-linking, site-directed mutagenesis, and spectroscopy [112–115]. These experimental data provide a starting point for the computational modeling of GPCR functional complexes.

A semiempirical model of rhodopsin oligomerization based on AFM-derived geometry was proposed that includes dimer interface contacts formed by TM4 and TM5, as well as residues in ICL2 and in the C-terminal region [111]. Also, an attempt to build a complete model of the rhodopsin signaling complex was performed by combining G_t heterotrimer structure with the predicted functional tetramer of rhodopsin [105], though the stoichiometry and helical arrangement of this model are questionable (e.g., an axial rotation of TM6 by 90°). Fanelli and coworkers generated putative models of dopamine (D2)/adenosine (A_{2a}) GPCR heterodimers [102] and lutropin receptor homodimers [116], by rigid body docking of predicted monomer models. The initial docking models were clustered, filtered, and ranked according to an empirical index defined by membrane topology, described by the normalized tilt angle and z-offset for the receptor [117]. The homodimer models generated by the method suggest a binding interface formed by TMs 4, 5, and 6 without involvement of loop regions, while heterodimer modeling could not reliably distinguish between two alternative configurations. Other studies from Fanelli and coworkers combined mutation data with structural modeling to suggest a mechanism of G protein binding and activation for several GPCRs. The common feature of these models, that is, binding of the C-terminus of $G\alpha_{q\alpha}$ to the conserved D(E)RY motif, accessible only in the "open" conformation of the activated receptor [103, 104, 118, 119], has been recently confirmed in high-resolution crystal structure of the Ops*–Gα-CT complex [98].

While these models deviate significantly in the details of intermolecular contacts and conformations of protein subunits, they provide tangible hypotheses for stoichiometry, orientation, and binding interfaces in GPCR signaling complexes useful for the design of further experimental tests. Additional spectroscopy, NMR, cryo-electron microscopy, and crystallography experiments are needed to resolve ambiguous structural and functional details of these complexes. More accurate methods for conformational modeling of membrane proteins [120, 121], which combine energy-based global optimization techniques with EM and other experimental restraints, may augment this analysis and improve its accuracy.

15.5. BEYOND CLASS A: MODELING OF OTHER GPCR FAMILIES

A multitude of clinically relevant GPCRs can be found outside of the Class A GPCR family. The calcitonin, glucagon, and parathyroid hormone receptors in the *secretin* (Class B) family and calcium-sensing, metabotropic glutamate, and GABA binding receptors in the *glutamate* (Class C) family are validated drug discovery targets [122]. Several additional GPCRs in these families are considered putative therapeutic targets, including several recently deorphanized receptors implicated in cancer and other pathologies [123]. Unlike Class A (rhodopsin like) receptors, the majority of Classes B and C receptors bind their endogenous ligands at the N-terminal extracellular domain. However, additional allosteric sites have been identified within the TM domains of these receptors that at least partially overlap with the retinal binding cavity in the rhodopsin [124]. Allosteric modulators may exhibit improved drug selectivity, as these allosteric sites are expected to be more structurally diverse than the N-terminal orthosteric ligand binding pocket [125, 126]. Additionally, overall structural similarity in the TM domains of all GPCRs may facilitate the application of rational design approaches developed for Class A receptors to other GPCR families and orphan GPCRs.

15.5.1. Modeling Secretin (Class B) Family GPCRs

The *secretin* (Class B) family comprises at least 15 GPCRs that share significant sequence similarity and an extracellular hormone-binding amino-terminal domain. Structures of isolated amino-termini for corticotropin-releasing factor (CRF), pituitary adenylate cyclase-activating polypeptide (hPAC1), gastric inhibitory polypeptide (GIP), and parathyroid hormone (PTH1) receptors have been recently solved by NMR and crystallography [127–129]. Though highly divergent in amino acid sequence, the Class B extracellular domains share a similar protein fold of about ~90 amino acids, stabilized by three conserved disulfide bonds, allowing each to interact with their respective hormone peptides in a relatively shallow binding groove. The latter three structures were characterized in complex with peptide hormones and may serve as structural templates for the rational design of orthosteric agonists and antagonists.

Computational modeling of the full-length secretin receptor was recently performed to investigate interactions among the receptor 7TM core, receptor N-terminus, and peptide ligand [130–132]. The complete model was constructed using the crystallographic N-terminal domain structures of CRF and GIP receptors [127, 128], homology models of the TM helical bundle, and distance constraints derived from photoaffinity labeling and FRET experiments [131–133]. The modeling indicates that the extracellular domain of the secretin receptor directs the peptide N-terminus toward the opening in the TM7 helical bundle. The refined model also suggests details of the receptor 7TM core interaction with a Trp-Asp-Asn epitope ($W^{48}D^{49}N^{50}$) in the receptor N-terminal domain, known to be involved in activation of several Class B GPCRs (Fig. 15.7).

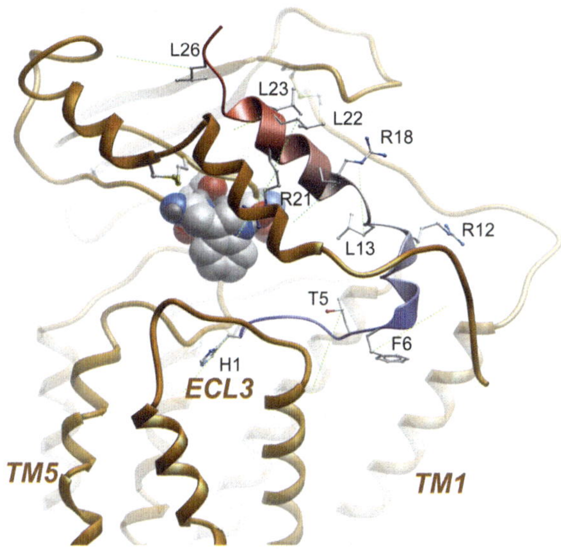

Isoproterenol

Figure 15.7 A secretin receptor model, generated by energy-based peptide docking. The model accommodates 10 photoaffinity labeling constraints and the three disulfide bonds demonstrated to exist in the secretin receptor. The full receptor is represented in orange ribbon, the secretin peptide is colored blue–red from the amino-terminus to the carboxyl terminus, and the 10 photoaffinity labeling constraints are represented as green dotted lines. The cross-linked residues on the secretin peptide are displayed in thin sticks. Residues of the secretin peptide are labeled in black. The proposed N-terminal domain epitope, implicated in secretin receptor activation (amino acids $W^{48}D^{49}N^{50}$), is shown in CPK.

Though the open shape and flexibility of the orthosteric pocket favor design of peptide-like ligands, small molecule modulators are sought that target allosteric sites in either the TM helical bundle or the interface between the extracellular and TM domains. In the past decade, a number of small molecule allosteric modulators of Class B GPCRs have been reported, most of them antagonists of the CRF (CRF1) [134–138]. Some of these orally available compounds have shown promising results in preclinical models of anxiety and depression [139], and initial clinical studies [140], suggesting the general utility of targeting allosteric sites in Class B GPCR modulation.

15.5.2. Glutamate/Class C

The Class C GPCR family comprises 22 human receptors [141], including such therapeutically important targets as the metabotropic glutamate receptors (GRM1-8) [142–144], GABA binding receptors (GABABRs), and the calcium-sensing receptor (CaR). For example, the $GABA_B$ agonist baclofen is used in the clinic as an anticonvulsant [122], while the allosteric CASR modulator cinacalcet is known to normalize calcium level in patients with hyperparathyroidism [145].

Class C receptors bind their endogenous ligands within a large (>500 amino acids) extracellular N-terminal domain. Crystal structures of this domain have recently been solved for GRM1, GRM3, and GRM7 [146–148]. This series of structures provides a high-resolution description of small molecule agonist and antagonist binding at an orthosteric pocket located between two subdomains (LB1 and LB2) connected by a flexible hinge. These structures provide an accurate template for structure-based design of small molecule ligands targeting the orthosteric site; however, an abundance of glutamate binding receptors in the CNS makes ligand specificity a serious concern. Nonetheless, homology modeling of the extracellular domain of other Class C receptors finds substantial diversity in the size and shape of the ligand binding pocket, suggesting that molecules substantially larger than glutamate can be used as selective Class C GPCR ligands [149, 150]. In one recent study, docking of tripeptides into the models based on crystal structure of the mGluR1 extracellular domain allowed identification of a specific orthosteric agonist of the CaR [151]. Docking in the orthosteric site of GRM1 has also been employed for rationalizing the agonistic or antagonistic action of a newly identified ligand series [152, 153]. Rational design strategies exploiting the new high-resolution structure of an extended glutamate orthosteric site [148] are likely to bring about other highly selective ligands with improved pharmacokinetic features and superior therapeutic potential.

While endogenous ligands bind the N-terminal domain of Class C GPCRs, many synthetic ligands identified in binding assays are noncompetitive. These compounds target an allosteric site between helices TM3-TM5-TM6-TM7, approximately corresponding to the binding cavity of rhodopsin-like GPCRs [142, 154–157]. Allosteric modulation has become the de facto approach to targeting Class C GPCRs, with cinacalcet (NPS-1493) being the first clinically approved drug of its kind [158]. Though most previous applications to discovery of Class C allosteric modulators employed a ligand-based approach, 3D modeling and docking is critical for understanding ligand interaction patterns with the allosteric site and the design of more selective inhibitors [159, 160]. Encouraging results were obtained in studies using a "ligand supported" homology model of GRM5 and a fingerprint-based scoring function, where a reasonable enrichment factor (~25-fold at 1% cutoff) was achieved in retrospective screening of a diverse compound library [161]. In another study, selective retrieval of agonist and antagonists in a small benchmark test was reported using AutoDock (http://autodock.scripps.edu/) and ArgusLab (http://www.arguslab.com/) docking; models of active and inactive GRMs in this study were generated from the bRho structure, though details of homology modeling and binding pocket refinement were not reported [162].

15.5.3. Orphan GPCRs

Over 120 GPCRs (60 in Class A) remain classified as orphans [123], that is, their endogenous ligands (and in many cases, synthetic ligands) have not been identified. There is a great interest in the functional and structural analysis of orphan receptors, which often leads to identification of new therapeutic targets. Limited structural information and lack of ligand binding data make orphan GPCRs especially hard targets for structure-based modeling.

Chemogenomics approaches allow clustering of orphan GPCRs with well-described receptors and attempt to extrapolate known SAR correlations between ligand activity and receptor/ligand structure to the orphan receptor [163]. Such approaches are not always useful in the search for endogenous ligands, as the pharmacological and phylogenetic classification of GPCRs is often paradoxically different. For example, GPCRs that recognize small molecule ligands versus peptides are not clearly separated in a phylogenetic tree. However, a detailed analysis of residue conservation in the 7TM binding cavity can greatly facilitate a focused search for synthetic, often allosteric, agonists and antagonists [3, 4]. Knowledge-based models have been trained on large sets of GPCR ligand binding data using machine learning techniques (e.g., support vector machines, or SVM) and generalized protein sequence and "ligand feature" representation [164]. Note that these bioinformatics approaches could greatly benefit from understanding the 3D context of the binding pocket, even for relatively low-resolution orphan GPCR models. For example, the accuracy of the knowledge-based models may be improved by using information about side chains lining the binding pocket and spatial distances between them. Though the direct application of small molecule docking to GPCR deorphanization has yet to be reported, accumulation of relevant structural templates and improved modeling techniques can eventually lead to more accurate orphan GPCR models for use in in silico ligand prediction. Induced fit modeling techniques, such as SCARE, may prove especially useful for predicting the bound conformations of new ligands [50].

15.6. SUMMARY AND CONCLUSIONS

The solution of GPCR crystal structures is a technically challenging, financially costly, and labor-intensive process. As additional structures are determined, it is critical that computational methodologies be in place to maximize the benefit of the experimental data to the drug discovery community. Energy-based conformational refinement of existing crystallographic data is an important first step toward generating a protonated, full-atom model suitable for docking. LGM with judicious use of experimental restraints can extend a given structure to describe interactions for other ligand/receptor complexes and GPCR types. As discussed here, very accurate conformational predictions may be made that enable selective VLS for agonists or antagonists. Computational modeling for orphan GPCRs and the Classes B and C receptor families provides a lower resolution view of the receptor structure and allosteric binding pocket; these models may facilitate drug discovery when paired with pharmacophore or SAR data. Finally, though significant gaps remain in our understanding of the mechanisms underlying receptor activation and signaling, computational modeling of these functional features provides a platform for examining experimental data and developing new hypotheses.

ACKNOWLEDGMENTS

The authors would like to especially thank our long-term collaborators Professors Patrick Sexton, Arthur Christopoulos, and Laurence J. Miller as well as their lab members, John Simms and Nathan Hall for in-depth discussions about GPCRs and GPCR modeling. Polo Lam and Irina Kufareva are also thanked for their help and contributions.

REFERENCES

1. Fredriksson, R., Lagerstrom, M.C., Lundin, L.-G., Schioth, H.B. (2003) The G-protein-coupled receptors in the human genome form five main families. Phylogenetic analysis, paralogon groups, and fingerprints. *Mol Pharmacol.* 63, 1256–1272.
2. Strachan, R.T., Ferrara, G., Roth, B.L. (2006) Screening the receptorome: An efficient approach for drug discovery and target validation. *Drug Discov Today.* 11, 708–716.
3. Surgand, J.S., Rodrigo, J., Kellenberger, E., Rognan, D. (2006) A chemogenomic analysis of the transmembrane binding cavity of human G-protein-coupled receptors. *Proteins.* 62, 509–538.
4. Jacoby, E. (2006) Designing compound libraries targeting GPCRs. *Ernst Schering Found Symp Proc.* 2, 93–103.
5. Oliveira, L., Hulsen, T., Hulsik, D.L., Paiva, A.C.M., Vriend, G. (2004) Heavier-than-air flying machines are impossible. *FEBS Lett.* 564, 269–273.

6. Bissantz, C., Wolfgang, C.S., Stahl, G.M. (2005) Focused library design in GPCR projects on the example of 5-HT2c agonists: Comparison of structure-based virtual screening with ligand-based search methods. *Proteins*. 61, 938–952.
7. Evers, A., Hessler, G., Matter, H., Klabunde, T. (2005) Virtual screening of biogenic amine-binding G-protein coupled receptors: Comparative evaluation of protein- and ligand-based virtual screening protocols. *J Med Chem*. 48, 5448–5465.
8. Sale, K., Faulon, J.-L., Gray, G.A., Schoeniger, J.S., Young, M.M. (2004) Optimal bundling of transmembrane helices using sparse distance constraints. *Protein Sci*. 13, 2613–2627.
9. Shacham, S., Marantz, Y., Bar-Haim, S., Kalid, O., Warshaviak, D., Avisar, N., Inbal, B., Heifetz, A., Fichman, M., Topf, M., Naor, Z., Noiman, S., Becker, O.M. (2004) PREDICT modeling and in-silico screening for G-protein coupled receptors. *Proteins*. 57, 51–86.
10. Vaidehi, N., Floriano, W.B., Trabanino, R., Hall, S.E., Freddolino, P., Choi, E.J., Zamanakos, G., Goddard III, W.A. (2002) Prediction of structure and function of G protein-coupled receptors. *Proc Natl Acad Sci U S A*. 99, 12622–12627.
11. Zhang, Y., DeVries, M.E., Skolnick, J. (2006) Structure modeling of all identified G protein coupled receptors in the human genome. *PLoS Comput Biol*. 2, e13.
12. Rosenbaum, D.M., Cherezov, V., Hanson, M.A., Rasmussen, S.G.F., Thian, F.S., Kobilka, T.S., Choi, H.-J., Yao, X.-J., Weis, W.I., Stevens, R.C., Kobilka, B.K. (2007) GPCR engineering yields high-resolution structural insights into β2-adrenergic receptor function. *Science*. 318, 266–273.
13. Rasmussen, S.G.F., Choi, H.-J., Rosenbaum, D.M., Kobilka, T.S., Thian, F.S., Edwards, P.C., Burghammer, M., Ratnala, V.R.P., Sanishvili, R., Fischetti, R.F., Schertler, G.F.X., Weis, W.I., Kobilka, B.K. (2007) Crystal structure of the human β2 adrenergic G-protein-coupled receptor. *Nature*. 450, 383–387.
14. Cherezov, V., Rosenbaum, D.M., Hanson, M.A., Rasmussen, S.G.F., Thian, F.S., Kobilka, T.S., Choi, H.-J., Kuhn, P., Weis, W.I., Kobilka, B.K., Stevens, R.C. (2007) High-resolution crystal structure of an engineered human β2-adrenergic G protein coupled receptor. *Science*. 318, 1258–1265.
15. Warne, T., Serrano-Vega, M.J., Baker, J.G., Moukhametzianov, R., Edwards, P.C., Henderson, R., Leslie, A.G., Tate, C.G., Schertler, G.F. (2008) Structure of a β1-adrenergic G-protein-coupled receptor. *Nature*. 454, 486–491.
16. Jaakola, V.P., Griffith, M.T., Hanson, M.A., Cherezov, V., Chien, E.Y., Lane, J.R., Ijzerman, A.P., Stevens, R.C. (2008) The 2.6 angstrom crystal structure of a human A2A adenosine receptor bound to an antagonist. *Science*. 322, 1211–1217.
17. Audet, M., Bouvier, M. (2008) Insights into signaling from the β2-adrenergic receptor structure. *Nat Chem Biol*. 4, 397–403.
18. Hanson, M., Cherezov, V., Roth, C., Griffith, M., Jaakola, V., Chien, E., Velasquez, J., Kuhn, P., Stevens, R. (2008) A specific cholesterol binding site is established by the 2.8 Å structure of the human β2-adrenergic receptor. *Structure*. 16, 897–905.
19. Kleywegt, G.J. (2007) Separating model optimization and model validation in statistical cross-validation as applied to crystallography. *Acta Crystallogr D Biol Crystallogr*. 63, 939–940.
20. Kleywegt, G.J. (2007) Crystallographic refinement of ligand complexes. *Acta Crystallogr D Biol Crystallogr*. 63, 94–100.

21. Weichenberger, C.X., Sippl, M.J. (2007) NQ-Flipper: Recognition and correction of erroneous asparagine and glutamine side-chain rotamers in protein structures. *Nucleic Acids Res*. 35, W403–W406.
22. Katritch, V., Reynolds, K.A., Cherezov, V., Hanson, M.A., Roth, C.B., Yeager, M., Abagyan, R. (2009) Analysis of full and partial agonists binding to b2-adrenergic receptor suggests a role of transmembrane helix V in agonist-specific conformational changes. *J Mol Recognit*. 22, 307–318.
23. Abagyan, R., Totrov, M. (1994) Biased probability Monte Carlo conformational searches and electrostatic calculations for peptides and proteins. *J Mol Biol*. 235, 983–1002.
24. Kusnetzow, A.K., Altenbach, C., Hubbell, W.L. (2006) Conformational states and dynamics of rhodopsin in micelles and bilayers. *Biochemistry*. 45, 5538–5550.
25. Ward, S.D., Hamdan, F.F., Bloodworth, L.M., Siddiqui, N.A., Li, J.H., Wess, J. (2006) Use of an in situ disulfide cross-linking strategy to study the dynamic properties of the cytoplasmic end of transmembrane domain VI of the M3 muscarinic acetylcholine receptor. *Biochemistry*. 45, 676–685.
26. Salom, D., Lodowski, D.T., Stenkamp, R.E., Le Trong, I., Golczak, M., Jastrzebska, B., Harris, T., Ballesteros, J.A., Palczewski, K. (2006) Crystal structure of a photoactivated deprotonated intermediate of rhodopsin. *Proc Natl Acad Sci U S A*. 103, 16123–16128.
27. Lodowski, D.T., Salom, D., Le Trong, I., Teller, D.C., Ballesteros, J.A., Palczewski, K., Stenkamp, R.E. (2007) Crystal packing analysis of rhodopsin crystals. *J Struct Biol*. 158, 455–462.
28. Brylinski, M., Skolnick, J. (2008) Q-Dock: Low-resolution flexible ligand docking with pocket-specific threading restraints. *J Comput Chem*. 29, 1574–1588.
29. Erickson, J.A., Jalaie, M., Robertson, D.H., Lewis, R.A., Vieth, M. (2004) Lessons in molecular recognition: The effects of ligand and protein flexibility on molecular docking accuracy. *J Med Chem*. 47, 45–55.
30. Totrov, M., Abagyan, R. (2008) Flexible ligand docking to multiple receptor conformations: A practical alternative. *Curr Opin Struct Biol*. 18, 178–184.
31. Kufareva, I., Laudet, B., Cochet, C., Abagyan, R. (2008) Structure-based discovery of small molecules that modulate kinase activity by disrupting the subunit interaction: Application to CK2. *Protein Sci*. 17, 265.
32. Totrov, M., Abagyan, R. (1997) Flexible protein-ligand docking by global energy optimization in internal coordinates. *Proteins*. 29, 215–220.
33. Abagyan, R., Totrov, M., Kuznetsov, D. (1994) ICM—A new method for protein modeling and design: Applications to docking and structure prediction from the distorted native conformation. *J Comput Chem*. 15, 488–506.
34. Totrov, M., Abagyan, R. (1994) Detailed ab initio prediction of lysozyme-antibody complex with 1.6 A accuracy. *Nat Struct Mol Biol*. 1, 259–263.
35. Cavasotto, C.N., Abagyan, R.A. (2004) Protein flexibility in ligand docking and virtual screening to protein kinases. *J Mol Biol*. 337, 209–225.
36. Totrov, M., Abagyan, R. (1999) Derivation of sensitive discrimination potential for virtual ligand screening. In *Proceedings of the 3rd Annual International Conference of Computational Molecular Biology (RECOMB 99)*. ACM Press, Lyon, France.
37. Bisson, W.H., Cheltsov, A.V., Bruey-Sedano, N., Lin, B., Chen, J., Goldberger, N., May, L.T., Christopoulos, A., Dalton, J.T., Sexton, P.M., Zhang, X.-K., Abagyan, R.

(2007) Discovery of antiandrogen activity of nonsteroidal scaffolds of marketed drugs. *Proc Natl Acad Sci U S A*. 104, 11927–11932.

38. Cavasotto, C.N., Orry, A.J.W., Murgolo, N.J., Czarniecki, M.F., Kocsi, S.A., Hawes, B.E., Neill, K.A., Hine, H., Burton, M.S., Voigt, J.H., Abagyan, R.A., Bayne, M.L., Monsma, F.J. (2008) Discovery of novel chemotypes to a G-protein-coupled receptor through ligand-steered homology modeling and structure-based virtual screening. *J Med Chem*. 51, 581–588.
39. Christopoulos, A. (2008) Structure-function meets ligand-directed modeling: Towards novel muscarinic receptor chemotypes. *Recent Advances in Muscarinic Receptor Pharmacology and Therapeutics*. ASPET Symposium, San Diego, CA.
40. Michino, M., Abola, E., Brooks III, C., Dixon, J.S., Moult, J., Stevens, R.C. (2009) Critical assessment of GPCR structure modeling and docking 2008. *Nat Rev Drug Discov*. 8, 455–463.
41. Katritch, V., Rueda, M., Lam, P. C., Yeager, M., Abagyan, R. (2010) GPCR 3D homology models for ligand screening: lessons learned from blind predictions of adenosine A2a receptor complex. *Proteins*. 78, 197–211.
42. Kobilka, B.K. (2007) G protein coupled receptor structure and activation. *Biochim Biophys Acta*. 1768, 794–807.
43. Park, P.S., Lodowski, D.T., Palczewski, K. (2008) Activation of G protein-coupled receptors: Beyond two-state models and tertiary conformational changes. *Annu Rev Pharmacol Toxicol*. 48, 107–141.
44. Park, J.H., Scheerer, P., Hofmann, K.P., Choe, H.-W., Ernst, O.P. (2008) Crystal structure of the ligand-free G-protein-coupled receptor opsin. *Nature*. 454, 183–187.
45. de Graaf, C., Rognan, D. (2008) Selective structure-based virtual screening for full and partial agonists of the beta2 adrenergic receptor. *J Med Chem*. 51, 4978–4985.
46. Strader, C.D., Candelore, M.R., Hill, W.S., Sigal, I.S., Dixon, R.A. (1989) Identification of two serine residues involved in agonist activation of the β-adrenergic receptor. *J Biol Chem*. 264, 13572–13578.
47. Ambrosio, C., Molinari, P., Cotecchia, S., Costa, T. (2000) Catechol-binding serines of β (2)-adrenergic receptors control the equilibrium between active and inactive receptor states. *Mol Pharmacol*. 257, 198–210.
48. Liapakis, G., Ballesteros, J.A., Papachristou, S., Chan, W.C., Chen, X., Javitch, J.A. (2000) The forgotten serine. A critical role for Ser-2035.42 in ligand binding to and activation of the β2-adrenergic receptor. *J Biol Chem*. 275, 37779–37788.
49. Liapakis, G., Chan, W.C., Papadokostaki, M., Javitch, J.A. (2004) Synergistic contributions of the functional groups of epinephrine to its affinity and efficacy at the β2 adrenergic receptor. *Mol Pharmacol*. 65, 1181–1190.
50. Bottegoni, G., Kufareva, I., Totrov, M., Abagyan, R. (2008) A new method for ligand docking to flexible receptors by dual alanine scanning and refinement (SCARE). *J Comput Aided Mol Des*. 22, 311–325.
51. Bissantz, C., Bernard, P., Hibert, M., Rognan, D. (2003) Protein-based virtual screening of chemical databases. II. Are homology models of g-protein coupled receptors suitable targets? *Proteins*. 50, 5–25.
52. Reynolds, K.A., Katritch, V., Abagyan, R. (2009) Identifying conformational changes of the beta(2) adrenoceptor that enable accurate prediction of ligand/

receptor interactions and screening for GPCR modulators. *J Comput Aided Mol Des*. 23, 273–288.

53. Topiol, S., Sabio, M. (2008) Use of the X-ray structure of the beta2-adrenergic receptor for drug discovery. *Bioorg Med Chem Lett*. 18, 1598–1602.
54. Donnelly, D., Overington, J.P., Blundell, T.L. (1994) The prediction and orientation of {alpha}-helices from sequence alignments: The combined use of environment-dependent substitution tables, Fourier transform methods and helix capping rules. *Protein Eng*. 7, 645–653.
55. Freddolino, P.L., Kalani, M.Y.S., Vaidehi, N., Floriano, W.B., Hall, S.E., Trabanino, R.J., Kam, V.W.T., Goddard III, W.A. (2004) Predicted 3D structure for the human {beta}2 adrenergic receptor and its binding site for agonists and antagonists. *Proc Natl Acad Sci U S A*. 101, 2736–2741.
56. Kalani, M.Y.S., Vaidehi, N., Hall, S.E., Trabanino, R.J., Freddolino, P.L., Kalani, M.A., Floriano, W.B., Kam, V.W.T., Goddard III, W.A. (2004) The predicted 3D structure of the human D2 dopamine receptor and the binding site and binding affinities for agonists and antagonists. *Proc Natl Acad Sci U S A*. 101, 3815–3820.
57. Becker, O.M., Marantz, Y., Shacham, S., Inbal, B., Heifetz, A., Kalid, O., Bar-Haim, S., Warshaviak, D., Fichman, M., Noiman, S. (2004) G protein-coupled receptors: In silico drug discovery in 3D. *Proc Natl Acad Sci U S A*. 101, 11304–11309.
58. Vaidehi, N., Schlyer, S., Trabanino, R.J., Floriano, W.B., Abrol, R., Sharma, S., Kochanny, M., Koovakat, S., Dunning, L., Liang, M., Fox, J.M., de Mendonca, F.L., Pease, J.E., Goddard III, W.A., Horuk, R. (2006) Predictions of CCR1 chemokine receptor structure and BX 471 antagonist binding followed by experimental validation. *J Biol Chem*. 281, 27613–27620.
59. Elofsson, A., Heijne, G.V. (2007) Membrane protein structure: Prediction versus reality. *Annu Rev Biochem*. 76, 125–140.
60. Ralf, J. (2007) HCYPRKNDC. Assessment of CASP7 structure predictions for template free targets. *Proteins*. 69, 57–67.
61. Trabanino, R.J., Hall, S.E., Vaidehi, N., Floriano, W.B., Kam, V.W.T., Goddard III, W.A. (2004) First principles predictions of the structure and function of G-protein-coupled receptors: Validation for bovine rhodopsin. *Biophys J*. 86, 1904–1921.
62. Zhang, Y., Arakaki, A.K., Skolnick, J. (2005) TASSER: An automated method for the prediction of protein tertiary structures in CASP6. *Proteins*. 61, 91–98.
63. Palczewski, K., Kumasaka, T., Hori, T., Behnke, C.A., Motoshima, H., Fox, B.A., Trong, I.L., Teller, D.C., Okada, T., Stenkamp, R.E., Yamamoto, M., Miyano, M. (2000) Crystal structure of rhodopsin: A G protein-coupled receptor. *Science*. 289, 739–745.
64. Li, J., Edwards, P.C., Burghammer, M., Villa, C., Schertler, G.F.X. (2004) Structure of bovine rhodopsin in a trigonal crystal form. *J Mol Biol*. 343, 1409–1438.
65. Murakami, M., Kouyama, T. (2008) Crystal structure of squid rhodopsin. *Nature*. 453, 363–367.
66. Okada, T., Sugihara, M., Bondar, A.-N., Elstner, M., Entel, P., Buss, V. (2004) The retinal conformation and its environment in rhodopsin in light of a new 2.2 Å crystal structure. *J Mol Biol*. 342, 571–583.
67. Standfuss, J., Xie, G., Edwards, P.C., Burghammer, M., Oprian, D.D., Schertler, G.F.X. (2007) Crystal structure of a thermally stable rhodopsin mutant. *J Mol Biol*. 372, 1179–1188.

68. White, S. (ed.) (2008) *Membrane Proteins of Known 3D Structure*, Vol. 2008. Irvine: Laboratory at UC Irvine.
69. Patny, A., Desai, P.V., Avery, M.A. (2006) Homology modeling of G-protein-coupled receptors and implications in drug design. *Curr Med Chem*. 13, 1667–1691.
70. An, J., Totrov, M., Abagyan, R. (2005) Pocketome via comprehensive identification and classification of ligand binding envelopes. *Mol Cell Proteomics*. 4, 752–761.
71. Costanzi, S. (2008) On the applicability of GPCR homology models to computer-aided drug discovery: A comparison between in silico and crystal structures of the beta2-adrenergic receptor. *J Med Chem*. 51, 2907–2914.
72. Avlani, V.A., Gregory, K.J., Morton, C.J., Parker, M.W., Sexton, P.M., Christopoulos, A. (2007) Critical role for the second extracellular loop in the binding of both orthosteric and allosteric G protein-coupled receptor ligands. *J Biol Chem*. 282, 25677–25686.
73. Shi, L., Javitch, J.A. (2004) The second extracellular loop of the dopamine D2 receptor lines the binding-site crevice. *Proc Natl Acad Sci U S A*. 101, 440–445.
74. Rossi, K.A., Weigelt, C.A., Nayeem, A., Krystek Jr., S.R. (2007) Loopholes and missing links in protein modeling. *Protein Sci*. 16, 1999–2012.
75. Mehler, E.L., Hassan, S.A., Kortagere, S., Weinstein, H. (2006) Ab initio computational modeling of loops in G-protein-coupled receptors: Lessons from the crystal structure of rhodopsin. *Proteins*. 64, 673–690.
76. Kortagere, S., Roy, A., Mehler, E. (2006) Ab initio computational modeling of long loops in G-protein coupled receptors. *J Comput Aided Mol Des*. 20, 427–436.
77. Cavasotto, C.N., Orry, A.J.W., Abagyan, R.A. (2003) Structure-based identification of binding sites, native ligands and potential inhibitors for G-protein coupled receptors. *Proteins*. 51, 423–433.
78. de Graaf, C., Foata, N., Engkvist, O., Rognan, D. (2008) Molecular modeling of the second extracellular loop of G-protein coupled receptors and its implication on structure-based virtual screening. *Proteins*. 71, 599–620.
79. Kane, B.E., Svensson, B., Ferguson, D.M. (2006) Molecular recognition of opioid receptor ligands. *AAPS J*. 8, E126–E137.
80. Nowak, M., Kolaczkowski, M., Pawlowski, M., Bojarski, A.J. (2006) Homology modeling of the serotonin 5-HT1A receptor using automated docking of bioactive compounds with defined geometry. *J Med Chem*. 49, 205–214.
81. Okuno, Y., Tamon, A., Yabuuchi, H., Niijima, S., Minowa, Y., Tonomura, K., Kunimoto, R., Feng, C. (2008) GLIDA: GPCR ligand database for chemical genomics drug discovery database and tools update. *Nucleic Acids Res*. 36, D907–D912.
82. Roth, B.L., Kroeze, W.K., Patel, S., Lopez, E. (2000) The multiplicity of serotonin receptors: Uselessly diverse molecules or an embarrassment of riches? *Neuroscientist*. 6, 252–262.
83. Wishart, D.S., Knox, C., Guo, A.C., Cheng, D., Shrivastava, S., Tzur, D., Gautam, B., Hassanali, M. (2008) DrugBank: A knowledgebase for drugs, drug actions and drug targets. *Nucleic Acids Res*. 36, D901–D906.
84. Chen, J.Z., Wang, J., Xie, X.Q. (2007) GPCR structure-based virtual screening approach for CB2 antagonist search. *J Chem Inf Model*. 47, 1626–1637.
85. Evers, A., Klabunde, T. (2005) Structure-based drug discovery using GPCR homology modeling: Successful virtual screening for antagonists of the alpha1A adrenergic receptor. *J Med Chem*. 48, 1088–1097.

86. Evers, A., Klebe, G. (2004) Successful virtual screening for a submicromolar antagonist of the neurokinin-1 receptor based on a ligand-supported homology model. *J Med Chem*. 47, 5381–5392.

87. Varady, J., Wu, X., Fang, X., Min, J., Hu, Z., Levant, B., Wang, S. (2003) Molecular modeling of the three-dimensional structure of dopamine 3 (D3) subtype receptor: Discovery of novel and potent D3 ligands through a hybrid pharmacophore- and structure-based database searching approach. *J Med Chem*. 46, 4377–4392.

88. Jacoby, E., Bouhelal, R., Gerspacher, M., Seuwen, K. (2006) The 7 TM G-protein-coupled receptor target family. *ChemMedChem*. 1, 761–782.

89. Emorine, L., Marullo, S., Briend-Sutren, M., Patey, G., Tate, K., Delavier-Klutchko, C., Strosberg, A. (1989) Molecular characterization of the human β3-adrenergic receptor. *Science*. 245, 1118–1121.

90. Kikkawa, H., Kurose, H., Isogaya, M., Sato, Y., Nagao, T. (1997) Differential contribution of two serine residues of wild type and constitutively active β2-adrenoceptors to the interaction with beta2-selective agonists. *Br J Pharmacol*. 121, 1059–1064.

91. Kikkawa, H., Isogaya, M., Nagao, T., Kurose, H. (1998) The role of the seventh transmembrane region in high affinity binding of a beta 2-selective agonist TA-2005. *Mol Pharmacol*. 53, 128–134.

92. Cacciari, B., Bolcato, C., Spalluto, G., Klotz, K., Bacilieri, M., Deflorian, F., Moro, S. (2007) Pyrazolo-triazolo-pyrimidines as adenosine receptor antagonists: A complete structure-activity profile. *Purinergic Signal*. 3, 183–193.

93. Jacobson, K.A., Gao, Z.-G. (2006) Adenosine receptors as therapeutic targets. *Nat Rev Drug Discov*. 5, 247–264.

94. Hoffmann, C., Zurn, A., Bunemann, M., Lohse, M.J. (2008) Conformational changes in G-protein-coupled receptors—The quest for functionally selective conformations is open. *Br J Pharmacol*. 153, S358–S366.

95. Kenakin, T. (2003) Ligand-selective receptor conformations revisited: The promise and the problem. *Trends Pharmacol Sci*. 24, 346–354.

96. Schwartz, T.W., Frimurer, T.M., Holst, B., Rosenkilde, M.M., Elling, C.E. (2006) Molecular mechanism of 7TM receptor activation—A global toggle switch model. *Annu Rev Pharmacol Toxicol*. 46, 481–519.

97. Kobilka, B.K., Deupi, X. (2007) Conformational complexity of G-protein-coupled receptors. *Trends Pharmacol Sci*. 28, 397–406.

98. Scheerer, P., Park, J.H., Hildebrand, P.W., Kim, Y.J., Krausz, N., Choe, H.-W., Hofmann, K.P., Ernst, O.P. (2008) Crystal structure of opsin in its G-protein-interacting conformation. *Nature*. 455, 497–502.

99. Shi, L., Liapakis, G., Xu, R., Guarnieri, F., Ballesteros, J.A., Javitch, J.A. (2002) Beta2 adrenergic receptor activation. Modulation of the proline kink in transmembrane 6 by a rotamer toggle switch. *J Biol Chem*. 277, 40989–40996.

100. McAllister, S.D., Hurst, D.P., Barnett-Norris, J., Lynch, D., Reggio, P.H., Abood, M.E. (2004) Structural mimicry in class A G protein-coupled receptor rotamer toggle switches: The importance of the F3.36(201)/W6.48(357) interaction in cannabinoid CB1 receptor activation. *J Biol Chem*. 279, 48024–48037.

101. Crocker, E., Eilers, M., Ahuja, S., Hornak, V., Hirshfeld, A., Sheves, M., Smith, S.O. (2006) Location of Trp265 in metarhodopsin II: Implications for the activation mechanism of the visual receptor rhodopsin. *J Mol Biol*. 357, 163–72.

102. Fanelli, F., De Benedetti, P.G. (2006) Inactive and active states and supramolecular organization of GPCRs: Insights from computational modeling. *J Comput Aided Mol Des*. 20, 449–461.

103. Angelova, K., Fanelli, F., Puett, D. (2008) Contributions of intracellular loops 2 and 3 of the lutropin receptor in Gs coupling. *Mol Endocrinol*. 22, 126–138.

104. Raimondi, F., Seeber, M., Benedetti, P.G., Fanelli, F. (2008) Mechanisms of inter- and intramolecular communication in GPCRs and G proteins. *J Am Chem Soc*. 130, 4310–4325.

105. Filipek, S., Krzysko, K.A., Fotiadis, D., Liang, Y., Saperstein, D.A., Engel, A., Palczewski, K. (2004) A concept for G protein activation by G protein-coupled receptor dimers: The transducin/rhodopsin interface. *Photochem Photobiol Sci*. 3, 628–638.

106. Sexton, P.M., Morfis, M., Tilakaratne, N., Hay, D.L., Udawela, M., Christopoulos, G., Christopoulos, A. (2006) Complexing receptor pharmacology: Modulation of family B G protein-coupled receptor function by RAMPs. *Ann N Y Acad Sci*. 1070, 90–104.

107. Milligan, G. (2008) A day in the life of a G protein-coupled receptor: The contribution to function of G protein-coupled receptor dimerization. *Br J Pharmacol*. 153, S216–S229.

108. Whorton, M.R., Jastrzebska, B., Park, P.S., Fotiadis, D., Engel, A., Palczewski, K., Sunahara, R.K. (2008) Efficient coupling of transducin to monomeric rhodopsin in a phospholipid bilayer. *J Biol Chem*. 283, 4387–4394.

109. Gurevich, V.V., Gurevich, E.V. (2008) GPCR monomers and oligomers: It takes all kinds. *Trends Neurosci*. 31, 74–81.

110. Fotiadis, D., Liang, Y., Filipek, S., Saperstein, D.A., Engel, A., Palczewski, K. (2003) Atomic-force microscopy: Rhodopsin dimers in native disc membranes. *Nature*. 421, 127–128.

111. Fotiadis, D., Liang, Y., Filipek, S., Saperstein, D.A., Engel, A., Palczewski, K. (2004) The G protein-coupled receptor rhodopsin in the native membrane. *FEBS Lett*. 564, 281–288.

112. Milligan, G., Canals, M., Pediani, J.D., Ellis, J., Lopez-Gimenez, J.F. (2006) The role of GPCR dimerisation/oligomerisation in receptor signalling. *Ernst Schering Found Symp Proc*. 2, 145–161.

113. Casado, V., Cortes, A., Ciruela, F., Mallol, J., Ferre, S., Lluis, C., Canela, E.I., Franco, R. (2007) Old and new ways to calculate the affinity of agonists and antagonists interacting with G-protein-coupled monomeric and dimeric receptors: The receptor-dimer cooperativity index. *Pharmacol Ther*. 116, 343–354.

114. Franco, R., Casado, V., Cortes, A., Mallol, J., Ciruela, F., Ferre, S., Lluis, C., Canela, E.I. (2008) G-protein-coupled receptor heteromers: Function and ligand pharmacology. *Br J Pharmacol*. 153, S90–S98.

115. Harikumar, K.G., Morfis, M.M., Sexton, P.M., Miller, L.J. (2008) Pattern of intra-family hetero-oligomerization involving the G-protein-coupled secretin receptor. *J Mol Neurosci*. 36, 279–285.

116. Fanelli, F. (2007) Dimerization of the lutropin receptor: Insights from computational modeling. *Mol Cell Endocrinol*. 260–262, 59–64.

117. Casciari, D., Seeber, M., Fanelli, F. (2006) Quaternary structure predictions of transmembrane proteins starting from the monomer: A docking-based approach. *BMC Bioinformatics*. 7, 340.

118. Fanelli, F., Dell'orco, D. (2008) Dark and photoactivated rhodopsin share common binding modes to transducin. *FEBS Lett.* 582, 991–996.

119. Puett, D., Li, Y., DeMars, G., Angelova, K., Fanelli, F. (2007) A functional transmembrane complex: The luteinizing hormone receptor with bound ligand and G protein. *Mol Cell Endocrinol.* 260–262, 126–136.

120. Kovacs, J.A., Baker, K.A., Altenberg, G.A., Abagyan, R., Yeager, M. (2007) Molecular modeling and mutagenesis of gap junction channels. *Prog Biophys Mol Biol.* 94, 15–28.

121. Kovacs, J.A., Yeager, M., Abagyan, R. (2007) Computational prediction of atomic structures of helical membrane proteins aided by EM maps. *Biophys J.* 93, 1950–1959.

122. Tyndall, J., Sandilya, R. (2005) GPCR agonists and antagonists in the clinic. *Med Chem.* 1, 405–421.

123. Lagerstrom, M.C., Schioth, H.B. (2008) Structural diversity of G protein-coupled receptors and significance for drug discovery. *Nat Rev Drug Discov.* 7, 339–357.

124. Soudijn, W., Van Wijngaarden, I., Ap, I.J. (2004) Allosteric modulation of G protein-coupled receptors: Perspectives and recent developments. *Drug Discov Today.* 9, 752–758.

125. Langmead, C.J., Christopoulos, A. (2006) Allosteric agonists of 7TM receptors: Expanding the pharmacological toolbox. *Trends Pharmacol Sci.* 27, 475–481.

126. Raddatz, R., Schaffhauser, H., Marino, M.J. (2007) Allosteric approaches to the targeting of G-protein-coupled receptors for novel drug discovery: A critical assessment. *Biochem Pharmacol.* 74, 383–391.

127. Grace, C.R., Perrin, M.H., DiGruccio, M.R., Miller, C.L., Rivier, J.E., Vale, W.W., Riek, R. (2004) NMR structure and peptide hormone binding site of the first extracellular domain of a type B1 G protein-coupled receptor. *Proc Natl Acad Sci U S A.* 101, 12836–12841.

128. Parthier, C., Kleinschmidt, M., Neumann, P., Rudolph, R., Manhart, S., Schlenzig, D., Fanghanel, J., Rahfeld, J.U., Demuth, H.U., Stubbs, M.T. (2007) Crystal structure of the incretin-bound extracellular domain of a G protein-coupled receptor. *Proc Natl Acad Sci U S A.* 104, 13942–13947.

129. Pioszak, A.A., Xu, H.E. (2008) Molecular recognition of parathyroid hormone by its G protein-coupled receptor. *Proc Natl Acad Sci U S A.* 105, 5034–5039.

130. Dong, M., Ding, X.Q., Thomas, S.E., Gao, F., Lam, P.C., Abagyan, R., Miller, L.J. (2007) Role of lysine187 within the second extracellular loop of the type A cholecystokinin receptor in agonist-induced activation. Use of complementary charge-reversal mutagenesis to define a functionally important interdomain interaction. *Biochemistry.* 46, 4522–4531.

131. Dong, M., Lam, P.C., Pinon, D.I., Sexton, P.M., Abagyan, R., Miller, L.J. (2008) Spatial approximation between secretin residue five and the third extracellular loop of its receptor provides new insight into the molecular basis of natural agonist binding. *Mol Pharmacol.* 74, 413–422.

132. Harikumar, K.G., Lam, P.C., Dong, M., Sexton, P.M., Abagyan, R., Miller, L.J. (2007) Fluorescence resonance energy transfer analysis of secretin docking to its receptor: Mapping distances between residues distributed throughout the ligand pharmacophore and distinct receptor residues. *J Biol Chem.* 282, 32834–32843.

133. Dong, M., Lam, P.C., Gao, F., Hosohata, K., Pinon, D.I., Sexton, P.M., Abagyan, R., Miller, L.J. (2007) Molecular approximations between residues 21 and 23 of secretin and its receptor: Development of a model for peptide docking with the amino terminus of the secretin receptor. *Mol Pharmacol.* 72, 280–290.

134. Chen, C., Dagnino Jr., R., De Souza, E.B., Grigoriadis, D.E., Huang, C.Q., Kim, K.I., Liu, Z., Moran, T., Webb, T.R., Whitten, J.P., Xie, Y.F., McCarthy, J.R. (1996) Design and synthesis of a series of non-peptide high-affinity human corticotropin-releasing factor1 receptor antagonists. *J Med Chem.* 39, 4358–4360.

135. Chen, C., Dagnino Jr., R., Huang, C.Q., McCarthy, J.R., Grigoriadis, D.E. (2001) 1-Alkyl-3-amino-5-aryl-1H-[1, 2, 4]triazoles: Novel synthesis via cyclization of N-acyl-S-methylisothioureas with alkylhydrazines and their potent corticotropin-releasing factor-1 (CRF(1)) receptor antagonist activities. *Bioorg Med Chem Lett.* 11, 3165–3168.

136. Chen, C., Wilcoxen, K.M., Huang, C.Q., McCarthy, J.R., Chen, T., Grigoriadis, D.E. (2004) Optimization of 3-phenylpyrazolo[1, 5-a]pyrimidines as potent corticotropin-releasing factor-1 antagonists with adequate lipophilicity and water solubility. *Bioorg Med Chem Lett.* 14, 3669–3673.

137. Hoare, S.R., Fleck, B.A., Gross, R.S., Crowe, P.D., Williams, J.P., Grigoriadis, D.E. (2008) Allosteric ligands for the corticotropin releasing factor type 1 receptor modulate conformational states involved in receptor activation. *Mol Pharmacol.* 73, 1371–1380.

138. Huang, C.Q., Wilcoxen, K.M., Grigoriadis, D.E., McCarthy, J.R., Chen, C. (2004) Design and synthesis of 3-(2-pyridyl)pyrazolo[1, 5-a]pyrimidines as potent CRF1 receptor antagonists. *Bioorg Med Chem Lett.* 14, 3943–3947.

139. St-Denis, Y., Di Fabio, R., Bernasconi, G., Castiglioni, E., Contini, S., Donati, D., Fazzolari, E., Gentile, G., Ghirlanda, D., Marchionni, C., Messina, F., Micheli, F., Pavone, F., Pasquarello, A., Sabbatini, F.M., Zampori, M.G., Arban, R., Vitulli, G. (2005) Substituted tetraazaacenaphthylenes as potent CRF1 receptor antagonists for the treatment of depression and anxiety. *Bioorg Med Chem Lett.* 15, 3713–3716.

140. Ising, M., Zimmermann, U.S., Kunzel, H.E., Uhr, M., Foster, A.C., Learned-Coughlin, S.M., Holsboer, F., Grigoriadis, D.E. (2007) High-affinity CRF1 receptor antagonist NBI-34041: Preclinical and clinical data suggest safety and efficacy in attenuating elevated stress response. *Neuropsychopharmacology.* 32, 1941–1949.

141. Bjarnadottir, T.K., Fredriksson, R., Schioth, H.B. (2005) The gene repertoire and the common evolutionary history of glutamate, pheromone (V2R), taste(1) and other related G protein-coupled receptors. *Gene.* 362, 70–84.

142. Gasparini, F., Kuhn, R., Pin, J.P. (2002) Allosteric modulators of group I metabotropic glutamate receptors: Novel subtype-selective ligands and therapeutic perspectives. *Curr Opin Pharmacol.* 2, 43–49.

143. Lavreysen, H., Dautzenberg, F.M. (2008) Therapeutic potential of group III metabotropic glutamate receptors. *Curr Med Chem.* 15, 671–684.

144. Recasens, M., Guiramand, J., Aimar, R., Abdulkarim, A., Barbanel, G. (2007) Metabotropic glutamate receptors as drug targets. *Curr Drug Targets.* 8, 651–681.

145. Shoback, D.M., Bilezikian, J.P., Turner, S.A., McCary, L.C., Guo, M.D., Peacock, M. (2003) The calcimimetic cinacalcet normalizes serum calcium in subjects with primary hyperparathyroidism. *J Clin Endocrinol Metab.* 88, 5644–5649.

146. Kunishima, N., Shimada, Y., Tsuji, Y., Sato, T., Yamamoto, M., Kumasaka, T., Nakanishi, S., Jingami, H., Morikawa, K. (2000) Structural basis of glutamate recognition by a dimeric metabotropic glutamate receptor. *Nature*. 407, 971–977.

147. Tsuchiya, D., Kunishima, N., Kamiya, N., Jingami, H., Morikawa, K. (2002) Structural views of the ligand-binding cores of a metabotropic glutamate receptor complexed with an antagonist and both glutamate and Gd3+. *Proc Natl Acad Sci U S A*. 99, 2660–2665.

148. Muto, T., Tsuchiya, D., Morikawa, K., Jingami, H. (2007) Structures of the extracellular regions of the group II/III metabotropic glutamate receptors. *Proc Natl Acad Sci U S A*. 104, 3759–3764.

149. Belenikin, M.S., Baskin, I.I., Costantino, G., Palyulin, V.A., Pellicciari, R., Zefirov, N.S. (2002) Comparative analysis of the ligand-binding sites of the metabotropic glutamate receptors mGluR1-mGluR8. *Dokl Biochem Biophys*. 386, 251–256.

150. Tropsha, A., Wang, S.X. (2006) QSAR modeling of GPCR ligands: Methodologies and examples of applications. *Ernst Schering Found Symp Proc*. 2, 49–73.

151. Wang, M., Yao, Y., Kuang, D., Hampson, D.R. (2006) Activation of family C G-protein-coupled receptors by the tripeptide glutathione. *J Biol Chem*. 281, 8864–8870.

152. Pellicciari, R., Filosa, R., Fulco, M.C., Marinozzi, M., Macchiarulo, A., Novak, C., Natalini, B., Hermit, M.B., Nielsen, S., Sager, T.N., Stensbol, T.B., Thomsen, C. (2006) Synthesis and preliminary biological evaluation of 2′-substituted 2-(3′-carboxybicyclo[1.1.1]pentyl)glycine derivatives as group I selective metabotropic glutamate receptor ligands. *ChemMedChem*. 1, 358–365.

153. Pellicciari, R., Marinozzi, M., Macchiarulo, A., Fulco, M.C., Gafarova, J., Serpi, M., Giorgi, G., Nielsen, S., Thomsen, C. (2007) Synthesis, molecular modeling studies, and preliminary pharmacological characterization of all possible 2-(2′-sulfonocyclopropyl)glycine stereoisomers as conformationally constrained L-homocysteic acid analogs. *J Med Chem*. 50, 4630–4641.

154. Lavreysen, H., Janssen, C., Bischoff, F., Langlois, X., Leysen, J.E., Lesage, A.S. (2003) [3H]R214127: A novel high-affinity radioligand for the mGlu1 receptor reveals a common binding site shared by multiple allosteric antagonists. *Mol Pharmacol*. 63, 1082–1093.

155. Malherbe, P., Kratochwil, N., Zenner, M.T., Piussi, J., Diener, C., Kratzeisen, C., Fischer, C., Porter, R.H. (2003) Mutational analysis and molecular modeling of the binding pocket of the metabotropic glutamate 5 receptor negative modulator 2-methyl-6-(phenylethynyl)-pyridine. *Mol Pharmacol*. 64, 823–832.

156. Malherbe, P., Kratochwil, N., Muhlemann, A., Zenner, M.T., Fischer, C., Stahl, M., Gerber, P.R., Jaeschke, G., Porter, R.H. (2006) Comparison of the binding pockets of two chemically unrelated allosteric antagonists of the mGlu5 receptor and identification of crucial residues involved in the inverse agonism of MPEP. *J Neurochem*. 98, 601–615.

157. Muhlemann, A., Ward, N.A., Kratochwil, N., Diener, C., Fischer, C., Stucki, A., Jaeschke, G., Malherbe, P., Porter, R.H. (2006) Determination of key amino acids implicated in the actions of allosteric modulation by 3,3′-difluorobenzaldazine on rat mGlu5 receptors. *Eur J Pharmacol*. 529, 95–104.

158. Ray, K., Tisdale, J., Dodd, R.H., Dauban, P., Ruat, M., Northup, J.K. (2005) Calindol, a positive allosteric modulator of the human Ca(2+) receptor, activates an extra-

cellular ligand-binding domain-deleted rhodopsin-like seven-transmembrane structure in the absence of Ca(2+). *J Biol Chem*. 280, 37013–37020.

159. Kratochwil, N.A., Malherbe, P., Lindemann, L., Ebeling, M., Hoener, M.C., Muhlemann, A., Porter, R.H., Stahl, M., Gerber, P.R. (2005) An automated system for the analysis of G protein-coupled receptor transmembrane binding pockets: Alignment, receptor-based pharmacophores, and their application. *J Chem Inf Model*. 45, 1324–1336.
160. Noeske, T., Jirgensons, A., Starchenkovs, I., Renner, S., Jaunzeme, I., Trifanova, D., Hechenberger, M., Bauer, T., Kauss, V., Parsons, C.G., Schneider, G., Weil, T. (2007) Virtual screening for selective allosteric mGluR1 antagonists and structure-activity relationship investigations for coumarine derivatives. *ChemMedChem*. 2, 1763–1773.
161. Radestock, S., Weil, T., Renner, S. (2008) Homology model-based virtual screening for GPCR ligands using docking and target-biased scoring. *J Chem Inf Model*. 48, 1104–1117.
162. Yanamala, N., Tirupula, K.C., Klein-Seetharaman, J. (2008) Preferential binding of allosteric modulators to active and inactive conformational states of metabotropic glutamate receptors. *BMC Bioinformatics*. 9, S16.
163. Jacoby, E., Schuffenhauer, A., Floersheim, P. (2003) Chemogenomics knowledge-based strategies in drug discovery. *Drug News Perspect*. 16, 93–102.
164. Bock, J.R., Gough, D.A. (2005) Virtual screen for ligands of orphan G protein-coupled receptors. *J Chem Inf Model*. 45, 1402–1414.

CHAPTER 16

X-Ray Structure Developments for GPCR Drug Targets

MICHAEL SABIO and SIDNEY W. TOPIOL
Lundbeck Research USA, Inc., Paramus, NJ

16.1. OVERVIEW

Recently published X-ray structures of Class A and Class C G protein-coupled receptors (GPCRs) may afford the direct use of GPCR X-ray structures in structure-based drug design, provide more reliable homology modeling templates for many other ligand-mediated GPCR proteins, and help reveal the mechanism of GPCR activation. In this chapter, we assess the potential of some of the recent GPCR X-ray structures to provide insights into the mechanism of activation, and we provide a perspective of the emerging opportunities and possible limitations in the use of the new X-ray structures in structure-based drug design.

16.2. INTRODUCTION

G protein-coupled receptors (GPCRs) span six subtypes and constitute the single largest protein target class of all marketed drugs [1, 2]. All GPCRs contain a transmembrane-spanning region comprised of seven α-helices, linked by alternating intracellular and extracellular loops (ICLs and ECLs). X-ray crystallography-derived structures would facilitate the application of structure-based drug-design approaches to these targets. Due to low concentrations, conformational flexibility, and instability in detergent solutions, the

GPCR Molecular Pharmacology and Drug Targeting: Shifting Paradigms and New Directions, Edited by Annette Gilchrist

purification and crystallization of GPCR targets have been challenging. Consequently, the only GPCR X-ray structures of the 7TM region available until 2007 have been those of bovine rhodopsin (a Class A GPCR), which is not ligand-activated and generally has less than 20% sequence identity relative to other Class A GPCR proteins. Thus, the use of a bovine rhodopsin X-ray structure as a template for homology modeling of other GPCR targets, particularly non-Class A GPCRs, may provide questionable results due to uncertainties in the alignment in loop regions (where the homology is generally very low) and expected variations between different proteins in binding-site relaxation (which may require the rearrangement of α-helices).

The first high-resolution X-ray structures of GPCR proteins began to emerge in 2000 (see Table 16.1 and references therein). Important insights into the mechanism of rhodopsin activation were facilitated [3] by studies that followed the publication of the bovine rhodopsin X-ray structures. However, it soon became apparent (e.g., References 4–6) that the construction and use of homology models for ligand-mediated Class A GPCRs would be challenging and likely produce results with limited accuracy. This is in contrast to other target classes such as kinases and proteases, where the use of X-ray structures and homology models based on closely related proteins is often possible. For structure-based drug design, the recently published X-ray structures of ligand-activated GPCR proteins should provide a distinct advantage over previously published X-ray structures of bovine rhodopsin, a light-activated Class A GPCR to which the endogenous chromophore retinal is covalently bound in the "dark" (inactive) state. The now commonly used structure-based drug-design methods, including the construction and use of homology models, may be applied to GPCR targets, especially in the more homologous, common seven-transmembrane structural core. Early questions to be addressed include the following: (1) Would GPCR structures within the same class or across classes demonstrate significant differences? (2) Could reliable models be built for other states of GPCRs, especially because it is believed that GPCR proteins exist in multiple states that depend on the nature and function of the target's ligands [7–13]? Most likely, separate models would have to be generated to analyze agonists, antagonists, and inverse agonists. In addition, as multiple conformations of an active state may exist (possibly leading to ligand-biased trafficking [14]), separate models for different activating ligands may be needed. (3) Should a GPCR target be modeled as a monomer, homodimer, heterodimer, or oligomer? These potential differences could profoundly affect structure-based drug design. (4) Do the available X-ray structures provide enough information to understand the molecular mechanisms of GPCR activation and G protein coupling?

In this chapter, we summarize the impact of published Class A and Class C GPCR X-ray structures that help to address several pertinent questions, and we examine early studies that may provide partial answers. More detailed information on the Class A GPCR X-ray structures is available [15–17].

TABLE 16.1 X-Ray Diffraction GPCR Structures Released by the PDB

PDB ID	Resolution (Å)	R-Free	Date Released	Protein and Chromophore/Ligand	Literature Reference
			A. (Rhod)opsin (a Class A GPCR)		
1F88	2.8	0.238	August 4, 2000	Bovine rhodopsin with retinal	[85]
1HZX	2.8	0.212	July 4, 2001	Bovine rhodopsin with retinal	[86]
1L9H	2.6	0.225	May 15, 2002	Bovine rhodopsin with retinal	[3]
1GZM	2.65	0.235	November 20, 2003	Bovine rhodopsin with retinal	[87]
1U19	2.2	0.222	October 12, 2004	Bovine rhodopsin with retinal	[88]
2HPY	2.8	0.238	August 22, 2006	Bovine lumirhodopsin with retinal	[89]
2G87	2.6	0.181	September 2, 2006	Bovine bathorhodopsin with retinal	[90]
2I35	3.8	0.418	October 17, 2006	Bovine rhodopsin with retinal	[91]
2I36	4.1	0.412	October 17, 2006	Bovine rhodopsin with retinal[a]	[91]
2I37	4.15	0.382	October 17, 2006	Bovine rhodopsin with retinal[a]	[91]
2J4Y	3.4	0.330	September 25, 2007	Bovine rhodopsin with retinal	[20]
2PED	2.95	0.289	October 30, 2007	Bovine rhodopsin with 9-*cis*-retinal	[92]
2ZIY	3.7	0.330	May 6, 2008	Squid rhodopsin with retinal	[49]
2Z73	2.5	0.206	May 13, 2008	Squid rhodopsin with retinal	[50]
3CAP	2.9	0.266	June 24, 2008	Bovine opsin, ligand-free rhodopsin	[31]
3C9L	2.65	0.216	August 5, 2008	Bovine rhodopsin with retinal	[93]
3C9M	3.4	0.219	August 5, 2008	Bovine rhodopsin with retinal	[93]
3DQB	3.2	0.248	September 23, 2008	Bovine opsin, ligand-free rhodopsin	[51]
			B. Other Class A GPCRs		
2RH1	2.4	0.232	October 30, 2007	Human β_2AR with carazolol	[22]
2R4R	3.4	0.270	November 6, 2007	Human β_2AR with carazolol[a]	[21]
2R4S	3.4	0.280	November 6, 2007	Human β_2AR with carazolol[a]	[21]

TABLE 16.1 (*Continued*)

PDB ID	Resolution (Å)	R-Free	Date Released	Protein and Chromophore/Ligand	Literature Reference
3D4S	2.8	0.273	June 17, 2008	Human β_2AR with timolol	[27]
2VT4	2.7	0.268	June 24, 2008	Turkey β_1Ar with cyanopindolol	[28]
3EML	2.6	0.231	October 14, 2008	Human A_{2a} adenosine receptor with ZM241385	[29]
			C. Class C GPCRs		
1EWK	2.2	0.227	December 18, 2000	Rat mGluR1 EC domain with glutamate	[73]
1EWT	3.7	0.287	December 18, 2000	Rat mGluR1 EC domain, free form I	[73]
1EWV	4.0	0.328	December 18, 2000	Rat mGluR1 EC domain, free form II	[73]
1ISR	4.0	0.259	March 13, 2002	Rat mGluR1 EC domain with glutamate/Gd^{3+}	[74]
1ISS	3.3	0.314	March 13, 2002	Rat mGluR1 EC domain with *S*-MCPG	[74]
2E4U	2.35	0.267	February 27, 2007	Rat mGluR3 EC domain with glutamate	[75]
2E4V	2.4	0.268	February 27, 2007	Rat mGluR3 EC domain with DCG-IV	[75]
2E4W	2.4	0.267	February 27, 2007	Rat mGluR3 EC domain with 1*S*,3*S*-ACPD	[75]
2E4X	2.75	0.265	February 27, 2007	Rat mGluR3 EC domain with 1*S*,3*R*-ACPD	[75]
2E4Y	3.4	0.284	February 27, 2007	Rat mGluR3 EC domain with 2*R*,4*R*-APDC	[75]
2E4Z	3.3	0.324	February 27, 2007	Rat mGluR7 EC domain with 2-MES	[75]

[a]The resolution in the active site was insufficient to determine the chromophore's or ligand's coordinates.

16.3. CLASS A GPCRs

16.3.1. Sequence Homology

The X-ray structures of bacteriorhodopsin and bovine rhodopsin have been used as templates for the construction of homology models of GPCR targets. However, because both proteins are only distantly related to GPCRs of drug-discovery interest, and bacteriorhodopsin is not a GPCR, the use of their X-ray structures as templates is challenging. For example, PSI-Blast [18] sequence alignments indicate that bacteriorhodopsin shares only 17%, 12%, 12%, 14%, and 14% sequence identity with the human β_1 adrenergic receptor (β_1AR), the human β_2 adrenergic receptor (β_2AR), the human A_{2a} adenosine receptor, the human dopamine D_2 receptor, and the human metabotropic glutamate receptor 2 (mGluR2), respectively. Bovine rhodopsin's sequence identity percentages relative to the same GPCRs marginally improve to 20%, 19%, 21%, 21%, and 17%, respectively. Interestingly, bacteriorhodopsin and bovine rhodopsin share only 17% sequence identity. The percentage identity values derived by the use of PSI-Blast for these selected examples are likely to represent upper limits, because PSI-Blast alignments tend to emphasize individual residue matches. Thus, the percentage identity values may differ when sequence alignments are performed manually while incorporating the knowledge of conserved motifs or biophysical data. Also, one would expect that the percentage identity would be lower in the loop regions.

The low levels of identity and overall homology create uncertainty in aligning [12] GPCR sequences of interest with that of bovine rhodopsin or bacteriorhodopsin. If an alignment is in error by even a single residue, the resulting model may have an inappropriately sized binding cavity (preventing a meaningful fitting of ligands) or have displaced critical residues so they are no longer capable of interacting with ligands. Such a model would be unusable for drug design, and without an adjustment to the alignment, the model cannot generally be corrected by relaxation afforded by molecular dynamics or other computational techniques.

16.3.2. Stabilization of X-Ray Structures

Due to its high abundance in bovine retina [19] and the limitations in the purification and crystallization of other GPCR proteins [20], bovine rhodopsin, until recently, was the only GPCR for which a three-dimensional (3D) structure was determined. Creating challenges to crystallization, other GPCR proteins are found in low concentrations, are conformationally flexible, and are unstable in detergent solutions. Since the bovine rhodopsin X-ray determinations, the first GPCR structures published were those of the β_2-adrenoreceptor (Protein Data Bank [PDB] accession IDs 2R4R, 2R4S, and 2RH1) [21–23] complexed with the picomolar-affinity inverse agonist carazolol (2RH1). These β_2-adrenoreceptor X-ray structures, which are also the first published ligand-mediated GPCR X-ray structures, were solved by two different approaches, both of which utilized a companion protein to stabilize the flexible intracellular loop 3 (ICL3). The authors used a number of other techniques, including ligand-affinity chromatography and embedding in a lipid cubic phase, to obtain crystallizable protein [24–26].

In the first β_2-adrenoreceptor X-ray determinations [21], which resulted in two 3.4-Å resolution structures with a poorly resolved active site, a β_2AR–Fab5 complex helped stabilize the two almost identical proteins and facilitate crystallization. In an attempt to improve the packing of the extracellular domains and thereby improve the electron density, the authors inserted a TEV (tobacco etch virus) cleavage site after the 24th amino acid of the N-terminus; however, this effort did not lead to a significant crystallographic result. In the second approach [22, 23], the authors replaced the flexible ICL3 loop, which may contribute to unrestricted helical movements, with residues from T4 lysozyme (T4L), and the fusion helped stabilize and crystallize the protein. T4L adds a polar surface that is important for forming crystal lattice contacts and restricting helical motions. To simulate a membrane-like environment and thus help stabilize the protein, the authors grew the crystals in a cholesterol-doped lipidic cubic phase. These innovative efforts were rewarded by a 2.4-Å resolution structure of β_2AR–T4L complexed with the partial inverse agonist carazolol. Because the authors observed only minor differences when they compared the β_2AR–T4L X-ray structure with the lower-resolution wild-type β_2AR–Fab5 X-ray complex, they concluded that the β_2AR–T4L X-ray structure, except for ICL3, is probably sufficiently similar in conformation to that of the native protein. A second high-resolution β_2AR X-ray structure was reported [27], this time in complex with the partial inverse agonist timolol. Splicing of T4L into the ICL3 region was again successful, without significant changes to the overall structure (relative to that observed in the first β_2AR–T4L X-ray determination). Consistent with the introduction of a different ligand, only small conformational changes in the binding site were observed. The crystal packing is different in the two high-resolution β_2AR structures, in that the β_2AR/timolol complex contains two cholesterol binding sites that are not involved in crystal packing. The cholesterol binding sites may be involved in cholesterol-mediated thermal stabilization, allosteric modulation of the high-affinity agonist binding state, and receptor trafficking (refer to Reference 27 and references therein).

An X-ray structure of a protein that is closely related to β_2AR, the turkey β_1 receptor complexed with the antagonist cyanopindolol, was published recently [28]. Employing a strategy to identify a combination of mutations in β_1AR that result in thermal stability, the authors found a set of six point mutations that sufficiently stabilized the complex without the introduction of a companion (Fab5 or T4L) protein. It should be noted that residues in ICL3 and the C-terminus were also deleted. The mutated protein was stable in many detergents used for crystallization and remained in an antagonist conformation even when the ligand was no longer present. Interestingly, the high-resolution β_1AR and β_2AR X-ray structures are very similar, indicating an absence of significant artifacts introduced by the use of a companion protein in the two approaches used for the β_2AR structures. The β_1AR and β_2AR ligand-binding cavities are very similar, showing differences of the scale expected for different complexed ligands, with small conformational variations in the binding-site residues. The source of β_1-versus-β_2 selectivity is not obvious, because the amino acid residues in the vicinity of the ligand differ by very little.

The reported ligand-mediated GPCR X-ray structures described above are the monoaminergic amine systems of β_1AR and β_2AR. A recently published X-ray structure is now available for a more distantly related Class A GPCR, the A_{2a} adenosine receptor in complex with the high-affinity antagonist ZM241385 [29]. The issue of thermal instability of this GPCR was addressed mainly by the T4L fusion approach, in which the ICL3 segment of Leu209 to Ala221 was replaced with T4L. Also, to improve the likelihood of crystallization, the C-terminus of the receptor, from Ala317 to Ser412, was deleted. The ZM241385 antagonist binding in this complex is markedly different from the binding of ligands in the β_1AR and β_2AR structures (vide infra).

Motivated by an earlier approach developed [30] for the selective extraction of rhodopsin, Park et al. [31] selectively extracted bovine opsin from rod cell disc membranes, thereby obtaining crystals of opsin in its native state at 2.9-Å resolution, without further purification steps and without protein structural modifications.

16.3.3. The Overall Topology of the 7TM Region

Electron microscopy experiments [32] with a resolution of 7 Å first showed that the transmembrane region of bacteriorhodopsin was, to a large extent, comprised of α-helices that are approximately perpendicular to the plane of the membrane. Later electron microscopy and electron diffraction efforts [33] with a resolution of about 6.5 Å demonstrated that bacteriorhodopsin was comprised of seven rod-shaped features (i.e., α-helices) with a possible linking of adjacent rods by polypeptide chains, which were later recognized to be alternating intracellular and extracellular loops. Electron cryomicroscopy experiments [34] showed that bovine rhodopsin structure also adopts a 7TM α-helical configuration. However, unlike the nearly parallel α-helical arrangement in the bacteriorhodopsin structure, only four α-helices in bovine rhodopsin are nearly perpendicular to the plane of the membrane, with an unexpected tilting of one of the four helices; the other three helices are more highly tilted. Moreover, the transmembrane helices are arranged differently in bacteriorhodopsin and bovine rhodopsin.

The determination of a low-resolution [35] and, later, a higher-resolution [36] bacteriorhodopsin X-ray structure motivated their use in drug design [6, 37] as templates to model GPCRs of interest. To refine the bacteriorhodopsin-based models, information derived from site-directed mutagenesis data, affinity labeling, and structure–activity relationships have been utilized. Examples of the use of bacteriorhodopsin as a drug-design template have been reviewed [12, 38–41].

The first X-ray diffraction structure of a GPCR, namely, bovine rhodopsin at 2.8-Å resolution, became available in 2000. Subsequently, as enumerated in Table 16.1, an additional 17 (bovine and squid) (rhod)opsin X-ray structures were elucidated. Especially with respect to the transmembrane region, the 18 (rhod)opsin X-ray structures are very superimposable, and the covalently bound chromophore, retinal (when present), varies minimally in position.

In the development of homology models, the X-ray structures of bovine rhodopsin were expected to provide a significant advantage relative to the bacteriorhodopsin X-ray structures, because bacteriorhodopsin is not a GPCR and bovine rhodopsin has at least a marginal increase in homology to GPCRs of interest for drug discovery efforts. In addition, the α-helices of the X-ray structures of bovine rhodopsin and bacteriorhodopsin differ significantly [40] in their positions, orientations, and packing [39, 42, 43]. Moreover, α-helix kinks appear in bovine rhodopsin, but the α-helices are more regularly shaped in bacteriorhodopsin.

Despite low sequence identity and the presence of a different chromophore/ligand when comparing bovine rhodopsin to the human β_2AR, the turkey β_1AR, and the human A_{2a} adenosine receptor, their overall topology is quite similar in the transmembrane region, as observed in X-ray structures (see Table 16.1). However, one notices changes in the tertiary structure and the positions of helices I, III, IV, V, and VI (e.g., when comparing the β_2 receptor and rhodopsin) [21–23].

16.3.4. The Binding Site

In X-ray structures of bovine or squid rhodopsin (see Table 16.1 and Fig. 16.1a,b), retinal is covalently bound and tightly enclosed in a mainly lipophilic binding pocket in the inactive state. In addition, the bovine rhodopsin binding site is blocked by the ECL2, which helps completely enclose retinal by folding into the receptor (see Fig. 16.1a,b). The topology of the rhodopsin binding site raises questions about whether rhodopsin X-ray structures can serve as reliable templates to model other GPCRs of interest. Are gross changes to the position, orientation, and kinking of α-helices necessary to accommodate ligands of different sizes, or are the requisite conformational changes (e.g., in rotational states of the residues) mainly localized to the binding site? Also, how would one expand rhodopsin's tight binding cavity to create a more reliable template from which to construct a homology model, whether for antagonists or agonists? For example, if one used an inactive-state bovine rhodopsin X-ray structure as a template, the construction of a reliable active-state homology model would be very difficult and may involve attempts to expand the binding site and rearrange the 7TM α-helices; these conformational changes are difficult to predict.

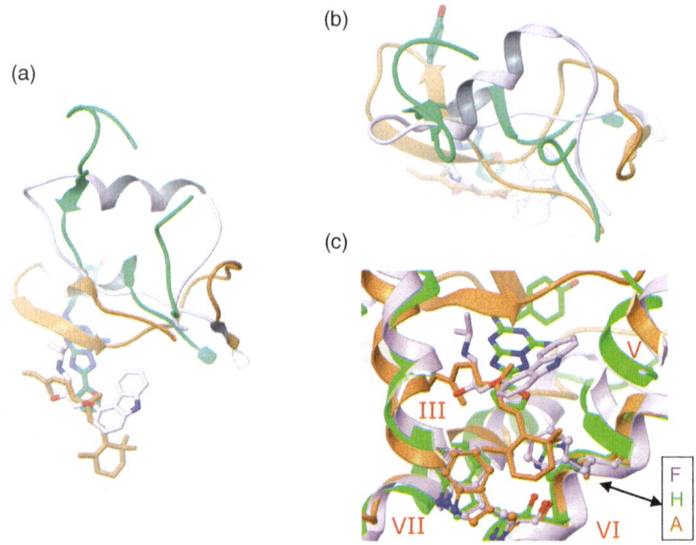

Figure 16.1 (a,b) Two orthogonal views of an overlay of the ECL2 region of PDB entries 1U19 (bovine rhodopsin in brown with retinal), 2RH1 (the human β_2AR in pink with carazolol), and 3EML (the human A_{2a} adenosine receptor in green with ZM241385); (c) the "toggle switch" region in the same proteins as in (a) and (b) viewed from inside the core toward the EC side. Note that the residues corresponding to Phe290 in the human β_2AR are Ala in rhodopsin and His in A_{2a}. All components of Figs. 16.1, 16.2, 16.3, and 16.5 were created in Maestro [94].

The binding sites in the X-ray structures of the human β_2AR and the turkey β_1AR are more open than in the rhodopsin X-ray structures, due to changes in the ECL2 architecture (see below) and a kinking of helix I toward the binding pocket in rhodopsin. Some of the transmembrane helices of the A_{2a} adenosine receptor X-ray structure are shifted or kinked relative to either rhodopsin or the adrenergic receptors. The binding of the antagonist ZM241385 in the A_{2a} adenosine receptor X-ray structure [29] is very different from that of retinal or the adrenergic receptor ligands (see Fig. 16.1a,b). Relative to the binding of retinal or the adrenergic receptor ligands, ZM241385 is positioned closer to the extracellular region and shifted toward helices VI and VII, providing very little overlap with the other X-ray ligands. In the A_{2a} adenosine receptor X-ray structure, the binding cavity is easily accessible to ligands. The observed differences in binding mode and ligand orientation in the A_{2a} adenosine receptor X-ray structure may indicate that selectivity differences among other GPCRs may be due to receptor plasticity, rather than being derived from the variation of amino acids on a nearly conformationally invariant backbone. These binding-site differences suggest severe challenges in the use of X-ray structures as homology modeling templates for remotely related GPCRs (also see Reference 22). An excellent illustration of this concern are recently published studies [44] demonstrating that the binding mode of ZM241385 is not accurately predicted when X-ray structures of remotely related GPCRs, the β_2AR and bovine rhodopsin, are used as templates for homology models of the A_{2a} adenosine receptor. In the A_{2a} homology model based on the high-resolution carazolol β_2 X-ray structure, the binding of ZM241385 is similar to that of carazolol in its respective X-ray structure. The bovine rhodopsin-based model produces a similar discrepancy.

16.3.5. The ECL2

In the rhodopsin X-ray structures, the ECL2 region is very long and contains β-sheets that help enclose the active site, but in the high-resolution X-ray structure of the β_2AR, the ECL2 has an unexpected α-helix (see Fig. 16.1a,b) with two cysteine bridges, one of which is found within the ECL2 and the other is linked to transmembrane helix 3. The α-helix and rigidity afforded by two cysteine bridges hold ECL2 away from the binding cavity of the transmembrane region, thus providing more accessible ligand entry. In the high-resolution X-ray structure of the β_2AR, Phe193 of the extended loop component of the ECL2 comes into contact with the bound ligand, carazolol.

Contrary to the extended β-sheet conformation in rhodopsin or the α-helical structure of the adrenergic receptors, the ECL2 of the A_{2a} adenosine receptor X-ray structure adopts a random coil conformation. However, in analogy to the β_2AR X-ray structure, the easily accessible binding cavity of the A_{2a} adenosine receptor X-ray structure is held open by three nearby disulfide bridges involving ECL1 and ECL2.

Near the extracellular region, the opsin X-ray structures possess an opening between TM5 and TM6, and another opening between TM1 and TM7, both of which arise from extracellular-side helical motions that alter the ligand binding cavity. The openings are necessary because ECL2 blocks the binding-site in opsin and rhodopsin. The authors suggest [31] that the two openings may allow the entry of 11-*cis*-retinal and the exit of all-*trans*-retinal, respectively. Due to differences in the ECL2 architecture, the modes of ligand entry and exit are expected to be different in the ligand-mediated GPCR X-ray structures. The two opsin X-ray structures advance the understanding of GPCR activation and may provide insights for developing homology models for agonist binding.

16.3.6. The Toggle Switch

Toward the intracellular side just below the covalently bound retinal in the rhodopsin X-ray structures, a conserved tryptophan (W6.48, using established nomenclature [45], see Fig. 16.1c) is part of a series of residues, in addition to a network of conserved water molecules, that help to propagate the activation/inactivation signal by interacting near the intracellular side along the inner transmembrane region. Believed to control this signal through a change of its rotational state, W6.48 is known as the "toggle switch" [46]. In rhodopsin, the β-ionone ring of retinal, which occupies a deep region of the binding pocket, interacts directly with W6.48, which is thereby locked into its inactive rotamer. The same inactivating conformation is achieved in the high-resolution β_2AR X-ray structure, even though the binding of carazolol does not interact directly with W6.48 due to the ligand's relatively shallow occupancy of the binding cavity. Carazolol, instead, interacts with Phe290 (corresponding to an alanine residue in rhodopsin), which forces W6.48 to adopt the same postulated inactivating conformation as in rhodopsin. Although the ligand of the A_{2a} adenosine X-ray structure is even further from W6.48, the protein is still found in the inactive state, where His250 plays the role of Phe290 in the β_2AR structure.

16.3.7. The Ionic Lock

In the inactive state of rhodopsin, located on the intracellular side, the "ionic lock," which constrains H6 inside the helical bundle, is comprised of an extended hydrogen-bonded network involving Arg135 (H3), Glu247 (H6), Glu134 (H3), and Thr251 (H6), where Glu134 and Arg135 are the first and second residues of the conserved "D(or E)RY" motif (see Fig. 16.2a–e). The presence of the "ionic lock" is understood to characterize the inactive state and is supported by biophysical data [47, 48]. Interestingly, the antibody-

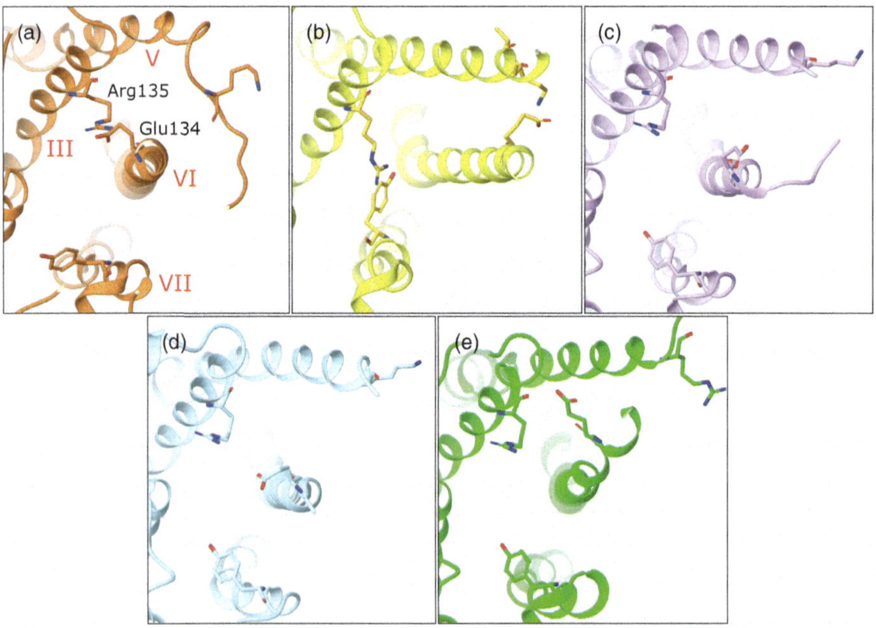

Figure 16.2 View of the "ionic lock" region from the IC side into the core in (a) bovine rhodopsin, (b) bovine opsin with Gα-CT, (c) the human β_2AR, (d) the human β_1AR, and (e) the human A_{2a} adenosine receptor.

complexed and T4L-spliced β_2AR X-ray structures lack the Glu(H6)-to-Arg(H3) interaction component of the "ionic lock." One could speculate that the absence of the "ionic lock" in these constructs is a consequence of introducing the companion proteins (antibody, T4L) into the crystal structures. Because of carazolol's comparable affinity as an inverse agonist for both the native protein and the β_2AR–T4L construct, one would assume inactive characteristics for the β_2AR–T4L construct. Alternatively, the ligand's function as an inverse agonist (vs. antagonist) may induce a protein conformation unlike that of the inactive state. Experiments with fluorescent probes demonstrated [23] that agonists can induce protein conformational changes that are consistent with receptor activation. Moreover, considerable experimental evidence shows that β_2AR has multiple conformations that correspond to different degrees of activation [7–11, 13].

In the β_1AR X-ray structure [28], the "ionic lock" observed in rhodopsin is also not formed. The β_1AR and β_2AR X-ray structures differ by the presence of a short α-helix, preserved in all four molecules of the unit cell, within ICL2 in β_1AR, allowing a hydrogen bond involving Tyr149 on ICL2 and H3's Asp138 of the "D(E)RY" motif. In contrast, in the β_2AR structures, this ICL2 secondary structure is not present. By analyzing mutational data, the authors conclude [28] that the β_1AR X-ray structure represents the physiologically relevant conformation. The authors also conclude that the inactivity of the complex of β_1AR with the antagonist cyanopindolol suggests that there is no evidence of the "ionic lock." In an alternative explanation, one may consider the Asp138-to-Tyr149 hydrogen bond as the cause of full antagonism and the lack of activity of the cyanopindolol/β_1AR complex, while the two β_2AR complexes with inverse agonists demonstrate residual basal activity.

The conformation of the "ionic lock" region in the A_{2a} adenosine receptor X-ray structure [29] is similar to that of the β_1AR X-ray structure, in that an ICL2 α-helix presents Tyr112$^{3.60}$ for interaction with Asp101$^{3.49}$ in the A_{2a} adenosine receptor X-ray structure. Thus, the ICL2 hydrogen bond can be achieved in a T4L construct. The authors therefore suggest [29] that the residual basal activity in the β_2AR X-ray structures reflects the absence of the ICL2 α-helix and the associated Tyr112$^{3.60}$-to-Asp101$^{3.49}$ interaction afforded by an ICL2 α-helix.

16.3.8. The ICL3 Region and Activation

Up until recently, the many X-ray structures of rhodopsin and the ligand-mediated GPCRs had not revealed very much direct information on GPCR activation. In contrast, recent publications [31, 49–52] now provide tangible details of the GPCR activation mechanism. Two X-ray structures of invertebrate squid rhodopsin, which couples to Gq, were published [49, 50] in May and June of 2008, and demonstrate noteworthy features in the intracellular region of helices 5 and 6, and their ICL3 tether, which is 12 residues longer than bovine rhodopsin's ICL3. On the intracellular side, H5 and H6 are longer and rigid, relative to their bovine rhodopsin counterparts, and extend far from

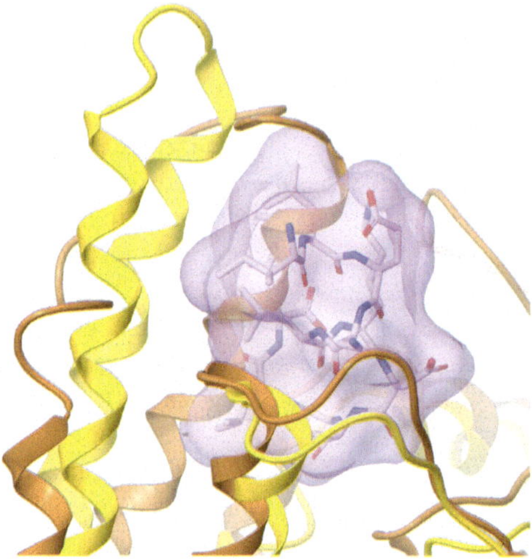

Figure 16.3 View near the ICL3 of bovine rhodopsin (brown) and of bovine opsin (yellow) bound to Gα-CT (pink with its surface).

the core. In one X-ray structure [50], H5 and H6 comprise one boundary of a binding region for an octylglucoside in close proximity. The authors speculated that the region around the extended helices and the ICL3 tether may be involved in binding the G protein.

Within the next few months, two publications [31, 51] on bovine opsin X-ray structures, both of which are in the active conformation, appeared in July and September. The second structure provides direct support of the active state in that it is complexed with an 11-amino acid peptide, Gα-CT, derived from the C-terminus of the transducin $G\alpha_t$ protein (ILENLKDCGLF, $G\alpha_t$ [(340–350) K341L]) (see Fig. 16.3).

Relative to the earlier bovine rhodopsin structures with which it shares overall topological similarity, the opsin X-ray structure demonstrates changes mainly in the intracellular region. For example, there is a short helical turn in ICL1. More importantly, because the N-terminal region of opsin's ICL3 adopts an α-helical structure, the intracellular side of H5 is elongated by 1.5–2.5 helical turns relative to the rhodopsin X-ray structures (see Fig. 16.3). Also, H5 is tilted toward the other helices of the 7TM core. As expected from earlier experimental evidence [53–55], the intracellular side of H6 is tilted by 6–7 Å away from the 7TM helical bundle, such that H5 and H6 are almost parallel and protrude from the intracellular region.

The highly conserved E(D)R135Y and NPxxY306$(x)_{5,6}$F motifs help effect the stabilization of H5 and H6 in their intracellular positions within opsin. In rhodopsin, Arg135 of the E(D)RY motif within H3 interacts with Glu247 of H6. However, in opsin, this salt bridge is broken, H6 moves away from H3 so that H6 is further from the center of the helical core but approaches H5, H5 approaches H3 in a direction toward the helical bundle's center, and H5 and H6 are stabilized in their rearranged positions by interactions involving (1) Lys231 of H5 and Glu247 of H6, and (2) Tyr223 of H5 and Arg135 of H3. In opsin, Tyr306 of the NPxxY306$(x)_{5,6}$F motif within H7 blocks the return of H6 to its position in rhodopsin by this residue's rotation into the helical core.

Stabilized by Tyr223, Arg135 is also involved in the binding of the Gα-CT peptide through hydrogen-bond formation involving the backbone carbonyl oxygen atom of Cys347 of Gα-CT. Gα-CT occupies a crevice bounded by Arg135, lipophilic residues of H5 and H6, and a hydrogen-bonding network involving H7 and H8. Conformational changes in the H7–H8 kink, which moves away from the helical core, and the rearrangement of H5 and H6 are required for Gα-CT binding [50]. Detailed structural analysis in the ICL3 region would benefit from comparisons with the ligand-mediated X-ray structures. However, the full, native ICL3 loop is not contained in any of the ligand-mediated X-ray structures.

Relative to its position in the inactive state in rhodopsin, Trp265 of the "toggle switch" is shifted in opsin but does not display the expected rotameric change; however, the lack of the rotameric change may be due to an empty binding pocket. In the Gα-CT-free opsin X-ray structure, electron density was not observed for Lys296, which is covalently bound to retinal in rhodopsin. However, electron density is observed in the opsin/Gα-CT complex, demonstrating that the binding of Gα-CT affects the chromophore-binding site. The authors suggest [51] that this observation supports the idea that the chromophore-binding site and the Gα-CT binding cleft are coupled during activation.

16.3.9. Computational Chemistry Successes and Limitations

Despite the uncertainties in aligning sequences of remotely related GPCRs to that of bovine rhodopsin, successful outcomes based on the resulting homology models have been cited (see, e.g., a recent review [6]). For example, using an X-ray structure (PDB's 1F88) of bovine rhodopsin as a structural template, Bissantz et al. [5] created antagonist-bound homology models of three human GPCRs (dopamine D3, muscarinic M1, and vasopressin V1a) and agonist-bound homology models of three human GPCRs (dopamine D3, β_2-adrenergic, and δ-opioid). Using three docking algorithms and seven scoring functions, the authors screened six databases, each containing 3D structures of 990 random analogs plus 10 known antagonists or agonists for each target. As the authors concluded, the homology models that were based on bovine rhodopsin were effective in identifying known antagonists that they seeded in the database but were not sufficiently accurate for identifying known agonists. To develop improved agonist models, the authors also utilized their own knowledge- and pharmacophore-based modeling protocol. Successful construction and validation of models have relied heavily on experimental results (e.g., mutational data, structure-activity relationships [SAR], etc.) [6]. In one example, Xie et al.

[56] developed a homology model, which agreed well with known biochemical and structural data, of the human CB_2 receptor, again using the X-ray structure of bovine rhodopsin as a template. Methodology that avoids the use of bovine rhodopsin as a template for constructing GPCR models has also been reported. Vaidehi et al., using the ab initio structure prediction algorithms, MembStruk and HierDock, reproduced [57] the bovine rhodopsin X-ray crystal structure within a root-mean-square (RMS) difference of 3.1 Å spanning the transmembrane region. In addition, the authors predicted protein structure and function without knowledge of the experimental structures for four classes of GPCR targets (β_1AR, EDG6, human sweet receptor, and mouse/rat I7 olfactory receptors). Shacham et al. developed [58] PREDICT, an algorithm that combines knowledge of the amino acid sequence with properties of the membrane environment without utilizing information about the 3D structure of rhodopsin. Using this methodology, the authors reproduced the X-ray geometry of rhodopsin within an RMS difference of 3.87 Å spanning the transmembrane region. The methodology also showed potential in creating other GPCR homology models for structure-based drug discovery, including the screening of virtual libraries. In other studies, hybrid approaches, including the use of receptor–ligand pharmacophores [59] and ligand-based homology modeling [60], were developed to enhance [43] the success of GPCR homology modeling efforts when a bovine rhodopsin X-ray structure is used as the template. For example, the use of structure–activity relationships, site-directed mutagenesis data, and affinity labeling studies have been integrated with various ligand-based approaches that resulted in a few encouraging examples [12, 39–41]. The lack of broad implementation demonstrates that GPCR modeling has not yet reached the stage of other successful drug-discovery approaches, due to the unavailability of high-resolution ligand-mediated GPCR X-ray structures.

The potential impact of using X-ray structures of ligand-activated GPCR proteins directly has recently been assessed with the high-resolution β_2AR/carazolol X-ray structure as a prototype for drug discovery. High-throughput docking alone efficiently extracted [61] low-nanomolar compounds from very large databases, demonstrating that GPCR X-ray structures could indeed play as critical a role in a drug-discovery setting as X-ray structures of other targets. Furthermore, the same X-ray structure was used for the docking of a number of known β_2AR ligands [61], including timolol. The accuracy of the predicted binding modes was validated with a later publication that reported a β_2AR X-ray structure complexed with the antagonist timolol [27, 62]. The predicted binding mode [61, 62] of timolol in the β_2AR/carazolol X-ray structure and the β_2AR/timolol X-ray structure are in very good agreement.

The limitations of using a remotely related GPCR as a template for homology modeling have been demonstrated in recent reports. Costanzi has shown [63] that docking carazolol into a rhodopsin-based β_2AR homology model produced poorer results than those obtained by the direct use of the β_2AR X-ray structure. Similarly, when epinephrine was docked to another rhodopsin-based β_2AR homology model, the hydroxy alkylamine moiety of the predicted binding mode [64] was involved in interactions different from those observed for the identical structural component within carazolol. The difference in interactions may be an artifact of the model or, for example, a reflection of different binding modes adopted by an inverse agonist (carazolol) and an agonist (epinephrine). In other efforts [44, 65], a ligand-mediated GPCR template, the β_2AR X-ray structure, was used for predicting other Class A GPCR structures. The binding of ZM241385 is very different in mode and location in the A_{2a} adenosine receptor X-ray structure [29] and the β_2AR-based homology model [44] of the A_{2a} adenosine receptor. Similarly, the β_2AR X-ray structure is now being used for agonist drug-discovery efforts [66–68]. For example, improved efficiency in database mining for agonists is observed when the β_2AR X-ray structure is modified to contain a binding pocket that more closely represents a closed (active) form that is expected for agonist binding [66].

16.4. CLASS C GPCRs

While the 7TM domain is the common component of GPCRs and is directly involved in the intracellular interaction with G proteins for all classes of GPCRs, large extracellular (N-terminal) extensions play key roles in Class B and Class C GPCRs. For Class C GPCRs, with regard to the extracellular regions, recent X-ray structures also have provided new opportunities that parallel those described above for the 7TM region. Indeed, conceptually analogous insights and questions are emerging.

16.4.1. Global Architecture

Class C GPCRs include the eight different mGluRs, which are subdivided into three subclasses (mGluR1 and 5 in subclass I; mGluR2 and 3 in subclass II; and mGluR4, 6, 7, and 8 in subclass III), as well as the sweet taste receptors, umami receptors, and the $GABA_B$ receptors. We focus on the mGluR receptors. The extracellular, N-terminal portion of these receptors contains the so called Venus flytrap (VFT) domain and the cysteine-rich (C-rich) domain (see Fig. 16.4). Unlike the Class A receptors whose ligand-binding site is the 7TM

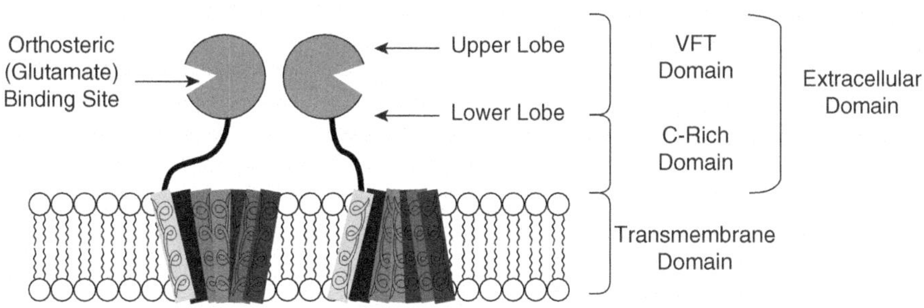

Figure 16.4 Schematic representation of the Class C GPCRs.

domain, as described above, the mGluR binding site of the endogenous ligand, glutamate, is the VFT domain. The C-rich domain lies between the N-terminal VFT and C-terminal 7TM domains. It is now well established that these proteins exist as dimers in the cell membrane.

From a drug-discovery perspective, by analogy to the binding sites of other receptors or enzymes, the VFT region would be the site to target for new ligands (agonists or antagonists) as it represents the endogenous ligand's binding site. Drug-discovery efforts, particularly before the availability of the VFT X-ray structures described below, tended to be built around the glutamate template. The belief that the VFT binding site required highly polar ligands, coupled with the argument that evolutionary pressure preserves common recognition features in the VFT to bind the common glutamate ligand, resulting in less subtype selectivity, together with pharmacological arguments, has led to the search for ligands that bind in the 7TM domain. These ligands act as "allosteric" modulators on the activity of the endogenous ligand, glutamate. Examples of positive and negative allosteric modulators are now common [69–71].

Fully understanding the mechanism of action of these multi-domain dimers is daunting, especially when considered in light of the challenges of just the 7TM domain discussed earlier for Class A GPCRs. While the site for endogenous ligand binding may be different for Class C GPCRs, ultimately, they too couple to G proteins and would be expected to have common coupling features. It is therefore intriguing, and arguably understandable, that a 7TM-acting positive allosteric modulator of mGluR5 was found to function as a full agonist in an N-terminal truncated (no extracellular [EC] region) receptor [72]. In spite of the complexity, here too, recent X-ray structures are leading to valuable developments.

16.4.2. The VFT Domain

The VFT domain is comprised of two lobes with α/β topology. In three articles [73–75] between 2000 and 2007, the same group published X-ray structures of VFT regions of mGluR1, mGluR3, and mGluR7, thereby spanning the I, II, and III subgroups, respectively (see Table 16.1). Included in these structures are examples of different domain conformations (open and closed), inter-domain orientations, and active-site occupations that can now serve as a broad suite of templates for direct use in drug design or homology modeling of other Class C targets of interest that are not exemplified. Examination of some of these structures illustrates the emerging picture. For mGluR1, X-ray structures are available for the two lobes in closed and open conformations (see Figs. 16.4 and 16.5). Both the closed and open conformations are found with the active site bound with glutamate or an empty (apo) active site. This supports earlier evidence that there is an equilibrium between the two conformations, and the role of the agonist is to shift the equilibrium toward the more active conformation, believed to be the closed conformation. Consistent

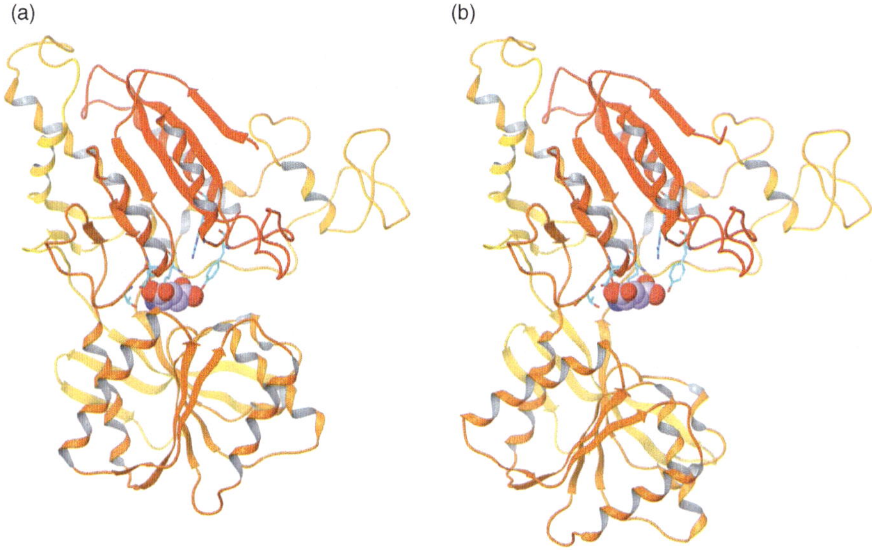

Figure 16.5 View of the rat mGluR1 EC domain in the (a) closed and (b) open conformations with glutamate bound (taken from chains A and B of PDB entry 1EWK, respectively).

with this, the open conformation is found with the antagonist (*S*)-(α)-methyl-4-carboxyphenylglycine (*S*-MCPG) bound. Closing of the VFT of one monomer (protomer) is believed to play a role in activation, but the relative structures between the VFT regions of the protomers in the dimer, as well as their relative orientations, were also found to change, and these changes are believed to play a role in the activation process [73, 75, 76]. For example, it has been suggested that these EC changes could affect the interactions, and thereby possibly the conformation, of the 7TM regions.

As expected, the active site is very polar. Comparisons of the active sites show that despite the use of a common endogenous ligand, there are differences between the X-ray structures. Thus, even glutamate binds slightly differently to the mGluR1 and mGluR3 VFT domains. The highly polar active site makes varying use of water molecules to bind different agonists in the X-ray structures of mGluR3 complexes. Also, comparisons of the X-ray structures showed that differences between residues in the active site explain the mGluR3 selectivity of DCG-IV [75].

16.4.3. The C-rich Domain

For mGluR3, there is now an available X-ray structure of the EC region including both the VFT and C-rich domains. The extensive cysteine bonds in this small domain render it fairly rigid. This suggests [75] that, to the extent that structural/conformational changes in the VFT are propagated via the C-rich region, the latter probably functions like a rigid connecting rod. In the related, Class C, sweet taste receptors, there is evidence suggesting that receptor activation can be achieved by interaction with the C-rich region.

16.4.4. Computational Studies

Prior to the publication of the X-ray structures of the mGluR EC regions described above, the leucine/isoleucine/valine binding protein (LIVBP) was used as a template for homology modeling (see, e.g., Reference 77 and references therein). LIVBP is a member of the bacterial periplasmic binding proteins (PBP), which are involved in ligand recognition and transport, and which share significant homology [78] with the EC domains of the mGluRs. Homology models were developed, generally in conjunction with mutational data, to help understand the binding of known ligands. For example, using LIVBP as a template for homology modeling, Costantino explored [77] the binding of 4MCPG [74] to mGluR1. Similarly, Hampson et al. developed [79] a model of L-serine O-phosphate (L-SOP) binding to mGluR4. With the X-ray publication of the EC domain, the mGluR1 structure has been used in studying subclass II (mGluR2 and 3) receptors. Malherbe et al. developed [80] models for the selective agonist LY354740, as did Monn et al. [81, 82], who also considered LY404040. Bertrand et al. [83] docked eight different ligands into the X-ray structure of mGluR1 and homology models of mGluR2 and mGluR4. Yao et al. studied [84] DCG-IV binding to an mGluR3 homology model. The subsequent publication of the X-ray structure of the mGluR3/DCG-IV complex revealed that the proposed [84] direct interaction between Arg277 and the ligand is not present. These findings underscore the value of the availability of X-ray structures spanning the various mGluR subclasses. Moreover, the full potential of the suite of EC X-ray structures for mGluRs in a drug-discovery environment has yet to be evaluated.

16.5. CONCLUSIONS

The importance of GPCR proteins in signal transduction in almost all physiological processes lead to their predominance as therapeutic targets. The long-awaited emergence of GPCR X-ray structures in the last 9 years is affording a better understanding of the structure/function relationships of GPCRs, including the mechanism of activation, and more effective drug design for these proteins.

For the transmembrane region, starting in 2000, crystallographic structural determinations were reported for rhodopsin, a light-activated Class A GPCR, and after 7 years, it was followed by the 2007 publication of the first non-rhodopsin/opsin (β_2AR) Class A GPCR X-ray structure. Since the reporting of the first β_2AR X-ray structure, several X-ray structures of ligand-mediated

Class A GPCRs and related publications have appeared. Due to significantly different ligand-binding cavities, for example, as observed in the β_1AR and β_2AR versus A_{2a} adenosine X-ray structures, it is clear that additional X-ray structures will be necessary for efficient structure-based drug design at many targets. Also, reliable homology models for remotely related GPCRs, such as those of Class B or Class C, will be difficult to construct if the models are based on Class A GPCR X-ray structures. However, the reported adaptation of different approaches to facilitate purification and crystallization suggests that other GPCR X-ray structures will be determined at an accelerated rate. The value of GPCR X-ray structures for use in drug discovery already has been demonstrated and validated, for example, by quickly identifying nanomolar β_2AR inhibitors by using a β_2AR X-ray structure. In addition, promising results from homology modeling have been reported. Other anticipated advancements in structure-based drug design driven by X-ray crystallography include an understanding of the structural differences of other GPCR classes and subclasses, ligands of other character (e.g., agonists, allosteric modulators, etc.), other protein states, and dimeric structures. Based on the progress observed in GPCR X-ray crystallography since 2007, one would expect an acceleration in the rate of structure determination, advances in understanding of GPCR structure/function relationships, and more successful GPCR structure-based drug design.

For the VFT domain of Class C GPCRs, a broad palette of X-ray structures are now available to be used directly or as templates for homology models. Some structure-based studies have already appeared, but extensive evaluation of the potential impact in a drug-discovery environment is still needed.

REFERENCES

1. Overington, J.P., Al-Lazikani, B., Hopkins, A.L. (2006) How many drug targets are there? *Nat Rev Drug Discov*. 5, 993–996.
2. Parrill, A.L. (2008) Crystal structures of a second G protein-coupled receptor: Triumphs and implications. *ChemMedChem*. 3, 1021–1023.
3. Okada, T., Fujiyoshi, Y., Silow, M., Navarro, J., Landau, E.M., Shichida, Y. (2002) Functional role of internal water molecules in rhodopsin revealed by X-ray crystallography. *Proc Natl Acad Sci U S A*. 99, 5982–5987.
4. Archer, E., Maigret, B., Escrieut, C., Pradayro, L., Fourmy, D. (2003) Rhodopsin crystal: New template yielding realistic models of G-protein-coupled receptors? *Trends Pharmacol Sci*. 24, 36–40.
5. Bissantz, C., Bernard, P., Hibert, M., Rognan, D. (2003) Protein-based virtual screening of chemical databases. II. Are homology models of G-protein coupled receptors suitable targets? *Proteins.* 50, 5–25.
6. Barton, N., Blaney, F.E., Garland, S., Tehan, B., Wall, I. (2007) Seven transmembrane G protein-coupled receptors: Insights for drug design from structure and modeling. In *Comprehensive Medicinal Chemistry II*, ed. J.B. Taylor, D.J. Triggle. Amsterdam: Elsevier, pp. 669–701.

7. Barak, L.S., Ménard, L., Ferguson, S.S.G., Colapietro, A.-M., Caron, M.G. (1995) The conserved seven-transmembrane sequence $NP(X)_{2,3}Y$ of the G-protein-coupled receptor superfamily regulates multiple properties of the β_2-adrenergic receptor. *Biochemistry*. 34, 15407–15414.
8. Ghanouni, P., Gryczynski, Z., Steenhuis, J.J., Lee, T.W., Farrensi, D.L., Lakowicz, J.R., Kobilka, B.K. (2001) Functionally different agonists induce distinct conformations in the G protein coupling domain of the β2 adrenergic receptor. *J Biol Chem*. 276, 24433–24436.
9. Swaminath, G., Xiang, Y., Lee, T.W., Steenhuis, J., Parnot, C., Kobilka, B.K. (2004) Sequential binding of agonists to the β_2 adrenoceptor. Kinetic evidence for intermediate conformational states. *J Biol Chem*. 279, 686–691.
10. Swaminath, G., Deupi, X., Lee, T.W., Zhu, W., Thian, F.S., Kobilka, T.S., Kobilka, B. (2005) Probing the β2 adrenoceptor binding site with catechol reveals differences in binding and activation by agonists and partial agonists. *J Biol Chem*. 280, 22165–22171.
11. Yao, X., Parnot, C., Deupi, X., Ratnala, V.R.P., Swaminath, G., Farrens, D., Kobilka, B. (2006) Coupling ligand structure to specific conformational switches in the β_2-adrenoceptor. *Nat Chem Biol*. 2, 417–422.
12. Reggio, P.H. (2006) Computational methods in drug design: Modeling G protein-coupled receptor monomers, dimers, and oligomers. *AAPS J*. 8, E322–E336.
13. Kobilka, B.K., Deupi, X. (2007) Conformational complexity of G-protein-coupled receptors. *Trends Pharmacol Sci*. 28, 397–406.
14. Violin, D.J., Lefkowitz, R.J. (2007) β-arrestin-biased ligands at seven-transmembrane receptors. *Trends Pharmacol Sci* 28, 416–421.
15. Weis, W.I., Kobilka, B.K. (2008) Structural insights into G-protein-coupled receptor activation. *Curr Opin Struct Biol*. 18, 734–740.
16. Mustafi, D., Palczewski, K. (2009) Topology of class A G protein-coupled receptors: Insights gained from crystal structures of rhodopsins, adrenergic and adenosine receptors. *Mol Pharmacol*. 75, 1–12.
17. Topiol, S., Sabio, M. (2009) X-ray structure breakthroughs in the GPCR transmembrane region. *Biochem Pharmacol*. 78, 11–20.
18. Altschul, S.F., Madden, T.L., Schaffer, A.A., Zhang, J., Zhang, Z., Miller, W., Lipman, D.J. (1997) Gapped BLAST and PSI-BLAST: A new generation of protein database search programs. *Nucleic Acids Res*. 25, 3389–3402.
19. Filipek, S., Teller, D.C., Palczewski, K., Stenkamp, R. (2003) The crystallographic model of rhodopsin and its use in studies of other G protein-coupled receptors. *Annu Rev Biophys Biomol Struct*. 32, 375–397.
20. Standfuss, J., Xie, G., Edwards, P.C., Burghammer, M., Oprian, D.D., Schertler, G.F.X. (2007) Crystal structure of a thermally stable rhodopsin mutant. *J Mol Biol*. 372, 1179–1188.
21. Rasmussen, S.G.F., Choi, H.-J., Rosenbaum, D.M., Kobilka, T.S., Thian, F.S., Edwards, P.C., Burghammer, M., Ratnala, V.R.P., Sanishvili, R., Fischetti, R.F., Schertler, G.F.X., Weis, W.I., Kobilka, B.K. (2007) Crystal structure of the human β_2 adrenergic G-protein-coupled receptor. *Nature*. 450, 383–388.
22. Cherezov, V., Rosenbaum, D.M., Hanson, M.A., Rasmussen, S.G.F., Thian, F.S., Kobilka, T.S., Choi, H.-J., Kuhn, P., Weis, W.I., Kobilka, B.K., Stevens, R.C. (2007)

High-resolution crystal structure of an engineered human β_2-adrenergic G protein-coupled receptor. *Science*. 318, 1258–1265.

23. Rosenbaum, D.M., Cherezov, V., Hanson, M.A., Rasmussen, S.G.F., Thian, F.S., Kobilka, T.S., Choi, H.-J., Yao, X.-J., Weis, W.I., Stevens, R.C., Kobilka, B.K. (2007) GPCR engineering yields high-resolution structural insights into β_2-adrenergic receptor function. *Science*. 318, 1266–1273.
24. Faham, S., Boulting, G.L., Massey, E.A., Yohannan, S., Yang, D., Bowie, J.U. (2005) Crystallization of bacteriorhodopsin from bicelle formulations at room temperature. *Protein Sci*. 14, 836–840.
25. Landau, E.M., Pebay-Peyroula, E., Neutze, R. (2003) Structural and mechanistic insight from high resolution structures of archaeal rhodopsins. *FEBS Lett*. 555, 51–56.
26. Cherezov, V., Peddi, A., Muthusubramaniam, L., Zheng, Y.F., Caffrey, M. (2004) A robotic system for crystallizing membrane and soluble proteins in lipidic mesophases. *Acta Crystallogr*. D60, 1795–1807.
27. Hanson, M.A., Cherezov, V., Griffith, M.T., Roth, C.B., Jaakola, V.-P., Chien, E.Y.T., Velasquez, J., Kuhn, P., Stevens, R.C. (2008) A specific cholesterol binding site is established by the 2.8 Å structure of the human β_2-adrenergic receptor. *Structure*. 16, 897–905.
28. Warne, T., Serrano-Vega, J., Baker, J.G., Moukhametzianov, R., Edwards, P.C., Henderson, R., Leslie, A.G.W., Tate, C.G., Schertler, G.F.X. (2008) Structure of a β1-adrenergic G-protein coupled receptor. *Nature* 454, 486–492.
29. Jaakola, V.-P., Griffith, M.T., Hanson, M.A., Cherezov, V., Chien, E.Y.T., Lane, J.R., Ijzerman, A.P., Stevens, R.C. (2008) The 2.6 angstrom crystal structure of a human A_{2A} adenosine receptor bound to an antagonist. *Science*. 322, 1211–1217.
30. Okada, T., Takeda, K., Kouyama, T. (1998) Highly selective separation of rhodopsin from bovine rod outer segment membranes using combination of divalent cation and alkyl(thio)glucoside. *Photochem Photobiol*. 67, 495–499.
31. Park, J.H., Scheerer, P., Hofmann, K.P., Choe, H.-W., Ernst, O.P. (2008) Crystal structure of the ligand-free G-protein-coupled receptor opsin. *Nature*. 454, 183–188.
32. Unwin, P.N.T., Henderson, R. (1975) Molecular structure determination by electron microscopy of unstained crystalline specimens. *J Mol Biol*. 94, 425–440.
33. Leifer, D., Henderson, R. (1983) Three-dimensional structure of orthorhombic purple membrane at 6.5 Å resolution. *J Mol Biol*. 163, 451–466.
34. Unger, V.M., Schertler, G.F.X. (1995) Low resolution structure of bovine rhodopsin determined by electron cryo-microscopy. *Biophys J*. 68, 1776–1786.
35. Pebay-Peyroula, E., Rummel, G., Rosenbusch, J.P., Landau, E.M. (1997) X-ray structure of bacteriorhodopsin at 2.5 Å from microcrystals grown in lipidic cubic phases. *Science*. 277, 1676–1681.
36. Luecke, H., Schobert, B., Richter, H.-T., Cartailler, J.-P., Lanyi, J.K. (1999) Structure of bacteriorhodopsin at 1.55 Å resolution. *J Mol Biol*. 291, 899–911.
37. Röper, D., Jacoby, E., Krüger, P., Engels, M., Grötzinger, J., Wollmer, A., Straßburger, W. (1994) Modeling of G-protein coupled receptors with bacteriorhodopsin as a template. A novel approach based on interaction energy differences. *J Recept Res*. 14, 167–186.

38. Gershengorn, M.C., Osman, R. (2001) Minireview: Insights into G protein-coupled receptor function using molecular models. *Endocrinology*. 142, 2–10.
39. Klabunde, T., Hessler, G. (2002) Drug design strategies for targeting G-protein-coupled receptors. *Chem Bio Chem*. 3, 928–944.
40. Becker, O.M., Shacham, S., Marantz, Y., Noiman, S. (2003) Modeling the 3D structure of GPCRs: Advances and application to drug discovery. *Curr Opin Drug Discov Devel*. 6, 353–361.
41. Mobarec, J.C., Filizola, M. (2008) Advances in the development and application of computational methodologies for structural modeling of G-protein-coupled receptors. *Expert Opin Drug Discov*. 3, 343–355.
42. Schertler, G.F.X., Villa, C., Henderson, R. (1993) Projection structure of rhodopsin. *Nature*. 362, 770–772.
43. Patny, A., Desai, P.V., Avery, M.A. (2006) Homology modeling of G-protein-coupled receptors and implications in drug design. *Curr Med Chem*. 13, 1667–1691.
44. Yuzlenko, O., Kiec'-Kononowicz, K. (2009) Molecular modeling of A_1 and A_{2A} adenosine receptors: Comparison of rhodopsin- and β2-adrenergic-based homology models through the docking studies. *J Comp Chem*. 30, 14–32.
45. Ballesteros, J.A., Weinstein, H. (1995) Integrated methods for the construction of three-dimensional models and computational probing of structure-function relations in G protein-coupled receptors. *Methods Neurosci*. 25, 366–428.
46. Schwartz, T.W., Frimurer, T.M., Holst, B., Rosenkilde, M.M., Elling, C.E. (2006) Molecular mechanism of 7TM receptor activation—A global toggle switch model. *Annu Rev Pharmacol Toxicol*. 46, 481–519.
47. Ballesteros, J., Kitanovic, S., Guarnieri, F., Davies, P., Fromme, B.J., Konvicka, K., Chi, L., Millar, R.P., Davidson, J.S., Weinstein, H., Sealfon, S.C. (1998) Functional microdomains in G-protein-coupled receptors. The conserved arginine-cage motif in the gonadotropin-releasing hormone receptor. *J Biol Chem*. 273, 10445–10453.
48. Ballesteros, J.A., Jensen, A.D., Liapakis, G., Rasmussen, S.G.F., Shi, L., Gether, U., Javitch, J.A. (2001) Activation of the β_2-adrenergic receptor involves disruption of an ionic lock between the cytoplasmic ends of transmembrane segments 3 and 6. *J Biol Chem*. 276, 29171–29177.
49. Shimamura, T., Hiraki, K., Takahashi, N., Hori, T., Ago, H., Masuda, K., Takio, K., Ishiguro, M., Miyano, M. (2008) Crystal structure of squid rhodopsin with intracellularly extended cytoplasmic region. *J Biol Chem*. 283, 17753–17756.
50. Murakami, M., Kouyama, T. (2008) Crystal structure of squid rhodopsin. *Nature*. 453, 363–368.
51. Scheerer, P., Park, J.H., Hildebrand, P.W., Kim, Y.J., Krauss, N., Choe, H.-W., Hofmann, K.P., Ernst, O.P. (2008) Crystal structure of opsin in its G-protein-interacting conformation. *Nature*. 455, 497–503.
52. Schwartz, T.W., Hubbell, W.L. (2008) A moving story of receptors. *Nature*. 455, 473–474.
53. Cohen, G.B., Oprian, D.D., Robinson, P.R. (1992) Mechanism of activation and inactivation of opsin: Role of Glu^{113} and Lys^{296}. *Biochemistry*. 31, 12592–12601.
54. Vogel, R., Siebert, F. (2001) Conformations of the active and inactive states of opsin. *J Biol Chem*. 276, 38487–38493.

55. Lamb, T.D., Pugh Jr., E.N. (2004) Dark adaptation and the retinoid cycle of vision. *Prog Retin Eye Res*. 23, 307–380.

56. Xie, X.-Q., Chen, J.-Z., Billings, E.M. (2003) 3D structural model of the G-protein-coupled cannabinoid CB_2 receptor. *Proteins*. 53, 307–319.

57. Vaidehi, N., Floriano, W.B., Trabanino, R., Hall, S.E., Freddolino, P., Choi, E.J., Zamanakos, G., Goddard III, W.A. (2002) Prediction of structure and function of G protein-coupled receptors. *Proc Natl Acad Sci U S A*. 99, 12622–12627.

58. Shacham, S., Topf, M., Avisar, N., Glaser, F., Marantz, Y., Bar-Haim, S., Noiman, S., Naor, Z., Becker, O.M. (2001) Modeling the 3D structure of GPCRs from sequence. *Med Res Rev*. 21, 472–483.

59. Kortagere, S., Welsh, W.J. (2006) Development and application of hybrid structure based method for efficient screening of ligands binding to G-protein coupled receptors. *J Comput Aided Mol Des*. 20, 789–802.

60. Moro, S., Deflorian, F., Bacilieri, M., Spalluto, G. (2006) Ligand-based homology modeling as attractive tool to inspect GPCR structural plasticity. *Curr Pharm Des*. 12, 2175–2185.

61. Topiol, S., Sabio, M. (2008) Use of the X-ray structure of the beta2-adrenergic receptor for drug discovery. *Bioorg Med Chem Lett*. 18, 1598–1602.

62. Sabio, M., Jones, K., Topiol, S. (2008) Use of the X-ray structure of the β2-adrenergic receptor for drug discovery. Part 2: Identification of active compounds. *Bioorg Med Chem Lett*. 18, 5391–5395.

63. Costanzi, S. (2008) On the applicability of GPCR homology models to computer-aided drug discovery: A comparison between in silico and crystal structures of the β_2-adrenergic receptor. *J Med Chem*. 51, 2907–2914.

64. Freddolino, P.L., Kalani, M.Y.S., Vaidehi, N., Floriano, W.B., Hall, S.E., Trabanino, R.J., Kam, V.W.T., Goddard III, W.A. (2004) Predicted 3D structure for the human β_2 adrenergic receptor and its binding site for agonists and antagonists. *Proc Natl Acad Sci U S A*. 101, 2736–2741.

65. Selent, J., López, L., Sanz, F., Pastor, M. (2008) Multi-receptor binding profile of clozapine and olanzapine: A structural study based on the new β_2 adrenergic receptor template. *ChemMedChem*. 3, 1194–1198.

66. de Graaf, C., Rognan, D. (2008) Selective structure-based virtual screening for full and partial agonists of the β2 adrenergic receptor. *J Med Chem*. 51, 4978–4985; erratum 51, 6620.

67. Baker, J.G., Proudman, R.G.W., Hawley, N.C., Fischer, P.M., Hill, S.J. (2008) Role of key transmembrane residues in agonist and antagonist actions at the two conformations of the human β1-adrenoceptor. *Mol Pharm*. 74, 1246–1260.

68. Lee, T., Schwandner, R., Swaminath, G., Weiszmann, J., Cardozo, M., Greenberg, J., Jaeckel, P., Ge, H., Wang, Y., Jiao, X., Liu, J., Kayser, F., Tian, H., Li, Y. (2008) Identification and functional characterization of allosteric agonists for the G protein-coupled receptor FFA2. *Mol Pharm*. 74, 1599–1609.

69. Kew, J.N.C. (2004) Positive and negative allosteric modulation of metabotropic glutamate receptors: Emerging therapeutic potential. *Pharmacol Ther*. 104, 233–244.

70. Ritzén, A., Mathiesen, J.M., Thomsen, C. (2005) Molecular pharmacology and therapeutic prospects of metabotropic glutamate receptor allosteric modulators. *Basic Clin Pharmacol Toxicol*. 97, 202–213.

71. Marino, M.J., Conn, P.J. (2006) Glutamate-based therapeutic approaches: Allosteric modulators of metabotropic glutamate receptors. *Curr Opin Pharmacol.* 6, 98–102.

72. Goudet, C., Gaven, F., Kniazeff, J., Vol, C., Liu, J., Cohen-Gonsaud, M., Acher, F., Prézeau, L., Pin, J.P. (2004) Heptahelical domain of metabotropic glutamate receptor 5 behaves like rhodopsin-like receptors. *Proc Natl Acad Sci U S A.* 101, 378–383.

73. Kunishima, N., Shimada, Y., Tsuji, Y., Sato, T., Yamamoto, M., Kumasaka, T., Nakanishi, S., Jingami, H., Morikawa, K. (2000) Structural basis of glutamate recognition by a dimeric metabotropic glutamate receptor. *Nature.* 407, 971–977.

74. Tsuchiya, D., Kunishima, N., Kamiya, N., Jingami, H., Morikawa, K. (2002) Structural views of the ligand-binding cores of a metabotropic glutamate receptor complexed with an antagonist and both glutamate and Gd^{3+}. *Proc Natl Acad Sci U S A.* 99, 2660–2665.

75. Muto, T., Tsuchiya, D., Morikawa, K., Jingami, H. (2007) Structures of the extracellular regions of the group II/III metabotropic glutamate receptors. *Proc Natl Acad Sci U S A.* 104, 3759–3764.

76. Parnot, C., Kobilka, B. (2004) Toward understanding GPCR dimers. *Nature Struct Mol Biol.* 11, 691–692.

77. Costantino, G. (2006) In silico approaches towards the understanding of the structure-function relationships in metabotropic glutamate receptors (mGluRs) and other family C GPRCs. *Curr Pharm Des.* 12, 2159–2173.

78. O'Hara, P.J., Sheppard, P.O., Thogersen, H., Venezia, D., Haldeman, B.A., McGrane, V., Houamed, K.M., Thomsen, C., Gilbert, T.L., Mulvihill, E.R. (1993) The ligand-binding domain in metabotropic glutamate receptors is related to bacterial periplasmic binding proteins. *Neuron.* 11, 41–52.

79. Hampson, D.R., Huang, X.-P., Pekhletski, R., Peltekova, V., Hornby, G., Thomsen, C., Thøgersen, H. (1999) Probing the ligand-binding domain of the mGluR4 subtype of metabotropic glutamate receptor. *J Biol Chem.* 274, 33488–33495.

80. Malherbe, P., Knoflach, F., Broger, C., Ohresser, S., Kratzeisen, C., Adam, G., Stadler, H., Kemp, J.A., Mutel, V. (2001) Identification of essential residues involved in the glutamate binding pocket of the group II metabotropic glutamate receptor. *Mol Pharmacol.* 60, 944–954.

81. Dominguez, C., Prieto, L., Valli, M.J., Massey, S.M., Bures, M., Wright, R.A., Johnson, B.G., Andis, S.L., Kingston, A., Schoepp, D.D., Monn, J.A. (2005) Methyl substitution of 2-sminobicyclo[3.1.0]hexane 2,6-dicarboxylate (LY354740) determines functional activity at metabotropic glutamate receptors: Identification of a subtype selective mGlu2 receptor agonist. *J Med Chem.* 48, 3605–3612.

82. Monn, J.A., Massey, S.M., Valli, M.J., Henry, S.S., Stephenson, G.A., Bures, M., Hérin, M., Catlow, J., Giera, D., Wright, R.A., Johnson, B.G., Andis, S.L., Kingston, A., Schoepp, D.D. (2007) Synthesis and metabotropic glutamate receptor activity of S-oxidized variants of (-)-4-amino-2-thiabicyclo-[3.1.0]hexane-4,6-dicarboxylate: Identification of potent, selective, and orally bioavailable agonists for mGlu2/3 receptors. *J Med Chem.* 50, 233–240.

83. Bertrand, H.O., Bessis, A.S., Pin, J.P., Acher, F.C. (2002) Common and selective molecular determinants involved in metabotropic glutamate receptor agonist activity. *J Med Chem.* 45, 3171–3183.

84. Yao, Y., Pattabiraman, N., Michne, W.F., Huang, X.P., Hampson, D.R. (2003) Molecular modeling and mutagenesis of the ligand-binding pocket of the mGlu3 subtype of metabotropic glutamate receptor. *J Neurochem.* 86, 947–957.

85. Palczewski, K., Kumasaka, T., Hori, T., Behnke, C.A., Motoshima, H., Fox, B.A., Trong, I.L., Teller, D.C., Okada, T., Stenkamp, R.E., Yamamoto, M., Miyano, M. (2000) Crystal structure of rhodopsin: A G protein-coupled receptor. *Science.* 289, 739–745.

86. Teller, D.C., Okada, T., Behnke, C.A., Palczewski, K., Stenkamp, R.E. (2001) Advances in determination of a high-resolution three-dimensional structure of rhodopsin, a model of G-protein-coupled receptors (GPCRs). *Biochemistry.* 40, 7761–7772.

87. Li, J., Edwards, P.C., Burghammer, M., Villa, C., Schertler, G.F.X. (2004) Structure of bovine rhodopsin in a trigonal crystal form. *J Mol Biol.* 343, 1409–1438.

88. Okada, T., Sugihara, M., Bondar, A.-N., Elstner, M., Entel, P., Buss, V. (2004) The retinal conformation and its environment in rhodopsin in light of a new 2.2 Å crystal structure. *J Mol Biol.* 342, 571–583.

89. Nakamichi, H., Okada, T. (2006) Local peptide movement in the photoreaction intermediate of rhodopsin. *Proc Natl Acad Sci U S A.* 103, 12729–12734.

90. Nakamichi, H., Okada, T. (2006) Crystallographic analysis of primary visual photochemistry. *Angew Chem Int Ed Engl.* 45, 4270–4273.

91. Salom, D., Lodowski, D.T., Stenkamp, R.E., Trong, I.L., Golczak, M., Jastrzebska, B., Harris, T., Ballesteros, J.A., Palczewski, K. (2006) Crystal structure of a photoactivated deprotonated intermediate of rhodopsin. *Proc Natl Acad Sci U S A.* 103, 16123–16128.

92. Nakamichi, H., Buss, V., Okada, T. (2007) Photoisomerization mechanism of rhodopsin and 9-cis-rhodopsin revealed by X-ray crystallography. *Biophys J.* 92, L106–L108.

93. Stenkamp, R.E. (2008) Alternative models for two crystal structures of bovine rhodopsin. *Acta Crystallogr D Biol Crystallogr.* D64, 902–904.

94. Schrödinger, L.L.C. (2008) *Maestro v8.5.* Portland, OR.

CHAPTER 17

Pharmacological Chaperones: Potential for the Treatment of Hereditary Diseases Caused by Mutations in G Protein-Coupled Receptors

KENNETH J. VALENZANO,[1] ELFRIDA R. BENJAMIN,[1] PATRICIA RENÉ,[2] and MICHEL BOUVIER[2]

[1]Amicus Therapeutics, Cranbury, NJ

[2]Département de Biochimie, Université de Montréal, Montréal, Quebec, Canada

17.1. OVERVIEW

Many human diseases result from mutations in specific genes. Once translated, the resulting aberrant proteins are often functionally competent and produced at normal levels. However, because of the mutations, the proteins are recognized as less stable by the quality control system of the endoplasmic reticulum (ER) and, as such, are not processed and trafficked correctly, ultimately leading to cellular dysfunction and disease. Small molecule pharmacological chaperones are a promising new therapeutic approach to treat these genetic disorders. Pharmacological chaperones selectively bind to the mutant proteins and stabilize a near-native conformation. This stabilization promotes normal trafficking of the mutant protein and allows passage through the ER quality control system, ultimately increasing protein levels and activity in relevant cellular locations and reducing ER accumulation, aggregation, and associated cell stress. Partial or complete restoration of normal function by pharmacological chaperones has been shown for numerous types of mutant proteins, including enzymes, secreted proteins, transcription factors, ion channels, and,

GPCR Molecular Pharmacology and Drug Targeting: Shifting Paradigms and New Directions, Edited by Annette Gilchrist

importantly, G protein-coupled receptors (GPCRs). This review highlights human diseases that are known to result from genetic mutations that lead to less stable or misfolded intracellularly retained GPCRs, and hence have the potential for pharmacological chaperone therapy. Particular emphasis is given to those diseases that have established proof of concept for the beneficial effects of pharmacological chaperones on the underlying mutant GPCRs, as demonstrated in cells, animals, and/or humans, including X-linked nephrogenic diabetes insipidus, retinitis pigmentosa, and idiopathic hypogonadotropic hypogonadism. In addition, considerations for the successful use and development of novel pharmacological chaperone therapies will be discussed.

17.2. INTRODUCTION

G protein-coupled receptors (GPCRs) represent the largest family of integral membrane receptors, comprised of over 1200 distinct members. GPCRs are structurally conserved, sharing a common architecture of seven transmembrane (TM) domains connected by three extracellular and three intracellular loops. In addition, the GPCR superfamily has been further divided into seven subclasses, denoted A, B, C, large N-terminal Family B-7TM, frizzled/smoothened, taste 2, and vomeronasal 1, based upon their native ligands, phylogenetic analysis of primary sequences, clustering analysis of genes in the human genome, and analysis of globular domains and motifs in the amino-terminus of large N-terminal Family B-7TM receptors (for review, see Reference 1). GPCRs are intimately involved in intercellular and intracellular communication, and either directly or indirectly control and modulate a variety of physiological functions, including, but not limited to, olfaction, gustation, vision, nociception, inflammation, neurotransmission, sex organ development/function, and cardiovascular, renal, and calcium homeostasis. Not surprisingly, over 60% of marketed drugs target this class of proteins. Furthermore, it has become evident that a variety of human diseases result from mutations in the genes coding for members of the GPCR superfamily, which can lead to constitutive activity (activating mutations) or loss of function (inactivating mutations). To date, more than 30 monogenic diseases have been described that are caused by mutant forms of GPCRs (for review, see Reference 2). As the focus of this review is restoration of function of mutated GPCRs that have been implicated in human disease, subsequent discussions will focus on those diseases that result from inactivating mutations that can lead to trafficking-defective receptors (Table 17.1).

Inactivating mutations in GPCRs can alter receptor structure and function to varying degrees. Large deletions, truncations, or frameshift mutations can lead to loss of entire receptor domains that grossly alter structure and function, and may even result in the complete loss of receptor expression. Similarly, splice site mutations can lead to incorrect processing of mRNA precursors, including exon skipping or splicing at cryptic splice points, often leading to

TABLE 17.1 Trafficking-Defective Mutant GPCRs Implicated in Human Diseases

Disease or Condition	GPCR	Class	Reference
RP	Rhodopsin	A	[111–114]
Hypogonadotropic hypogonadism	Gonadotropin-releasing hormone receptor	A	[4, 150]
X-linked NDI	V2R	A	[27]
Hirschsprung disease	ETbR	A	[213]
Male pseudohermaphroditism	LHR	A	[201, 202, 244]
Ovarian dysgenesis	FSH receptor	A	[208, 211]
Congenital hypothyroidism	TSHR	A	[199, 245]
Red head color and fair skin phenotype and propensity to skin cancer	MC1R	A	[189, 190]
FGD	Melanocortin 2 (ACTH) receptor	A	[246–248]
Congenital morbid obesity	MC3R	A	[180, 181]
Congenital morbid obesity	MC4R	A	[180, 249, 250]
Resistance to HIV-1 infection	CCR5 chemokine receptor	A	[251, 252]
Dominant inherited bleeding disorder	Thromboxane A2 receptor	A	[253]
Blomstrand chondrodysplasia	PTH-R1	B	[180, 220]
Dwarfism	GHRHR	B	[180, 254]
FHH/NSHPT	CaR	C	[223, 255]
FEVR	Frizzled-4 receptor	Frizzled/smoothened	[232, 233]

gross alterations in the structure and function of the receptor. Small in-frame deletions and insertions, or missense mutations, which result in a single base pair substitution in the coding sequence, lead to more subtle changes in receptor structure that can influence mRNA expression, protein folding, intracellular trafficking, ligand binding, signal transduction, internalization, and/or receptor turnover. Approximately two-thirds of the mutations in GPCRs that cause human disease are missense, and over 40% of the missense mutations and small in-frame insertions and deletions are estimated to result in less stable or trafficking-defective receptors (for review, see References 2–4).

Folding and maturation of GPCRs are monitored by the quality control system of the endoplasmic reticulum (ER), which permits correctly folded and assembled proteins to leave this cellular compartment and progress through the secretory pathway to their final destinations [5, 6]. The primary quality control mechanisms rely on ubiquitous *molecular* chaperones and folding factors, like BiP, calnexin, calreticulin, thiol-disulfide oxidoreductases, protein disulfide isomerase, ERp57, and ERdj5, that recognize common structural features to distinguish stable native protein conformations from unstable nonnative ones (for review, see References 7–9). Important features for recognition include exposure of hydrophobic regions, unpaired cysteine residues, and tendency to aggregate. The mechanisms that distinguish native receptor conformations from nonnative ones and assist in folding are undoubtedly crucial for biogenesis and cell surface expression, but they have remained largely uncharacterized. Three molecular chaperones, calnexin, calreticulin, and BiP, which are known to be involved in the primary ER quality control, have been shown to interact with GPCRs (for review; see References 10–14). In addition, cytosolic chaperones, such as Hsp40 proteins, HSJ1a and HSJ1b, and Hsp70, have been shown to interact with GPCRs [15, 16]. Furthermore, more specialized chaperones and escort proteins are involved in the folding and trafficking of specific GPCRs (for review, see References 14, 17). If, despite the action of the chaperones, folding of the nascent protein fails, they are recognized by the ER quality control system as aberrant and targeted for degradation. This ER-associated degradation (ERAD) involves polyubiquitination and retrotranslocation through the Sec61 translocon to the cytosol, where the less stable or misfolded proteins are degraded by the 26S proteasomes (for review, see Reference 18).

Although highly efficient in most cases, the quality control systems are imperfect. In some cases, proteins harboring mutations that do not, or only modestly, compromise their functional integrity may be recognized as less stable or misfolded, leading to unnecessary intracellular retention and premature degradation. As a consequence, proteins that have slight modifications in their stability or conformation may not be released from the ER and are degraded, resulting in a loss-of-function phenotype due to absence at their normal site of action.

In some cases, the pathogenesis associated with inherited mutations in GPCRs has been linked to enhanced susceptibility of the mutated protein to degradation by the ubiquitin–proteasome system as just described, leading to loss of function (for review, see Reference 19), while in other cases, the mutated proteins accumulate in cells, forming aggregates that can have toxic consequences, often referred to as gain of function (for review, see References 20, 21). Efforts to address these underlying molecular defects have led to the development of numerous types of interventions that have the potential to rescue mutant proteins from aggregation and/or degradation. One promising new strategy utilizes cell-permeant small molecules that have the potential to specifically bind folding intermediates of mutated proteins, conferring enhanced thermodynamic stability and facilitating proper transport through the secretory pathway to the correct site of action (for review, see References 4, 22). These molecules appear to act primarily in the ER early during biosynthesis, where they facilitate the release of the mutated proteins from the ER quality control machinery, preventing premature degradation by the ubiquitin–proteasome system and promoting transport through the Golgi apparatus and, ultimately, to the cell surface (for review, see References 19, 23, 24).

Facilitated protein folding and restoration of normal cellular trafficking are also expected to relieve stress on the ER associated with protein accumulation and to minimize the toxic consequences of protein aggregation (for review, see Reference 21). Because these small molecules are designed to bind specifically to the protein of interest, thereby promoting proper cellular trafficking, they have been termed *pharmacological* chaperones. It is important to differentiate pharmacological chaperones from *chemical* chaperones, which are small molecules that nonspecifically bind and generally increase protein stability (e.g., dimethyl sulfoxide [DMSO], glycerol, trimethylamine oxide [TMAO]) (for review, see Reference 25). In addition to GPCRs, restoration of partial or complete function by pharmacological chaperones has been shown for other types of mutated proteins, including enzymes, secreted proteins, transcription factors, ion channels, and transporters (see Table 17.2), that lead to diseases, such as cystic fibrosis, hypercholesteremia, cataracts, Huntington's, Alzheimer's, and Parkinson's diseases, cancer, and numerous lysosomal storage disorders.

While many GPCRs that harbor missense mutations or small in-frame insertions or deletions fail to exit the ER efficiently, introduction of a pharmacological chaperone may not only increase cell surface expression but may also restore the ability of the receptor to bind endogenous ligand and activate intracellular signaling pathways. To this end, proof of concept for pharmacological chaperones at the cellular level has been established for mutant forms of several Class A receptors, including the vasopressin V2 receptor (V2R), rhodopsin, and the gonadotropin-releasing hormone receptor, which underlie the molecular bases for X-linked nephrogenic diabetes insipidus (NDI), autosomal dominant forms of retinitis pigmentosa (RP), and idiopathic hypogonadotropic hypogonadism (IHH), respectively. Furthermore, pharmacological chaperone proof of concept in animals has been established for RP and in the clinic for both RP and NDI. More recently, proof of concept at the cellular level has also been established for a Class C GPCR, the calcium-sensing receptor (CaR). This review will first detail the use of pharmacological chaperones for NDI, RP, and IHH, followed by brief discussions of their potential to treat other diseases that result from GPCR loss of function, and will close with some important considerations for their therapeutic use.

17.3. NDI AND THE V2R

Congenital NDI is a hereditary disease associated with renal tubular resistance to the antidiuretic hormone vasopressin [26]. NDI is most common in its acquired form, due to electrolyte disturbances, urinary tract obstruction, or as a side effect of lithium salts used to treat bipolar disorder [27]. In its congenital form, two primary molecular defects can cause NDI: Mutations in the gene encoding the aquaporin-2 water channel (AQP2) lead to the autosomal dominant and recessive forms of NDI [28–30]; mutations in the gene (*AVPR2*) encoding the V2R lead to the X-linked recessive form of NDI and account for

TABLE 17.2 Representative Diseases/Mutated Proteins for Which Pharmacological Chaperone Proof of Concept Has Been Established

Disease or Condition	Protein/Class	Pharmacological Chaperone	Reference
RP	Rhodopsin/GPCR	9-*cis*-retinal	[124, 135, 136]
		11-*cis*-retinal	
		11-*cis*-7-ring-retinal	
Hypogonadotropic hypogonadism	Gonadotropin-releasing hormone receptor/GPCR	Indoles	[154, 163]
		Quinolones	
		Erythromycin-derived macrolides	
X-linked NDI	V2R/GPCR	SR121463	[27, 58, 61, 62, 64]
		SR49059	
		VPA-985	
		YM087	
		OPC41061	
		OPC31260	
FHH/NSHPT	CaR/GPCR	NPS R-568	[193]
Cancer	Smo/GPCR	Cyclopamine	[256]
Long QT syndrome	hERG/potassium channel	Cisapride	[257–259]
		E-4031	
		Astemizole	
Cystic fibrosis	CFTR/chloride transporter	Benzo(c)quinoliziniums	[260–264]
		Curcumin	
		Aminobenzothiazoles	
		Aminoarylthiazoles	
		Quinazolinylaminopyrimidinones	
		Bisaminomethylbithiazoles	
		4-cyclohexyloxy-2-{1-[4-(4-methoxy-benzenesulfonyl)piperazin-1-yl]ethyl} quinazoline	
		Sildenafil	
Hyperinsulinemic hypoglycemia	SUR1 (sulfonylurea receptor)/potassium channel-associated protein	Diazoxide	[265, 266]
		Sulfonylureas	

TABLE 17.2 *(Continued)*

Disease or Condition	Protein/Class	Pharmacological Chaperone	Reference
Phenylketonuria	Phenylalanine hydroxylase/enzyme	(3-amino-2-benzyl-7-nitro-4-(2-quinolyl)-1,2-dihydroisoquinolin-1-one) (5,6-dimethyl-3-(4-methyl-2-pyridinyl)-2-thioxo-2,3-dihydrothieno[2,3-d]pyrimidin-4(1H)-one)	[267]
Gaucher disease	β-glucocerebrosidase/lysosomal enzyme	*N*-(*n*-nonyl)deoxynojirimycin Isofagomine	[268, 269]
β-galactosidosis	β-galactosidase/lysosomal enzyme	*N*-octyl-4-epi-β-valienamine 1-deoxy-galactonojirimycin *N*-(*n*-butyl)-deoxy-galactonojirimycin	[270, 271]
Fabry disease	α-galactosidase A/lysosomal enzyme	1-deoxygalactonojirimycin Galactose	[272, 273]
Tay–Sachs disease	β-hexosamidase/lysosomal enzyme	*N*-acetyl-glucosamine-thiazoline Pyrimethamine Bisnaphthalimide Nitro-indan-1-one Pyrrolo[3,4d]pyridazin-1-one	[274–276]
Pompe disease	Acid α-glucosidase/lysosomal enzyme	*N*-butyl-deoxynojirimycin	[277, 278]
Cancer	p53/transcription factor	PRIMA-1 Peptide Fl-CDB3	[279, 280]
Menkes disease	MNK/copper-transporting ATPase	Copper	[281]
α1-antitrypsin deficiency-mediated cirrhosis	α1-antitrypsin/secreted serine proteinase inhibitor	6-mer peptide Ac-FLEAIG 6-mer peptide non-Ac-FLEAIG Numerous small molecule polymerization inhibitors	[282–284]
Prion disease	PrP^{sc}/membrane-bound copper binding protein	IPrP13 Quinacrine Chlorpromazine	[285, 286]

CFTR, cystic fibrosis transmembrane conductance regulator.

~75% of all cases. X-linked NDI is very rare, occurring in ~1:250,000 individuals. Affected patients are unable to efficiently concentrate their urine, resulting in the excretion of large volumes of hypotonic urine (up to 25 L/day in adults). If sufficient quantities of water are not ingested, this polyuria can lead to severe episodes of dehydration with clinical manifestations, including hypernatremia, hyperthermia, vomiting, and constipation [31]. The occurrence of repeated episodes of dehydration in children often leads to mental retardation and to growth disorders that can, in extreme cases, be fatal as a result of hypertonic encephalopathy.

In normal subjects, the action of vasopressin on water reabsorption and urine concentration by the kidney results from its interaction with V2R located on the basolateral surface of the principal cells of the collecting ducts. Ligand stimulation of the V2R promotes intracellular cyclic adenosine monophosphate (cAMP) accumulation and leads to the activation of the cAMP-dependent protein kinase (PKA). Upon activation of this pathway, intracellular vesicles containing high concentrations of AQP2 migrate toward and fuse with the apical membrane of the principal cells, thus dramatically increasing their water permeability [32–34]. The reabsorbed water then exits the cell through AQP3 and AQP4 that are constitutively expressed on the basolateral plasma membrane [35]. PKA-mediated phosphorylation of AQP2 itself [36] and interaction with PKA anchoring proteins [37, 38] have been shown to play essential roles in the vesicular trafficking of this water channel. Accumulating evidence, however, indicates the existence of other yet ill-defined trafficking mechanisms that could also contribute to the response [39, 40]. In addition to its rapid translocation to the plasma membrane, longer term regulation of AQP2 expression has been shown to occur in response to vasopressin stimulation, with both PKA-dependent [41] and PKA-independent [42] processes being involved. Interestingly, V2R was recently found to activate mitogen-activated protein (MAP) kinase independent of G protein and PKA activation by a mechanism relying on β-arrestin recruitment and involving c-Src as well as receptor tyrosine kinase transactivation events [43, 44]. MAP kinase activation has also been shown to potentiate vasopressin-promoted AQP2 expression [45]; however, the potential role of MAP kinase in AQP2 translocation remains to be investigated.

Patients with NDI fail to have an antidiuretic response to the administration of either vasopressin or the strong antidiuretic vasopressin agonist 1-desamino[8-D-arginine] vasopressin [46, 47]. It has been recognized for some time that renal tubular resistance to the action of vasopressin in NDI patients results from a defect in the signal transduction network mediating the cellular actions of vasopressin [48]. In 1992, the identification of two mutations in *AVPR2* of two unrelated males suffering from X-linked NDI confirmed that the genetic defect responsible for this form of the pathology resides in the V2R [49]. In humans, the V2R is comprised of 371 amino acids. To date, more than 200 distinct mutations have been identified in more than 250 unrelated affected families [4, 50, 51]. Approximately 50% of the mutations in the coding region of the V2R are missense and lead to a single amino acid substitution. The remainder are deletion and nonsense mutations that lead to frameshift and premature termination of the coding sequence [4]. In all cases, the mutation leads to a nonfunctional receptor unable to relay the antidiuretic action of vasopressin. Although a small number of mutations were found that either affected vasopressin binding [52–56] or G protein coupling [52, 55, 57], the vast majority (75 out of the 95) of missense mutations identified were found to be deficient in their cell surface expression [4]. These mutations are not clustered but rather spread throughout the V2R sequence [5, 22]. In most cases, the deficient cell surface expression was attributed to intracellular trapping of the receptor resulting from its retention by the ER quality control system.

For the V2R, association with calnexin was shown to be prolonged in the case of NDI mutants [11], indicating that calnexin may be intimately involved in V2R folding and ER retention of the V2R mutants. Biochemical analyses were also consistent with ER entrapment of the NDI V2R mutants. Indeed, intracellularly retained mutants were found to be immature core-glycosylated receptors that are sensitive to the action of the endoglycosidase H (Endo H), indicating that they did not undergo the glycosylation processing that normally occurs during Golgi transit [11, 58]. Immunofluorescence microscopy studies are also consistent with this notion since, in most cases, the intracellularly retained NDI V2R mutants were found to colocalize with the ER-restricted protein calnexin but not with Golgi markers [11, 58].

For some mutations, however, the receptor seems to be able to proceed through the ER to some extent, with subsequent retention in the Golgi [59]. The molecular basis leading to the retention of different V2R mutants in distinct compartments of the secretory pathway remains to be determined. It was suggested that some NDI V2R mutations can also lead to the inappropriate targeting of the receptor to endosomal compartments [60]. For instance, the R137H mutant was largely found in endocytotic vesicles as a result of its constitutive phosphorylation, stable interaction with β-arrestin, and internalization. However, as is found with most other NDI-causing mutations, the R137H V2R is also poorly processed through the ER. Thus, while a large proportion of the synthesized R137H mutant is retained in the ER similar to other V2R mutants, the small fraction that does traffic to the cell surface is rapidly and constitutively endocytosed [61].

Because many of the mutations leading to intracellular retention of mutant forms of V2R are single amino acid substitutions and would not be predicted to grossly affect receptor function, it was hypothesized that promoting the escape of mutant proteins from the ER quality control system could be sufficient to restore cell surface expression and thus enhance receptor function. In the first study assessing this hypothesis, cells heterologously expressing intracellularly retained NDI V2R mutants were exposed for several hours to the membrane-permeable V2R antagonist SR121463 [58]. The treatment-rescued cell surface expression of the eight mutants tested and restored their ability to stimulate cAMP production in response to vasopressin. Although the functional recovery was very significant for most mutant receptors, in some cases, the recovery was more modest despite the restored trafficking to the cell surface, indicating that these mutations may have detrimental effects on the activity of the receptor that cannot be corrected even if the receptors reach the cell surface. Since then, the ability of SR121463 and five other V2R antagonists (VPA-985, SR49059, OPC31260, OPC41061, and YM087) to rescue cell surface expression and function of various NDI V2R mutants has been confirmed in various cellular backgrounds [58, 61–66], including studies performed in MDCK kidney cells, where the polarized targeting of the receptors to the basolateral membrane of the cells was confirmed [27, 67]. To date, more than 25 NDI-causing V2R mutations have been tested, with most showing cell surface and functional rescue after treatment with cell-permeable V2R antagonists.

It was proposed that the vasopressin antagonists may function as selective pharmacological chaperones by binding to the precursor form of the mutant receptor, most likely in the ER, thus stabilizing the receptor in a conformation that allows for release from the quality control system. In agreement with this mechanism of action, only lipohilic antagonists that can penetrate biological membranes (and thus reach the protein secretory apparatus), and not membrane-impermeable peptide antagonists, could rescue cell surface expression of the intracellularly trapped NDI V2R mutants [58]. Also consistent with an action in the early phase of receptor ontogeny, pulse–chase metabolic labeling experiments showed that the mutant receptors could proceed from their immature core-glycosylated (Endo H-sensitive) form to the fully mature form harboring complex carbohydrates only following treatment with the membrane-permeable antagonists. The pharmacological selectivity of action was confirmed by the observation that treatment with other membrane-permeable GPCR antagonists that do not bind V2R did not rescue cell surface expression of the NDI V2R mutants, confirming that the compounds need to bind the mutant receptor to promote its release from the quality control system. These results, which point to a stabilization of a more native conformation as a result of binding of a ligand, are consistent with the observation that other types of interventions that facilitate folding into the native conformation, such as low-temperature incubation [67] or chemical chaperones [64], also rescued cell surface expression and function of some NDI V2R mutants. Interestingly, the action of different chemical chaperones and low temperature appeared to be mutation dependent, with distinct NDI V2R mutants being rescued by the different treatments. In contrast, less mutational variability, in response, was observed for the pharmacological chaperones, with the same broad subset of mutations being rescued by the different molecules tested. To date, none of the mutations that could be rescued by low temperature or chemical chaperones have been resistant to pharmacological chaperone action, indicating that the spectrum of action of the latter must be broader. Although it is generally assumed that V2R pharmacological chaperones act by binding to the mutant receptors during folding in the ER, a recent study showed that a membrane-permeable V2R antagonist could also rescue an NDI V2R mutant shown to be retained in the Golgi, suggesting that pharmacological chaperones could also favor escape of mutant receptors from quality control steps downstream of ER exit [66].

The observations that treatment with pharmacologically selective small molecules can restore cell surface expression and function of NDI V2R mutants have clear potential therapeutic implications for NDI patients. The possible use of pharmacological chaperones for the treatment of NDI was recently validated by a pilot clinical trial that tested whether SR49059 administration could restore vasopressin responsiveness in five NDI patients harboring three distinct mutations that were found to be rescued in heterologous expression systems (del62-64, W164S, and R137H). A 3-day treatment with SR49059 led to a 30% reduction in urine output and to a 40% increase in urine osmolality without affecting sodium, potassium, and creatinine excretion, or plasma sodium levels [62]. Unfortunately, the trial could not be extended for longer times or to other patients as a result of potential hepatotoxicity observed in a separate clinical trial being conducted with SR49059 for a different indication. Nevertheless, this study provided clear proof of concept that pharmacological chaperones represent a viable avenue for the treatment of NDI and, most likely, for other diseases resulting from decreased protein stability and misfolding. The advent of new vasopressin ligands available for clinical testing should allow the initiation of new clinical trials for NDI. The fact that a large proportion of patients carry missense mutations leading to intracellular retention and that many of these mutations can be rescued by pharmacological chaperones in heterologous cell systems allow us to be hopeful that, if efficacious, the treatment could be used for a significant proportion of NDI patients.

17.4. RP AND THE RHODOPSIN RECEPTOR

RP is a heterogeneous group of inherited disorders that lead to retinal degeneration [68]. Patients afflicted with RP show progressive night blindness, loss of central vision, rod cell degeneration often accompanied by subsequent cone loss, and progressive decreases in electroretinogram (ERG) potentials (for review, see Reference 69). The progressive death of rod and cone cells precipitates other pathological symptoms in the retina, including attenuation of retinal vasculature, a pale optic disk, and the accumulation of intra-retinal pigment deposits from which the disease derives its name (for review, see Reference 70). RP is estimated to affect 2 million people worldwide, with an incidence rate of approximately 1:3500 [71]. Mutations leading to RP have been found in over 20 different genes (for review, see References 69, 72–74]), with mutations in the rhodopsin gene representing the most common single cause of autosomal dominant RP (adRP) and accounting for up to 50% of all adRP cases (for review, see References 69, 75). To date, over 100 missense mutations and over 20 other insertion, deletion, nonsense, and splice site mutations that are associated with RP have been identified in the rhodopsin gene (http://www.sph.uth.tmc.edu/Retnet/disease.htm; http://www.retina-international.org/sci-news/rhomut.htm). P23H is the most common single mutant form of rhodopsin, accounting for up to 10% of all adRP cases [75]. Currently, there is no effective therapy for RP, though supplementation with vitamin A (the precursor of 11-*cis*-retinal; see below) has shown some efficacy in a subset of RP patients [76].

Rhodopsin is the dim light-activated photoreceptor located in vertebrate rod cells of the retina, where it functions to absorb light and initiate the visual photoexcitation response (for review, see Reference 77). Rhodopsin accounts for >70% of the total protein in the membranes of rod cell outer segment stacked disks. While rhodopsin is a member of the GPCR superfamily, it uniquely is comprised of two components: opsin, the membrane-bound polypeptide, and 11-*cis*-retinal, a chromophore component that is covalently bound to opsin via a protonated Schiff base. 11-*Cis*-retinal is derived from vitamin A and acts as an inverse agonist when occupying opsin's binding pocket that yields a distinct UV-visible spectrum with a λ_{max} of ~500 nm (for review, see Reference 78). Once bound to opsin, 11-*cis*-retinal is stable in the absence of light. Rhodopsin captures photons of light to catalyze the isomerization of 11-*cis*-retinal to all-*trans*-retinal. Upon photoisomerization, both 11-*cis*-retinal and opsin undergo a series of conformational changes that result in the conversion of rhodopsin to metarhodopsin II, the active form of rhodopsin. Once activated by rhodopsin, $G_t(\alpha)$ relieves inhibition of cyclic guanosine monophosphate (cGMP) phosphodiesterase, resulting in the hydrolysis of cGMP to 5′-GMP. The subsequent decrease in cytosolic cGMP triggers the closure of cGMP-gated calcium channels, lowering cytosolic calcium levels and leading to hyperpolarization of the cell, which is transmitted as a neural signal.

The gene for human rhodopsin is comprised of five exons, spans a total of ~7000 bases, and codes for a glycoprotein of 348 amino acids with a molecular mass of approximately 40 kDa [79]. Nascent rhodopsin molecules are co-translationally inserted into the ER membrane of the rod cell inner segment, where they undergo *N*-linked glycosylation and conformational maturation prior to export to the Golgi apparatus and vesicular transport to the stacked disk membranes in the rod cell outer segment (for review, see Reference 80). Like other GPCRs, rhodopsin is comprised of three domains: extracellular (intradiscal), TM, and intracellular (cytoplasmic). The integrity of these three domains is critical for proper function, as evidenced by adRP-associated mutations having been identified in all three regions. The X-ray crystal structure for bovine rhodopsin in the dark state has been solved at 2.8-Å resolution [81]. Mapping of point-mutated residues onto the structure for rhodopsin has been instrumental in providing a molecular understanding for many of the defects caused by mutations in rhodopsin's three domains.

The extracellular (intradiscal) domain is comprised of the amino-terminus (residues 1–33) as well as extracellular loops I (residues 101–105), II ("plug" residues 173–198), and III (residues 277–285). The amino-terminal domain has two *N*-linked glycosylation sites at Asn2 and Asn15, which are important for proper folding and trafficking of rhodopsin. Mutations in this region of the protein result in improper folding and poor 11-*cis*-retinal binding (for review, see Reference 82). Extracellular loop II contains a twisted β-hairpin and Cys187 that forms a disulfide bond with Cys110 in α-helix III of the TM domain; this is critical for proper folding, stability, and function [83–86]. Additionally, extracellular loop II forms a "plug" upon which 11-*cis*-retinal lies, with Glu181 and Cys187 in close proximity to the C12 of retinal [81, 86]. Similarly, residues of the amino-terminus form a series of β-strands that are also an integral part of this amino-terminal plug [78], which keeps 11-*cis*-retinal in the proper position.

The TM domain, comprised of the seven 20–33 residue membrane-spanning α-helices (I–VII), is structurally coupled to the extracellular domain, such that mutations in the extracellular domain can impact opsin's ability to bind 11-*cis*-retinal in the TM domain on the ε-amino group of Lys296 within α-helix VII [87–89]. Substitution of Lys296, as is found in some patients with severe adRP, abolishes 11-*cis*-retinal binding and leads to constitutive activity of opsin in the absence of light [90, 91]. Once bound to wild-type opsin, the 11-*cis*-retinal chromophore is further stabilized by salt bridge formation with negatively charged Glu113 of α-helix III [92, 93]; other TM domain residues are also important for chromophore stabilization, photoisomerization efficiency, opsin conformational changes, and optimal wavelength absorption [94–96]. Charged residues on the cytoplasmic side of the TM domain, including the conserved Glu134-Arg135-Tyr136 triplet, are critical for interaction with, and activation of, $G_t(\alpha)$ [97]. Lastly, mutations in the TM domain can lead to formation of an abnormal disulfide bond between Cys185 and Cys187 in the extracellular domain, preventing proper function and leading to irreversible protein misfolding [84, 85, 96, 98, 99].

The intracellular (cytoplasmic) domain consists of intracellular loops I, II, and III, an additional α-helical loop formed between the end of membrane-spanning α-helix VII and the palmitoyl attachment sites at Cys322 and Cys323 [100, 101], as well as the carboxy-terminal tail. It is now clear that the intracellular loops are critical for G_t binding as mutations in loops II and III impair rhodopsin signaling [97, 102–104]. In addition, rhodopsin kinase associates with the carboxy-terminal tail and phosphorylates Ser334, Ser338, and Ser343 upon photoactivation [105–107], thus promoting interaction with visual arrestin and termination of signaling [108]. Mutations in this region may also result in impaired trafficking to the rod outer segment (see below).

P23H was the first point mutation identified within rhodopsin that was associated with adRP [109] and is the most common mutation found in adRP, accounting for up to 40% of the cases in the United States [110, 111]. As mentioned above, over 100 missense mutations and approximately 20 other insertion, deletion, nonsense, and splice site mutations that are associated with adRP have now been identified and are randomly distributed throughout the sequence of rhodopsin. Overexpression of mutant rhodopsins that have been identified in adRP patients show a number of molecular impairments, including protein folding, 11-*cis*-retinal chromophore binding, G protein coupling/activation, and/or intracellular trafficking. On the basis of these biochemical defects, mutant rhodopsins have generally been classified into three categories [111–115]. *In vitro*, Class I mutants are similar to wild-type rhodopsin in that they can form functional photopigment upon reconstitution with 11-*cis*-retinal in the dark and show plasma membrane localization when expressed in HEK293 or COS cells [112–114]. Class I mutants typically cluster near the carboxy-terminal domain of rhodopsin and are defective in trafficking to the rod outer segment *in vivo* [112, 116–119]. Some Class I mutants also inefficiently couple to G_t [82, 120, 121]. Class II mutants are the most prevalent, showing defects in 11-*cis*-retinal binding, glycosylation, and diminished cell surface localization due to various extents of ER retention [111, 114]. Compared to wild-type rhodopsin, several members of this mutant class (e.g., P23H, G188R) have extensive, long-lived interactions with ER-resident molecular chaperones (including Grp78, Hsp60, and Grp94) [75, 122] and are substrates for degradation by the ubiquitin–proteasome system [123, 124]. In addition, Class II mutants have been further subdivided, with Class IIa mutants showing less cell surface expression than Class IIb mutants [112]. Mutations leading to the Class II phenotype are found in all three domains [111, 113, 114, 125, 126]. Lastly, Class III mutants (characterized by mutations at Arg135) show low levels of retinal binding, are hyperphosphorylated, and are constitutively internalized when expressed in HEK293 cells due to prolonged interaction with visual arrestin [115]. These rhodopsin–arrestin complexes alter the morphology of endosomal compartments and severely damage receptor-mediated endocytic functions, suggesting that impaired endocytic activity may underlie the pathogenesis of adRP caused by Class III mutations [115].

Because more than 100 adRP mutations have been identified and these mutations are distributed throughout the different functional domains of opsin, it is likely that they lead to retinal degeneration through multiple pathways. Photoreceptor death in adRP is believed to occur by apoptosis, with sorting failures and prolonged rhodopsin activation having been postulated as key contributors to cell death in patients (for review, see Reference 127). In some cases, adRP and other neurodegenerative diseases appear to share a common etiology, having abundant production of folding-defective polypepetides that lead to overt inclusion bodies and ubiquitin immunoreactivity [123, 124, 128, 129]. The P23H rhodopsin is degraded by the ubiquitin–proteasome system and, unlike wild-type and the Class I mutant V345M rhodopsin, forms aggregates and results in a generalized impairment of the ubiquitin–proteasome system even when expressed at low levels. These data suggest that, in the case of Class II mutants, rhodopsin aggregation in the ER and Golgi may lead to a toxic gain of function, similar to other aggregation-prone proteins associated with neurodegenerative diseases. In the case of Class I mutants, which show attenuated targeting to the rod outer segment and subsequent accumulation in the plasma membrane, other possible mechanisms for the induction of apoptosis that are consistent with a sorting defect at or beyond the *trans*-Golgi network have been proposed [130]. Mutant rhodopsin may overwhelm the normal vesicular machinery of the plasma membrane-bound pathway, interfering with the routing of legitimate cargo. In addition, the physical presence of high levels of mutant opsin in the plasma membrane could interfere with normal cellular processes, such as neurotransmitter release. Lastly, the source of photoreceptor damage could be related to the buildup of protein in the cell body, perhaps by producing an excessive metabolic burden associated with its destruction.

Based on *in vitro* studies, it is now clear that the stability of opsin can be increased upon binding of 11-*cis*-retinal or other retinoids [131–134], and stability is a key property in the mechanism of a pharmacological chaperone. Thus, via stabilization and structural correction of less stable mutant forms of opsin, retinoids have the potential to act as pharmacological chaperones to increase the cellular content and cell surface localization of mutant opsins. To this end, the level of T17M rhodopsin isolated from cell membranes increased 10-fold when overexpressed in the presence of 11-*cis*-retinal [117]. Similarly, cellular expression levels and cell surface localization of P23H rhodopsin were increased when incubated with 9-*cis*-retinal, 11-*cis*-retinal, or a locked, non-photoisomerizable 7-membered ring form of retinal (11-*cis*-7-ring-retinal) [124, 135, 136]. Recent high-throughput screening efforts have identified additional molecules that can increase cell surface expression of P23H rhodopsin [137]. These molecules were shown to have low affinity for opsin and, as such, could be displaced with 11-*cis*-retinal after introduction to the culture media. Importantly, mutant rhodopsin was formed efficiently only when retinals were added during opsin expression, which facilitated trafficking of the mutant opsin through the secretory pathway to the cell surface [136], suggesting that the interaction and stabilization occur early in the biogenesis of opsin and most likely facilitate the folding and ER export of mutant forms of opsin. These data are supported by earlier studies showing that rhodopsin is present in the rough ER of rod cells, indicating that retinal binds to opsin in the inner segment early in its biogenesis [138]. Collectively, these studies demonstrate that retinoids can act as pharmacological chaperones to stabilize mutant forms of opsin, facilitating retinal binding and generating a functional rhodopsin. Furthermore, studies using the "locked" forms of retinals that are easily accepted into the binding site of opsin may provide additional information about contact sites that could be important to more efficiently facilitate folding and stabilization of mutant forms of rhodopsin [139, 140].

Several animal models of rhodopsin-associated RP have been generated. The rhodopsin knockout mouse, which lacks any functional rhodopsin, undergoes rapid retinal degeneration in the first 3 months of life, with reductions in the quantity of rod photoreceptors and thinning of the rod outer segment noted as early as postnatal day 24 [141]. More importantly, transgenic mouse lines that have different copy numbers of human P23H rhodopsin have been created on a wild-type background. The mice show a gene-dosing effect for disease severity that closely mimics the human condition, including diminished or absent ERG responses and profound loss of photoreceptor cells that begins in the central retina and progresses peripherally [125, 126]. A similar pattern of retinal degeneration is also seen in mice harboring a murine transgene containing P23H opsin as well as two other non-adRP-related point mutations located at the amino-terminus (V20G and P27L) [142]. Interestingly, overexpression of human wild-type rhodopsin at five times the normal levels also caused retinal degeneration, similar to that seen in mice carrying the mutant P23H transgene [125]. These studies indicate that photoreceptors, similar to HEK293 and COS cells, can express mutant forms of rhodopsin *in vivo* and that their expression can lead to disease, with loss-of-function, gain-of-function, and/or dominant-negative effects having been proposed. Other studies using transgenic mice harboring different mutant forms of human opsin (K296E, P347S, Q344X), as well as a transgenic rat model expressing Ser334X rhodopsin, have also shown gene-dosing effects, whereby the expression levels of the mutant opsins directly correlate with the rate of retinal degeneration [91, 116, 130, 143].

The utility of pharmacological chaperones in adRP was explored *in vivo* using high-dose dietary vitamin A supplementation (40 times higher than obtained in a traditional rodent diet) in two transgenic mouse lines [117]. Mice harboring either T17M (amino-terminal domain) or P347S (carboxy-terminal domain) mutant forms of rhodopsin were chosen as representative of Class II and Class I mutants, respectively. High doses of vitamin A significantly reduced the rate of decline as measured by electroretinography in mice carrying a T17M mutant form of rhodopsin. Corresponding histological evaluation showed that the treatment was associated with significantly longer photoreceptor inner and outer segments, and a thicker outer nuclear layer. These effects were corroborated *in vitro*, as inclusion of 11-*cis*-retinal in the culture media partially stabilized T17M mutant opsin expressed in HEK293 cells. No effect of vitamin A was seen in mice harboring P347S rhodopsin, though this mutant did show aberrant transport to the outer segments [143]. Similar observations have been made in pigs harboring a P347L opsin [144, 145] and in mice lacking the last five amino acids of opsin due to a premature stop codon (Q344X; [116]), pointing to a potentially important role of the carboxy-terminus in rod outer segment targeting. Similar to the observations in mice, adult RP patients with the common forms of the disease that were administered oral vitamin A supplementation showed, on average, a slowing of the rate of retinal degeneration as measured by decline in ERG amplitude [76]; however, response was not correlated to genotype in this study. These data suggest that vitamin A supplementation may confer therapeutic benefit by stabilizing mutant opsins through increased availability of chromophore.

17.5. IHH AND THE GONADOTROPIN-RELEASING HORMONE RECEPTOR

IHH is a disorder characterized by a complete or partial lack of sexual development that may be either congenital or acquired [146, 147]. IHH is a rare disease, with estimated incidence of 1:10,000 to 1:86,000 [147]. Patients afflicted with IHH have a deficient pituitary gonadotrope response to gonoadotropin-releasing hormone (GnRH), which results in reduced or apulsatile gonadotropin release, delayed onset or absence of puberty, and infertility (for review, see References 4, 146, 147). IHH patients may also present with a variety of anomalies, such as anosmia/hyposmia, which is usually diagnosed as Kallmann syndrome, cleft palate, dental agenesis, visual abnormalities, deafness, and mental retardation, among others [146]. The disease may present at birth, adolescence, or adulthood. IHH is genetically heterogeneous, with X-linked recessive, autosomal recessive, autosomal dominant, and apparent sporadic inheritance patterns. Mutations in one of three genes are most commonly associated with IHH and account for 15–20% of all cases: *KAL1*, *GNRHR*, and *FGFR1*. Mutations in *KAL1*, the gene encoding a cell surface protein of unknown function, are transmitted as an X-linked recessive mode of inheritance and commonly lead to IHH with anosmia, or Kallman syndrome. Mutations in *FGFR1*, the gene encoding the fibroblast growth factor-1 receptor, are transmitted as an autosomal dominant mode of inheritance and lead to either anosmic/hyposmic or normosmic IHH. *GNRHR* mutations are transmitted as an autosomal recessive mode of inheritance and are the most common cause of normosmic IHH [146]. The prevalence of *GNRHR* mutations in all IHH patients ranges between 1.6% and 4.6% [148]. The prevalence of *GNRHR* mutations in normosmic IHH patients is 3.5–10.6%, and in IHH families showing an autosomal recessive mode of inheritance, it is between 6% and 11%. No *GNRHR* mutations have been identified in hyposmic or anosmic IHH patients [148]. As the present chapter is focused on GPCR targets that may be amenable to pharmacological chaperones, only *GNRHR* will be discussed further.

GNRHR encodes the 328-amino acid type 1 gonadotropin-releasing hormone receptor (GnRHR) [4]. GnRHR is expressed on the surface of gonadotrope cells of the anterior pituitary, as well as in lymphocytes, breast, ovary, placenta, and prostate [149]. GnRHR mediates responses to its ligand GnRH, a decapeptide that is synthesized and secreted in a pulsatile fashion by neurons located in the medial basal and anterior hypothalamus (for review, see Reference 4). Pulsatile GnRH release and cognate binding to GnRHR in pituitary gonadotropes result in the association of the receptor with G proteins that activate phosphatidylinositol–calcium second messenger signaling. The pituitary gonadotrope responds with pulsatile release of the gonadotropins, luteinizing hormone (LH) or follicle-stimulating hormone (FSH). LH release is favored by a more rapid GnRH pulse frequency, and FSH release is favored by a slower pulse frequency. These gonadotropins bind to specific GPCRs on the gonads to stimulate the production of sex steroids and peptide hormones that, in turn, modulate GnRH and gonadotropin release. Loss-of-function mutations in GnRHR can disrupt this cycle of GnRH frequency modulation, gonadotropin release, and regulatory hormonal feedback central to the normal function of the hypothalamic–pituitary–gonadal axis [4]. GnRHR mutations prevent response of pituitary gonadotropes to GnRH, decreased or apulsatile gonadotropin release, and, ultimately, a partial or complete lack of sexual development that is characteristic of IHH.

To date, 21 mutations in the human *GNRHR* gene have been described that lead to IHH (for review, see References 4, 150). Nineteen result from the substitution of a single amino acid, and 17 of these have been shown to be defective in trafficking to the plasma membrane [4, 150]. Point mutations occur throughout most domains of the receptor, including the extracellular amino-terminus, TM domains II–VII, extracellular loops I and II, and intracellular loop III [150, 151]. More than 50% of GnRHR mutations that lead to IHH are mediated by two common point substitutions: Q106R and R262Q [146]. When individually expressed in heterologous cell systems, these mutations show reduced ligand binding and/or intracellular signaling compared to the wild-type receptor [152, 153]. Similarly, heterologous expression of 11 other GnRHRs with point mutations that have been described in IHH also showed decreased or absent ligand binding and intracellular signaling [154–156]. Traditionally, these mutations had been thought to directly interfere with ligand binding or effector coupling, thereby reducing the intrinsic functionality of the receptor. However, several lines of structural- and genetic-based evidence were inconsistent with this hypothesis. First, none of the naturally occurring mutations directly involved residues from the defined ligand binding pocket [157]. Second, IHH disease-causing mutations occur throughout the receptor and are not limited to known functional domains [150, 151, 157]. Finally, ligand binding and intracellular signaling for the E90K mutant receptor could be genetically rescued to near normal levels by either removal of a primate-specific residue that limits cell surface expression levels (Lys191) or by addition of a carboxy-terminal extension present in fish/bird GnRHRs that increases cell surface expression [158–161]. These results demonstrate that the mutant receptor is intrinsically functional and supports the alternative hypothesis that the defect may be the result of decreased protein stability or misfolding with increased intracellular degradation and reduced trafficking to the cell surface [150, 151, 157, 162].

Consistent with the idea that GnRHR mutations result in decreased stability and misfolding, and reduced trafficking, a high-affinity cell-permeant peptidomimetic antagonist of human GnRHR, referred to as IN3, was shown to exhibit pharmacological chaperone activity for a majority of missense mutant forms of GnRHR (for review, see References 4, 150, 151, 162). Ligand binding and agonist-stimulated intracellular signaling in COS7 cells that were transiently transfected with each of the common Q106R and R262Q mutant forms of GnRHR were restored to near wild-type levels after incubation with IN3 followed by a period of compound washout [155]. Moreover, IN3 significantly rescued, though to variable degrees, ligand binding and receptor function in nine other less prevalent mutant forms of GnRHR [154–156]. The rescue effect of IN3 was reproduced by other structurally related cell membrane-permeant peptidomimetics with high affinity for GnRHR, but not by the low-affinity congener IN4 [154, 163]. Peptide antagonists that bind to the same site as IN3, but cannot permeate cells, were unable to rescue mutant forms of GnRHR [154]. Such data indicate that IN3 and its high-affinity analogs act intracellularly as folding templates, stabilizing mutant forms of GnRHR to increase trafficking to the plasma membrane, thereby restoring function [154, 155, 163]. However, the activity of IN3 is not applicable to all known IHH disease-causing mutations, as IN3 did not exhibit any significant rescue for two missense mutant forms of GnRHR, S168R and S217R, or the truncation mutant 205X [154, 155]. It is possible that these mutations may affect binding, processing, and/or function too severely for pharmacological chaperone rescue. Interestingly, IN3 treatment also increased the expression level of the wild-type human GnRHR [154, 155]. This observation suggests that a significant proportion of the newly synthesized wild-type human GnRHR is inefficiently processed by the cell, being retained in the ER and eventually degraded, analogous to that which has been observed for the wild-type human δ-opioid receptor [164].

Many patients with IHH are compound heterozygotes, expressing two different mutant forms of GnRHR. Genotype–phenotype correlation studies that compare compound heterozygous patients with patients homozygous for either mutant form suggest that clinical phenotype and response to GnRH are dictated primarily by the mutant form of GnRHR that shows the less severe loss of function *in vitro* [165]. Similarly, cotransfection experiments that used pairwise combinations of nine naturally occurring mutant forms of GnRHR found in IHH patients showed partial or full restoration of ligand binding and function after incubation with the pharmacological chaperone IN3 [166]. However, the magnitude of the response was unpredictable, in that for some combinations, it was similar (active-predominant effect), higher (additive effect), or lower (dominant-negative effect) than that exhibited by the less severe mutant form alone [166].

Most of the studies of pharmacological chaperone rescue of IHH disease-causing mutant forms of GnRHR have been conducted *in vitro*. For the purpose of drug development of a pharmacological chaperone, an appropriate animal model is also necessary. One existing model is the *N*-ethyl-*N*-nitrosourea-induced *Gnrhr* gene mutation mouse model of IHH [167]. The genetic defect is an L117P point mutation in murine *Gnrhr*. Phenotypic characterization of mice homozygous for this mutation revealed abnormalities in sexual differentiation and sterility in male and female mice that were transmitted as an autosomal recessive mode of inheritance. However, *in vitro* heterologous expression of the L117P mutant receptor showed that it exhibits no ligand binding or intracellular signaling, exerts a dominant-negative effect on wild-type receptor function, and, unfortunately, is not rescuable with any current GnRHR pharmacological chaperone [167]. Thus, this animal model would not be applicable for testing the *in vivo* effects of pharmacological chaperones for mutant GnRHR. Other models would need to be established where animals exhibit a phenotype relevant to IHH *and* express a mutant GnRHR, or combination of mutant receptors, that is rescuable by pharmacological chaperones. Relevant tissue-specific expression of a human mutant GnRHR or combination of human mutant receptors on a null background is preferable, as cellular trafficking and regulation of plasma membrane expression of the rodent receptor is significantly different from that of the human [4, 150]. In this manner, the extensive proof of concept that has been demonstrated *in vitro* for pharmacological chaperone rescue of mutant forms of GnRHR associated with IHH in humans could be tested and optimized for dosing paradigms that maximize efficacy *in vivo*.

17.6. OTHER HUMAN DISEASES CAUSED BY INACTIVATING MUTATIONS IN GPCRs

In the preceding sections, the success of the pharmacological chaperone strategy to restore activity to three different Class A GPCRs harboring loss-of-function mutations (V2R, rhodopsin, GnRHR) suggests that this approach may also be applicable to other mutated Class A receptors and raises the possibility that it may also be viable for other non-Class A receptors, as recently demonstrated for the Class C CaR (see below). This is certainly an important point given the continuous identification of diseases caused by mutant GPCRs across the various classes (Table 17.1). Below, we present summaries of other currently identified diseases that result from inactivating mutations in GPCRs that belong to Classes A, B, C, and frizzled/smoothened. Because many of these mutations lead to intracellularly retained receptors, a pharmacological chaperone approach may be applicable.

17.6.1. Class A GPCRs

Melanocortin Receptors The melanocortin system consists of five GPCRs that all stimulate the adenylyl cyclase signal transduction pathway. The endogenous agonists are derived by posttranslational processing of the proopiomelanocortin gene transcript by the prohormone convertases PC1 and PC2 to generate adrenocorticotropin (ACTH) and the α-, β-, and γ-melanocyte-stimulating hormones (MSH). The melanocortin system also has the only two known endogenous GPCR antagonists, termed agouti and agouti-related protein (AgRP) (for review, see Reference 168). Inactivating mutations in the genes of four of these receptors have now been linked to human diseases.

Pharmacological studies in rodents demonstrated a pivotal role of central melanocortin 4 receptors (MC4R) in the regulation of energy homeostasis (for review, see References 169, 170). Further evidence came from the observation that in MC4R null mice, obesity results from the combined effects of increased food intake (hyperphagia) and decreased energy expenditure [171–174]. Subsequently, genetic studies in severely obese patients confirmed that the MC4R is also critical for maintaining energy homeostasis in humans [175–178]. The prevalence of MC4R mutations in the human population now makes this the most common monogenic cause of early-onset morbid obesity [179]. To date, 29 trafficking-defective mutant forms of MC4R have been reported among the 80 mutations described in the human population, suggesting that pharmacological chaperones could be utilized for some forms of this disease [180].

Strong evidence for a causative role of melanocortin 3 receptor (MC3R) mutations in obesity has also recently emerged. In MC3R knockout mice, high feed efficiency (i.e., mice gain more fat per calorie of food consumed compared to wild-type littermates), rather than hyperphagia, seems to contribute to increased fat mass. The first potential human obesity-linked mutation in MC3R from two related obese patients in Singapore was reported in 2002 [181]. This mutation, I183N, changes a highly conserved Ile at the junction between TM domain III and intracellular loop II, residing within the highly conserved DRYxxI/V motif. Two groups have independently reported the functional properties of I183N MC3R [182, 183]. Both studies showed that this mutation indeed results in a complete loss of function, but there was a discrepancy regarding its cell surface expression. This discrepancy could be due to the use a green fluorescent protein (GFP)-tagged MC3R [182], which might affect intracellular trafficking of the receptor. Recently, three additional mutations have been identified in obese patients from Italy, including A293T in TM domain VI, I335S in TM domain VII, and X361S (which changes the stop codon to Ser, resulting in the addition of seven amino acids before a downstream stop codon) [184]. *In vitro* expression demonstrated a loss of function of I335S MC3R caused by intracellular retention. This Ile is part of the conserved N/DPxxY motif (DPLIY) in TM domain VII and is fully conserved in all five melanocortin receptors. The amino acid Ile335 in MC3R corresponds to Ile301 in MC4R. The I301T mutation in MC4R has been described as a loss-of-function mutation, suggesting a critical role of this amino acid position for full biological function of both MC3R and MC4R [177]. The *in vitro* analysis of the A293T and X361S mutant receptors showed normal cell surface expression and signaling capacity in the presence of the superagonist NDP-MSH and the endogenous agonist α-MSH [184, 185]. However, a more complete signaling study with γ-MSH and AgRP would be required in order to draw any conclusion regarding the role of these mutations in the development of obesity.

Mutations in the melanocortin 2 receptor (MC2R), or ACTH receptor, are responsible for familial glucocorticoid deficiency (FGD) syndrome and account for approximately 25% of FGD cases. Thirty distinct mutations associated with FGD have been reported, including 27 missense mutations (for review, see Reference 180). Functional studies of the MC2R mutants have been hindered by the fact that MC2R cell surface expression depends on expression of a melanocortin receptor accessory protein, which is only expressed in cells derived from the adrenal gland [186, 187]. In most studies, ligand binding and localization experiments were not done, preventing conclusions from being drawn on whether the mutant MC2Rs are defective in cell surface expression, ligand binding, and/or signaling.

Lastly, mutations in the melanocortin 1 receptor (MC1R) are responsible for pigmentation defects in humans. The MC1R is a major factor in controlling skin and hair pigmentation, with activation increasing intracellular cAMP levels and producing eumelanin that leads to a darkening of skin and hair [188]. Mutant forms of MC1R that are responsible for the red hair color (RHC) phenotype in humans (red hair, fair skin, and poor ability to tan) have been reported [189]. To date, more than 60 mutant forms have been identified in MC1R with various skin and hair abnormalities, including RHC and an increased susceptibility to melanoma and other skin cancers. Recently, four of these MC1R mutants were shown to have greatly decreased cell surface expression (D84E in TM domain II, R151C, I155T, and R160W in intracellular loop II), with two others showing a smaller decrease in cell surface expression (V60L in TM domain I and R163Q in TM domain IV) [190]. A good correlation was seen between those alleles strongly associated with the RHC phenotype and significantly decreased cell surface expression of the MC1R.

Glycoprotein Hormone Receptors Loss-of-function mutations have also been described for a number of glycoprotein hormone receptors. This receptor family is unique among the GPCRs, because in contrast to other receptors, this contains a large amino-terminal, extracellular (ecto-) domain (ECD) containing leucine-rich repeats that are important for recognition and binding of its glycoprotein ligands (for review, see Reference 191). A priori, this specific feature could present a challenge for the design of pharmacological chaperones. However, it has been demonstrated that highly conserved residues in the TM domains of these receptors can modulate constitutive activity of the receptor, thereby also regulating the efficiency of hormone recognition by the ECD. This observation potentially opens up the possibility of using allosteric compounds that bind in the TM domains as pharmacological chaperones for glycoprotein hormone receptors with mutations in the ECD. Allosteric modulators have been reported for one member of this receptor subfamily, the FSH receptor [192], and have proven their utility as pharmacological chaperones for the related Family C CaR [193] (see below).

Thyroid-stimulating hormone (TSH) binds and activates the cell surface TSH receptor (TSHR), resulting in thyroid hormone synthesis and secretion as well as cell proliferation and differentiation within the thyroid gland. Inactivating mutations in the TSHR cause congenital hypothyroidism, which has an autosomal recessive mode of inheritance, with patients either homozygous or compound heterozygous for TSHR mutations. The first inactivating mutations in TSHR causing TSH resistance were reported in 1995 [194]. To date, 32 distinct mutations have been reported in this gene, and among them, six result in intracellular retention of the receptor. Specifically, four missense mutations, I167N and L252P in the ECD, T447I in extracellular loop I, and A553T in TM domain IV [195–198], as well as two frameshift mutations [199], have been characterized as intracellularly retained or partially expressed at the cell surface.

Loss-of-function mutations in the LH receptor (LHR) lead to Leydig cell hypoplasia, a rare form of male pseudohermaphroditism. In females, LHR inactivation results in hypergonatropic hypogonadism and primary amenorrhea. Like congenital hypothyroidism, the pattern of inheritance is autosomal recessive, with homozygous or compound heterozygous patients. To date, 22 distinct mutations have been described in this receptor (for review, see Reference 180). The mutant forms V144F, F194V, and C343S in the ECD, C543R in TM domain V, A593P in TM domain VI, S616Y and I625K in TM domain VII, and two frameshift mutations, 33 base pairs in exon 1 and a microdeletion in TM domain VII (ΔL608/V609), impair trafficking and cell surface expression [200–206].

Lastly, inactivating mutations in the FSH receptor (FSHR) have been reported. FSH is important for spermatogenesis in males and is absolutely required for follicle growth in females. Mutations in the FSHR result in ovarian dysgenesis, with amenorrhea and infertility in women. Since the first missense mutation, A189T, in the ECD of the receptor was reported in 1995 [207], eight additional loss-of-function missense mutations have been described. Four of those nine mutants have been characterized as trafficking defective: A189T, I160T, and D224V in the ECD and P519T of extracellular loop II [208–211].

Endothelin b Receptor The endothelin b receptor (ETbR) is derived from one of the primary gene targets (*ETRB*) involved in Hirschprung disease. Hirschsprung disease (HSCR) is a congenital disorder characterized by an absence of ganglion cells in the nerve plexuses of the lower digestive tract. Although mutations in eight different genes (*ETRB*, *EDN3*, *ECE1*, *SOX10*, *RET*, *GDNF*, *NTN*, and *SIP1*) have been identified in affected individuals, it is now clear that *RET* and *ETRB* are the primary genes implicated in the etiology of HSCR (for review, see Reference 212). All eight genes are involved in the early development of the enteric nervous system, and most act through two distinct biochemical pathways mediated by the products of *RET* and *ETRB*. Mutations in *ETRB* account for 5–10% of the HSCR cases in the general population. To date, only two of the 22 mutations thus far reported, C109R in TM domain I and P383L in TM domain VII of the N/DPxxY conserved motif, have been described as trafficking-defective mutants of ETbR [213, 214].

17.6.2. Class B GPCRs

Growth Hormone-Releasing Hormone Receptor Nine distinct inactivating mutations have been described in the growth hormone-releasing hormone receptor (GHRHR). Loss-of-function of GHRHR leads to severe growth retardation, associated with isolated growth hormone deficiency. Six of the nine currently identified mutations are missense and were found to be defective in signaling, though the mechanism for the dysfunction is not clear due to the lack of detailed binding and subcellular localization studies [215].

Parathyroid Hormone-Related Peptide Type 1 Receptor Loss-of-function mutations in the parathyroid hormone-related peptide type 1 receptor (PTH-R1) result in Blomstrand osteochondrodysplasia (BOCD), a rare autosomal recessive disorder characterized by advanced maturation and premature ossification of all skeletal elements (for review, see Reference 216). Inactivating mutations in PTH-R1 responsible for BOCD were first described in 1998, with a homozygous P132L mutation in the ECD reported in two independent studies [217, 218]. Importantly, P132L PTH-R1 was well expressed at the cell surface but showed low levels of ligand binding and signaling. This mutation was again reported in two families living in the same region of England in 2007, suggesting derivation from a common ancestor [219]. Expression of this mutant receptor confirmed and extended previous observations, that is, compromised activation of the cAMP/PKA pathway, but not of the phospholipase C (PLC)/protein kinase C (PKC) pathway. A second inactivating mutation was reported that deleted 11 amino acids in TM domain V [220]. Also, in this patient, the paternal allele was not expressed. Expression of the truncated receptor confirmed a loss-of-function phenotype but did not shed light on cell surface expression or ligand binding. Subsequently, two additional patients were found to be homozygous for two different inactivating mutations. The first was a frameshift mutation (deletion of G at nucleotide 1122 modifying the coding sequence after codon 364) in TM domain V [221], and the second was a truncation mutation at the carboxy-terminus deleting at least 108 amino acids [222]. More recently, two additional patients homozygous for inactivating point mutations were identified. The first contained a premature stop codon at position R104 of the ECD; the second resided in the intron between the exons coding for TM domain IV and extracellular loop II [219]. This intronic mutation creates a highly conserved consensus sequence for exon–intron splicing that has higher homology than the native splice site, leading to preferential use of the mutant splice site with concomitant expression of wild-type receptor at low levels. Utilization of this novel splice site results in an aberrant transcript coding for a receptor lacking TM domains V, VI, and VII, the intervening intracellular and extracellular loops, and the cytoplasmic tail. Due to the severity of the mutations identified thus far in this disease, the potential ability for pharmacological chaperone correction of BOCD-associated mutant forms of PTH-R1 appears low.

17.6.3. Class C GPCRs

CaR Recently, the potential for pharmacological chaperone activity has been demonstrated *in vitro* for the CaR, which belongs to the Class C GPCR family and contains a large calcium-binding ECD. The CaR plays a critical role in calcium homeostasis, translating changes in plasma calcium levels into reciprocal changes in parathyroid hormone secretion from the parathyroid glands. Inactivating mutations in CaR produce a benign form of hypercalcemia when present in the heterozygous state, termed familial hypocalciuric hypercalcemia (FHH), while homozygous mutations produce a much more severe hypercalcemic disorder resulting from marked hyperparathyroidism, called neonatal severe hyperparathyroidism (NSHPT) (for review, see Reference 223). These diseases are characterized by generalized resistance to extracellular calcium and increased circulating parathyroid hormone levels. Since the first report describing inactivating mutations in the CaR [224], more than 225 missense mutations have been identified, with at least 30 resulting in a loss-of-function phenotype (http://www.casrdb.mcgill.ca/). A majority of the mutations occur in the ECD, with others found in TM domains II and VI, extracellular loops I and II, intracellular loop III, and the carboxy-terminal tail of the receptor. A number of these mutations have been shown to result in severe ER retention, including R66C, R680C, R795W, and G549R, while others showed partial ER retention, including R185Q and V817I (for review, see References 180, 193).

A series of allosteric modulators (agonist and antagonist) were recently tested against the wild-type CaR, as well as against receptors with activating and inactivating mutations [193]. By Western blot analysis, increased levels of mature wild-type receptor and four inactive mutants (R185Q in the ECD, R680C in TM domain III, R795W in intracellular loop III, and V817I in TM domain VI) were seen after incubation with the allosteric *agonist* NPS-568. Similarly, increased plasma membrane localization of the wild-type and mutant receptors was also observed. Importantly, functional rescue was proportional to the level of plasma membrane expression. In contrast, the allosteric *antagonist* NPS-2143 showed no effect against these mutant forms with inactivating mutations and decreased the activity of these mutants. This is in contrast to what has been observed for the V2R and GnRHR, for which antagonists were shown to act as pharmacological chaperones (described above). Also, for the δ-opioid receptor, both agonists and antagonists were found to have pharmacological chaperone properties [225]. Whether this difference results from the allosteric nature of the compounds tested for the CaR or is a characteristic of the receptor itself is unclear.

It should be noted that in clinical studies, NPS-568 effectively lowered circulating levels of parathyroid hormone in dialysis patients with secondary hyperparathyroidism [226], suggesting that long-term treatment with calcimimetics can upregulate the levels of wild-type CaR, thus providing possibilities for intervention in diseases affecting calcium homeostasis.

17.6.4. Family Frizzled/Smoothened GPCRs

Frizzled-4 Mutations in the gene *FZD4* cause familial exudative vitreoretinopathy (FEVR). FEVR is an autosomal dominant hereditary ocular disorder characterized by impaired development of retinal vessels and various secondary complications, including retinal folds and retinal detachments. The gene product of *FZD4*, frizzled-4, is a GPCR that binds the Wnt family of signaling peptides. To date, 12 mutations have been described (five deletions, one nonsense, and six missense) [227–231]. Only one frameshift mutation, L501 fs X533 that results in deletion of the receptor's carboxy-terminal tail, has been characterized *in vitro* [232]. This truncated receptor is retained in the ER and exerts a dominant-negative effect on wild-type frizzled-4, preventing its ability to signal [233]. Whether this mutant form could be rescued by a pharmacological chaperone approach remains to be investigated.

The above examples provide evidence that loss-of-function mutations linked to human diseases can be found in many GPCRs and across multiple GPCR subclasses. There is little doubt that this list will continue to grow in the coming years. Additional studies will be needed to determine which of these mutations lead to GPCRs with decreased stability, misfolding, and intracellular retention that would be amenable to pharmacological chaperone therapy.

17.7. CONSIDERATIONS FOR THE THERAPEUTIC USE OF PHARMACOLOGICAL CHAPERONES

17.7.1. Pharmacogenetics

Mutational heterogeneity represents a significant barrier to development of therapies for many inherited diseases. In several instances, the outcome of the mutation is fairly easy to predict. For example, the presence of a premature stop codon that leads to a truncated receptor after the first TM domain certainly leads to a protein that, even if inserted in the plasma membrane, is unlikely to bind ligand or transduce signal. In contrast, missense mutations that result in the substitution of a single amino acid in the encoded protein theoretically could have three major effects: (1) mutations that cause loss of stability or misfolding of the protein, and thus interfere with the normal targeting and insertion of the receptor into the plasma membrane; (2) mutations that affect residues involved in the formation of the binding site, and thus prevent binding of the ligand or drastically reduce the receptor's affinity for it; and (3) mutations that affect domains involved in transducing the signal from the receptor to signaling partners, thus leading to a receptor that can bind the ligand but cannot propagate the signal to downstream effectors. As approximately two-thirds of the mutations in GPCRs that cause human disease are missense, it is important to determine the functional consequences of these more subtle mutations. In recent years, this has been accomplished by expressing the DNA encoding the missense mutated receptors in defined heterologous cellular systems that have little to no endogenous receptor. The ability of the mutant receptors to insert in the plasma membrane, bind agonist, interact with G proteins and stimulate downstream effectors, and undergo regulation can then be compared with the characteristics of the wild-type receptors expressed in the same system. This approach allows one to determine which aspects of receptor function are impaired by specific mutations. As mentioned earlier, over 40% of the missense mutations, and small in-frame insertions and deletions are predicted to result in less stable or misfolded trafficking-defective GPCRs (for review, see References 2, 3) but do not necessarily affect other receptor functions, such as ligand binding and signal transduction. Mutated GPCRs of this type are the ones most likely to be responsive to a pharmacological chaperone that specifically binds to and stabilizes the receptor, promoting folding and efficient trafficking to the plasma membrane, where the otherwise functional mutant receptor can carry out its physiological function. Missense mutant GPCRs that are *unlikely* to respond to a pharmacological chaperone include those that fold and traffic normally, but are deficient in ligand binding or coupling, those with severe folding deficiencies that cannot be effectively stabilized by a small molecule, or those with gross alterations in structure. In diseases caused by a large number of different missense mutations in different domains of a GPCR, it is important to clearly understand the nature of the molecular defect *in vitro*. Efforts should then be focused on those mutant forms that are trafficking defective and functionally rescuable, and hence define the potentially responsive patient population, prior to testing a pharmacological chaperone in a clinical setting.

Determining the mutant forms of a GPCR that are most likely amenable to pharmacological chaperone therapy is relatively straightforward as responsive mutant forms will show increased cell surface expression, ligand binding, and/or intracellular signaling after incubation with the chaperone in cell culture. However, a period of compound washout after exposure is required prior to measuring ligand binding or intracellular signaling as many of the pharmacological chaperones for GPCRs that have been described are competitive or reversible antagonists (see below). In contrast, increases in cell surface expression can be measured even in the presence of the pharmacological chaperone by using epitope-tagged receptors. As described above, 11 of 13 missense mutant forms of GnRHR with confirmed or suspected trafficking defects were responsive to IN3, as shown by increased ligand binding and intracellular signaling after incubation [154–156]. The responses of mutant forms of GnRHRs to IN3 were of variable magnitude, with the level of functional rescue ranging from fully restored to near wild-type levels to nonresponsive, meaning that no increases in ligand binding or intracellular signaling were detected after incubation with the pharmacological chaperone. Similarly, various levels of both cell surface expression and function for 25 mutant forms of the V2R that are found in NDI were responsive to pharmacological chaperones (for review, see References 58, 62).

17.7.2. Dominant-Negative Effects

Loss of receptor function often becomes apparent only when both alleles are affected. In some cases, however, impaired trafficking of the wild-type receptor to the plasma membrane or defects in signal transduction is caused by dominant-negative effects of coexpressed mutated receptors in the heterozygous state (for review, see Reference 4). This has recently been postulated to explain the dominant mode of inheritance in adRP [234]. Expression of the Class II rhodopsin mutants P23H and G188R was shown to interfere with the biosynthetic maturation of coexpressed wild-type rhodopsin, leading to retention in the ER. In contrast, ER retention of wild-type rhodopsin was not seen when coexpressed with the Class I mutant, V345M rhodopsin. Direct, physical interaction between mutant and wild-type rhodopsins was responsible for ER retention and premature proteasome degradation of the wild-type protein [234]. These observations confirmed and extended earlier observations seen in *Drosophila* and mammalian expression systems [124, 235, 236], and further support the potential for rhodopsin dimerization during biosynthesis, trafficking, and signal transduction [237, 238], similar to the role played by dimerization with other GPCRs (for review, see Reference 239). While compelling, the dominant-negative effect is not necessarily mutually exclusive with other mechanisms that may also be at play in RP, such as the toxic gain of function shown for some mutant forms of rhodopsin that lead to photoreceptor death as described above. Similar to rhodopsin, GnRHR has been shown to dimerize as part of the process of receptor activation [240]. Cotransfection of mutant and wild-type GnRHRs *in vitro* revealed that the mutated receptors exert a dominant-negative effect on wild-type receptor with respect to ligand binding and intracellular signaling [163, 241, 242] as a result of wild-type receptor retention in the ER by association with trafficking-defective mutant receptors [241]. This phenomenon has also been observed for other GPCRs, including a naturally occurring mutant form of MC4R that has been shown to exert a dominant-negative effect on wild-type receptor, potentially through dimerization [243]. It will be necessary to better understand the effects of pharmacological chaperones on dominant-negative effects, such that facilitation of mutant receptor export from the ER also permits trafficking of wild-type receptors to the cell surface.

Similar to the dominant-negative effects on wild-type receptors, evidence now suggests that coexpression of two *mutant* forms of a GPCR are functionally unique and may respond differently to a pharmacological chaperone than either mutant form expressed alone. This has been clearly shown by the different effects of pharmacological chaperones on coexpressed pairs of GnRHR mutants *in vitro* as discussed above [165, 166]. In these cases, the *in vitro* response of the *combination* of mutant receptors to a pharmacological chaperone would be expected to be more predictive of the *in vivo* response to a pharmacological chaperone than the *in vitro* response of each individual mutant receptor alone. Thus, a pharmacogenetic approach that involves screening for responsive mutations in a well-designed *in vitro* assay, whether alone or in combination with wild-type or other mutant receptor forms, is likely to help select patients that harbor mutations or combinations of mutations that may be responsive *in vivo*. Ideally, for *homozygous* autosomal recessive or autosomal dominant diseases caused by mutated GPCRs, an *in vitro* response will likely enrich for the population of patients harboring mutations that may be responsive *in vivo*. Similarly, for autosomal recessive diseases in which patients are most likely to be *heterozygous* for two different GPCR mutations, the pharmacogenetic approach would be most useful if the *in vitro* response of the combination of mutant receptors was assessed. Finally, the correlation of the *in vitro* and *in vivo* responses to a pharmacological chaperone may apply only to the biochemical measures of receptor cell surface expression, ligand binding, and intracellular signaling. Predictive value for clinical improvement could be achieved only if the *in vitro* response was positively correlated with improved clinical symptoms or decreased disease severity.

17.7.3. Function of Rescued GPCRs with Missense Mutations

As just discussed, it is important to consider the relationship of the *in vitro* and *in vivo* responses of a mutant GPCR to a pharmacological chaperone with respect to clinical outcome. It is expected that a mutated GPCR that shows no response *in vitro* would not respond *in vivo*. However, predicting the *in vivo* response of mutated GPCR that *does* respond *in vitro* is more difficult, as cultured cells differ significantly from a living organism. For example, given that mutated receptors tend to be less stable than their wild-type counterparts, increased turnover rates could limit the time a rescued receptor is available to interact with endogenous ligand at the cell surface. In addition, the pharmacokinetic profile of the pharmacological chaperone, including its tissue distribution and clearance rate, could influence the net response, especially if the elimination half-life is long or tissue penetration is poor (see below). Furthermore, physiological factors need to be considered. For instance, in the case of IHH, pulsatile release of GnRH and subsequent binding to GnRHR in pituitary gonadotropes is critical for downstream effects, with LH release favored by a more rapid GnRH pulse frequency and FSH release favored by a slower pulse frequency. Hence, both temporal and quantitative rescue of GnRHR may be necessary to restore full physiological function. Similarly, in the case of adRP, interaction of opsin with 11-*cis*-retinal is not only critical for receptor stabilization, but also for achieving the optimal spectral properties characterized by a λ_{max} of 500 nm. It has been shown that rescue of P23H rhodopsin with 11-*cis*-7-ring-retinal leads to *transient* chromophore binding and a λ_{max} of 490 nm, properties that may provide only limited function of the receptor in the retina [135].

17.7.4. Biophysical Requirements of Pharmacological Chaperones

The success of a pharmacological chaperone depends critically on biophysical properties that dictate affinity, specificity, and reversibility for its target, as well as cell permeability. With respect to permeability, it is imperative that the pharmacological chaperone penetrate both the plasma and ER membranes to bind and stabilize the newly synthesized protein during biosynthesis, thus facilitating ER export. With respect to affinity and specificity, the chaperone must bind the mutant form of the GPCR at concentrations that can be achieved in cells and tissues after administration of clinically relevant doses, as the magnitude of ligand-mediated rescue is correlated with binding affinity [22]. In addition, the concentration of pharmacological chaperone required to bind and stabilize the mutated GPCR must be relatively low in order to avoid off-target interactions with other proteins. With respect to reversibility, pharmacological chaperones that are antagonists or partial agonists must be able to dissociate quickly from the receptor. In the case of molecules with high affinity and/or slow receptor off rates, there would be a risk that prolonged binding could interfere with the interaction of the rescued receptor with its endogenous hormone or neurotransmitter, thereby minimizing therapeutic benefit. Receptor agonists are also capable of promoting trafficking of mutant receptors to the cell surface. Indeed, for GPCRs such as the δ-opioid receptor, membrane-permeable agonists have been found to act as pharmacological chaperones [225]. However, functional rescue may be countered by their ability to activate the receptor and subsequently elicit internalization and desensitization, diminishing overall long-term efficacy. Partial agonists may address the issues of rapid activation and subsequent desensitization but, like antagonists, could inhibit binding of the natural ligand. Hence, reversible, moderate-affinity competitive antagonists or partial agonists with a limited propensity to induce internalization and desensitization may provide the best opportunity for clinical efficacy. As discussed above, low-affinity pharmacological chaperones that target P23H opsin have been identified. These pharmacological chaperones can be displaced by 11-*cis*-retinal after introduction to the culture media, thus restoring mutant rhodopsin functionality *in vitro* [137]. In the case of the NDI clinical trial, SR49059 was chosen as a tool for proof of concept because it has a 10-fold lower affinity for the V2R than the other V2R pharmacological chaperones that were tested *in vitro* [62]. This lower affinity was seen as an advantage in the initial trial, and it remains to be seen if high-affinity V2R antagonists will be as efficacious as the lower affinity SR49059.

Lastly, noncompetitive allosteric modulators may offer an alternative to competitive ligands. As discussed above, prolonged incubation with the allosteric agonist NPS-568 increased the levels of cell surface wild-type and mutant CaR, most likely by favoring an active, more stable conformation, with reduced ubiquitination and premature degradation [193]. These results support the hypothesis that activation of the CaR can facilitate trafficking through the secretory pathway, that is, receptors in an active conformation are more stable than those in an inactive conformation, allowing passage through the ER quality control system and avoidance of ERAD. As such, it could be envisioned that mutant forms of the CaR, with point mutations in the calcium-binding ECD, could be stabilized with allosteric agonists that favor an active conformation of the receptor in the calcium-rich environment of the ER. This strategy might also prove useful in the case of point mutations that affect the ability of the endogenous ligand to interact with the rescued cell surface receptor. Proof of concept for this approach with other mutant GPCRs with large ECDs remains to be established.

17.7.5. Safety

Compared to small molecule chemical chaperones, such as glycerol, DMSO, and TMAO, it is envisioned that specific pharmacological chaperones will have a safety advantage in the clinical setting. While both strategies are effective in promoting protein folding in the ER and subsequent trafficking through the secretory pathway *in vitro*, very high concentrations of chemical chaperones are typically required to see an effect, often too high to be practical at all. Furthermore, chemical chaperones act nonspecifically on many proteins, raising the possibility that they could lead to premature ER release of folding intermediates for many other normal proteins, some of which could lack stability and have a propensity for aggregation and toxicity in the post-ER environment [20]. Pharmacological chaperones, however, specifically target the protein of interest and are designed to elicit little to no global perturbation of the ER quality control system and the general protein folding environment. Because pharmacological chaperones specifically bind to their target proteins and can be selected to have suitably high affinity, lower concentrations may be sufficient to lead to therapeutic benefit, reducing or preventing off-target side effects (for review, see Reference 4).

17.8. CONCLUDING REMARKS

Diseases that result from mutations in the genes that encode GPCRs represent a large and growing unmet medical need, with current therapies often insufficient to treat associated symptoms and underlying molecular defects. Pharmacological chaperones represent an alternative therapeutic approach aimed at restoring activity to mutant proteins that, although harbor molecular defects, are often functionally competent but not trafficked to their appropriate cellular location to carry out their intended physiological function. Compelling evidence for the utility of this approach has been provided *in vitro*, *in vivo*, and in the clinic for several diseases that result from mutations in Class A GPCRs. Importantly, restoration of activity for diseases with both autosomal recessive and autosomal dominant modes of inheritance highlights the ability of this approach to correct both loss-of-function and gain-of-function mutations. In order to exert their beneficial effects on receptor stability and trafficking, however, pharmacological chaperones often bind the same sites as the endogenous ligands, potentially hindering full therapeutic benefit due to competitive inhibition. As such, this strategy presents challenges for clinical development that necessitate a thorough understanding and appropriate utilization of the biophysical and pharmaceutical properties of the chaperone as well as the biological properties of the rescued protein. The extension of this approach to allosteric modulators may alleviate some of these concerns and possibly extend the therapeutic possibilities beyond Class A GPCRs and to other types of mutant receptors that are compromised in their ability to bind endogenous ligands.

ACKNOWLEDGMENTS

The authors would like to thank Dr. David Lockhart for critical review of the manuscript and Ms. Tammy Allen for her assistance in its preparation.

REFERENCES

1. Kristiansen, K. (2004) Molecular mechanisms of ligand binding, signaling, and regulation within the superfamily of G-protein-coupled receptors: Molecular modeling and mutagenesis approaches to receptor structure and function. *Pharmacol Ther.* 103, 21–80.
2. Schoneberg, T., Schulz, A., Biebermann, H., Hermsdorf, T., Rompler, H., Sangkuhl, K. (2004) Mutant G-protein-coupled receptors as a cause of human diseases. *Pharmacol Ther.* 104, 173–206.
3. Rompler, H., Staubert, C., Thor, D., Schulz, A., Hofreiter, M., Schoneberg, T. (2007) G protein-coupled time travel: Evolutionary aspects of GPCR research. *Mol Intervention.* 7, 17–25.
4. Conn, P., Ulloa-Aguirre, A., Ito, J., Janovick, J. (2007) G protein-coupled receptor trafficking in health and disease: Lessons learned to prepare for therapeutic mutant rescue in vivo. *Pharmacol Rev.* 59, 225–250.
5. Ellgaard, L., Helenius, A. (2003) Quality control in the endoplasmic reticulum. *Nat Rev Mol Cell Biol.* 4, 181–191.
6. Trombetta, E., Parodi, A. (2003) Quality control and protein folding in the secretory pathway. *Annu Rev Cell Dev Biol.* 19, 649–676.
7. Herbert, D., Molinari, M. (2007) In and out of the ER: Protein folding, quality control, degradation, and related human diseases. *Physiol Rev.* 87, 1377–1408.
8. Anelli, T., Sitia, R. (2008) Protein quality control in the early secretory pathway. *Embo J.* 27, 315–327.
9. Ushioda, R., Hoseki, J., Araki, K., Jansen, G., Thomas, D., Nagata, K. (2008) ERdj5 is required as a disulfide reductase for degradation of misfolded proteins in the ER. *Science.* 321, 569–572.
10. Rozell, T., Davis, D., Chai, Y., Segaloff, D. (1998) Association of gonadotropin receptor precursors with the protein folding chaperone calnexin. *Endocrinology.* 139, 1588–1593.
11. Morello, J., Salahpour, A., Petaja-Repo, U., Laperriere, A., Lonergan, M., Arthus, M., Nabi, I.R., Bichet, D., Bouvier, M. (2001) Association of calnexin with wild type and mutant AVPR2 that causes nephrogenic diabetes insipidus. *Biochemistry.* 40, 6766–6775.

12. Siffroi-Fernandez, S., Giraud, A., Lanet, J., Franc, J. (2002) Association of the thyrotropin receptor with calnexin, calreticulin and BiP. Effects on the maturation of the receptor. *Eur J Biochem.* 269, 4930–4937.
13. Lu, M., Echeverri, F., Moyer, B. (2003) Endoplasmic reticulum retention, degradation, and aggregation of olfactory G-protein coupled receptors. *Traffic.* 4, 416–433.
14. Dong, C., Filipeanu, C., Duvernay, M., Wu, G. (2007) Regulation of G protein-coupled receptor export trafficking. *Biochim Biophys Acta.* 1768, 853–870.
15. Chapple, J., Cheetham, M. (2003) The chaperone environment at the cytoplasmic face of the endoplasmic reticulum can modulate rhodopsin processing and inclusion formation. *J Biol Chem.* 278, 19087–19094.
16. Lanctot, P., Leclerc, P., Escher, E., Guillemette, G., Leduc, R. (2006) Role of N-glycan-dependent quality control in the cell-surface expression of the AT1 receptor. *Biochem Biophys Res Commun.* 340, 395–402.
17. Achour, L., Labbe-Jullie, C., Scott, M., Marullo, S. (2008) An escort for GPCRs: Implications for regulation of receptor density at the cell surface. *Trends Pharmacol Sci.* 29, 528–535.
18. Bonifacino, J., Weissman, A. (1998) Ubiquitin and the control of protein fate in the secretory and endocytic pathways. *Annu Rev Cell Dev Biol.* 14, 19–57.
19. Sitia, R., Braakman, I. (2003) Quality control in the endoplasmic reticulum protein factory. *Nature.* 426, 891–894.
20. Selkoe, D. (2003) Folding proteins in fatal ways. *Nature.* 426, 900–904.
21. Chaudhuri, T., Paul, S. (2006) Protein-misfolding diseases and chaperone-based therapeutic approaches. *FEBS J.* 273, 1331–1349.
22. Bernier, V., Bichet, D., Bouvier, M. (2004) Pharmacological chaperone action on G-protein-coupled receptors. *Curr Opin Pharmacol.* 4, 528–533.
23. Kopito, R., Ron, D. (2000) Conformational disease. *Nat Cell Biol.* 2, E207–E209.
24. Chiti, F., Dobson, C. (2006) Protein misfolding, functional amyloid, and human disease. *Annu Rev Biochem.* 75, 333–366.
25. Papp, E., Csermely, P. (2006) Chemical chaperones: Mechanisms of action and potential use. In *Molecular Chaperones in Health and Disease*, ed. M. Gaestal, pp. 405–416. Berlin Heidelberg: Springer.
26. Culpepper, R., Hebert, S., Andreoli, T. (1983) Nephrogenic diabetes insipidus. In *The Metabolic Basis of Inherited Diseases*, ed. J. Stanbury, J. Wyngaarden, D. Frederickson, J. Goldstein, M. Brown, pp. 1867–188. New York: McGraw-Hill, New York.
27. Robben, J., Deen, P. (2007) Pharmacological chaperones in nephrogenic diabetes insipidus: Possibilities for clinical application. *BioDrugs.* 21, 157–166.
28. Deen, P., Croes, H., van Aubel, R., Ginsel, L., van Os, C. (1995) Water channels encoded by mutant aquaporin-2 genes in nephrogenic diabetes insipidus are impaired in their cellular routing. *J Clin Invest.* 95, 2291–2296.
29. Mulders, S., Knoers, N., Van Lieburg, A., Monnens, L., Leumann, E., Wuhl, E., Schober, E., Rijss, J., Van Os, C., Deen, P. (1997) New mutations in the AQP2 gene in nephrogenic diabetes insipidus resulting in functional but misrouted water channels. *J Am Soc Nephrol.* 8, 242–248.
30. Mulders, S., Bichet, D., Rijss, J., Kamsteeg, E., Arthus, M., Lonergan, M., Fujiwara, M., Morgan, K., Leijendekker, R., van der Sluijs, P., van Os, C., Deen, P. (1998) An

aquaporin-2 water channel mutant which causes autosomal dominant nephrogenic diabetes insipidus is retained in the Golgi complex. *J Clin Invest.* 102, 57–66.

31. Reeves, W., Andreoli, T. (1989) Nephrogenic diabetes insipidus. In *The Metabolic Basis of Inherited Disease*, ed. S. Scriver, A. Beaudet, W. Sly, D. Valle, pp. 1985–2001. New York: McGraw-Hill.
32. Brown, D. (1989) Membrane recycling and epithelial cell function. *Am J Physiol.* 256, F1–F12.
33. Fushimi, K., Uchida, S., Hara, Y., Hirata, Y., Marumo, F., Sasaki, S. (1993) Cloning and expression of apical membrane water channel of rat kidney collecting tubule. *Nature.* 361, 549–552.
34. Hayashi, M., Sasaki, S., Tsuganezawa, H., Monkawa, T., Kitajima, W., Konishi, K., Fushimi, K., Marumo, F., Saruta, T. (1994) Expression and distribution of aquaporin of collecting duct are regulated by vasopressin V2 receptor in rat kidney. *J Clin Invest.* 94, 1778–1783.
35. Nielsen, S., Smith, B., Christensen, E., Agre, P. (1993) Distribution of the aquaporin CHIP in secretory and resorptive epithelia and capillary endothelia. *Proc Natl Acad Sci U S A.* 90, 7275–7279.
36. van Balkom, B., Savelkoul, P., Markovich, D., Hofman, E., Nielsen, S., van der Sluijs, P., Deen, P. (2002) The role of putative phosphorylation sites in the targeting and shuttling of the aquaporin-2 water channel. *J Biol Chem.* 277, 41473–41479.
37. Klussmann, E., Maric, K., Wiesner, B., Beyermann, M., Rosenthal, W. (1999) Protein kinase A anchoring proteins are required for vasopressin-mediated translocation of aquaporin-2 into cell membranes of renal principal cells. *J Biol Chem.* 274, 4934–4938.
38. Henn, V., Edemir, B., Stefan, E., Wiesner, B., Lorenz, D., Theilig, F., Schmitt, R., Vossebein, L., Tamma, G., Beyermann, M., Krause, E., Herberg, F., Valenti, G., Bachmann, S., Rosenthal, W., Klussmann, E. (2004) Identification of a novel A-kinase anchoring protein 18 isoform and evidence for its role in the vasopressin-induced aquaporin-2 shuttle in renal principal cells. *J Biol Chem.* 279, 26654–26665.
39. Christensen, B., Zelenina, M., Aperia, A., Nielsen, S. (2000) Localization and regulation of PKA-phosphorylated AQP2 in response to V(2)-receptor agonist/antagonist treatment. *Am J Physiol Renal Physiol.* 278, F29–F42.
40. Nejsum, L., Zelenina, M., Aperia, A., Frokiaer, J., Nielsen, S. (2005) Bidirectional regulation of AQP2 trafficking and recycling: Involvement of AQP2-S256 phosphorylation. *Am J Physiol Renal Physiol.* 288, F930–F938.
41. Hozawa, S., Holtzman, E., Ausiello, D. (1996) cAMP motifs regulating transcription in the aquaporin 2 gene. *Am J Physiol.* 270, C1695–C1702.
42. Umenishi, F., Narikiyo, T., Vandewalle, A., Schrier, R. (2006) cAMP regulates vasopressin-induced AQP2 expression via protein kinase A-independent pathway. *Biochim Biophys Acta.* 1758, 1100–1105.
43. Charest, P., Bouvier, M. (2003) Palmitoylation of the V2 vasopressin receptor carboxyl tail enhances beta-arrestin recruitment leading to efficient receptor endocytosis and ERK1/2 activation. *J Biol Chem.* 278, 41541–41551.
44. Charest, P., Oligny-Longpre, G., Bonin, H., Azzi, M., Bouvier, M. (2007) The V2 vasopressin receptor stimulates ERK1/2 activity independently of heterotrimeric G protein signalling. *Cell Signal.* 19, 32–41.

45. Bustamante, M., Hasler, U., Kotova, O., Chibalin, A., Mordasini, D., Rousselot, M., Vandewalle, A., Martin, P., Feraille, E. (2005) Insulin potentiates AVP-induced AQP2 expression in cultured renal collecting duct principal cells. *Am J Physiol Renal Physiol.* 288, F334–F344.

46. Schwartz, J., Liard, J., Ott, C., Cowley Jr., A. (1985) Hemodynamic effects of neurohypophyseal peptides with antidiuretic activity in dogs. *Am J Physiol.* 249, H1001–H1008.

47. Williams, T., Lightman, S., Leadbeater, M. (1986) Hormonal and cardiovascular responses to DDAVP in man. *Clin Endocrinol (Oxf).* 24, 89–96.

48. Bichet, D., Razi, M., Arthus, M., Lonergan, M., Tittley, P., Smiley, R., Rock, G., Hirsch, D. (1989) Epinephrine and dDAVP administration in patients with congenital nephrogenic diabetes insipidus. Evidence for a pre-cyclic AMP V2 receptor defective mechanism. *Kidney Int.* 36, 859–866.

49. Rosenthal, W., Seibold, A., Antaramian, A., Lonergan, M., Arthus, M., Hendy, G., Birnbaumer, M., Bichet, D. (1992) Molecular identification of the gene responsible for congenital nephrogenic diabetes insipidus. *Nature.* 359, 233–235.

50. Fujiwara, T., Morgan, K., Bichet, D. (1995) Molecular biology of diabetes insipidus. *Annu Rev Med.* 46, 331–343.

51. Bichet, D. (2008) Vasopressin receptor mutations in nephrogenic diabetes insipidus. *Semin Nephrol.* 28, 245–251.

52. Birnbaumer, M., Gilbert, S., Rosenthal, W. (1994) An extracellular congenital nephrogenic diabetes insipidus mutation of the vasopressin receptor reduces cell surface expression, affinity for ligand, and coupling to the Gs/adenylyl cyclase system. *Mol Endocrinol.* 8, 886–894.

53. Pan, Y., Wilson, P., Gitschier, J. (1994) The effect of eight V2 vasopressin receptor mutations on stimulation of adenylyl cyclase and binding to vasopressin. *J Biol Chem.* 269, 31933–31937.

54. Yokoyama, K., Yamauchi, A., Izumi, M., Itoh, T., Ando, A., Imai, E., Kamada, T., Ueda, N. (1996) A low-affinity vasopressin V2-receptor gene in a kindred with X-linked nephrogenic diabetes insipidus. *J Am Soc Nephrol.* 7, 410–414.

55. Sadeghi, H., Robertson, G., Bichet, D., Innamorati, G., Birnbaumer, M. (1997) Biochemical basis of partial nephrogenic diabetes insipidus phenotypes. *Mol Endocrinol.* 11, 1806–1813.

56. Ala, Y., Morin, D., Mouillac, B., Sabatier, N., Vargas, R., Cotte, N., Dechaux, M., Antignac, C., Arthus, M., Lonergan, M., Turner, M., Balestre, M., Alonso, G., Hibert, M., Barberis, C., Hendy, G., Bichet, D., Jard, S. (1998) Functional studies of twelve mutant V2 vasopressin receptors related to nephrogenic diabetes insipidus: Molecular basis of a mild clinical phenotype. *J Am Soc Nephrol.* 9, 1861–1872.

57. Rosenthal, W., Antaramian, A., Gilbert, S., Birnbaumer, M. (1993) Nephrogenic diabetes insipidus. A V2 vasopressin receptor unable to stimulate adenylyl cyclase. *J Biol Chem.* 268, 13030–13033.

58. Morello, J., Salahpour, A., Laperriere, A., Bernier, V., Arthus, M., Lonergan, M., Petaja-Repo, U., Angers, S., Morin, D., Bichet, D., Bouvier, M. (2000) Pharmacological chaperones rescue cell-surface expression and function of misfolded V2 vasopressin receptor mutants. *J Clin Invest.* 105, 887–895.

59. Hermosilla, R., Oueslati, M., Donalies, U., Schonenberger, E., Krause, E., Oksche, A., Rosenthal, W., Schulein, R. (2004) Disease-causing V(2) vasopressin receptors are retained in different compartments of the early secretory pathway. *Traffic.* 5, 993–1005.

60. Barak, L., Oakley, R., Laporte, S., Caron, M. (2001) Constitutive arrestin-mediated desensitization of a human vasopressin receptor mutant associated with nephrogenic diabetes insipidus. *Proc Natl Acad Sci U S A.* 98, 93–98.

61. Bernier, V., Lagace, M., Lonergan, M., Arthus, M., Bichet, D., Bouvier, M. (2004) Functional rescue of the constitutively internalized V2 vasopressin receptor mutant R137H by the pharmacological chaperone action of SR49059. *Mol Endocrinol.* 18, 2074–2084.

62. Bernier, V., Morello, J., Zarruk, A., Debrand, N., Salahpour, A., Lonergan, M., Arthus, M., Laperriere, A., Brouard, R., Bouvier, M., Bichet, D. (2006) Pharmacologic chaperones as a potential treatment for X-linked nephrogenic diabetes insipidus. *J Am Soc Nephrol.* 17, 232–243.

63. Cheong, H., Cho, H., Park, H., Ha, I.S., Choi, Y. (2007) Molecular genetic study of congenital nephrogenic diabetes insipidus and rescue of mutant vasopressin V2 receptor by chemical chaperones. *Nephrology (Carlton).* 12, 113–117.

64. Robben, J., Sze, M., Knoers, N., Deen, P. (2006) Rescue of vasopressin V2 receptor mutants by chemical chaperones: Specificity and mechanism. *Mol Biol Cell.* 17, 379–386.

65. Wuller, S., Wiesner, B., Loffler, A., Furkert, J., Krause, G., Hermosilla, R., Schaefer, M., Schulein, R., Rosenthal, W., Oksche, A. (2004) Pharmacochaperones post-translationally enhance cell surface expression by increasing conformational stability of wild-type and mutant vasopressin V2 receptors. *J Biol Chem.* 279, 47254–47263.

66. Oueslati, M., Hermosilla, R., Schonenberger, E., Oorschot, V., Beyermann, M., Wiesner, B., Schmidt, A., Klumperman, J., Rosenthal, W., Schulein, R. (2007) Rescue of a nephrogenic diabetes insipidus-causing vasopressin V2 receptor mutant by cell-penetrating peptides. *J Biol Chem.* 282, 20676–20685.

67. Tan, C., Nickols, H., Limbird, L. (2003) Appropriate polarization following pharmacological rescue of V2 vasopressin receptors encoded by X-linked nephrogenic diabetes insipidus alleles involves a conformation of the receptor that also attains mature glycosylation. *J Biol Chem.* 278, 35678–35686.

68. Milam, A., Li, Z., Fariss, R. (2003) Histopathology of the human retina in retinitis pigmentosa. *Prog Retin Eye Res.* 17, 175–205.

69. Phelan, J., Bok, D. (2000) A brief review of retinitis pigmentosa and the identified retinitis pigmentosa genes. *Mol Vis.* 6, 116–124.

70. Kalloniatis, M., Fletcher, E. (2004) Retinitis pigmentosa: Understanding the clinical presentation, mechanisms and treatment options. *Clin Exp Optom.* 87, 65–80.

71. Inglehearn, C. (1998) Molecular genetics of human retinal dystrophies. *Eye.* 12, 571–579.

72. Dryja, T., Li, T. (1995) Molecular genetics of retinitis pigmentosa. *Hum Mol Genet.* 4, 1739–1743.

73. Wang, Q., Chen, Q., Zhao, K., Wang, L., Wang, L., Traboulsi, E. (2001) Update on the molecular genetics of retinitis pigmentosa. *Ophthalmic Genet.* 22, 133–154.

74. Farrar, G., Kenna, P., Humphries, P. (2002) On the genetics of retinitis pigmentosa and on mutation-independent approaches to therapeutic intervention. *Embo J.* 21, 857–864.

75. Chapple, J., Grayson, C., Hardcastle, A., Saliba, R., van der Spuy, J., Cheetham, M. (2001) Unfolding retinal dystrophies: A role for molecular chaperones? *Trends Mol Med.* 7, 414–421.

76. Berson, E., Rosner, B., Sandberg, M., Hayes, K., Nicholson, B., Weigel-DiFrano, C., Willett, W. (1993) Vitamin A supplementation for retinitis pigmentosa. *Arch Ophthalmol.* 111, 1456–1459.

77. Stojanovic, A., Hwa, J. (2002) Rhodopsin and retinitis pigmentosa: Shedding light on structure and function. *Receptors Channels.* 8, 33–50.

78. Filipek, S., Stenkamp, R., Teller, D., Palczewski, K. (2003) G protein-coupled receptor rhodopsin: A prospectus. *Annu Rev Physiol.* 65, 851–879.

79. Nathans, J., Hogness, D. (1984) Isolation and nucleotide sequence of the gene encoding human rhodopsin. *Proc Natl Acad Sci U S A.* 81, 4851–4855.

80. Sung, C., Tai, A. (2000) Rhodopsin trafficking and its role in retinal dystrophies. *Int Rev Cytol.* 195, 215–267.

81. Palczewski, K., Kumasaka, T., Hori, T., Behnke, C., Motoshima, H., Fox, B., Le Trong, I., Teller, D., Okada, T., Stenkamp, R., Yamamoto, M., Miyano, M. (2000) Crystal structure of rhodopsin: A G protein-coupled receptor. *Science.* 289, 739–745.

82. Rattner, A., Sun, H., Nathans, J. (1999) Molecular genetics of human retinal disease. *Annu Rev Genet.* 33, 89–131.

83. Karnik, S., Sakmar, T., Chen, H., Khorana, H. (1988) Cysteine residues 110 and 187 are essential for the formation of correct structure in bovine rhodopsin. *Proc Natl Acad Sci U S A.* 85, 8459–8463.

84. Hwa, J., Reeves, P., Klein-Seetharaman, J., Davidson, F., Khorana, H. (1999) Structure and function in rhodopsin: Further elucidation of the role of the intradiscal cysteines, Cys-110, -185, and -187, in rhodopsin folding and function. *Proc Natl Acad Sci U S A.* 96, 1932–1935.

85. Hwa, J., Klein-Seetharaman, J., Khorana, H. (2001) Structure and function in rhodopsin: Mass spectrometric identification of the abnormal intradiscal disulfide bond in misfolded retinitis pigmentosa mutants. *Proc Natl Acad Sci U S A.* 98, 4872–4876.

86. Teller, D., Okada, T., Behnke, C., Palczewski, K., Stenkamp, R. (2001) Advances in determination of a high-resolution three-dimensional structure of rhodopsin, a model of G-protein-coupled receptors (GPCRs). *Biochemistry.* 40, 7761–7772.

87. Bownds, D. (1967) Site of attachment of retinal in rhodopsin. *Nature.* 216, 1178–1181.

88. Wang, J., McDowell, J., Hargrave, P. (1980) Site of attachment of 11-cis retinal in bovine rhodopsin. *Biochemistry.* 19, 5111–5117.

89. Hargrave, P., Bownds, D., Wang, J., McDowell, J. (1982) Retinyl peptide isolation and characterization. *Methods Enzymol.* 81, 211–214.

90. Robinson, P., Buczylko, J., Ohguro, H., Palczewski, K. (1994) Opsins with mutations at the site of chromophore attachment constitutively activate transducin but are not phosphorylated by rhodopsin kinase. *Proc Natl Acad Sci U S A.* 91, 5411–5415.

91. Li, T., Franson, W., Gordon, J., Berson, E., Dryja, T. (1995) Constitutive activation of phototransduction by K296E opsin is not a cause of photoreceptor degeneration. *Proc Natl Acad Sci U S A.* 92, 3551–3555.
92. Sakmar, T., Franke, R., Khorana, H. (1989) Glutamic acid-113 serves as the retinylidene Schiff base counterion in bovine rhodopsin. *Proc Natl Acad Sci U S A.* 86, 8309–8313.
93. Nathans, J. (1990) Determinants of visual pigment absorbance: Identification of the retinylidene Schiff's base counterion in bovine rhodopsin. *Biochemistry.* 29, 9746–9752.
94. Hwa, J., Garriga, P., Liu, X., Khorana, H. (1997) Structure and function in rhodopsin: Packing of the helices in the transmembrane domain and folding to a tertiary structure in the intradiscal domain are coupled. *Proc Natl Acad Sci U S A.* 94, 10571–10576.
95. Creemers, A., Kiihne, S., Bovee-Geurts, P., DeGrip, W., Lugtenburg, J., de Groot, H. (2002) (1)H and (13)C MAS NMR evidence for pronounced ligand-protein interactions involving the ionone ring of the retinylidene chromophore in rhodopsin. *Proc Natl Acad Sci U S A.* 99, 9101–9106.
96. Stojanovic, A., Hwang, I., Khorana, H., Hwa, J. (2003) Retinitis pigmentosa rhodopsin mutations L125R and A164V perturb critical interhelical interactions: New insights through compensatory mutations and crystal structure analysis. *J Biol Chem.* 278, 39020–39028.
97. Arnis, S., Fahmy, K., Hofmann, K., Sakmar, T. (1994) A conserved carboxylic acid group mediates light-dependent proton uptake and signaling by rhodopsin. *J Biol Chem.* 269, 23879–23881.
98. Garriga, P., Liu, X., Khorana, H. (1996) Structure and function in rhodopsin: Correct folding and misfolding in point mutants at and in proximity to the site of the retinitis pigmentosa mutation Leu-125→Arg in the transmembrane helix C. *Proc Natl Acad Sci U S A.* 93, 4560–4564.
99. Andres, A., Kosoy, A., Garriga, P., Manyosa, J. (2001) Mutations at position 125 in transmembrane helix III of rhodopsin affect the structure and signalling of the receptor. *Eur J Biochem.* 268, 5696–5704.
100. Ovchinnikov Yu, A., Abdulaev, N., Bogachuk, A. (1988) Two adjacent cysteine residues in the C-terminal cytoplasmic fragment of bovine rhodopsin are palmitylated. *FEBS Lett.* 230, 1–5.
101. Papac, D., Thornburg, K., Bullesbach, E., Crouch, R., Knapp, D. (1992) Palmitylation of a G-protein coupled receptor. Direct analysis by tandem mass spectrometry. *J Biol Chem.* 267, 16889–16894.
102. Konig, B., Arendt, A., McDowell, J., Kahlert, M., Hargrave, P., Hofmann, K. (1989) Three cytoplasmic loops of rhodopsin interact with transducin. *Proc Natl Acad Sci U S A.* 86, 6878-6882.
103. Franke, R., Sakmar, T., Graham, R., Khorana, H. (1992) Structure and function in rhodopsin. Studies of the interaction between the rhodopsin cytoplasmic domain and transducin. *J Biol Chem.* 267, 14767–14774.
104. Ernst, O., Hofmann, K., Sakmar, T. (1995) Characterization of rhodopsin mutants that bind transducin but fail to induce GTP nucleotide uptake. Classification of mutant pigments by fluorescence, nucleotide release, and flash-induced light-scattering assays. *J Biol Chem.* 270, 10580–10586.

105. McDowell, J., Nawrocki, J., Hargrave, P. (1993) Phosphorylation sites in bovine rhodopsin. *Biochemistry.* 32, 4968–4974.
106. Ohguro, H., Palczewski, K., Ericsson, L., Walsh, K., Johnson, R. (1993) Sequential phosphorylation of rhodopsin at multiple sites. *Biochemistry.* 32, 5718–5724.
107. Papac, D., Oatis Jr., J., Crouch, R., Knapp, D. (1993) Mass spectrometric identification of phosphorylation sites in bleached bovine rhodopsin. *Biochemistry.* 32, 5930–5934.
108. Chen, J., Makino, C., Peachey, N., Baylor, D., Simon, M. (1995) Mechanisms of rhodopsin inactivation in vivo as revealed by a COOH-terminal truncation mutant. *Science.* 267, 374–377.
109. Dryja, T., McGee, T., Reichel, E., Hahn, L.B., Cowley, G., Yandell, D., Sandberg, M., Berson, E. (1990) A point mutation of the rhodopsin gene in one form of retinitis pigmentosa. *Nature.* 343, 364–366.
110. Dryja, T., Hahn, L., Cowley, G., McGee, T., Berson, E. (1991) Mutation spectrum of the rhodopsin gene among patients with autosomal dominant retinitis pigmentosa. *Proc Natl Acad Sci U S A.* 88, 9370–9374.
111. Sung, C., Davenport, C., Hennessey, J., Maumenee, I., Jacobson, S., Heckenlively, J., Nowakowski, R., Fishman, G., Gouras, P., Nathans, J. (1991) Rhodopsin mutations in autosomal dominant retinitis pigmentosa. *Proc Natl Acad Sci U S A.* 88, 6481–6485.
112. Sung, C., Schneider, B., Agarwal, N., Papermaster, D., Nathans, J. (1991) Functional heterogeneity of mutant rhodopsins responsible for autosomal dominant retinitis pigmentosa. *Proc Natl Acad Sci U S A.* 88, 8840–8844.
113. Sung, C., Davenport, C., Nathans, J. (1993) Rhodopsin mutations responsible for autosomal dominant retinitis pigmentosa. Clustering of functional classes along the polypeptide chain. *J Biol Chem.* 268, 26645–26649.
114. Kaushal, S., Khorana, H. (1994) Structure and function in rhodopsin. 7. Point mutations associated with autosomal dominant retinitis pigmentosa. *Biochemistry.* 33, 6121–6128.
115. Chuang, J., Vega, C., Jun, W., Sung, C. (2004) Structural and functional impairment of endocytic pathways by retinitis pigmentosa mutant rhodopsin-arrestin complexes. *J Clin Invest.* 114, 131–140.
116. Sung, C., Makino, C., Baylor, D., Nathans, J. (1994) A rhodopsin gene mutation responsible for autosomal dominant retinitis pigmentosa results in a protein that is defective in localization to the photoreceptor outer segment. *J Neurosci.* 14, 5818–5833.
117. Li, T., Sandberg, M., Pawlyk, B., Rosner, B., Hayes, K., Dryja, T., Berson, E. (1998) Effect of vitamin A supplementation on rhodopsin mutants threonine-17$\rightarrow$ methionine and proline-347$\rightarrow$serine in transgenic mice and in cell cultures *Proc Natl Acad Sci U S A.* 95, 11933–11938.
118. Tai, A., Chuang, J., Bode, C., Wolfrum, U., Sung, C. (1999) Rhodopsin's carboxy-terminal cytoplasmic tail acts as a membrane receptor for cytoplasmic dynein by binding to the dynein light chain Tctex-1. *Cell.* 97, 877–887.
119. Tam, B., Moritz, O., Hurd, L., Papermaster, D. (2000) Identification of an outer segment targeting signal in the COOH terminus of rhodopsin using transgenic *Xenopus laevis*. *J Cell Biol.* 151, 1369–1380.

120. Min, K., Zvyaga, T., Cypess, A., Sakmar, T. (1993) Characterization of mutant rhodopsins responsible for autosomal dominant retinitis pigmentosa. Mutations on the cytoplasmic surface affect transducin activation. *J Biol Chem.* 268, 9400–9404.
121. Papermaster, D. (2002) The birth and death of photoreceptors: The Friedenwald lecture. *Invest Ophthalmol Vis Sci.* 43, 1300–1309.
122. Anukanth, A., Khorana, H. (1994) Structure and function in rhodopsin. Requirements of a specific structure for the intradiscal domain. *J Biol Chem.* 269, 19738–19744.
123. Illing, M., Rajan, R., Bence, N., Kopito, R. (2002) A rhodopsin mutant linked to autosomal dominant retinitis pigmentosa is prone to aggregate and interacts with the ubiquitin proteasome system. *J Biol Chem.* 277, 34150–34160.
124. Saliba, R., Munro, P., Luthert, P., Cheetham, M. (2002) The cellular fate of mutant rhodopsin: Quality control, degradation and aggresome formation. *J Cell Sci.* 115, 2907–2918.
125. Olsson, J., Gordon, J., Pawlyk, B., Roof, D., Hayes, A., Molday, R., Mukai, S., Cowley, G., Berson, E., Dryja, T. (1992) Transgenic mice with a rhodopsin mutation (Pro23His): A mouse model of autosomal dominant retinitis pigmentosa. *Neuron.* 9, 815–830.
126. Roof, D., Adamian, M., Hayes, A. (1994) Rhodopsin accumulation at abnormal sites in retinas of mice with a human P23H rhodopsin transgene. *Invest Ophthalmol Vis Sci.* 35, 4049–4062.
127. Marigo, V. (2007) Programmed cell death in retinal degeneration: Targeting apoptosis in photoreceptors as potential therapy for retinal degeneration. *Cell Cycle.* 6, 652–655.
128. Tam, B., Moritz, O. (2006) Characterization of rhodopsin P23H-induced retinal degeneration in a *Xenopus laevis* model of retinitis pigmentosa. *Invest Ophthalmol Vis Sci.* 47, 3234–3241.
129. Lin, J., Li, H., Yasumura, D., Cohen, H., Zhang, C., Panning, B., Shokat, K., Lavail, M., Walter, P. (2007) IRE1 signaling affects cell fate during the unfolded protein response. *Science.* 318, 944–949.
130. Green, E., Menz, M., LaVail, M., Flannery, J. (2000) Characterization of rhodopsin mis-sorting and constitutive activation in a transgenic rat model of retinitis pigmentosa. *Invest Ophthalmol Vis Sci.* 41, 1546–1553.
131. De Grip, W., Lester, P. (1982) Thermal stability of rhodopsin and opsin in some novel detergents. *Methods Enzymol.* 81, 256–265.
132. Khan, S., Bolen, W., Hargrave, P., Santoro, M., McDowell, J. (1991) Differential scanning calorimetry of bovine rhodopsin in rod-outer-segment disk membranes. *Eur J Biochem.* 200, 53–59.
133. Okada, T., Le Trong, I., Fox, B., Behnke, C., Stenkamp, R., Palczewski, K. (2000) X-ray diffraction analysis of three-dimensional crystals of bovine rhodopsin obtained from mixed micelles. *J Struct Biol.* 130, 73–80.
134. Landin, J., Katragadda, M., Albert, A. (2001) Thermal destabilization of rhodopsin and opsin by proteolytic cleavage in bovine rod outer segment disk membranes. *Biochemistry.* 40, 11176–11183.
135. Noorwez, S., Kuksa, V., Imanishi, Y., Zhu, L., Filipek, S., Palczewski, K., Kaushal, S. (2003) Pharmacological chaperone-mediated in vivo folding and stabilization

of the P23H-opsin mutant associated with autosomal dominant retinitis pigmentosa. *J Biol Chem.* 278, 14442–14450.

136. Noorwez, S., Malhotra, R., McDowell, J., Smith, K., Krebs, M., Kaushal, S. (2004) Retinoids assist the cellular folding of the autosomal dominant retinitis pigmentosa opsin mutant P23H. *J Biol Chem.* 279, 16278–16284.
137. Noorwez, S., Ostrov, D., McDowell, J., Krebs, M., Kaushal, S. (2008) A combinatorial high-throughput screening method for small molecule pharmacological chaperones of misfolded rhodopsin. *Invest Ophthalmol Vis Sci.* 49, 3224–3230.
138. St. Jules, R., Wallingford, J., Smith, S., O'Brien, P. (1989) Addition of the chromophore to rat rhodopsin is an early post-translational event. *Exp Eye Res.* 48, 653–665.
139. Lou, J., Tan, Q., Karnaukhova, E., Berova, N., Nakanishi, K., Crouch, R., Krzystof, P. (2000) Synthetic retinals: Convenient probes of rhodopsin and visual transduction process. *Methods Enzymol.* 315, 219–237.
140. Kuksa, V., Bartl, F., Maeda, T., Jang, G., Ritter, E., Heck, M., Van Hooser, J., Liang, Y., Filipek, S., Gelb, M., Hofmann, K., Palczewski, K. (2002) Biochemical and physiological properties of rhodopsin regenerated with 11-cis-6-ring- and 7-ring-retinals. *J Biol Chem.* 277, 42315–42324.
141. Humphries, M., Rancourt, D., Farrar, G., Kenna, P., Hazel, M., Bush, R., Sieving, P., Sheils, D., McNally, N., Creighton, P., Erven, A., Boros, A., Gulya, K., Capecchi, M., Humphries, P. (1997) Retinopathy induced in mice by targeted disruption of the rhodopsin gene. *Nat Genet.* 15, 216–219.
142. Naash, M., Hollyfield, J., al-Ubaidi, M., Baehr, W. (1993) Simulation of human autosomal dominant retinitis pigmentosa in transgenic mice expressing a mutated murine opsin gene. *Proc Natl Acad Sci U S A.* 90, 5499–5503.
143. Li, T., Snyder, W., Olsson, J., Dryja, T. (1996) Transgenic mice carrying the dominant rhodopsin mutation P347S: Evidence for defective vectorial transport of rhodopsin to the outer segments. *Proc Natl Acad Sci U S A.* 93, 14176–14181.
144. Petters, R., Alexander, C., Wells, K., Collins, E., Sommer, J., Blanton, M., Rojas, G., Hao, Y., Flowers, W., Banin, E., Cideciyan, A., Jacobson, S., Wong, F. (1997) Genetically engineered large animal model for studying cone photoreceptor survival and degeneration in retinitis pigmentosa. *Nat Biotechnol.* 15, 965–970.
145. Li, Z., Wong, F., Chang, J., Possin, D., Hao, Y., Petters, R., Milam, A. (1998) Rhodopsin transgenic pigs as a model for human retinitis pigmentosa. *Invest Ophthalmol Vis Sci.* 39, 808–819.
146. Bhagavath, B., Layman, L. (2007) The genetics of hypogonadotropic hypogonadism. *Sem Reprod Med.* 25, 272–286.
147. Seminara, S., Hayes, F., Crowley, W. (1998) Gonadotropin-releasing hormone deficiency in the human (idiopathic hypogonadotropic hypogonadism and Kallmann's syndrome): Pathophysiological and genetic considerations. *Endocr Rev.* 19, 521–539.
148. Bhagavath, B., Ozata, M., Ozdemir, I., Bolu, E., Bick, D., Sherins, R., Layman, L. (2005) The prevalence of gonadotropin-releasing hormone receptor mutations in a large cohort of patients with hypogonadotropic hypogonadism. *Fertil Steril.* 84, 951–957.
149. Hapgood, J., Sadie, H., van Biljon, W., Ronacher, K. (2005) Regulation of expression of mammalian gonadotrophin-releasing hormone receptor genes. *J Neuroendocrinol.* 17, 619–638.

150. Ulloa-Aguirre, A., Janovick, J., Miranda, A., Conn, P. (2006) G-protein-coupled receptor trafficking: Understanding the chemical basis of health and disease. *ACS Chem Bio.* 1, 631–638.

151. Ulloa-Aguirre, A., Janovick, J., Brothers, S., Conn, P. (2004) Pharmacologic rescue of conformationally-defective proteins: Implications for the treatment of human disease. *Traffic.* 5, 821–837.

152. de Roux, N., Young, J., Misrahi, M., Genet, R., Chanson, P., Schaison, G., Milgrom, E. (1997) A family with hypogonadotropic hypogonadism and mutations in the gonadotropin-releasing hormone receptor. *N Engl J Med.* 337, 1597–1602.

153. Layman, L., Cohen, D., Jin, M., Xie, J., Li, Z., Reindollar, R., Bolbolan, S. (1998) Mutations in gonadotropin-releasing hormone receptor gene cause hypogonadotropic hypogonadism. *Nat Genet.* 18, 14–15.

154. Janovick, J., Maya-Nunez, G., Conn, P. (2002) Rescue of hypogonadotropic hypogonadism-causing and manufactured GnRH receptor mutants by a specific protein-folding template: Misrouted proteins as a novel disease etiology and therapeutic target. *J Clin Endocrinol Metab.* 87, 3255–3262.

155. Leanos-Miranda, A., Janovick, J., Conn, P. (2002) Receptor-misrouting: An unexpectedly prevalent and rescuable etiology in GnRHR-mediated hypogonadotropic hypogonadism. *J Clin Endocrinol Metab.* 87, 4825–4828.

156. Topaloglu, A., Lu, Z.-L., Farooqi, I., Mungan, N., Yuksel, B., O'Rahilly, S., Millar, R. (2006) Molecular genetic analysis of normosmic hypogonadotropic hypogonadism in a Turkish population: Identification and detailed functional characterization of a novel mutation in the gonadotropin-releasing hormone receptor gene. *Neuroendocrinol.* 84, 301–308.

157. Bedecarrats, G., Kaiser, U. (2007) Mutations in the human gonadotropin-releasing hormone receptor: Insights into receptor biology and function. *Semin Repro Med.* 25, 368–378.

158. Arora, K., Chung, H., Catt, K. (1999) Influence of a species-specific extracellular amino acid on expression and function of the human gonadotropin-releasing hormone receptor. *Mol Endocrinol.* 13, 890–896.

159. Lin, X., Janovick, J., Brothers, S., Blomenrohr, M., Bogerd, J., Conn, P. (1998) Addition of catfish gonadotropin-releasing hormone (GnRH) receptor intracellular carboxyl-terminal tail to rat GnRH receptor alters receptor expression and regulation. *Mol Endocrinol.* 12, 161–171.

160. Maya-Nunez, G., Janovick, J., Conn, P. (2000) Combined modification of intracellular and extracellular loci on human gonadotropin-releasing hormone receptor provides a mechanism for enhanced expression. *Endocrine.* 3, 401–409.

161. Maya-Nunez, G., Soderlund, D., Ulloa-Aguirre, A., Conn, P., Janovick, J., Mendez, J. (2002) Molecular basis of hypogonadotropic hypogonadism: Restoration of mutant (E90K) GnRH receptor function by a deletion at a distant site. *J Clin Endocrinol Metab.* 87, 2144–2149.

162. Conn, P., Leanos-Miranda, A., Janovick, J. (2002) Protein origami: Therapeutic rescue of misfolded gene products. *Mol Intervention.* 2, 308–316.

163. Janovick, J., Goulet, M., Bush, E., Greer, J., Wettlaufer, D., Conn, P. (2003) Structure-activity relations of successful pharmacologic chaperones for rescue of naturally occurring and manufactured mutants of the gonadotropin-releasing hormone receptor. *J Pharmacol Exp Ther.* 305, 608–614.

164. Petaja-Repo, U., Hogue, M., Laperriere, A., Bhalla, S., Walker, P., Bouvier, M. (2001) Newly synthesized human delta opioid receptors retained in the endoplasmic reticulum are retrotranslocated to the cytosol, deglycosylated, ubiquitinated, and degraded by the proteasome. *J Biol Chem.* 276, 4416–4423.

165. Karges, B., Karges, W., de Roux, N. (2003) Clinical and molecular genetics of the human GnRH receptor. *Hum Reprod Update.* 9 523–530.

166. Leanos-Miranda, A., Ulloa-Aguirre, A., Janovick, J., Conn, P. (2005) In vitro coexpression and pharmacological rescue of mutant gonadotropin-releasing hormone receptors causing hypogonadotropic hypogonadism in humans expressing compound heterozygous alleles. *J Clin Endocrinol Metab.* 90, 3001–3008.

167. Pask, A., Kanasaki, H., Kaiser, U., Conn, P., Janovick, J., Stockton, D., Hess, D., Justice, M., Behringer, R. (2005) A novel mouse model of hypogonadotrophic hypogonadism: N-ethyl-N-nitrosourea-induced gonadotropin-releasing hormone receptor gene mutation. *Mol Endocrinol.* 19, 972–981.

168. Cone, R. (2006) Studies on the physiological functions of the melanocortin system. *Endocr Rev.* 27, 736–749.

169. Cone, R. (2005) Anatomy and regulation of the central melanocortin system. *Nat Neurosci.* 8, 571–578.

170. Tao, Y. (2005) Molecular mechanisms of the neural melanocortin receptor dysfunction in severe early onset obesity. *Mol Cell Endocrinol.* 239, 1–14.

171. Balthasar, N., Dalgaard, L., Lee, C., Yu, J., Funahashi, H., Williams, T., Ferreira, M., Tang, V., McGovern, R., Kenny, C., Christiansen, L., Edelstein, E., Choi, B., Boss, O., Aschkenasi, C., Zhang, C., Mountjoy, K., Kishi, T., Elmquist, J., Lowell, B. (2005) Divergence of melanocortin pathways in the control of food intake and energy expenditure. *Cell.* 123, 493–505.

172. Chen, A., Marsh, D.J., Trumbauer, M., Frazier, E., Guan, X., Yu, H., Rosenblum, C., Vongs, A., Feng, Y., Cao, L., Metzger, J., Strack, A., Camacho, R., Mellin, T., Nunes, C., Min, W., Fisher, J., Gopal-Truter, S., MacIntyre, D., Chen, H., Van der Ploeg, L. (2000) Inactivation of the mouse melanocortin-3 receptor results in increased fat mass and reduced lean body mass. *Nat Genet.* 26, 97–102.

173. Huszar, D., Lynch, C., Fairchild-Huntress, V., Dunmore, J., Fang, Q., Berkemeier, L., Gu, W., Kesterson, R., Boston, B., Cone, R., Smith, F., Campfield, L., Burn, P., Lee, F. (1997) Targeted disruption of the melanocortin-4 receptor results in obesity in mice. *Cell.* 88, 131–141.

174. Ste. Marie, L., Miura, G., Marsh, D., Yagaloff, K., Palmiter, R. (2000) A metabolic defect promotes obesity in mice lacking melanocortin-4 receptors. *Proc Natl Acad Sci U S A.* 97, 12339–12344.

175. Vaisse, C., Clement, K., Guy-Grand, B., Froguel, P. (1998) A frameshift mutation in human MC4R is associated with a dominant form of obesity. *Nat Genet.* 20, 113–114.

176. Yeo, G., Farooqi, I., Aminian, S., Halsall, D., Stanhope, R., O'Rahilly, S. (1998) A frameshift mutation in MC4R associated with dominantly inherited human obesity. *Nat Genet.* 20, 111–112.

177. Vaisse, C., Clement, K., Durand, E., Hercberg, S., Guy-Grand, B., Froguel, P. (2000) Melanocortin-4 receptor mutations are a frequent and heterogeneous cause of morbid obesity. *J Clin Invest.* 106, 253–262.

178. Yeo, G., Lank, E., Farooqi, I., Keogh, J., Challis, B., O'Rahilly, S. (2003) Mutations in the human melanocortin-4 receptor gene associated with severe familial obesity disrupts receptor function through multiple molecular mechanisms. *Hum Mol Genet.* 12, 561–574.

179. Farooqi, I., Keogh, J., Yeo, G., Lank, E., Cheetham, T., O'Rahilly, S. (2003) Clinical spectrum of obesity and mutations in the melanocortin 4 receptor gene. *N Engl J Med.* 348, 1085–1095.

180. Tao, Y.X. (2006) Inactivating mutations of G protein-coupled receptors and diseases: Structure-function insights and therapeutic implications. *Pharmacol Ther.* 111, 949–973.

181. Lee, Y., Poh, L., Loke, K. (2002) A novel melanocortin 3 receptor gene (MC3R) mutation associated with severe obesity. *J Clin Endocrinol Metab.* 87, 1423–1426.

182. Rached, M., Buronfosse, A., Begeot, M., Penhoat, A. (2004) Inactivation and intracellular retention of the human I183N mutated melanocortin 3 receptor associated with obesity. *Biochim Biophys Acta.* 1689, 229–234.

183. Tao, Y., Segaloff, D. (2004) Functional characterization of melanocortin-3 receptor variants identify a loss-of-function mutation involving an amino acid critical for G protein-coupled receptor activation. *J Clin Endocrinol Metab.* 89, 3936–3942.

184. Mencarelli, M., Walker, G., Maestrini, S., Alberti, L., Verti, B., Brunani, A., Petroni, M., Tagliaferri, M., Liuzzi, A., Di Blasio, A. (2008) Sporadic mutations in melanocortin receptor 3 in morbid obese individuals. *Eur J Hum Genet.* 16, 581–586.

185. Tao, Y. (2007) Functional characterization of novel melanocortin-3 receptor mutations identified from obese subjects. *Biochim Biophys Acta.* 1772, 1167–1174.

186. Xu, A., Choi, K., Wang, Y., Permana, P., Xu, L., Bogardus, C., Cooper, G. (2002) Identification of novel putative membrane proteins selectively expressed during adipose conversion of 3T3-L1 cells. *Biochem Biophys Res Commun.* 293, 1161–1167.

187. Metherell, L., Chapple, J., Cooray, S., David, A., Becker, C., Ruschendorf, F., Naville, D., Begeot, M., Khoo, B., Nurnberg, P., Huebner, A., Cheetham, M., Clark, A. (2005) Mutations in MRAP, encoding a new interacting partner of the ACTH receptor, cause familial glucocorticoid deficiency type 2. *Nat Genet.* 37, 166–170.

188. Rouzaud, F., Kadekaro, A., Abdel-Malek, Z., Hearing, V. (2005) MC1R and the response of melanocytes to ultraviolet radiation. *Mutat Res.* 571, 133–152.

189. Valverde, P., Healy, E., Jackson, I., Rees, J., Thody, A. (1995) Variants of the melanocyte-stimulating hormone receptor gene are associated with red hair and fair skin in humans. *Nat Genet.* 11, 328–330.

190. Beaumont, K., Newton, R., Smit, D., Leonard, J., Stow, J., Sturm, R. (2005) Altered cell surface expression of human MC1R variant receptor alleles associated with red hair and skin cancer risk. *Hum Mol Genet.* 14, 2145–2154.

191. Caltabiano, G., Campillo, M., De Leener, A., Smits, G., Vassart, G., Costagliola, S., Pardo, L. (2008) The specificity of binding of glycoprotein hormones to their receptors. *Cell Mol Life Sci.* 65, 2484–2489.

192. Yanofsky, S., Shen, E., Holden, F., Whitehorn, E., Aguilar, B., Tate, E., Holmes, C., Scheuerman, R., MacLean, D., Wu, M., Frail, D., Lopez, F., Winneker, R., Arey, B., Barrett, R. (2006) Allosteric activation of the follicle-stimulating hormone (FSH) receptor by selective, nonpeptide agonists. *J Biol Chem.* 281, 13226–13233.

193. Huang, Y., Breitwieser, G. (2007) Rescue of calcium-sensing receptor mutants by allosteric modulators reveals a conformational checkpoint in receptor biogenesis. *J Biol Chem.* 282, 9517–9525.

194. Sunthornthepvarakui, T., Gottschalk, M., Hayashi, Y., Refetoff, S. (1995) Brief report: Resistance to thyrotropin caused by mutations in the thyrotropin-receptor gene. *N Engl J Med.* 332, 155–160.

195. Abramowicz, M., Duprez, L., Parma, J., Vassart, G., Heinrichs, C. (1997) Familial congenital hypothyroidism due to inactivating mutation of the thyrotropin receptor causing profound hypoplasia of the thyroid gland. *J Clin Invest.* 99, 3018–3024.

196. Costagliola, S., Sunthorntepvarakul, T., Migeotte, I., Van Sande, J., Kajava, A., Refetoff, S., Vassart, G. (1999) Structure-function relationships of two loss-of-function mutations of the thyrotropin receptor gene. *Thyroid.* 9, 995–1000.

197. Tonacchera, M., Agretti, P., Chiovato, L., Rosellini, V., Ceccarini, G., Perri, A., Viacava, P., Naccarato, A., Miccoli, P., Pinchera, A., Vitti, P. (2000) Activating thyrotropin receptor mutations are present in nonadenomatous hyperfunctioning nodules of toxic or autonomous multinodular goiter. *J Clin Endocrinol Metab.* 85, 2270–2274.

198. Tonacchera, M., Perri, A., De Marco, G., Agretti, P., Banco, M., Di Cosmo, C., Grasso, L., Vitti, P., Chiovato, L., Pinchera, A. (2004) Low prevalence of thyrotropin receptor mutations in a large series of subjects with sporadic and familial nonautoimmune subclinical hypothyroidism. *J Clin Endocrinol Metab.* 89, 5787–5793.

199. Biebermann, H., Schoneberg, T., Krude, H., Schultz, G., Gudermann, T., Gruters, A. (1997) Mutations of the human thyrotropin receptor gene causing thyroid hypoplasia and persistent congenital hypothyroidism. *J Clin Endocrinol Metab.* 82, 3471–3480.

200. Richter-Unruh, A., Korsch, E., Hiort, O., Holterhus, P., Themmen, A., Wudy, S. (2005) Novel insertion frameshift mutation of the LH receptor gene: Problematic clinical distinction of Leydig cell hypoplasia from enzyme defects primarily affecting testosterone biosynthesis. *Eur J Endocrinol.* 152, 255–259.

201. Gromoll, J., Schulz, A., Borta, H., Gudermann, T., Teerds, K., Greschniok, A., Nieschlag, E., Seif, F. (2002) Homozygous mutation within the conserved Ala-Phe-Asn-Glu-Thr motif of exon 7 of the LH receptor causes male pseudohermaphroditism. *Eur J Endocrinol.* 147, 597–608.

202. Martens, J., Lumbroso, S., Verhoef-Post, M., Georget, V., Richter-Unruh, A., Szarras-Czapnik, M., Romer, T., Brunner, H., Themmen, A., Sultan, C. (2002) Mutant luteinizing hormone receptors in a compound heterozygous patient with complete Leydig cell hypoplasia: Abnormal processing causes signaling deficiency. *J Clin Endocrinol Metab.* 87, 2506–2513.

203. Tao, Y., Johnson, N., Segaloff, D. (2004) Constitutive and agonist-dependent self-association of the cell surface human lutropin receptor. *J Biol Chem.* 279, 5904–5914.

204. Martens, J., Verhoef-Post, M., Abelin, N., Ezabella, M., Toledo, S., Brunner, H., Themmen, A. (1998) A homozygous mutation in the luteinizing hormone receptor causes partial Leydig cell hypoplasia: Correlation between receptor activity and phenotype. *Mol Endocrinol.* 12, 775–784.

205. Wu, S., Hallermeier, K., Laue, L., Brain, C., Berry, A., Grant, D., Griffin, J., Wilson, J., Cutler Jr., G., Chan, W. (1998) Inactivation of the luteinizing hormone/chorionic

gonadotropin receptor by an insertional mutation in Leydig cell hypoplasia. *Mol Endocrinol.* 12, 1651–1660.

206. Latronico, A., Chai, Y., Arnhold, I., Liu, X., Mendonca, B., Segaloff, D. (1998) A homozygous microdeletion in helix 7 of the luteinizing hormone receptor associated with familial testicular and ovarian resistance is due to both decreased cell surface expression and impaired effector activation by the cell surface receptor. *Mol Endocrinol.* 12, 442–450.
207. Aittomaki, K., Lucena, J., Pakarinen, P., Sistonen, P., Tapanainen, J., Gromoll, J., Kaskikari, R., Sankila, E., Lehvaslaiho, H., Engel, A., Nieschlag, E., Huhtaniemi, I., de la Chapelle, A. (1995) Mutation in the follicle-stimulating hormone receptor gene causes hereditary hypergonadotropic ovarian failure. *Cell.* 82, 959–968.
208. Rannikko, A., Pakarinen, P., Manna, P., Beau, I., Misrahi, M., Aittomaki, K., Huhtaniemi, I. (2002) Functional characterization of the human FSH receptor with an inactivating Ala189Val mutation. *Mol Hum Reprod.* 8, 311–317.
209. Beau, I., Touraine, P., Meduri, G., Gougeon, A., Desroches, A., Matuchansky, C., Milgrom, E., Kuttenn, F., Misrahi, M. (1998) A novel phenotype related to partial loss of function mutations of the follicle stimulating hormone receptor. *J Clin Invest.* 102, 1352–1359.
210. Touraine, P., Beau, I., Gougeon, A., Meduri, G., Desroches, A., Pichard, C., Detoeuf, M., Paniel, B., Prieur, M., Zorn, J., Milgrom, E., Kuttenn, F., Misrahi, M. (1999) New natural inactivating mutations of the follicle-stimulating hormone receptor: Correlations between receptor function and phenotype. *Mol Endocrinol.* 13, 1844–1854.
211. Meduri, G., Touraine, P., Beau, I., Lahuna, O., Desroches, A., Vacher-Lavenu, M., Kuttenn, F., Misrahi, M. (2003) Delayed puberty and primary amenorrhea associated with a novel mutation of the human follicle-stimulating hormone receptor: Clinical, histological, and molecular studies. *J Clin Endocrinol Metab.* 88, 3491–3498.
212. Amiel, J., Sproat-Emison, E., Garcia-Barcelo, M., Lantieri, F., Burzynski, G., Borrego, S., Pelet, A., Arnold, S., Miao, X., Griseri, P., Brooks, A., Antinolo, G., de Pontual, L., Clement-Ziza, M., Munnich, A., Kashuk, C., West, K., Wong, K., Lyonnet, S., Chakravarti, A., Tam, P., Ceccherini, I., Hofstra, R., Fernandez, R. (2008) Hirschsprung disease, associated syndromes and genetics: A review. *J Med Genet.* 45, 1–14.
213. Tanaka, H., Moroi, K., Iwai, J., Takahashi, H., Ohnuma, N., Hori, S., Takimoto, M., Nishiyama, M., Masaki, T., Yanagisawa, M., Sekiya, S., Kimura, S. (1998) Novel mutations of the endothelin B receptor gene in patients with Hirschsprung's disease and their characterization. *J Biol Chem.* 273, 11378–11383.
214. Fuchs, S., Amiel, J., Claudel, S., Lyonnet, S., Corvol, P., Pinet, F. (2001) Functional characterization of three mutations of the endothelin B receptor gene in patients with Hirschsprung's disease: Evidence for selective loss of Gi coupling. *Mol Med.* 7, 115–124.
215. Alba, M., Salvatori, R. (2005) Naturally-occurring missense mutations in the human growth hormone-releasing hormone receptor alter ligand binding. *J Endocrinol.* 186, 515–521.
216. Urena, P. (2003) [The PTH/PTHrP receptor: Biological implications]. *Nefrologia.* 23, 12–17.

217. Karaplis, A., He, B., Nguyen, M., Young, I., Semeraro, D., Ozawa, H., Amizuka, N. (1998) Inactivating mutation in the human parathyroid hormone receptor type 1 gene in Blomstrand chondrodysplasia. *Endocrinology.* 139, 5255–5258.

218. Zhang, P., Jobert, A., Couvineau, A., Silve, C. (1998) A homozygous inactivating mutation in the parathyroid hormone/parathyroid hormone-related peptide receptor causing Blomstrand chondrodysplasia. *J Clin Endocrinol Metab.* 83, 3365–3368.

219. Hoogendam, J., Farih-Sips, H., Wynaendts, L., Lowik, C., Wit, J., Karperien, M. (2007) Novel mutations in the parathyroid hormone (PTH)/PTH-related peptide receptor type 1 causing Blomstrand osteochondrodysplasia types I and II. *J Clin Endocrinol Metab.* 92, 1088–1095.

220. Jobert, A., Zhang, P., Couvineau, A., Bonaventure, J., Roume, J., Le Merrer, M., Silve, C. (1998) Absence of functional receptors for parathyroid hormone and parathyroid hormone-related peptide in Blomstrand chondrodysplasia. *J Clin Invest.* 102, 34–40.

221. Karperien, M., van der Harten, H., van Schooten, R., Farih-Sips, H., den Hollander, N., Kneppers, S., Nijweide, P., Papapoulos, S., Lowik, C. (1999) A frame-shift mutation in the type I parathyroid hormone (PTH)/PTH-related peptide receptor causing Blomstrand lethal osteochondrodysplasia. *J Clin Endocrinol Metab.* 84, 3713–3720.

222. Duchatelet, S., Ostergaard, E., Cortes, D., Lemainque, A., Julier, C. (2005) Recessive mutations in PTHR1 cause contrasting skeletal dysplasias in Eiken and Blomstrand syndromes. *Hum Mol Genet.* 14, 1–5.

223. Brown, E. (2007) The calcium-sensing receptor: Physiology, pathophysiology and CaR-based therapeutics. *Subcell Biochem.* 45, 139–167.

224. Pollak, M., Brown, E., Chou, Y., Hebert, S., Marx, S., Steinmann, B., Levi, T., Seidman, C., Seidman, J. (1993) Mutations in the human Ca(2+)-sensing receptor gene cause familial hypocalciuric hypercalcemia and neonatal severe hyperparathyroidism. *Cell.* 75, 1297–1303.

225. Petaja-Repo, U., Hogue, M., Bhalla, S., Laperriere, A., Morello, J., Bouvier, M. (2002) Ligands act as pharmacological chaperones and increase the efficiency of delta opioid receptor maturation. *Embo J.* 21, 1628–1637.

226. Goodman, W., Frazao, J., Goodkin, D., Turner, S., Liu, W., Coburn, J. (2000) A calcimimetic agent lowers plasma parathyroid hormone levels in patients with secondary hyperparathyroidism. *Kidney Int.* 58, 436–445.

227. Toomes, C., Bottomley, H., Scott, S., Mackey, D., Craig, J., Appukuttan, B., Stout, J., Flaxel, C., Zhang, K., Black, G., Fryer, A., Downey, L., Inglehearn, C. (2004) Spectrum and frequency of FZD4 mutations in familial exudative vitreoretinopathy. *Invest Ophthalmol Vis Sci.* 45, 2083–2090.

228. Yoshida, S., Arita, R., Yoshida, A., Tada, H., Emori, A., Noda, Y., Nakao, S., Fujisawa, K., Ishibashi, T. (2004) Novel mutation in FZD4 gene in a Japanese pedigree with familial exudative vitreoretinopathy. *Am J Ophthalmol.* 138, 670–671.

229. Nallathambi, J., Shukla, D., Rajendran, A., Namperumalsamy, P., Muthulakshmi, R., Sundaresan, P. (2006) Identification of novel FZD4 mutations in Indian patients with familial exudative vitreoretinopathy. *Mol Vis.* 12, 1086–1092.

230. Kondo, H., Qin, M., Tahira, T., Uchio, E., Hayashi, K. (2007) Severe form of familial exudative vitreoretinopathy caused by homozygous R417Q mutation in frizzled-4 gene. *Ophthalmic Genet.* 28, 220–223.

231. Muller, M., Kusserow, C., Orth, U., Klar-Dissars, U., Laqua, H., Gal, A. (2008) [Mutations of the frizzled-4 gene. Their impact on medical care of patients with autosomal dominant exudative vitreoretinopathy]. *Ophthalmologe.* 105, 262–268.
232. Robitaille, J., MacDonald, M., Kaykas, A., Sheldahl, L., Zeisler, J., Dube, M., Zhang, L., Singaraja, R., Guernsey, D., Zheng, B., Siebert, L., Hoskin-Mott, A., Trese, M., Pimstone, S., Shastry, B., Moon, R., Hayden, M., Goldberg, Y., Samuels, M. (2002) Mutant frizzled-4 disrupts retinal angiogenesis in familial exudative vitreoretinopathy. *Nat Genet.* 32, 326–330.
233. Kaykas, A., Yang-Snyder, J., Heroux, M., Shah, K., Bouvier, M., Moon, R. (2004) Mutant Frizzled 4 associated with vitreoretinopathy traps wild-type Frizzled in the endoplasmic reticulum by oligomerization. *Nat Cell Biol.* 6, 52–58.
234. Rajan, R., Kopito, R. (2005) Suppression of wild-type rhodopsin maturation by mutants linked to autosomal dominant retinitis pigmentosa. *J Biol Chem.* 280, 1284–1291.
235. Colley, N., Cassill, J., Baker, E., Zuker, C. (1995) Defective intracellular transport is the molecular basis of rhodopsin-dependent dominant retinal degeneration. *Proc Natl Acad Sci U S A.* 92, 3070–3074.
236. Kurada, P., Tonini, T., Serikaku, M., Piccini, J., O'Tousa, J. (1998) Rhodopsin maturation antagonized by dominant rhodopsin mutants. *Vis Neurosci.* 15, 693–700.
237. Liang, Y., Fotiadis, D., Filipek, S., Saperstein, D., Palczewski, K., Engel, A. (2003) Organization of the G protein-coupled receptors rhodopsin and opsin in native membranes. *J Biol Chem.* 278, 21655–21662.
238. Fotiadis, D., Liang, Y., Filipek, S., Saperstein, D., Engel, A., Palczewski, K. (2003) Atomic-force microscopy: Rhodopsin dimers in native disc membranes. *Nature.* 421, 127–128.
239. Bouvier, M. (2001) Oligomerization of G-protein-coupled transmitter receptors. *Nat Rev Neurosci.* 2, 274–286.
240. Conn, P., Rogers, D., Steward, J., Niedel, J., Sheffield, T. (1982) Conversion of a gonadotropin-releasing hormone antagonist to an agonist. *Nature.* 296, 653–655.
241. Brothers, S., Cornea, A., Janovick, J., Conn, P. (2004) Human loss-of-function gonadotropin-releasing hormone receptor mutants retain wild-type receptors in the endoplasmic reticulum: Molecular basis of the dominant-negative effect. *Mol Endocrinol.* 18, 1787–1797.
242. Leanos-Miranda, A., Ulloa-Aguirre, A., Ji, T., Janovick, J., Conn, P. (2003) Dominant-negative action of disease-causing gonadotropin-releasing hormone receptor (GnRHR) mutants: A trait that potentially coevolved with decreased plasma membrane expression of GnRHR in humans. *J Clin Endocrinol Metab.* 88, 3360–3367.
243. Biebermann, H., Krude, H., Elsner, A., Chubanov, V., Gudermann, T., Gruters, A. (2003) Autosomal-dominant mode of inheritance of a melanocortin-4 receptor mutation in a patient with severe early-onset obesity is due to a dominant-negative effect caused by receptor dimerization. *Diabetes.* 52, 2984–2988.
244. Piersma, D., Verhoef-Post, M., Berns, E., Themmen, A. (2007) LH receptor gene mutations and polymorphisms: An overview. *Mol Cell Endocrinol.* 260–262, 282–286.
245. Davies, T., Ando, T., Lin, R., Tomer, Y., Latif, R. (2005) Thyrotropin receptor-associated diseases: From adenomata to Graves disease. *J Clin Invest.* 115, 1972–1983.

246. Clark, A., McLoughlin, L., Grossman, A. (1993) Familial glucocorticoid deficiency associated with point mutation in the adrenocorticotropin receptor. *Lancet.* 341, 461–462.

247. Tsigos, C., Arai, K., Hung, W., Chrousos, G. (1993) Hereditary isolated glucocorticoid deficiency is associated with abnormalities of the adrenocorticotropin receptor gene. *J Clin Invest.* 92, 2458–2461.

248. Clark, A., Metherell, L., Cheetham, M., Huebner, A. (2005) Inherited ACTH insensitivity illuminates the mechanisms of ACTH action. *Trends Endocrinol Metab.* 16, 451–457.

249. Tao, Y., Segaloff, D. (2003) Functional characterization of melanocortin-4 receptor mutations associated with childhood obesity. *Endocrinology.* 144, 4544–4551.

250. Govaerts, C., Srinivasan, S., Shapiro, A., Zhang, S., Picard, F., Clement, K., Lubrano-Berthelier, C., Vaisse, C. (2005) Obesity-associated mutations in the melanocortin 4 receptor provide novel insights into its function. *Peptides.* 26, 1909–1919.

251. Rana, S., Besson, G., Cook, D., Rucker, J., Smyth, R., Yi, Y., Turner, J., Guo, H., Du, J., Peiper, S., Lavi, E., Samson, M., Libert, F., Liesnard, C., Vassart, G., Doms, R., Parmentier, M., Collman, R. (1997) Role of CCR5 in infection of primary macrophages and lymphocytes by macrophage-tropic strains of human immunodeficiency virus: Resistance to patient-derived and prototype isolates resulting from the delta ccr5 mutation. *J Virol.* 71, 3219–3227.

252. Blanpain, C., Lee, B., Tackoen, M., Puffer, B., Boom, A., Libert, F., Sharron, M., Wittamer, V., Vassart, G., Doms, R., Parmentier, M. (2000) Multiple nonfunctional alleles of CCR5 are frequent in various human populations. *Blood.* 96, 1638–1645.

253. Hirata, T., Kakizuka, A., Ushikubi, F., Fuse, I., Okuma, M., Narumiya, S. (1994) Arg60 to Leu mutation of the human thromboxane A2 receptor in a dominantly inherited bleeding disorder. *J Clin Invest.* 94, 1662–1667.

254. Salvatori, R., Fan, X., Phillips III, J., Espigares-Martin, R., Martin De Lara, I., Freeman, K., Plotnick, L., Al-Ashwal, A., Levine, M. (2001) Three new mutations in the gene for the growth hormone (gh)-releasing hormone receptor in familial isolated gh deficiency type ib. *J Clin Endocrinol Metab.* 86, 273–279.

255. Brown, E. (2007) Clinical lessons from the calcium-sensing receptor. *Nat Clin Pract Endocrinol Metab.* 3, 122–133.

256. Chen, J., Taipale, J., Cooper, M., Beachy, P. (2002) Inhibition of Hedgehog signaling by direct binding of cyclopamine to Smoothened. *Genes Dev.* 16, 2743–2748.

257. Curran, M., Splawski, I., Timothy, K., Vincent, G., Green, E., Keating, M. (1995) A molecular basis for cardiac arrhythmia: HERG mutations cause long QT syndrome. *Cell.* 80, 795–803.

258. Zhou, Z., Gong, Q., January, C. (1999) Correction of defective protein trafficking of a mutant HERG potassium channel in human long QT syndrome. Pharmacological and temperature effects. *J Biol Chem.* 274, 31123–31126.

259. Paulussen, A., Raes, A., Matthijs, G., Snyders, D., Cohen, N., Aerssens, J. (2002) A novel mutation (T65P) in the PAS domain of the human potassium channel HERG results in the long QT syndrome by trafficking deficiency. *J Biol Chem.* 277, 48610–48616.

260. Dormer, R., Derand, R., McNeilly, C., Mettey, Y., Bulteau-Pignoux, L., Metaye, T., Vierfond, J., Gray, M., Galietta, L., Morris, M., Pereira, M., Doull, I., Becq, F.,

McPherson, M. (2001) Correction of delF508-CFTR activity with benzo(c)quinolizinium compounds through facilitation of its processing in cystic fibrosis airway cells. *J Cell Sci.* 114, 4073–4081.

261. Egan, M., Pearson, M., Weiner, S., Rajendran, V., Rubin, D., Glockner-Pagel, J., Canny, S., Du, K., Lukacs, G., Caplan, M. (2004) Curcumin, a major constituent of turmeric, corrects cystic fibrosis defects. *Science.* 304, 600–602.
262. Pedemonte, N., Lukacs, G., Du, K., Caci, E., Zegarra-Moran, O., Galietta, L., Verkman, A. (2005) Small-molecule correctors of defective DeltaF508-CFTR cellular processing identified by high-throughput screening. *J Clin Invest.* 115, 2564–2571.
263. Loo, T., Bartlett, M., Clarke, D. (2005) Rescue of DeltaF508 and other misprocessed CFTR mutants by a novel quinazoline compound. *Mol Pharm.* 2, 407–413.
264. Dormer, R., Harris, C., Clark, Z., Pereira, M., Doull, I., Norez, C., Becq, F., McPherson, M. (2005) Sildenafil (Viagra) corrects DeltaF508-CFTR location in nasal epithelial cells from patients with cystic fibrosis. *Thorax.* 60, 55–59.
265. Partridge, C., Beech, D., Sivaprasadarao, A. (2001) Identification and pharmacological correction of a membrane trafficking defect associated with a mutation in the sulfonylurea receptor causing familial hyperinsulinism. *J Biol Chem.* 276, 35947–35952.
266. Yan, F., Lin, C.-W., Weisiger, E., Cartier, E., Taschenberger, G., Shyng, S.-L. (2004) Sulfonylureas correct trafficking defects of ATP-sensitive potassium channels caused by mutations in the sulfonylurea receptor. *J Biol Chem.* 279, 11096–11105.
267. Pey, A., Ying, M., Cremades, N., Velazquez-Campoy, A., Scherer, T., Thony, B., Sancho, J., Martinez, A. (2008) Identification of pharmacological chaperones as potential therapeutic agents to treat phenylketonuria. *J Clin Invest.* 118, 2858–2867.
268. Sawkar, A., Cheng, W., Beutler, E., Wong, C., Balch, W., Kelly, J. (2002) Chemical chaperones increase the cellular activity of N370S beta-glucosidase: A therapeutic strategy for Gaucher disease. *Proc Natl Acad Sci U S A.* 99, 15428–15433.
269. Steet, R., Chung, S., Wustman, B., Powe, A., Do, H., Kornfeld, S. (2006) The iminosugar isofagomine increases the activity of N370S mutant acid {beta}-glucosidase in Gaucher fibroblasts by several mechanisms. *Proc Natl Acad Sci U S A.* 109, 13813–13818.
270. Tominaga, L., Ogawa, Y., Taniguchi, M., Ohno, K., Matsuda, J., Oshima, A., Suzuki, Y., Nanba, E. (2001) Galactonojirimycin derivatives restore mutant human beta-galactosidase activities expressed in fibroblasts from enzyme-deficient knockout mouse. *Brain Dev.* 23, 284–287.
271. Matsuda, J., Suzuki, O., Oshima, A., Yamamoto, Y., Noguchi, A., Takimoto, K., Itoh, M., Matsuzaki, Y., Yasuda, Y., Ogawa, S., Sakata, Y., Nanba, E., Higaki, K., Ogawa, Y., Tominaga, L., Ohno, K., Iwasaki, H., Watanabe, H., Brady, R., Suzuki, Y. (2003) Chemical chaperone therapy for brain pathology in GM1-gangliosidosis. *Proc Natl Acad Sci U S A.* 100, 15912–15917.
272. Fan, J.-Q., Ishii, S., Asano, N., Suzuki, Y. (1999) Accelerated transport and maturation of lysosomal a-galactosidase A in Fabry lymphoblasts by an enzyme inhibitor. *Nature Med.* 5, 112–115.
273. Frustaci, A., Chimenti, C., Ricci, R., Natale, L., Russo, M., Pieroni, M., Eng, C., Desnick, R. (2001) Improvement in cardiac function in the cardiac variant of Fabry's disease with galactose-infusion therapy. *N Engl J Med.* 345, 25–32.

274. Tropak, M., Reid, S., Guiral, M., Withers, S., Mahuran, D. (2004) Pharmacological enhancement of beta-hexosaminidase activity in fibroblasts from adult Tay-Sachs and Sandhoff Patients. *J Biol Chem.* 279, 13478–13487.

275. Maegawa, G., Tropak, M., Buttner, J., Stockley, T., Kok, F., Clarke, J., Mahuran, D. (2007) Pyrimethamine as a potential pharmacological chaperone for late-onset forms of GM2 gangliosidosis. *J Biol Chem.* 282, 9150–9161.

276. Tropak, M., Blanchard, J., Withers, S., Brown, E., Mahuran, D. (2007) High-throughput screening for human lysosomal beta-N-acetyl hexosaminidase inhibitors acting as pharmacological chaperones. *Chem Biol.* 14, 153–164.

277. Parenti, G., Zuppaldi, A., Gabriela Pittis, M., Rosaria Tuzzi, M., Annunziata, I., Meroni, G., Porto, C., Donaudy, F., Rossi, B., Rossi, M., Filocamo, M., Donati, A., Bembi, B., Ballabio, A., Andria, G. (2007) Pharmacological enhancement of mutated alpha-glucosidase activity in fibroblasts from patients with Pompe disease. *Mol Ther.* 15, 508–514.

278. Okumiya, T., Kroos, M., Vliet, L., Takeuchi, H., Van der Ploeg, A., Reuser, A. (2007) Chemical chaperones improve transport and enhance stability of mutant alpha-glucosidases in glycogen storage disease type II. *Mol Genet Metab.* 90, 49–57.

279. Bykov, V., Issaeva, N., Shilov, A., Hultcrantz, M., Pugacheva, E., Chumakov, P., Bergman, J., Wiman, K., Selivanova, G. (2002) Restoration of the tumor suppressor function to mutant p53 by a low-molecular-weight compound. *Nat Med.* 8, 282–288.

280. Issaeva, N., Friedler, A., Bozko, P., Wiman, K., Fersht, A., Selivanova, G. (2003) Rescue of mutants of the tumor suppressor p53 in cancer cells by a designed peptide. *Proc Natl Acad Sci U S A.* 100, 13303–13307.

281. Kim, B.-E., Smith., K., Meagher, C., Petris, M. (2002) A conditional mutation affecting localization of the Menkes disease copper ATPase. Suppression by copper supplementation. *J Biol Chem.* 277, 44079–44084.

282. Mahadeva, R., Dafforn, T., Carrell, R., Lomas, D. (2002) 6-mer peptide selectively anneals to a pathogenic serpin conformation and blocks polymerization. Implications for the prevention of Z alpha(1)-antitrypsin-related cirrhosis. *J Biol Chem.* 277, 6771–6774.

283. Parfrey, H., Dafforn, T., Belorgey, D., Lomas, D., Mahadeva, R. (2004) Inhibiting polymerization: New therapeutic strategies for Z alpha1-antitrypsin-related emphysema. *Am J Respir Cell Mol Biol.* 31, 133–139.

284. Mallya, M., Phillips, R., Saldanha, S., Gooptu, B., LeighBrown, S., Termine, D., Shirvani, A., Wu, Y., Sifers, R., Abagyan, R., Lomas, D. (2007) Small molecules block the polymerization of Z alpha1-antitrypsin and increase the clearance of intracellular aggregates. *J Med Chem.* 50, 5357–5363.

285. Soto, C., Kascsak, R., Saborío, G., Aucouturier, P., Wisniewski, T., Prelli, F., Kascsak, R., Mendez, E., Harris, D., Ironside, J., Tagliavini, F., Carp, R., Frangione, B. (2000) Reversion of prion protein conformational changes by synthetic b-sheet breaker peptides. *Lancet.* 355, 192–197.

286. Korth, C., May, B., Cohen, F., Prusiner, S. (2001) Acridine and phenothiazine derivatives as pharmacotherapeutics for prion disease. *Proc Natl Acad Sci U S A.* 98, 9836–9841.

INDEX

GPCR Molecular Pharmacology and Drug Targeting: Shifting Paradigms and New Directions, Edited by Annette Gilchrist

Wiley The manufacturer's authorized representative according to the EU General Product Safety Regulation is Wiley-VCH GmbH, Boschstr. 12, 69469 Weinheim, Germany, e-mail: Product_Safety@wiley.com.

Printed and bound by CPI Group (UK) Ltd, Croydon, CR0 4YY
19/08/2026
02201077-0008